D0142232

INTRODUCTION TO INTEGRATED CIRCUIT ENGINEERING

INTRODUCTION TO INTEGRATED CIRCUIT ENGINEERING

D. K. Reinhard
Michigan State University

HOUGHTON MIFFLIN COMPANY Boston
Dallas Geneva, Illinois Lawrenceville, New Jersey Palo Alto

Art created by Boston Graphics, Inc.

Printed in the U.S.A.

Library of Congress Catalog Card Number: 86-80842

ISBN: 0-395-37068-X

ABCDEFGHIJ-H-89876

To F. O. Arntz
For his early encouragement and support

Contents

Preface

Although the engineering team responsible for the realization of large-scale engineering projects necessarily consists of specialists, it is useful for each team member to have exposure to all facets of the engineering project. This text is written from that point of view and, like integrated circuit engineering, covers a wide range of skills and knowledge. The text is designed to be used at the senior, or beginning graduate, level.

Integrated circuit engineering includes the process of devising circuits, chip-layouts, components, and fabrication processes to meet a given set of objectives. In this book, students who have had a previous introductory course in solid state materials, devices, and circuits are introduced to the field of integrated circuit engineering, with balanced emphasis placed on the design *and* fabrication of these devices. In brief, my book describes and compares metal-oxide-semiconductor (MOS) and bipolar technologies, presents layout principles, and demonstrates integrated circuit design methodologies with examples. In addition, the areas of testing, yields, and packaging are examined from both a technical and economic perspective.

Recognizing that further exploitation of very large scale integrated (VLSI) systems will require continued interplay between device technology and circuit design, this text equips students to understand *both* circuit design *and* the technologies used in the fabrication of these devices. This book establishes a foundation for the interplay of these two crucial factors in integrated circuit engineering at the senior level, before students move on to more specialized studies.

Another distinguishing feature of this book is the inclusion of laboratory experiments that cover basic integrated circuit fabrication processes and provide students with a unique opportunity to compare theoretical predictions with actual device and circuit performance since they know exactly what went into the making of the chip. The basic material in the first four chapters is based on a firm experimental foundation, and this is best appreciated in a "learning by doing" environment. Although the experiments are optional, in my experience they greatly enhance the learning process.

Although integrated circuit design was once the province of relatively few individuals, the wide availability of computer-aided design tools, also discussed here, has opened the area dramatically. Furthermore, the information is useful as background material to engineers who use, rather than design, integrated circuits to better understand the fundamental distinctions between various integrated circuit types. In this way, their relative advantages and disadvantages, as well as performance limitations are best appreciated.

Chapter 1 primarily gives the student a preview of the material in the following chapters. Salient points in the history of integrated circuits are briefly discussed, as well as fundamental limitations on future growth in circuit density. Chapter 2 covers the fundamentals of integrated circuit processing including diffusion, ion implantation, oxidation, and a variety of deposition and etching techniques. Chapters 3 and 4 cover the two main integrated circuit families, MOS and bipolar, and describe basic fabrication sequences, device models and parasitics, basic gate configurations, and speed considerations.

In Chapters 5 and 6 layout principles and design methodologies are discussed, respectively. Both MOS and bipolar, and analog as well as digital, examples are included. Because layout of integrated circuits is a design process, unique solutions do not exist. However, several general considerations for layout design are discussed and demonstrated by elementary examples in Chapter 5. Likewise, several circuit design methodologies have emerged in modern integrated circuit design, many of which are quite different from traditional design based on small scale integrated circuits, and Chapter 6 introduces a sampling of basic circuit design methods by example. In addition, the important role of computers in integrated circuit design is discussed from several perspectives.

In Chapter 7, three topics needed to round out the picture of integrated circuit engineering are covered, namely packaging, yields, and testing, all of which are of vital importance in integrated circuit fabrication. Packaging, testing, yields, and processing costs all combine to determine the production cost of an integrated circuit, and exercises in estimating the production cost of an integrated circuit are included in this chapter.

At the end of the book, Appendix A presents laboratory experiments that reinforce the major topics presented in the book's first four chapters. The first experiment is concerned with basic wafer characterization and the oxidation process. The next two experiments involve a series of mask steps in which a variety of devices and simple circuits are fabricated.

The chapters include problems, an important part of the learning process, and the experiments contain sets of questions, which may form the basis of reports.

The book has evolved from course material taught at Michigan State University over the past five years. It is presently taught in two courses, a design lecture course, and also in a combined lecture and laboratory course, to senior and first year-graduate students. The first two experiments in the Appendix have formed the standard portion of the combined lecture and laboratory course offered at Michigan State since 1982 and the third experiment has been used for special individual or group projects.

This text has benefited from suggestions and critiques from several classes of students who have used draft versions of this material, and I express gratitude for this to EE 478 and EE 871 students at Michigan State University. I thank Gary Anderson, Bob Matthews, and Gordon Priebe for their independent study work that led to mask designs used in the text, and also Martin Perrine, whose cooperative education student work is used as a design example in Chapter 6. Material from this book has also been taught by Thaddeus Roppel, Mahmoud Dahimene, Mohsen Alavi, and Vivek Joshi, and I have appreciated their supportive comments and helpful suggestions. Appreciation is expressed to Man Kuan Vai for applying his valuable expertise with computer aided layout to the preparations of illustrations for this text, and to the Case Center for Computer Aided Design at Michigan State University for supplying needed equipment. The text has also benefited from many helpful comments by the manuscript reviewers: Stanley Burns, Iowa State University; the late Basil Cochrun, Northeastern University; Barry J. Farbrother, Rose-Hulman Institute of Technology; Thomas Higgins, University of Wisconsin; Faquir C. Jain, University of Connecticut; and Gerold W. Neudeck, Purdue University.

D.K.R.

INTRODUCTION TO INTEGRATED CIRCUIT ENGINEERING

1 Basic Concepts of Integrated Circuits

1.1 Introduction

Modern integrated circuits may contain hundreds of thousands of components and require a multitude of highly critical fabrication steps. In spite of their small size, they may represent large-scale engineering projects involving several man-years of multidisciplinary design effort. Furthermore, the growth in integrated circuit complexity has been paralleled by the growth in the integrated circuit industry. More than three-fourths of the world semiconductor market is accounted for by integrated circuits, which means that the dollar value of the market is measured in billions of dollars per year. The design of electronic systems has been, and continues to be, radically changed by the availability of integrated circuits with increasingly powerful functions. If microelectronics has made possible a new industrial revolution, integrated circuits represent the technological cornerstone of that revolution.

The study of integrated circuits, in the broadest sense, encompasses a wide range of skills and knowledge. Consider the microprocessor, which is an excellent example of the modern integrated circuit. The microprocessor may be viewed from vastly different perspectives. Using a programmer's model, this integrated circuit may be considered as a computing machine with a variety of programming features. These might include several 32-bit data registers, address registers, a stack pointer, a program counter, and an arithmetic-logic unit with some control and timing circuitry. A corresponding instruction set allows the user to perform logic and arithmetic operations. From this perspective, the particular technology on which the integrated circuit is based, such as n-channel metal-oxide-semiconductor (NMOS) or complementary metal-oxide-semiconductor (CMOS), may appear to be relatively unimportant.

On the other hand, the same integrated circuit might be considered as a piece of single crystalline silicon, about 500 μm thick, of which the top 5 μm or so contain varying trace amounts of impurities along with holes and electrons of varying effective masses, mobilities, and concentrations. From this viewpoint, whether it is a microprocessor, a memory circuit, or an amplifier may seem to be of small consequence.

Therefore, by combining these perspectives, it may be argued that a full understanding of the microprocessor should span the spectrum from solid-state physics to computer science. Of course, no text could pretend to adequately cover such a range of topics. However, the fundamentals of integrated circuit fabrication, design, and performance limitations presented here should provide a useful foundation on which further, more specialized study can be based.

1.2 A Historical Perspective

Credit for the first public prophecy of integrated circuits generally goes to G. W. A. Dummer, who in 1952 suggested that electronic equipment could be constructed "in a solid block with no connecting wires. The block may consist of layers of insulating, conducting, rectifying and amplifying materials, the electrical functions being connected directly by cutting out areas of the various layers" [1]. Dummer's vision of the monolithic integrated electronic circuit is all the more remarkable considering that fabrication of the first transistor had been announced only four years earlier.

A full account of the many leading contributors to the development of integrated circuit technology, along with associated subtleties of the patent law, is beyond the scope of this section. However, the credit for an actual realization of the first monolithic semiconductor integrated circuit goes to Jack Kilby, who began working for Texas Instruments in 1958 as a young engineer and project leader. Working for 2 weeks in a nearly deserted plant shut down for summer vacation (for which he was not yet eligible), Kilby developed ideas for forming resistors, capacitors, and transistors in a single slice of semiconductor. By the fall of 1958 he had used these concepts to build a circuit with devices that were integral to a single piece of germanium and interconnected by thermally bonded gold wires to form a phase-shift oscillator. Since external wires were used to connect neighboring devices, the circuit was not entirely integrated. However, Kilby's 1959 patent described how the circuit elements could also be interconnected by an electrically conducting layer, such as gold, laid down on insulating material to make the necessary device connections.

Although the early monolithic germanium circuits described by Kilby were the first realization of Dummer's prediction, they were quite unlike modern integrated circuits in terms of their technical features. A few months later, at Fairchild Semiconductor Company, Robert Noyce independently announced a planar silicon integrated circuit in which many of the basic features of modern integrated circuit technology were clearly evident. Noyce and Kilby were joined in their pioneering work by several other contributors, such as Harwick Johnson at RCA, who did early work on integrated germanium structures; Kurt Lehovec at Sprague Electric Co., who patented the

Figure 1.1 Exponential growth of the number of components on a chip. (SSI = small-scale integration; MSI = medium-scale integration; LSI = large-scale integration; VLSI = very large scale integration.)
SOURCE: Adapted from S. M. Sze, in *Quick Reference Manual for Silicon Integrated Circuit Technology*, W. E. Beadle, J. C. C. Tsai, and R. D. Plummer, eds., New York: Copyright © 1985 by John Wiley & Sons, pp. 9-16. Reprinted by permission of John Wiley & Sons, Inc.

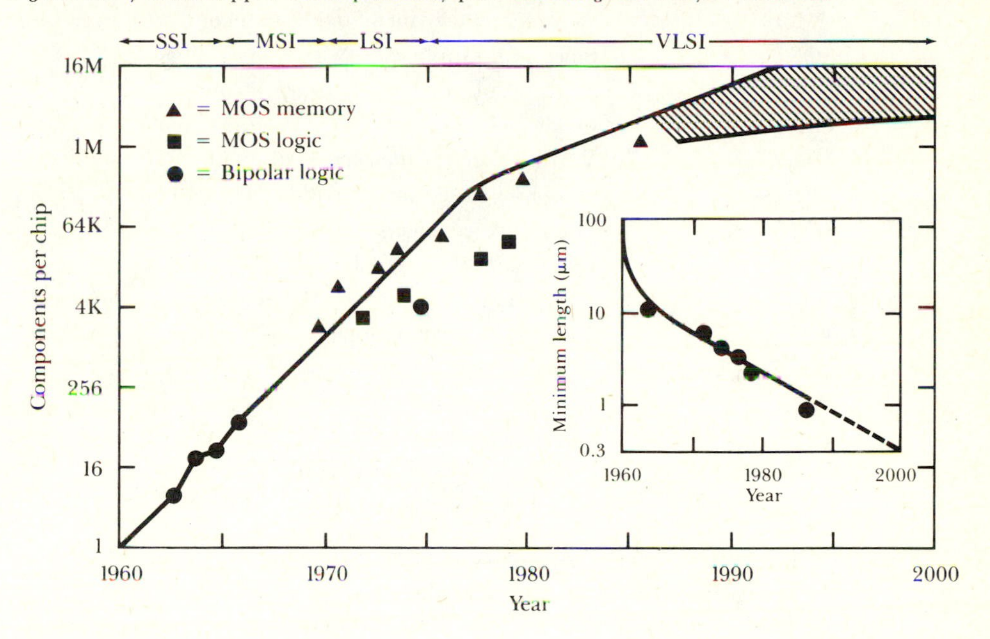

method of *p-n* junction isolation of integrated circuit devices; and Jean Hoerni at Fairchild, who invented the planar transistor process [2]. The combined efforts led to the first commercial integrated circuits, which were primarily simple bipolar logic circuits.

From these modest beginnings, with a few transistors and resistors per circuit, or per chip, the growth of integrated circuit technology proceeded at a rapid pace. For more than a decade, the number of components per chip increased by a factor of two every year, until by the early 1970s, approximately 10^4 components per chip were achievable. And when thousands of devices could be incorporated into a single integrated circuit, important new applications became practical. Most notable among them were the microprocessor and semiconductor memory, both of which spurred further developments in integrated circuitry. The exponential growth in the number of components per chip and projections for the future are shown in Figure 1.1 [3].

Several integrated circuit technologies have emerged over the past 3 decades. Consequently, many types of integrated circuits are described in professional journals such as the *IEEE Transactions on Electron Devices* and the *IEEE Transactions on Solid State Circuits*. As a sample, consider the 15

Table 1.1 A sampling of integrated circuit types

Abbreviation	Description
NMOS	*n*-Channel metal-oxide-semiconductor
PMOS	*p*-Channel metal-oxide-semiconductor
CMOS	Complementary metal-oxide-semiconductor
SOS	Silicon-on-sapphire
SOI	Silicon-on-insulator
OXIL	Oxide isolated logic
VMOS	Vertical metal-oxide-semiconductor
DMOS	Double-diffused metal-oxide-semiconductor
HMOS	High-performance metal-oxide-semiconductor
PSA	Polysilicon self-aligned
E/D MOS	Enhancement/depletion metal-oxide-semiconductor
I^2L	Integrated injection logic
ISL	Integrated Schottky logic
MTL	Merged transistor logic
CCD	Charge-coupled devices

different integrated circuit types listed in Table 1.1. The integrated circuit type distinctions may be due to different transistor types, to different circuit designs and organization, or to different fabrication processes. It is desirable to know not only what the letters in Table 1.1 stand for, but also to have a physical picture of the various integrated circuit types. In this way, the relative advantages and disadvantages of each, as well as the underlying performance limitations, are best appreciated. Integrated circuit performance limitations may be due to the processing technology, to an individual component design such as a transistor, to the circuit design itself, or to a combination of all three factors.

All of the entries in Table 1.1 refer to monolithic integrated circuits; that is, the substrate on which and in which the circuit resides is a solid block of material. Monolithic integrated circuits are generally fabricated in a single crystalline semiconductor, and most commonly silicon is the material of choice. Less specifically, however, the term *integrated circuit* may refer to any circuit in which the components are inseparably connected. In addition to monolithic integrated circuits, two other integrated circuit technologies with important commercial uses are thick-film hybrid integrated circuits and thin-film hybrid integrated circuits. In both technologies, the substrate is an insulator and serves a passive supporting role. The latter two technologies are hybrid in the sense that resistor and capacitor films are combined with miniature components such as discrete transistors or monolithic integrated circuits on a single insulating substrate. However, the bulk of the material in this text is concerned with variations on the monolithic integrated circuit theme rather than on hybrid integrated circuits.

1.3 A Basic Fabrication Sequence

Although the steps in the fabrication process of integrated circuits are covered in detail in later chapters, it is useful to consider a simple fabrication sequence at this point as a preview. As an example of several of the basic steps in the planar silicon integrated circuit process, consider the metal-oxide-semiconductor field-effect transistor (MOSFET) fabrication sequence illustrated in simplified form in Figure 1.2. For purposes of this discussion, the steps in this figure, shown by cross-sectional views of the transistor, require that we assume the ability to define patterned layers with critical dimensions both vertically and horizontally. (The horizontal and vertical dimensions in Figure 1.2 and subsequent figures are not to scale.) Also, for the moment, let us take for granted that oxide and metal layers can be added to a silicon wafer and also be selectively removed. As an aside, it is noted that although transistors are generally thought of as semiconductor devices, both a good conductor and a good insulator are also required for their actual realization in integrated circuits. Therefore, the three main classes of electrical materials, namely insulators, conductors, and semiconductors, are all represented in the integrated circuit. Furthermore, since the atomic order is amorphous in the insulator, single crystalline in the semiconductor, and polycrystalline in the conductor, the three broad classes of atomic order in solids are also represented. The oxide layer, most typically silicon dioxide, is a superb insulator, although it serves an even broader purpose in the example shown in Figure 1.2. Note that in step 2, the oxide serves as a mask that allows donor impurities to enter the wafer only in the areas desired for the source and drain regions. In this step it acts simply as a physical barrier. However, in step 7 the oxide layer plays an electrical role, acting as the gate insulator of the MOSFET. A variety of conducting materials would suit the purpose of the fabrication sequence shown in Figure 1.2, although the actual choice of material plays an important role in determining the threshold voltage of the MOSFET. A metal is shown here, but another alternative for the conducting material is a heavily doped and therefore highly conductive semiconductor layer.

Now that the fabrication sequence for a single device is seen, it is not hard to extrapolate to the fabrication sequence for a circuit composed of several MOSFETs. Neighboring devices may be fabricated simultaneously with the final metalization providing the interconnect between devices. By modifying the fabrication sequence, resistors, capacitors, diodes, and other transistor types may also be incorporated, as is described in later chapters.

Figure 1.2 Metal gate MOS fabrication—an example of planar silicon technology. A. Step 1: Initial oxidation. This will be an n channel device. B. Step 2: Oxide openings to define the source and drain regions. C. Step 3: After introduction of donors to establish the source and drain. D. Step 4: Gate oxide region definition. E. Step 5: Gate oxide growth. F. Step 6: Contact cuts. G. Step 7: Metalization.

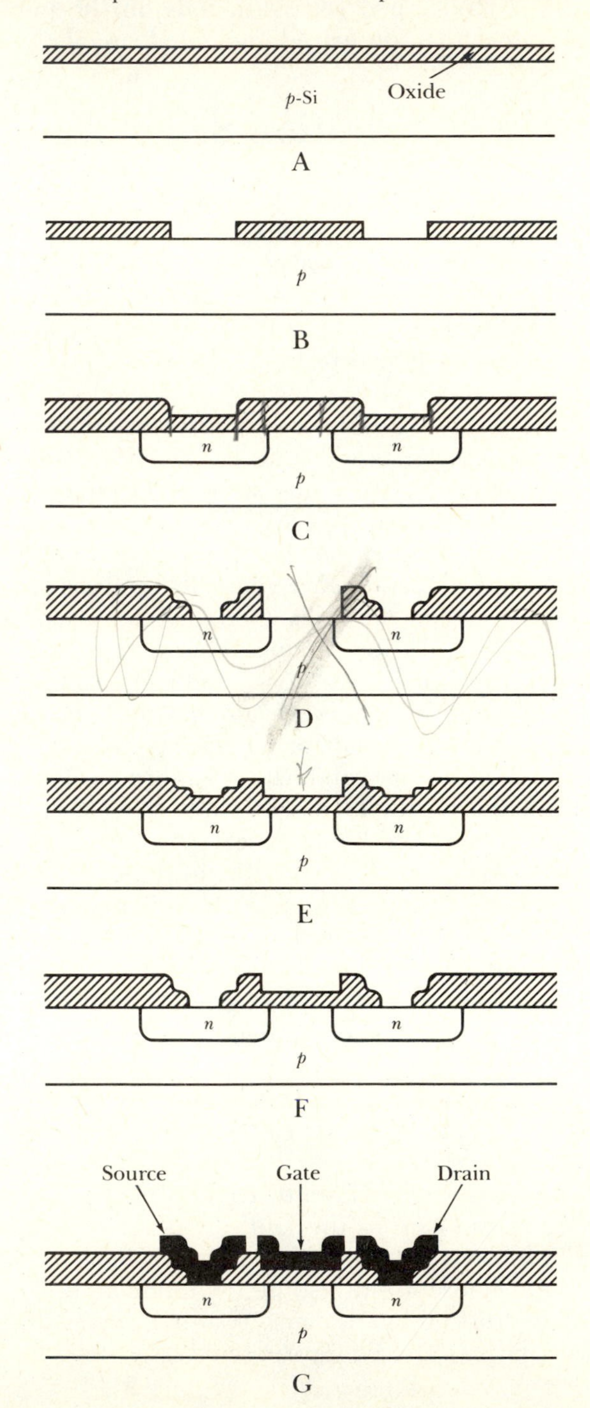

Only seven steps are shown for the MOSFET fabrication sequence in Figure 1.2, but two qualifying statements must be made. First, the steps shown represent one of the least complicated of many possible MOSFET fabrication schemes. Second, and more important, a multitude of hidden steps are not shown in Figure 1.2. In fact, a step-by-step set of instructions would involve more than 60 individual processing operations to go from step 1 to step 7 in the simple fabrication sequence shown in Figure 1.2.

As an example of the hidden steps, consider in more detail step 2 of the fabrication sequence. How can an opening of fixed size and shape be established in the oxide? Such patterning is generally carried out by selectively exposing a layer of sensitive material, commonly called resist, to ionizing radiation. The radiation may be ultraviolet light, x-rays, or electron beams. The process is referred to as microlithography, or alternatively, to indicate the type of radiation, photolithography, x-ray lithography, or electron beam lithography. When a polymer-based negative resist is exposed to, say, ultraviolet radiation, cross-polymerizing bonds in the resist are created by photoactivation such that the exposed resist is insoluble in an organic solvent. Unexposed photoresist, however, is soluble in the solvent and may be removed in a developing process, as shown in Figure 1.3. (Positive resist is also available, in which case the exposed resist is removed.) Selective exposure of the resist is most often achieved by placing a mask, with transparent and opaque regions corresponding to the desired pattern, between the radiation source and the resist-coated wafer. For photolithography and negative resist, the mask is a glass plate coated with an opaque material over areas where the resist is to be removed. After development of the resist, and with subsequent heat treatment, the remaining resist acts as a protective barrier to an

Figure 1.3 The microlithography sequence. A. The resist coated wafer. B. Selective resist exposure through a mask. C. Resist development. D. Etching. E. Resist strip.

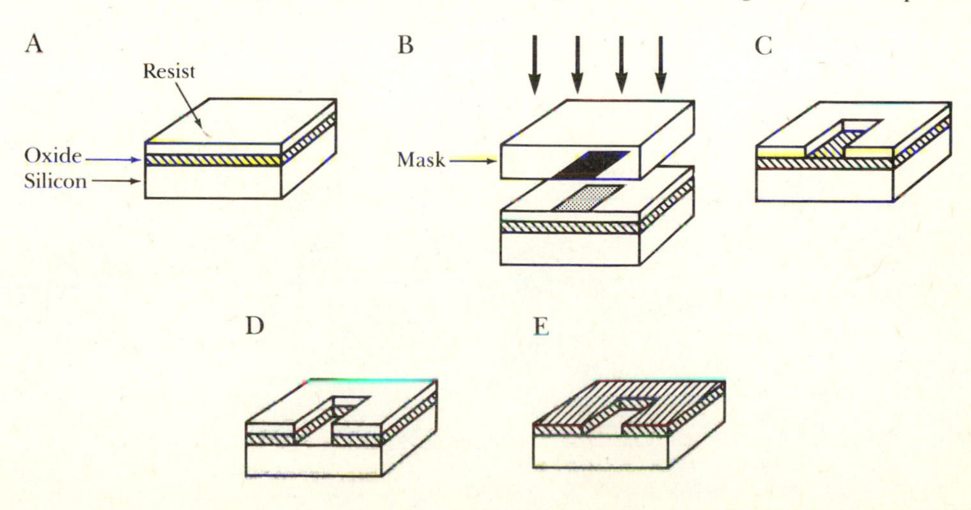

Figure 1.4 Layout and mask set for a MOSFET gate. A. Circuit schematic. B. Diffusion mask. C. Gate oxide mask. D. Contact cut mask. E. Metalization mask. F. Composite layout.

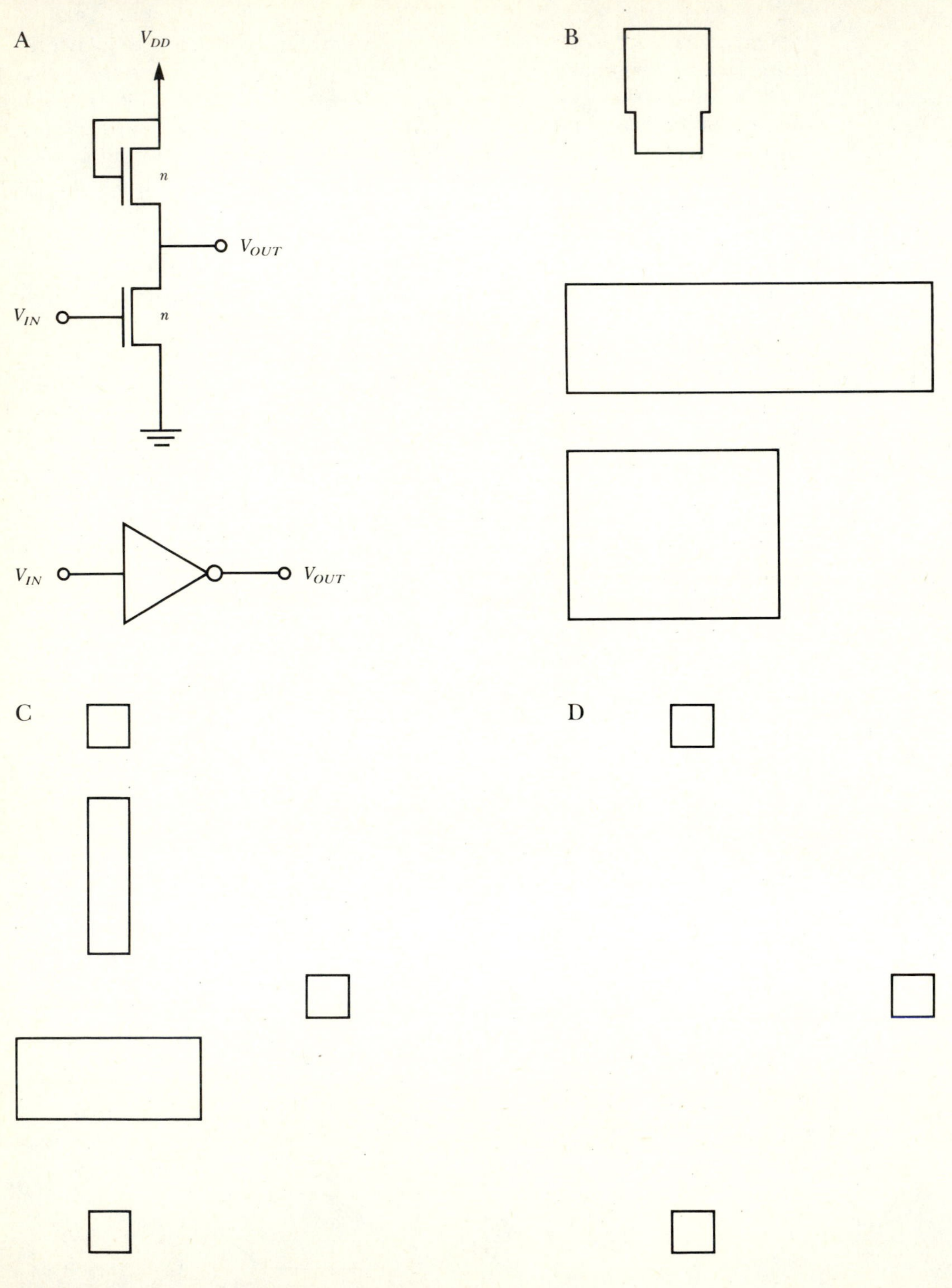

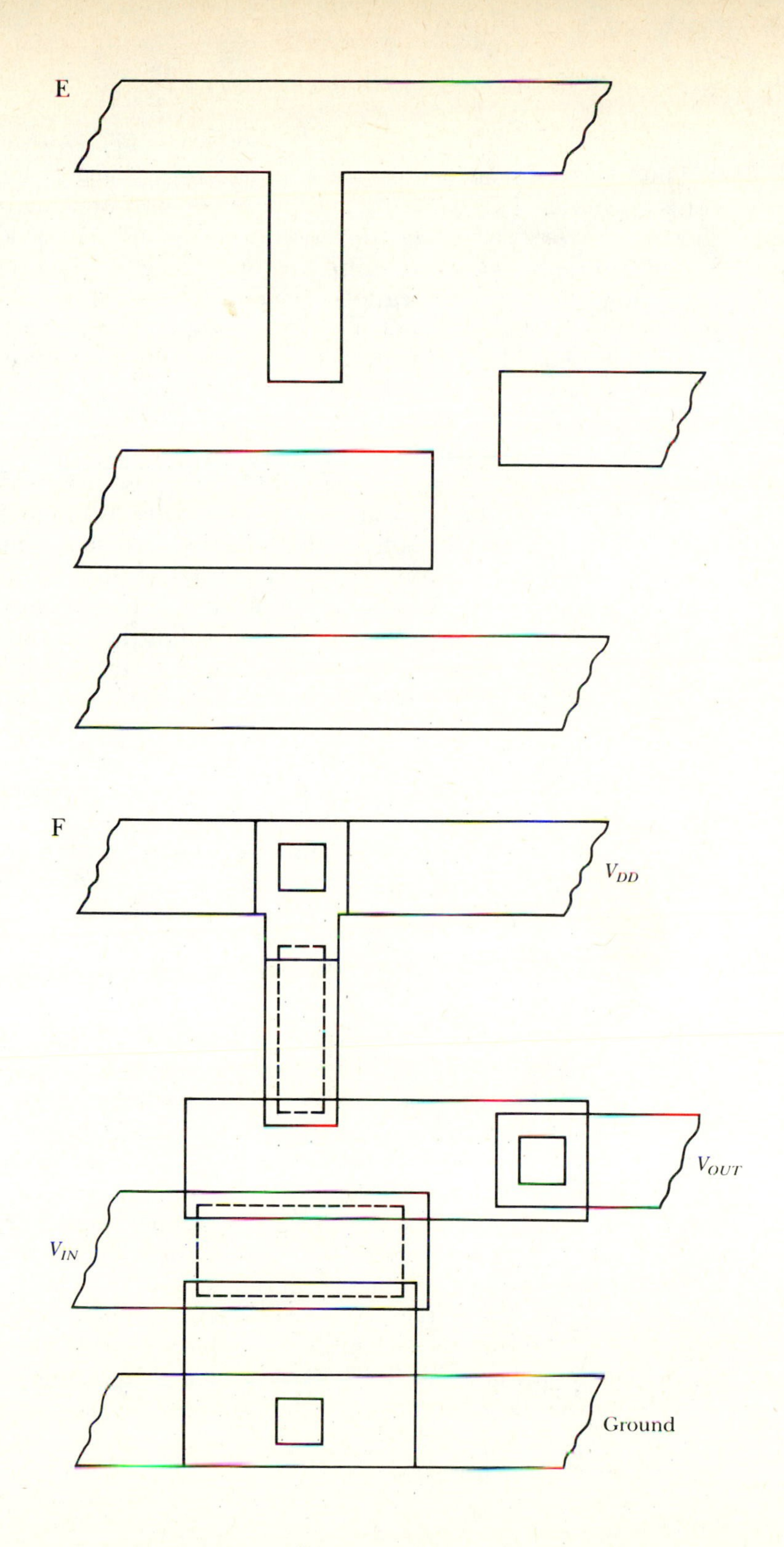
E
F
V_{DD}
V_{OUT}
V_{IN}
Ground

etching medium that removes the exposed oxide. Finally, the remaining resist is chemically stripped away and the result is an opening in the oxide, as shown in the last step of Figure 1.3 and the second step of Figure 1.2. The microlithography sequence is repeated once for each mask, typically from 5 to 15 times during the integrated circuit fabrication process, depending on the fabrication complexity. For the sequence shown in Figure 1.2, only 4 masks are used. The first mask defines the regions for source and drain diffusions, the second defines the gate oxide regions, the third defines the contact cuts through the oxide for source and drain contacts, and the fourth establishes the metal pattern.

The sequence shown in Figure 1.2 may also be used for a simple MOSFET circuit. A top view of a set of mask patterns for an NMOS inverter is shown in Figure 1.4, along with the circuit schematic and a composite drawing showing the relative positions of the diffused regions, the gate oxides, the contact cuts, and the metalization. For the metalization mask, assuming negative resist, the mask material within the enclosed areas is clear and the rest of the mask is opaque. For the other three masks, the enclosed areas are opaque and the rest of the mask area is transparent.

Finally, as an example of a more sophisticated fabrication process, cross-sectional views of a silicon-on-insulator MOSFET process are shown in Figure 1.5 [4]. Here additional patterning is seen both horizontally and vertically, and new materials have been added as well. Hundreds of separate processing steps are required, creating a trade-off between integrated circuit performance and processing complexity, cost, and yield.

1.4 Design Route for an Integrated Circuit

Depending on the complexity of the circuit, the effort that goes into an integrated circuit's design, from initial specifications to final design, may range from a group project over several months to a single designer's effort over several days. For example, the Intel 8086, a 16-bit microprocessor, is reported to have involved a layout design effort of 13 man-years [5]. The Motorola 68000, a 16-bit microprocessor with enhanced features and approximately 70,000 transistors, required 52 man-years of design time [5]. Both of these integrated circuits represent custom integrated circuit designs in which the designers essentially began with a blank piece of silicon. In contrast, a gate array integrated circuit entails a much more limited design effort. Because an array of basic logic circuits, or cells, preexists, the final design effort only needs to provide the proper cell interconnections to realize the desired circuit function. This section briefly considers the more general case of a custom integrated circuit design.

A block diagram flow chart for the design process is shown in Figure 1.6. The actual design of the circuit layout on the chip begins with the floor

Figure 1.5 Buried nitride silicon on insulator MOSFET fabrication process
SOURCE: G. Zimmer and H. Vogt, "CMOS on Buried Nitride—a VLSI SOI Technology," *IEEE Transactions on Election Devices,* ED-30 (1983), 1515–1520. © 1983 IEEE.

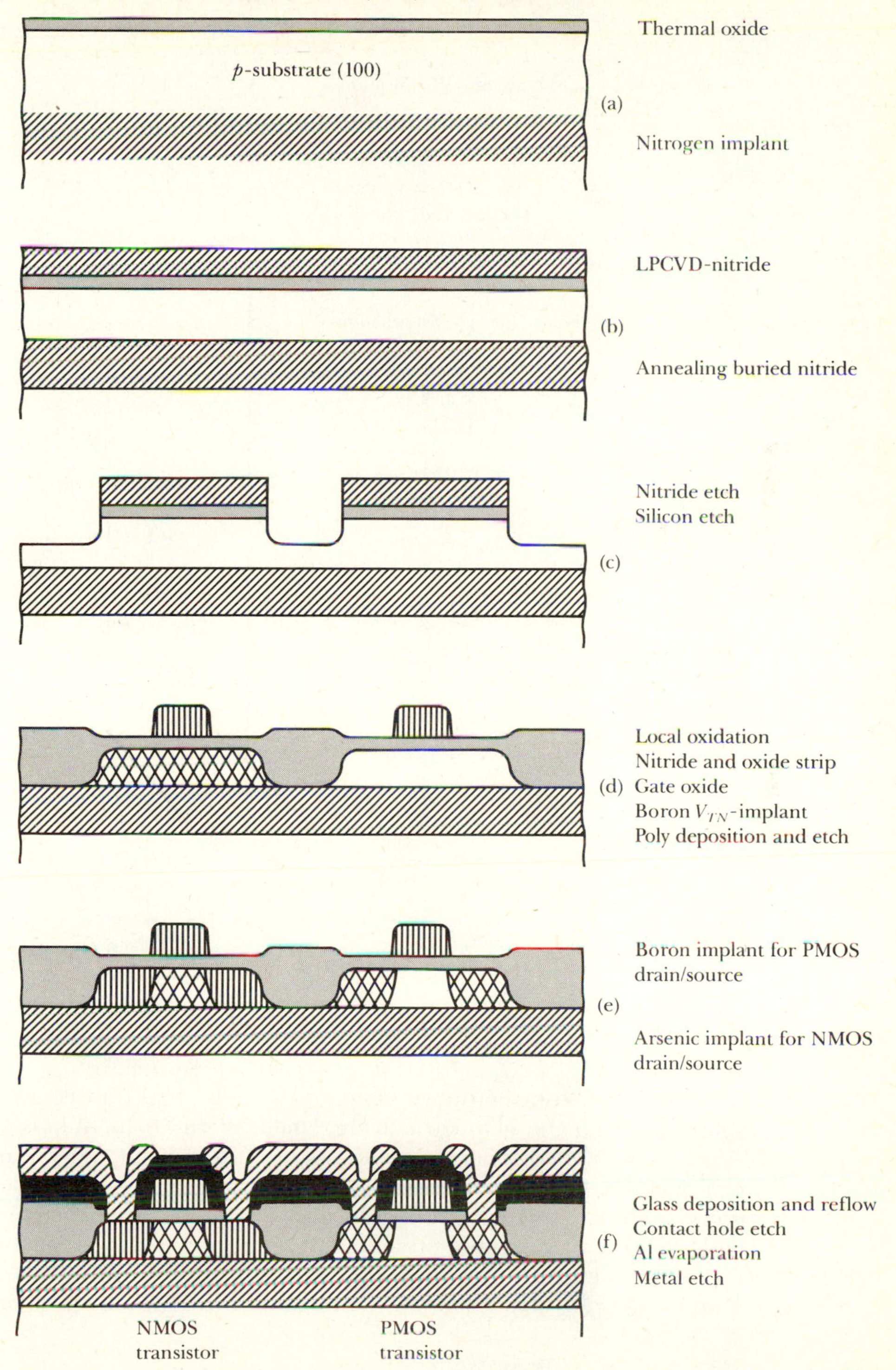

Figure 1.6 A design route for an integrated circuit.

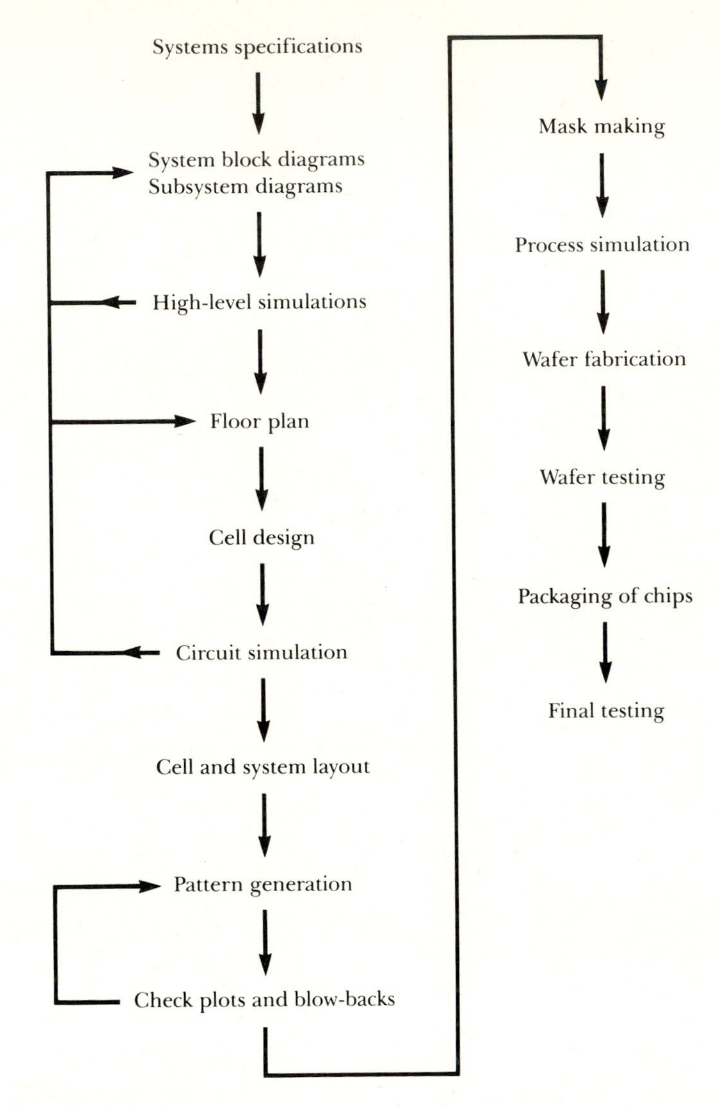

plan. At that point, the design is not at the individual transistor level, but is concerned with the geometric arrangement of the main subsystems of the circuit. A good floor plan simplifies the often formidable interconnect problem. Examples of subsystems in a digital integrated circuit include shift registers, counters, adders, and arithmetic-logic units. If the circuit is to be designed using an existing on-line fabrication process, as is usually the case, then a well-defined set of design rules guides the detailed layout of the subsystems and cells that make up the integrated circuit. Furthermore, the designer can frequently draw on an existing library of subsystem layouts. More rarely, the formulation of the design rules themselves is also part of the integrated circuit design effort.

A prototype layout is encoded digitally as a pattern-generation file from which the several masks that define patterns for the different integrated circuit levels can be created. However, before the pattern is fully committed to glass, the design group often generates check plots and blow-backs to provide a final check. The blow-backs are transparency photo enlargements of the pattern that will actually appear on the masks. Once the masks are approved, the design of the integrated circuit is applied to wafers and success or failure is documented at the wafer test point. Note that simulation at several levels plays a crucial role in the design process, as indicated in Figure 1.6.

1.5 A Look to the Future

Will fundamental limits on the packing density of devices on a chip eventually cause progress in integrated circuit development to plateau? Consider the limits imposed by microlithography. The theoretical limit on resolution is $\lambda/2$, where λ is the wavelength of the exposure radiation; however, practical considerations cause the resolution limit to be larger. For example, a diffraction limited projection aligner used for integrated circuit pattern printing may have a resolution limit of about 3λ. Consider a photon in the near ultraviolet range with an energy, E, of 3.5 eV. The wavelength is given by

$$\lambda = c/f = hc/E \tag{1.1}$$

where c is the speed of light, h is Planck's constant, and f is the frequency. Conveniently, when the energy is expressed in eV, Eq. (1.1) gives a λ value in micrometers, or microns, of approximately $(1.24/E)$. Therefore the wavelength for a 3.5-eV photon is approximately 0.35 μm. Considering the minimum resolution to be on the order of 3λ, let us assume that the pattern resolution is 1 μm and that this corresponds approximately to one feature of an integrated circuit transistor. If, as a consequence, the total transistor size is approximately 10 μm by 10 μm, and if the chip is 1 cm by 1 cm, then the number of transistors per chip is 10^6. Since this corresponds approximately to the state of the art (as of 1987), it appears that VLSI technology is pushing against the fundamental limits of optical microlithography. Indeed this is the case. An obvious solution is to use radiation with a smaller wavelength, such as deep ultraviolet light, electron beams, or x-rays.

Consider a 2-keV x-ray photon. The wavelength is

$$\lambda = [1.24/(2 \times 10^3)] \simeq 6.2 \times 10^{-4}\ \mu\text{m} = 6.2\ \text{Å}$$

or nearly 1000 times less than is the case for optical lithography. With this wavelength, in the absence of other limitations, the number of devices per square centimeter would exceed the remarkable value of 10^{11}. However, owing to practical limitations, the best line widths achieved with either x-ray or electron lithography are several hundred Angstroms.

Quite aside from the microlithographic ones, there are other important limits to consider. One is power dissipation [6]. The more components per unit area, the more problem there is with heat dissipation, particularly for bipolar devices. Another important limit relates to device scaling. X-ray lithography has been used to fabricate functioning FETs with channel lengths of less than 1000 Å. How small can a transistor be before it no longer behaves like a transistor? More fundamentally, how small can a conductor be and still conduct electricity?

The scaling of MOSFETs, which are the dominant VLSI device, is limited by quantum mechanical tunneling through the gate oxide as the gate oxide approaches 50 Å in thickness [7]. As a result, logic-gate scaling laws indicate a minimum silicon MOSFET channel length of about 0.2 μm, more or less, depending on the degree of nonideal effects considered tolerable in the device characteristics. The corresponding transistor density per square centimeter would be approximately 10^8. If the present VLSI chip size of approximately 1 cm^2 were also to increase, say, to 10 cm^2, then it may be speculated that MOSFET-based integrated circuits could, in principle, reach a billion devices per chip before encountering hard physical limits. However, if this level of integration is ever to be achieved in an economically viable framework, many challenging technological problems will need to be solved.

Looking further, future integrated circuits need not be limited to traditional transistor structures or to two-dimensional layouts. Several new devices are currently being investigated as alternatives to conventional transistors, including switching devices based on quantum effects in superlattices, which can be scaled to smaller dimensions than MOSFETs. Furthermore, three-dimensional monolithic structures in which multiple superimposed layers contain interconnected devices offer the possibility of achieving higher circuit densities by vertical integration. Whatever form future integrated circuits take, progress will continue to depend on synergetic efforts in materials and device technology and circuit and system design.

BIBLIOGRAPHY

An interesting history of microelectronics, including an account of key contributions by several individuals to integrated circuit development, is contained in R. M. Warner, Jr., and B. L. Grung. *Transistors, Fundamentals for the Integrated-Circuit Engineer.* New York: Wiley, 1983, Chap. 1.

REFERENCES

1. G. W. A. Dummer, *Proceedings of the Symposium of the IRE-AIEE-RTMA,* Washington, D.C., May 1952.

2. M. F. Wolff, "The Genesis of the Integrated Circuit," *IEEE Spectrum,* 13 (Aug. 1976), 44–53.
3. S. M. Sze, in *Quick Reference Manual for Silicon Integrated Circuit Technology,* eds. W. E. Beadle, J. C. C. Tsai, and R. D. Plummer (New York: Wiley, 1985), 9-16.
4. G. Zimmer and H. Vogt, "CMOS on Buried Nitride—a VLSI SOI Technology," *IEEE Transactions on Electron Devices,* ED-30 (1983), 1515–1520.
5. D. F. Barbe, "VHSIC Systems and Technology," *IEEE Computer Magazine* (Feb. 1981), 13–22.
6. R. W. Keyes, "Physical Limits in Digital Electronics," *IEEE Proceedings,* 63 (1975), 740–767.
7. B. Hoeneisen and C. Mead, "Fundamental Limitations in Microelectronics—I. MOS Technology," *Solid State Electronics,* 15 (1972), 819–829.

2 Processing Fundamentals

2.1 Introduction

The original success of the planar silicon integrated circuit hinged largely on the development of a mutually compatible and economically feasible set of fabrication processes. Today, fundamental processing steps—including oxidation, epitaxial growth, diffusion, ion implantation, film deposition, and etching—are blended in a variety of ways to produce several variations on the monolithic integrated circuit theme. These processes combine concepts from engineering, chemistry, physics, and metallurgy to provide methods for establishing and delineating the intricate patterns of materials that make up a modern integrated circuit. Indeed, the steady evolution toward increasingly powerful integrated circuits is due in large part to a corresponding evolution in processing capabilities.

The properties and performance limitations of transistors and transistor circuits depend strongly on fabrication details such as doping profiles, device-isolation techniques, and interconnect path properties. In turn, these fabrication details are a direct result of the processing steps used during the chip fabrication. Therefore, the fundamentals of integrated circuit processing are an important part of the study of integrated circuit engineering.

2.2 Semiconductor Alternatives

Integrated circuits require a semiconductor in which the active electronic devices may be realized. Semiconductors may be elemental, such as silicon or germanium, or compounds. A few examples of simple binary-compound semiconductors are GaAs, GaP, InP, InSb, CdS, and CdTe. The first four are examples of III–V compounds and the last two are II–VI compounds, so named because of the corresponding columns of the periodic table, shown in Figure 2.1, from which the elements are drawn. However, semiconductors may also exist as more complex materials. Consider, as one of many examples, $GaAs_xP_{1-x}$ where x may vary between 0 and 1. Since fundamental semi-

Period	Group I	Group II											Group III	Group IV	Group V	Group VI	Group VII	Group VIII
1	1.00 **H** 1																	4.00 **He** 2
2	6.94 **Li** 3	9.01 **Be** 4											10.82 **B** 5	12.01 **C** 6	14.01 **N** 7	16.00 **O** 8	19.00 **F** 9	20.18 **Ne** 10
3	22.99 **Na** 11	24.32 **Mg** 12											26.98 **Al** 13	28.09 **Si** 14	30.98 **P** 15	32.07 **S** 16	35.46 **Cl** 17	39.94 **Ar** 18
4	39.10 **K** 19	40.08 **Ca** 20	44.96 **Sc** 21	47.90 **Ti** 22	50.95 **V** 23	52.01 **Cr** 24	54.94 **Mn** 25	55.85 **Fe** 26	58.94 **Co** 27	58.71 **Ni** 28	63.54 **Cu** 29	65.38 **Zn** 30	69.72 **Ga** 31	72.60 **Ge** 32	74.91 **As** 33	78.96 **Se** 34	79.92 **Br** 35	83.8 **Kr** 36
5	85.48 **Rb** 37	87.63 **Sr** 38	88.92 **Y** 39	91.22 **Zr** 40	92.91 **Nb** 41	95.95 **Mo** 42	(99) **Tc** 43	101.1 **Ru** 44	102.91 **Rh** 45	106.7 **Pd** 46	107.88 **Ag** 47	112.41 **Cd** 48	114.82 **In** 49	118.70 **Sn** 50	121.76 **Sb** 51	127.61 **Te** 52	126.91 **I** 53	131.30 **Xe** 54
6	132.91 **Cs** 55	137.36 **Ba** 56	57-71 *	178.58 **Hf** 72	180.95 **Ta** 73	183.86 **W** 74	186.22 **Re** 75	190.2 **Os** 76	192.2 **Ir** 77	195.09 **Pt** 78	197.0 **Au** 79	200.61 **Hg** 80	204.39 **Tl** 81	207.21 **Pb** 82	209.00 **Bi** 83	210 **Po** 84	(211) **At** 85	(222) **Rn** 86
7	(223) **Fr** 87	226.05 **Ra** 88	89-102 **															

*Rare earths	138.92 **La** 57	140.13 **Ce** 58	140.91 **Pr** 59	144.27 **Nd** 60	(145) **Pm** 61	150.35 **Sm** 62	152.0 **Eu** 63	157.26 **Gd** 64	158.93 **Tb** 65	162.51 **Dy** 66	164.94 **Ho** 67	167.27 **Er** 68	168.94 **Tm** 69	173.04 **Yb** 70	174.99 **Lu** 71
**Actinides	227 **Ac** 89	232.05 **Th** 90	231 **Pa** 91	238.07 **U** 92	237 **Np** 93	(244) **Pu** 94	(243) **Am** 95	(247) **Cm** 96	(247) **Bk** 97	(251) **Cf** 98	(254) **Es** 99	(257) **Fm** 100	(258) **Md** 101	(259) **No** 102	(260) **Lr** 103

Figure 2.1 The periodic table. The number above the symbol of each element is its atomic weight, and that below is its atomic number. The elements whose atomic weights are given in parentheses do not occur in nature, but have been prepared artificially in nuclear reactions. The atomic weight in such a case is the mass number of the most long-lived radioactive isotope of the element.

conductor properties such as the energy gap and effective masses vary with the value of x in such compounds, there are in principle an infinite number of possible semiconductors. Nevertheless, at this time, the vast majority of integrated circuits employ silicon, and most of the remaining nonsilicon integrated circuits are GaAs. Although silicon is expected to remain the chief integrated circuit semiconductor for at least the near future, there is considerable research interest in other options.

There are two classes of criteria that determine the suitability of a semiconductor for general integrated circuit applications. The first broad class of criteria concerns the basic physical properties of the material. For example, the carrier mobilities should be high to allow high-speed device operation. Also, the energy gap should be sufficiently high to allow room temperature and near room temperature operation. More specifically, the intrinsic carrier concentration at near room temperature should be much smaller than the usual doping levels. These two conditions combine to require a semiconductor with an energy gap preferably greater than approximately 0.5 eV and with mobilities preferably greater than 1000 $cm^2/V \cdot s$.

The second broad class of criteria concerns the technology associated with the material. The material should be producible in high-purity form with impurities measured in parts per billion and with few crystalline defects. Furthermore, the production of such high-quality material should be economically feasible as well as technically possible. Also, it should be possible to dope the semiconductor, both n- and p-type, over a wide range of carrier concentrations. Finally, the material should lend itself well to standard integrated circuit fabrication processes such as oxidation and etching.

Early workers in integrated circuit development had essentially two choices at the time, silicon and germanium. Silicon won the contest quite easily, although applying the first set of criteria to these two materials does not yield an obvious choice. Germanium has substantially larger electron and hole mobilities than does silicon, but silicon has a larger energy gap and therefore lower junction-leakage currents. Also silicon can operate over a wider temperature range before intrinsic carrier effects become noticeable. However, the deciding factors for silicon over germanium came from the second class of criteria related to the technology. Chiefly, it is relatively easy to produce a high-quality oxide layer on silicon and difficult to do so with germanium. And as noted in Chapter 1, an oxide layer is extremely useful in planar ingegrated circuit processing. Silicon and germanium are included in Table 2.1, which lists mobility and energy gaps for several semiconductors that are of interest for comparative purposes.

Silicon has maintained its preeminence based on technological advantages rather than on inherent advantages in energy gap or mobilities. GaAs, for example, has a higher electron mobility and a larger energy gap than silicon, as seen in Table 2.1. However, oxides and insulator alternatives to

Table 2.1 A comparison of semiconductor properties

Material	Electron mobility ($cm^2/V{\cdot}s$)	Hole mobility ($cm^2/V{\cdot}s$)	Energy gap (eV)
Si[a]	1500	450	1.12
Ge[a]	3900	1900	0.66
GaAs[a]	8500	400	1.42
InP[a]	4600	150	1.35
Cryogenic MBE GaAs[b]	1.06×10^6	—	1.52

[a] Values are for 300 K and for relatively pure material. Doping will reduce mobility values from those given here.
[b] Value is for 4.2 K and refers to a molecular beam epitaxy (MBE)-grown lattice-matched system of GaAs/AlGaAs. (This mobility value from Heiblum et al. [1].)

oxides on GaAs are characterized by a relatively high concentration of defects. Consequently, metal-oxide-semiconductor field-effect transistor (MOSFET) applications are less practical at this time with GaAs integrated circuits, which generally emphasize non-MOS circuitry. InP, which is also listed in Table 2.1, has a high electron mobility and an attractive energy gap. It is considered a potential integrated circuit semiconductor because it appears more suitable for MOS applications than does GaAs.

Molecular beam epitaxy (MBE) offers further intriguing possibilities for advanced integrated circuit materials. In this method, the material is deposited, atomic layer by atomic layer, in an ultra-high vacuum environment. In addition to being well suited to preparation of compound semiconductors, MBE is also very useful for fabricating structures with more than one type of semiconductor. Devices with junctions of dissimilar semiconductors are called heterojunction devices and provide another potential degree of freedom in integrated circuit device design. For example, a bipolar transistor could consist of three different semiconductors, with one material being optimum for the emitter, a second for the base, and a third for the collector [2].

Furthermore, by separating highly doped material from high-purity material, extremely high mobilities have been reported for MBE semiconductors. This technique, which uses selectively doped heterojunctions, is sometimes referred to as modulation doping. For example, an AlGaAs layer may be doped n-type while a neighboring GaAs layer is grown as pure as possible. Since AlGaAs has a lower electron affinity than GaAs, free electrons liberated from donors in AlGaAs reside in the GaAs near the interface. Because there is very little impurity scattering in the highly pure GaAs, large mobilities are observed in the direction parallel to the interface. Electron mobilities over 10^6 $cm^2/V \cdot s$ have been observed at cryogenic temperatures, as noted in Table 2.1 [1].

Since the bulk of integrated circuits are expected to be silicon for the near future, most of the semiconductor-specific material in this text will be based on silicon. However, it may be expected that nonsilicon integrated circuits will find increased use in applications with special requirements, particularly those requiring very high speed and high frequency operation.

2.3 The Silicon Wafer

Silicon for integrated circuit fabrication is provided in the form of wafers, typically 10 to 20 cm in diameter and less than 1 mm in thickness. The raw material from which the silicon is produced is SiO_2 via the reaction

$$SiO_2 + 2C \rightarrow Si + 2CO \tag{2.1}$$

Pure SiO_2 in mineral form is quartz; however, the actual source of silicon dioxide for this reaction is typically quartzite, a granular rock consisting to a large degree of interlocking grains of quartz. The carbon in the reaction is provided by wood chips, coal, and coke. The reaction takes place via intermediate steps at high temperatures in large metallurgical electrode-arc furnaces operating with power sources of approximately 25 MW, and the resulting silicon is referred to as metallurgical-grade (MG) silicon. MG silicon is approximately 98% pure, and there are several approaches for further purification. The general method is to form a silicon-containing gas and then reduce the gas to obtain a pure silicon precipitate. For example, one pathway is

$$\text{MG Si} + 3HCl \rightarrow SiHCl_3 + H_2 \tag{2.2}$$

then,

$$SiHCl_3 + H_2 \rightarrow Si + 3HCl \tag{2.3}$$

(MG Si and Si are solid; the other elements are in the gaseous state.) The reaction in Eq. 2.3, which produces electronic-grade (EG) silicon, is carried out at high temperatures, and the EG silicon is deposited from the chemical vapor onto thin rods of single-crystalline silicon. Although EG silicon is polycrystalline rather than single-crystalline, it is extremely pure, with impurity levels typically expressed in the parts-per-billion range.

EG silicon must next be converted to single-crystalline silicon to achieve the mobilities listed in Table 2.1. This requires remelting the silicon and allowing it to cool in such a fashion that on solidification, the silicon atoms

Figure 2.2 The diamond lattice unit cell. The cube edge, or lattice constant, for silicon is 5.43 Å. In addition to the lattice sites, five interstitial sites large enough to contain a displaced silicon atom are also noted.
SOURCE: From S. K. Ghandhi, *The Theory and Practice of Microelectronics,* New York: John Wiley & Sons, copyright © 1968. Reprinted by permission of John Wiley & Sons, Inc.

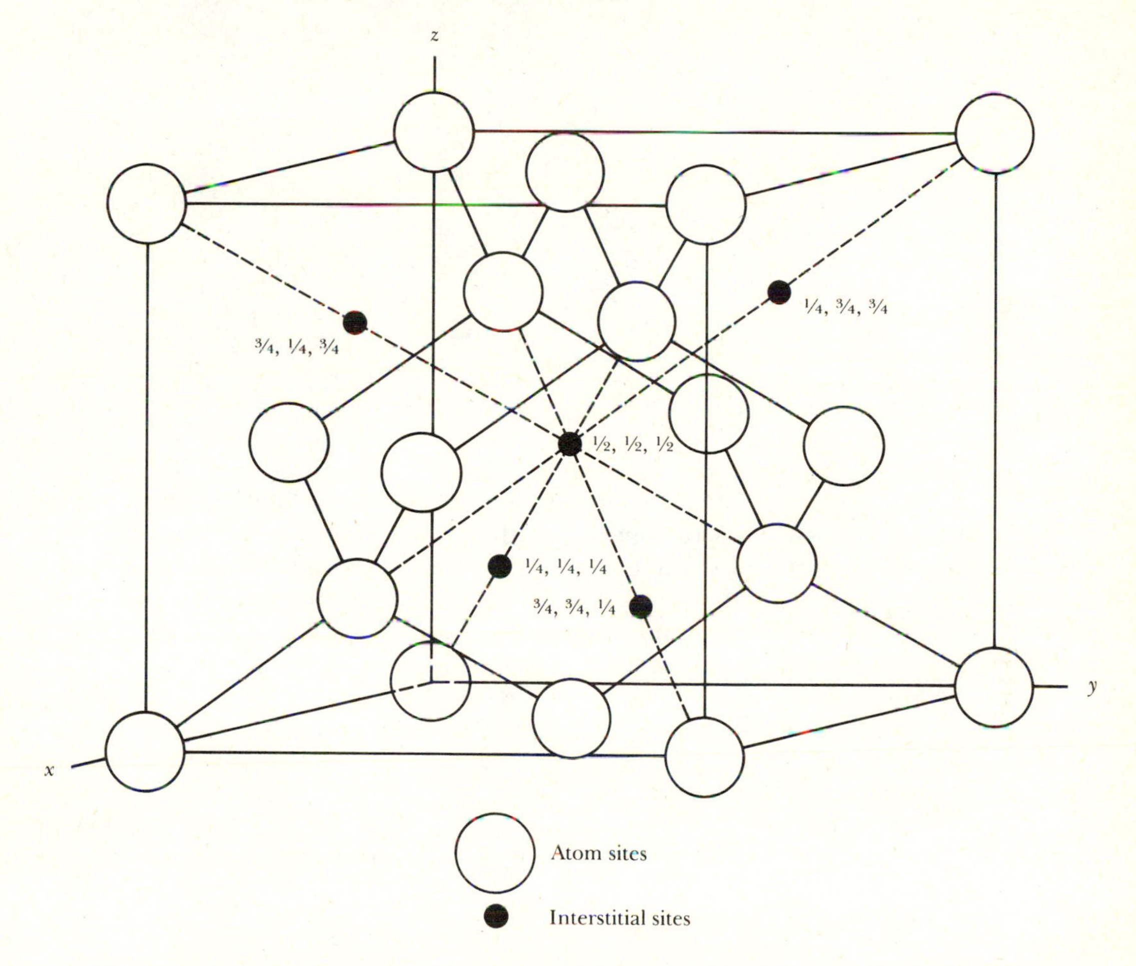

form in the diamond lattice shown in Figure 2.2, which is the lowest energy state for solid silicon. The Czochralski method for crystal growth is most appropriate for large-diameter wafers and is illustrated schematically in Figure 2.3. In this method the crystal is "pulled" from a silicon melt charge such that atoms from the liquid phase solidify at the constantly moving solid-liquid interface. A seed crystal is required to provide the initial growth sites. The combination of pull rate and speed of rotation determines the diameter of the crystalline rod, or ingot extracted from the melt. Since pull rates are typically about 1 mm/min and an ingot is usually more than a meter in length,

Figure 2.3 The Czochralski method of crystal growing.

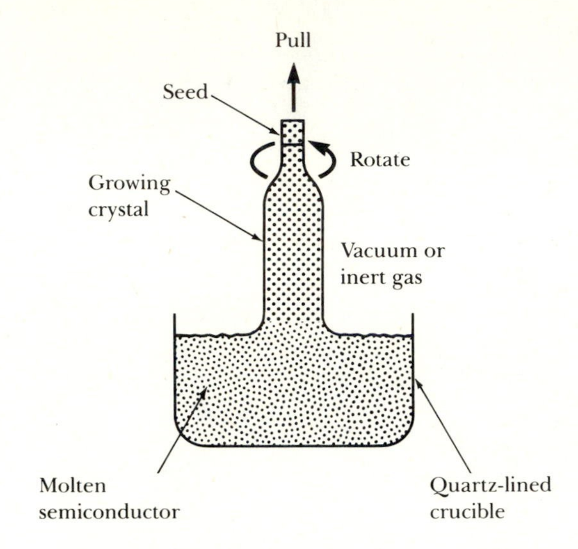

producing one ingot may take days. During this time precise mechanical tolerances are required to maintain controlled pull rates, and vibrations must be minimized. The actual growth of the crystal is monitored remotely by a video camera. Commercial Czochralski crystal growers are approximately 20 feet in height, weigh approximately 3 tons, and are mounted on vibration-reducing platforms to reduce crystal defects. A photograph of a modern crystal grower is shown in Figure 2.4, and a photograph of a silicon ingot grown by the Czochralski method is shown in Figure 2.5.

Since the silicon wafers are to be either *n*-type or *p*-type, it is necessary to add either donors or acceptors to the melt. In the crystal-growing process, impurities tend to be segregated in the melt. If C_S and C_L are the concentration of an impurity solute, by weight, in the solid and liquid phases respectively, then the distribution coefficient, K_d, is defined as

$$K_d = C_S/C_L \tag{2.4}$$

Values of K_d vary with the impurity. For gold, K_d is 2.25×10^{-5}, and for boron K_d is 0.72. The distribution coefficient must be accounted for in calculating the amount of dopant that should be added to the melt to achieve a final donor or acceptor concentration in the crystalline silicon. This calculation is illustrated by the example that follows.

EXAMPLE 2.1

It is desired that a silicon ingot be *p*-type and contain 10^{15} cm^{-3} acceptors. The ingot is drawn by the Czochralski method from a 30-kg charge of mol-

Figure 2.4 Photograph of a computer-controlled Czochralski method crystal puller.
SOURCE: Photo Courtesy of Siltec Corporation, Mountain View, California.

ten silicon and the acceptor is to be boron. How many grams of boron should be added to the crucible? The density of silicon is 2.33 g/cm^3 and the atomic weight of boron is 10.82.

Solution Since K_d has been given as 0.72, the concentration in the melt should be

$$10^{15}/0.72 = 1.39 \times 10^{15}\ \text{cm}^{-3}$$

Figure 2.5 Photograph of a Czochralski-grown silicon ingot.
SOURCE: Photo courtesy of Siltec Corporation, Mountain View, California.

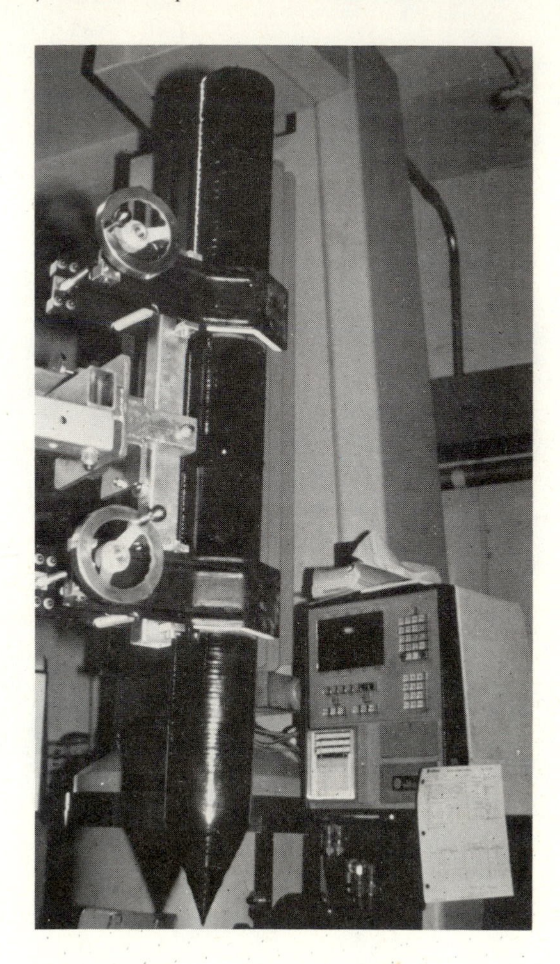

In calculating the volume of the melt, the contribution of the boron may be ignored since it is so low. Consequently, the volume of the melt is

$$\frac{30{,}000 \text{ g}}{2.33 \text{ g/cm}^3} = 1.29 \times 10^4 \text{ cm}^3$$

and the initial number of boron atoms in the melt is

$$(1.29 \times 10^4 \text{ cm}^3)(1.39 \times 10^{15} \text{ cm}^{-3}) = 1.79 \times 10^{19}$$

Therefore, the amount of boron to be added is

$$\frac{1.79 \times 10^{19} \text{ atoms} \times 10.82 \text{ g/mol}}{6.02 \times 10^{23} \text{ atoms/mol}} = 3.21 \times 10^{-4} \text{ g}$$

Since the distribution coefficient is less than unity, the melt will become richer as the crystal grows. Therefore the doping of the ingot increases near the end of the ingot. Also, as a practical note, the boron is not actually added to the melt in elemental form, but rather in the form of heavily doped silicon.

After production of the ingot, the silicon must be shaped into wafers. Thin, diamond-coated saw blades slice the ingot into wafers of a specified thickness and with a surface orientation corresponding to a specified crystal plane. In the process, from one-third to one-half of the carefully prepared ingot ends up as sawdust, or kerf loss. Next the wafers are polished mechanically, or "lapped," with an Al_2O_3 slurry to a high degree of flatness (within approximately 2 μm), and the edges are contoured. A chemical etch removes the top layers of mechanically damaged material, and as a final step the wafers are polished to a mirror finish using a slurry of very fine (100-Å diameter) SiO_2 particles in an aqueous NaOH solution.

2.4 Doping Principles

A pure semiconductor has equal numbers of holes and electrons. The carrier concentrations in equilibrium may be expressed as

$$n_0 = p_0 = n_i = \sqrt{N_C N_V} \exp\left(\frac{-E_G}{2kT}\right) \tag{2.5}$$

where n_0 is the equilibrium electron concentration, p_0 is the equilibrium hole concentration, n_i is the intrinsic concentration, N_C and N_V are the effective density of states for the valence and conduction band respectively, k is Boltzmann's constant, T is the temperature in degrees Kelvin, and E_G is the energy gap. Alternative expressions for n_0 and p_0 are

$$n_0 = N_C \exp[-(E_C - E_F)/kT] \tag{2.6}$$

and

$$p_0 = N_V \exp[-(E_F - E_V)/kT] \tag{2.7}$$

where E_C is the energy at the bottom of the conduction band, E_V is the energy at the top of the valence band, and E_F is the Fermi energy.

For integrated circuit applications, however, the semiconductor is intentionally doped with impurities to produce either p- or n-type material. Equa-

Figure 2.6 Semiconductor energy band diagrams. A. Intrinsic semiconductor. B. n-type semiconductor. C. p-type semiconductor.

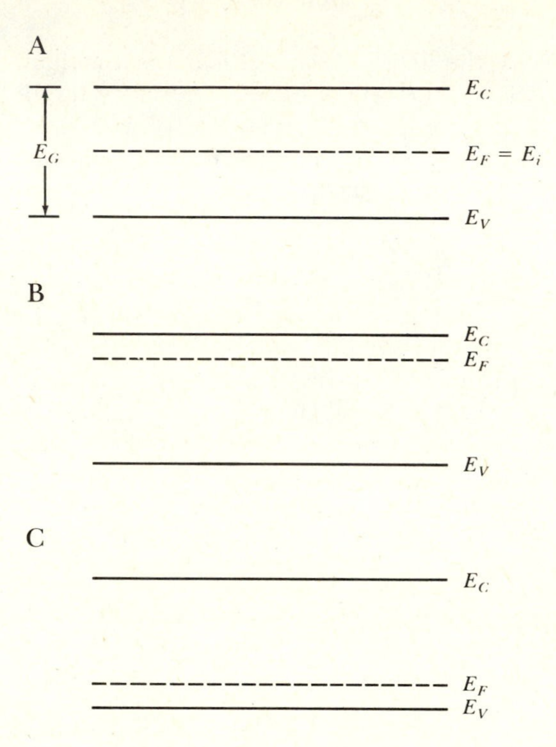

tions (2.6) and (2.7) still apply for doped materials, provided the impurity concentrations are somewhat less than the effective density of states. The energy band diagrams for the cases of intrinsic, n-type, and p-type semiconductors are shown in Figure 2.6.

In principle, the carrier concentrations can be solved for by finding the Fermi energy and applying Eqs. (2.6) and (2.7). In practice, though, it is often more convenient to use the procedure that follows.

As long as the temperature is sufficiently high to ionize all donors and acceptors, then by charge neutrality,

$$n_0 + N_A = p_0 + N_D \tag{2.8}$$

where N_A and N_D are the acceptor and doping concentrations, respectively. The impurity ionization requirement is met at room temperature, provided that doping levels are moderate. Also, by multiplying Eqs. (2.6) and (2.7) and comparing the result to Eq. (2.5), it follows that

$$n_0 p_0 = (n_i)^2 \tag{2.9}$$

Figure 2.7 The intrinsic carrier concentration, n_i, as a function of temperature for Ge, Si, GaAs, and GaP.
SOURCE: From C. D. Thurmond, "The Standard Thermodynamic Functions for the Formation of Electrons and Holes in Ge, Si, GaAs, and GaP," *Journal of the Electrochemical Society*, 122 (1975), 1133–1141. Reprinted by permission of the publisher, The Electrochemical Society.

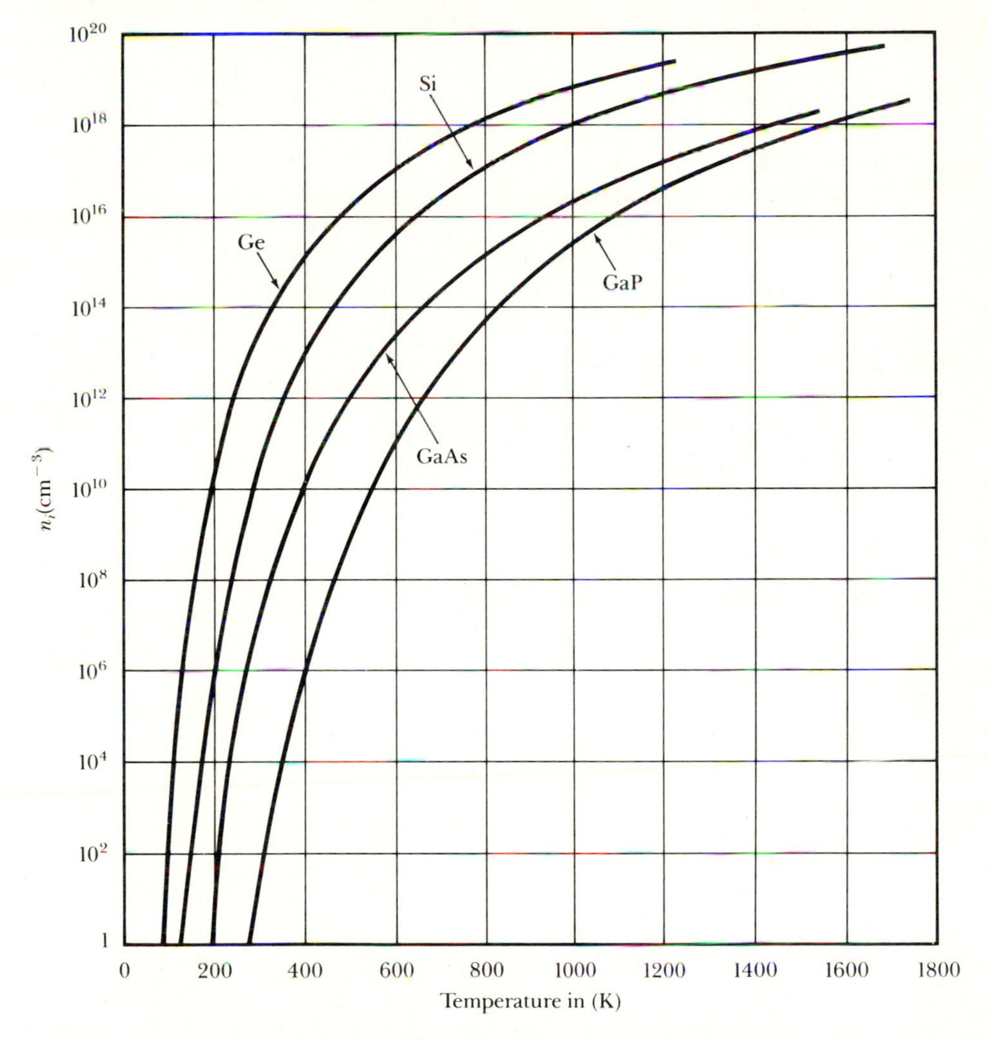

As shown in Figure 2.7, at a given temperature, for a given semiconductor, n_i is a constant except under exceptionally high doping conditions that cause energy gap narrowing [3]. Therefore, if the doping concentrations are given, Eqs. (2.8) and (2.9) represent two equations for the two unknown carrier concentrations n_0 and p_0. For the case of n-type material—that is, when $N_D > N_A$—if Eq. (2.8) is multiplied through by n_0 and Eq. (2.9) is used to eliminate the n_0p_0 term, then the resulting quadratic equation has the following solution, keeping only the root with physical meaning:

$$n_0 = \tfrac{1}{2}\left[(N_D - N_A) + \sqrt{(N_D - N_A)^2 + 4(n_i)^2}\,\right] \tag{2.10}$$

For p-type material—that is, where $N_A > N_D$—it is convenient to multiply Eq. (2.8) by p_0 and solve the resulting quadratic equation for p_0 as follows:

$$p_0 = \tfrac{1}{2}\left[(N_A - N_D) + \sqrt{(N_A - N_D)^2 + 4(n_i)^2}\,\right] \tag{2.11}$$

In any case, having solved for one of the carrier concentrations, the other may be found using Eq. (2.9).

EXAMPLE 2.2

Consider a silicon wafer with the following impurity densities, which are uniformly distributed throughout the wafer:

Phosphorus at 10^{16} cm^{-3}
Arsenic at 4×10^{16} cm^{-3}
Boron at 10^{17} cm^{-3}

Calculate the hole and electron concentrations at 300 K. The value of n_i at 300 K is 1.45×10^{10} cm^{-3}.

Solution Since phosphorus and arsenic are from column V of the periodic table, they have five outer electrons. Only four electrons are required for covalent bonding in the silicon lattice, so the fifth electron is weakly attached. Therefore phosphorus and arsenic act as donors in silicon. Boron is from column III of the periodic table and therefore has three outer electrons. As a result there is a vacant valence bond and boron acts as an acceptor.

Using Eq. (2.11), it may be seen that $(N_A - N_D)^2$ is much larger than $(n_i)^2$, so to a good approximation,

$$p_0 \simeq N_A - N_D = 5 \times 10^{16} \text{ cm}^{-3}$$

Therefore, from Eq. (2.9), n_0 is equal to

$$n_0 = \frac{(1.45 \times 10^{10})^2}{5 \times 10^{16}} = 4.2 \times 10^3 \text{ cm}^{-3}$$

For this example, the holes are the majority carriers and the electrons are the minority carriers.

Example 2.2 illustrates two important points. First, so long as the net doping concentration, donors minus acceptors or acceptors minus donors, is considerably greater than n_i, then the majority carrier concentration is essentially equal to the net doping concentration. Under these conditions, the semiconductor is in the extrinsic state. The second point to note is that when

Figure 2.8 Compensation. A. Compensation of acceptors by donors. B. Compensation of donors by acceptors.

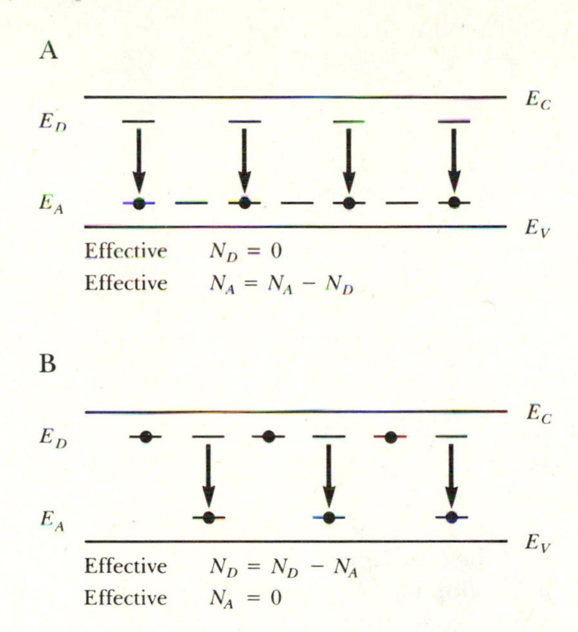

multiple dopants are present, compensation may take place. In Ex. 2.2, the effective acceptor concentration is $N_A - N_D$. The 5×10^{16} cm^{-3} donor impurities have effectively donated their electrons to an equal number of acceptors, thereby canceling the effect of 5×10^{16} cm^{-3} acceptors. The process of compensation is shown from an energy band point of view in Figure 2.8, with donor states near the conduction band and acceptor states near the valence band. Since much of the semiconductor in an integrated circuit contains multiple types of dopants, compensation is an important concept.

The minority carrier concentration is always strongly temperature dependent because n_i increases exponentially with temperature. However, in the extrinsic region, the majority carrier concentration is fairly constant with changing temperature. The low temperature end of the extrinsic region is determined by carrier freezeout, which refers to the case of incompletely ionized donors and acceptors. At temperatures corresponding to carrier freezeout in n-type material, for example, the majority carrier concentration decreases with decreasing temperature since there is insufficient thermal energy to release all the weakly bound fifth electrons on donor sites. The high temperature end of the extrinsic region is determined by the point at which the intrinsic carrier concentration is not negligible in Eqs. (2.10) and (2.11), and the majority carrier concentration begins to significantly increase with increasing temperature owing to broken Si–Si bonds. Temperature endpoints for the extrinsic region depend strongly on the doping levels.

Devices are generally operated at temperatures such that the extrinsic condition holds in regions of moderate doping concentrations. However, it is often important to know the carrier concentrations under high temperature processing conditions, such as diffusions, where the majority carrier concentration is not simply equal to the net doping concentration.

EXAMPLE 2.3

Consider the silicon wafer in Ex. 2.2, but at a temperature of 900°C, where n_i is 3.5×10^{18} cm^{-3}. Now what are the values of n_0 and p_0?

Solution In this case, the intrinsic carrier concentration is much greater than the net doping concentration so, from Eq. (2.11),

$$p_0 \simeq n_i = 3.5 \times 10^{18} \text{ cm}^{-3}$$

From Eq. (2.9),

$$n_0 \simeq n_i = 3.5 \times 10^{18} \text{ cm}^{-3}$$

Under the condition that the intrinsic carrier concentration is much greater than the net doping concentration, the semiconductor is in the intrinsic state. The intrinsic condition may correspond to pure material where the doping concentration is negligible, or to materials with perfect compensation, or as in Ex. 2.3, to doped materials at sufficiently high temperatures.

In addition to donors and acceptors that are added to the semiconductor to tailor the carrier concentrations to desired values, there are also impurities that act as neither donors or acceptors. Gold, for example, introduces deep-lying states near the middle of the energy gap. These energy levels have essentially no effect on n_0 or p_0, but they have a marked effect on the lifetime of excess carriers. Gold impurities may reduce carrier lifetimes in silicon from 10^{-5} seconds to 10^{-10} seconds, a fact sometimes used to good effect in high-speed switching devices.

2.5 Diffusion

2.5.1 Basic Concepts

Diffusion is one of the two principal ways of introducing impurities in a controlled fashion into the wafer (the second way is ion implantation). Furthermore, diffusion phenomena are also important in oxidation processes, in post–ion implant annealing steps, and in impurity motion during the growth of epitaxial layers. The diffusion method relies on the fact that a concentration gradient, in the absence of opposing forces, gives rise to a flux. This flux is simply the result of random thermal motion. For example, if a puff of hydrogen is released from a gas cylinder in one corner of a closed room, a net flux of H_2 follows until the concentration in the room is constant. Since the gas molecules have random thermal motion, hydrogen will first flow away from the initial volume, but not in; so the hydrogen spreads out, or diffuses. Equilibrium is reached when each volume element in the room has, on the average, the same inward flux of hydrogen as outward. The higher the temperature, the higher the thermal velocity and the faster diffusion takes place.

Diffusion also takes place in solids. If an atmosphere containing boron, for example, is maintained at the surface of a silicon wafer, then there will be a flux of boron into the silicon wafer. At room temperature, that flux would be so small that it would not be perceptible, since diffusion takes place more slowly in solids than in gases. Therefore, semiconductor diffusions are carried out in a temperature range of approximately 900 to 1100°C.

Referring to Figures 2.2 and 2.9A, the most obvious way for impurity atoms to move through the crystal is by squeezing between the crystal's atoms, moving from interstitial site to interstitial site. In silicon, this mechanism is indeed favored by impurities with small ionic radii, such as elements from columns I and VIII of the periodic table. However, impurities from columns III and V diffuse in silicon mainly by substitutional diffusion. Substitutional diffusion requires a point defect at one of the lattice sites, as illustrated in Figure 2.9B. In the first mechanism shown, the point defect is a silicon vacancy, Si_V, to which a neighboring impurity moves. If there are silicon vacancies, then interstitial silicon point defects, Si_I, are also expected. The second part of Figure 2.9B shows one way in which interstitial silicon defects can be involved in the diffusion process. Here the Si_I displaces an impurity atom, which travels interstitially to a silicon vacancy. Evidence exists that the first mechanism of Figure 2.9B dominates diffusion of donors and acceptors in silicon at low temperatures, but that the second mechanism dominates at temperatures greater than about 1100°C [4]. A vacancy that results from mising atoms is called a Schottky defect. A Frenkel defect refers to a vacancy-interstitial pair, with a silicon atom in an interstitial lattice site and a nearby associated vacancy.

Examples of interstitial diffusing elements are Li, S, Fe, Cu, Au, O, and Ni. Examples of substitutional impurities include P, B, Sb, As, In, Ga, Al, and Au. Note that gold employs both methods.

Interstitial diffusion occurs at a much faster rate than substitutional diffusion, as can be seen by further considering the mechanisms involved in the two cases. If a boron atom, for example, is to move substitutionally in silicon, it must find vacant lattice sites; in other words, there must be an Si_V. Such a defect is a result of thermal energy fluctuations in the crystal. The expected number of such defects can be calculated, based on a Boltzmann distribution in energy, according to

$$n_{\mathrm{DEFECT}} = N[\exp(-E_{\mathrm{DEFECT}}/kT)] \tag{2.12}$$

Figure 2.9 Diffusion mechanisms. A. Interstitial diffusion. B. Two mechanisms for substitutional diffusion.

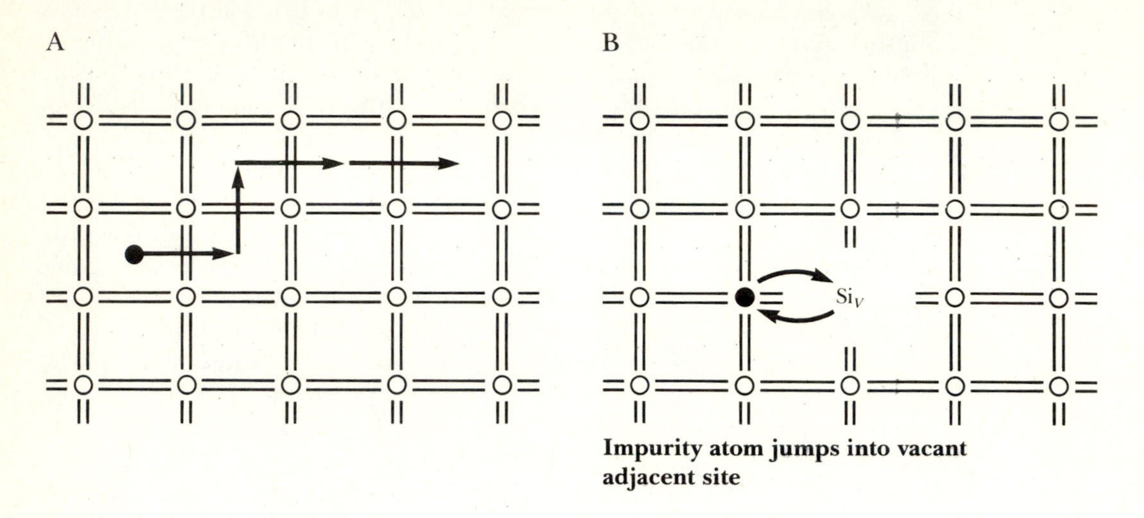

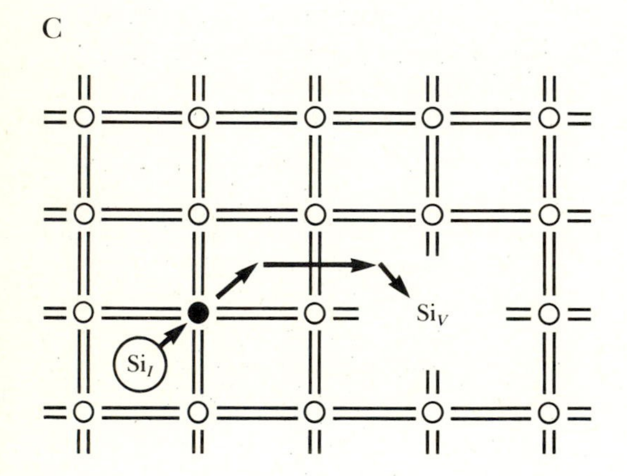

An interstitial silicon atom displaces a substitutional impurity that subsequently travels to a nearby vacant site

where N is the number of atoms in the crystal and the value of E_{DEFECT} is material dependent. For silicon, E_{DEFECT} is approximately 2.3 eV for Schottky defects.

EXAMPLE 2.4

Calculate the expected number of Schottky point-defect silicon vacancies per cubic centimeter at 20°C and at 1100°C.

Solution There are 5.0×10^{22} silicon atoms per cubic centimeter. Since k is equal to 8.62×10^{-5} eV/K, at 20°C, or 293 K,

$$n_{\text{DEFECT}} = 5.0 \times 10^{22} \exp\left[\frac{-2.3}{(8.62 \times 10^{-5})(293)}\right]$$

$$= 1.41 \times 10^{-17} \text{ cm}^{-3}$$

which is certainly negligible. The only Schottky defect vacancies expected at room temperatures are those that were "frozen in" at the time the ingot was grown or as a result of fast-cooling after a high-temperature wafer-processing procedure.

However, when 293 K is replaced by 1373 K in this calculation,

$$n_{\text{DEFECT}} = 1.82 \times 10^{14} \text{ cm}^{-3}$$

Clearly the concentration of point defects is strongly temperature dependent.

An impurity diffusing by substitution must have a neighboring vacancy. In addition to finding a vacancy, it must break the covalent bond holding it, which requires energy, E_{BOND}. The bond-breaking probability is also thermally activated, so that the rate of substitutional diffusions per second is given by

$$\nu_{\text{SUBST}} = 4\nu_o \exp(-E_{\text{BOND}}/kT) \exp(-E_{\text{DEFECT}}/kT) \tag{2.13}$$

where 4 is the number of neighbors in the diamond lattice, ν_o is the lattice vibration frequency, and the exponential terms represent probabilities of bond breaking and defect formation. A typical ν_o is 10^{14} s^{-1}. The value of E_{BOND} is dependent on both the type of impurity and the material in which the impurity is moving. The notation ν_{SUBST} may be interpreted as the frequency with which thermal energy fluctuations occur with sufficient magnitude to allow substitutional motion. In other words, it is the jump rate.

For interstitial impurities, on the other hand, there is always a plentiful supply of vacant interstitial sites, as shown in Figure 2.2. Also, there are no covalent bonds that must be broken to allow a jump. Nevertheless, the impurities must squeeze through constrictions in the lattice, which requires an interaction energy, E_{INTER}. The interstitial jump rate may be expressed as

$$\nu_{\text{INTER}} = 4\nu_o \exp(-E_{\text{INTER}}/kT) \tag{2.14}$$

since there are 4 neighboring interstitial sites. Values of E_{INTER} are approximately 1 to 2 eV, so at a given temperature, ν_{INTER} will be substantially larger than ν_{SUBST}.

Figure 2.10 Models for Fick's laws of diffusion. A. Model for the first law. B. Model for the second law.

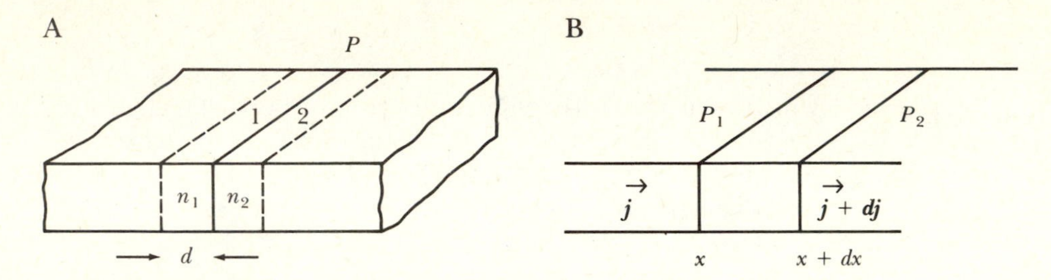

2.5.2 Fick's Equations

Diffusion may be treated quantitatively using Fick's first and second laws of diffusion [5]. Consider a crystal of cross-sectional area A. For simplicity's sake, consider a simple cubic lattice with a spacing between crystal planes of d, and consider two adjacent layers, 1 and 2, separated by d as shown in Figure 2.10. Let

$\mathcal{N}_1$ = number of impurity atoms in layer 1
$\mathcal{N}_2$ = number of impurity atoms in layer 2
ν = frequency with which impurity atoms jump from one layer to another

Then the net flow of impurity atoms across the dividing plane, P, is per time increment Δt,

$$\Delta\mathcal{N}/\Delta t = \nu(\mathcal{N}_1/2 - \mathcal{N}_2/2) \tag{2.15}$$

Here positive flow is taken from left to right. We have allowed that thermal motion is random; thus, at any time, half of the moving impurity atoms in a given layer will be moving to the left and half to the right.

The volume concentrations are $N_1 = \mathcal{N}_1/Ad$, and $N_2 = \mathcal{N}_2/Ad$, so

$$\Delta\mathcal{N}/\Delta t = (\nu Ad/2)(N_1 - N_2) \tag{2.16}$$

Also, the impurity concentration gradient may be expressed as

$$\Delta N/\Delta x = (N_2 - N_1)/d \tag{2.17}$$

Combining Eqs. (2.16) and (2.17) yields a relationship between the net impurity flow and the concentration gradient as follows:

$$\Delta\mathcal{N}/\Delta t = -(\nu Ad^2/2)(\Delta N/\Delta x) \tag{2.18}$$

Equation (2.18) may be written as

$$j = -D(\partial N/\partial x) \tag{2.19}$$

where $j = (\Delta N/\Delta t)A^{-1}$ is the impurity flux in atoms/cm$^2\cdot$s, $D = \nu d^2/2$ is the diffusion coefficient, and the concentration gradient is expressed as a derivative. Equation (2.19) is Fick's first law. In three dimensions,

$$j = -D\,\nabla N \tag{2.20}$$

From the previous discussion of the jump rate, ν, D is expected to be strongly temperature dependent and to be a function of both the impurity type and the host material. From Eq. (2.13), D may be expressed as

$$D = D_0 \exp(-E_A/kT) \tag{2.21}$$

where the prefactor, D_0, is given by

$$D_0 = 4\nu_0(d^2/2) \tag{2.22}$$

and the activation energy for substitutional impurities is

$$E_A = E_{\text{DEFECT}} + E_{\text{BOND}} \tag{2.23}$$

Actually, the single exponential relationship for D in Eq. (2.21) is considerably oversimplified because of some physical phenomena not considered in the jump rate discussion. These phenomena will be treated in the next section. However, for limited temperature ranges and moderate doping concentrations, Eq. (2.21) approximates D. For boron and phosphorus in silicon, for example, values of D_0 and E_A of 10.5 cm^2/s and 3.69 eV, respectively, give approximate average values of D between about 950°C and 1200°C [6].

EXAMPLE 2.5

Calculate the diffusion coefficient for boron at 1000°C and 1100°C, assuming the simple expression in Eq. (2.21) is appropriate for the doping concentrations that are of interest.

Solution At 1000°C,

$$D = 10.5 \exp\left[\frac{-3.69}{(8.62 \times 10^{-5})(1000 + 273)}\right]$$

$$D = 2.61 \times 10^{-14}\ \text{cm}^2/\text{s}$$

At 1100°C,

$$D = 10.5 \exp\left[\frac{-3.69}{(8.62 \times 10^{-5})(1100 + 273)}\right]$$
$$D = 3.03 \times 10^{-13}\ \text{cm}^2/\text{s}$$

Fick's first law provides the impurity flux given the concentration gradient and the diffusion coefficient. However, it does not provide the means to solve for $N(x,t)$ given the boundary conditions and temperature. For this solution, Fick's second law, which follows from an application of flow-continuity considerations, is required.

Refe ring to Figure 2.10B, the number of impurity atoms per second entering the region between x and $x + dx$ through plane P_1 is jA, and the number per second leaving that region through plane P_2 is $(j + dj)A$. Therefore, the net number of impurity atoms per second leaving the center region is Adj. This net rate of change may be expressed mathematically as

$$-Adj = (\partial N/\partial t)Adx \tag{2.24}$$

where N is the impurity concentration per unit volume and Adx is the volume of the center region of interest. Therefore,

$$-(\partial N/\partial t) = \partial j/\partial x \tag{2.25}$$

Combining Fick's first law, Eq. (2.19), with Eq. (2.25) yields

$$\partial N/\partial t = D(\partial^2 N/\partial x^2) \tag{2.26}$$

which is Fick's second diffusion law. In three dimensions,

$$\partial N/\partial t = D(\nabla^2 N) \tag{2.27}$$

This equation can be solved for $N(x,t)$ if the boundary conditions are known, that is, if the details of how the diffusion process is actually carried out are known.

2.5.3 A Second Look at D

The simple exponential form for D shown in Eq. (2.21) is a fairly crude approximation because of two important factors that are addressed in this section. The first is that impurities are often charged, and therefore, a built-in electric field is associated with the diffusion process. If there is a gradient of donors, for example, there will also be a gradient of electrons. Since electrons are more mobile than atoms, they have a much higher diffusion rate than do the donor atoms and "out-run" them. This creates a charge separation and an electric field that enhances motion of the positively charged donor atoms in the direction of the gradient.

The electric field may be accounted for by modifying Fick's first equation according to

$$j = -D\nabla N + \mu N\mathscr{E} \tag{2.28}$$

where $\mathscr{E}$ is the electric field vector and μ is the impurity mobility. Using the Einstein relationship between D and μ, Eq. (2.28) can be rewritten as

$$j = -D\nabla N + (qD/kT)N\mathscr{E} \tag{2.29}$$

However, since the electric field is related to the concentration gradient, a more convenient form of Eq. (2.29) may be derived, namely,

$$j = -hD\nabla N \tag{2.30}$$

where

$$h = 1 + [1 + 4(n_i/N)^2]^{-1/2} \tag{2.31}$$

The derivation of equations (2.30) and (2.31) is an exercise in the chapter problems.

The second consideration in achieving more accurate diffusion calculations is that point-defect vacancies in a semiconductor crystal may exist in various states of electrical charge. For example, V^0, V^+, V^-, and V^{--} refer to vacancies with neutral, single positive, single negative, and double negative charges, respectively. A different diffusion rate is associated with each type of vacancy. Furthermore, the number of each vacancy type present at a given temperature depends on the concentration of holes and electrons. For example, the number of positively charged vacancies increases as the hole concentration increases.

Taking into account both the electric field enhancement and the charged vacancy effects, the general expression for D may be written as

$$D = h[D^0 + D^-(n/n_i) + D^{--}(n/n_i)^2 + D^+(p/n_i)] \tag{2.32}$$

where n and p are the hole and electron concentrations, respectively [7]. Each of the superscripted diffusion coefficients is exponentially dependent on temperature with a prefactor and activation energy, as shown in Eq. (2.21). Values of the activation energies and prefactors, as described in the previous section, are listed in Table 2.2 for boron, phosphorus, arsenic, and antimony in silicon. Predictably, the boron diffusion rates associated with negatively charged vacancies are negligible, since the boron atoms are also negatively charged. Likewise, the donor impurities, which are positively charged, diffuse via either neutral vacancies or negatively charged vacancies. Arsenic and

Table 2.2 Activation energy and prefactor for diffusion coefficients in silicon[a]

Vacancy charge		Boron	Arsenic	Phosphorus	Antimony
Neutral	D_0	0.091	0.38	3.85	0.214
	E_A	3.36	3.58	3.66	3.65
Single negative	D_0	—	22.9	4.44	13
	E_A	—	4.1	4.0	4.0
Double negative	D_0	—	—	44.2	—
	E_A	—	—	4.37	—
Single positive	D_0	166.3	—	—	—
	E_A	4.08	—	—	—

[a] The units of E_A are eV, and the units of D_0 are cm^2/s.

Source: From R. B. Fair, "Recent Advances in Implantation and Diffusion Modeling for the Design and Process Control of Bipolar IC's," in *Semiconductor Silicon 1977*, eds. H. R. Huff and E. Sirtl (Princeton: Electrochemical Society, 1977), pp. 968–987. This table was originally presented at the Spring 1977 Meeting of the Electrochemical Society, Inc., held in Philadelphia, Pennsylvania.

antimony diffuse mainly by V^0 and V^-, but phosphorus diffuses by V^0, V^-, and V^{--}. The physical reasons for the difference between phosphorus and the other two donors may not be clear, but the results of that difference are significant.

EXAMPLE 2.6

Use Eq. (2.32) and the information in Table 2.2 to calculate the boron diffusion coefficient at 1000°C for boron concentrations of 10^{18} cm^{-3} and 10^{19} cm^{-3}. The intrinsic carrier concentration is 6.5×10^{18} cm^{-3}, and the wafer is n-type, with a uniform background doping of 10^{16} cm^{-3}. Compare the solution with the results of Ex. 2.5.

Solution For boron, equation (2.32) becomes

$$D = h[D^0 + D^+(p/n_i)]$$

where p is evaluated using Eq. (2.11). Using the tabulated values in Table 2.2,

$$D^0 = 0.091 \exp\left[\frac{-3.36}{(8.62 \times 10^{-5})(273 + 1000)}\right]$$

or

$$D^0 = 4.58 \times 10^{-15} \text{ cm}^2/\text{s}$$

Likewise,

$$D^+ = 166.3 \exp[-4.08/(8.62 \times 10^{-5})(273 + 1000)]$$

or

$$D^+ = 1.18 \times 10^{-14} \text{ cm}^2/\text{s}$$

In applying Eq. (2.11), note that $N_A = N_{\text{BORON}}$, and N_D is the background doping, 10^{16} cm^{-3}. The results are summarized in the following table.

N_{BORON} (cm^{-3})	p (cm^{-3})	h	D cm^2/s
10^{18}	7.01×10^{18}	1.08	1.87×10^{-14}
10^{19}	1.32×10^{19}	1.61	4.61×10^{-14}

The approximate approach of Ex. 2.5, which did not consider the dependence of D on N, produced a value for D of 2.61×10^{-14} cm^2/s, which lies between the more exact values arrived at in this example.

2.5.4 Solutions to the Diffusion Equation

Since the diffusion coefficient depends on the impurity concentration N, as well as on temperature and the wafer doping, a closed-form solution to Fick's second equation is usually not achieved. Therefore, numerical techniques are required for accurate solutions. However, it is instructive to consider the simpler case where D is assumed to be constant with N. In this case, closed-form solutions are obtainable and, although the results are only approximate, the solutions do provide an appreciation for the basic nature of diffusion profiles.

Solutions to Eq. (2.27) will be considered for two sets of boundary conditions, namely, the constant-source and limited-source conditions. Most integrated circuit diffusion processes can be treated approximately as either one of these two cases or a combination of the two. In both cases a semi-infinite wafer will be assumed, with the y-axis along the horizontal surface of the wafer and the x-axis vertical into the wafer. Since a typical wafer thickness is greater than 100 μm and a typical diffusion depth is less than 10 μm, the semi-infinite approximation is reasonable in the vertical direction. It is also acceptable in the horizontal direction if the point of interest is not near the edge of a diffusion mask.

The constant-source condition refers to the case where a constant number of impurity atoms per cubic centimeter, N_o, is maintained at the surface of the wafer. If the diffusion is carried out in an atmosphere sufficiently rich in the impurity, then N_o is the solid solubility limit, that is, the maximum density of impurity atoms that the host crystal can accommodate without a serious distortion of the crystalline structure. Figure 2.11 shows the values, as a function of temperature, of N_o as the solid solubility limit for several impurities in silicon.

If the impurity is phosphorus, for example, and the temperature is 1000°C, then N_o is approximately 9×10^{20} cm^{-3}. Suppose that a wafer is placed in such an atmosphere for a certain length of time. What is the resulting distribution of phosphorus atoms in the wafer? Taking $x = 0$ at the wafer surface and assuming that the wafer is placed in the atmosphere at time $t = 0$, the boundary conditions on $N(x,t)$ are

$$N(x,0^-) = 0 \tag{2.33}$$

and

$$N(0,t) = N_o u(t) \tag{2.34}$$

where $u(t)$ is the step function. Applying a Laplace transform to Eq. (2.26),

$$sN(x,s) - N(x,t = 0^-) = D\left[\frac{d^2N(x,s)}{dx^2}\right] \tag{2.35}$$

where $N(x,s)$ is the Laplace transform of $N(x,t)$.

The general solution to Eq. (2.35), considering the first boundary condition and that the wafer is semi-infinite, is

$$N(x,s) = B \exp\left(-x\sqrt{\frac{s}{D}}\right) \tag{2.36}$$

B is found from the second boundary condition, so that,

$$N(x,s) = (N_o/s)\exp[-(s/D)^{1/2}x] \tag{2.37}$$

Finally, taking the inverse transform,

$$N(x,t) = N_o \operatorname{erfc} \frac{x}{2\sqrt{Dt}} \tag{2.38}$$

where erfc is the complementary error function plotted in Figure 2.12.

EXAMPLE 2.7

A constant-source boron diffusion is carried out at 1050°C into an n-type silicon wafer with uniform background doping, N_B, where $N_B = 10^{16}$ cm^{-3}. The surface concentration is maintained at 4×10^{20} cm^{-3}, which is, incidentally, the solid solubility limit. A junction depth of 1 μm is desired. What should be the diffusion time?

Figure 2.11 Solid solubility of impurities in silicon.
SOURCE: Original art appeared in "Solid Solubilities of Impurity Elements in Germanium and Silicon," by F. A. Trumbore, *Bell System Technical Journal,* 39 (1960), 205–234. Reprinted with permission from the *AT&T Technical Journal,* copyright © 1960, AT&T. Adapted by S. K. Ghandhi, *The Theory and Practice of Microelectronics* (New York: John Wiley & Sons, copyright © 1968). Reprinted by permission of John Wiley & Sons, Inc.

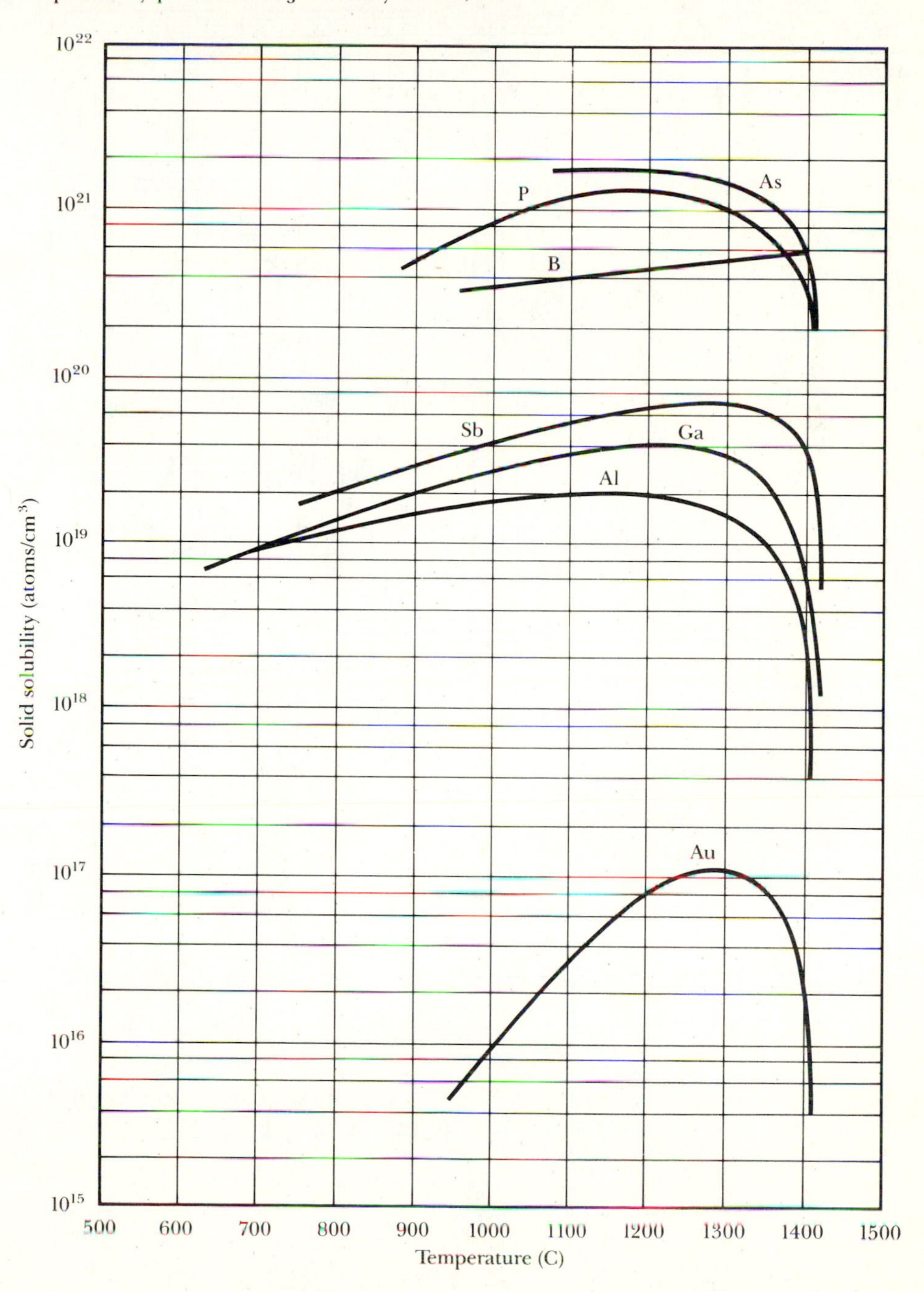

Figure 2.12 Gaussian and erfc diffusion profiles.

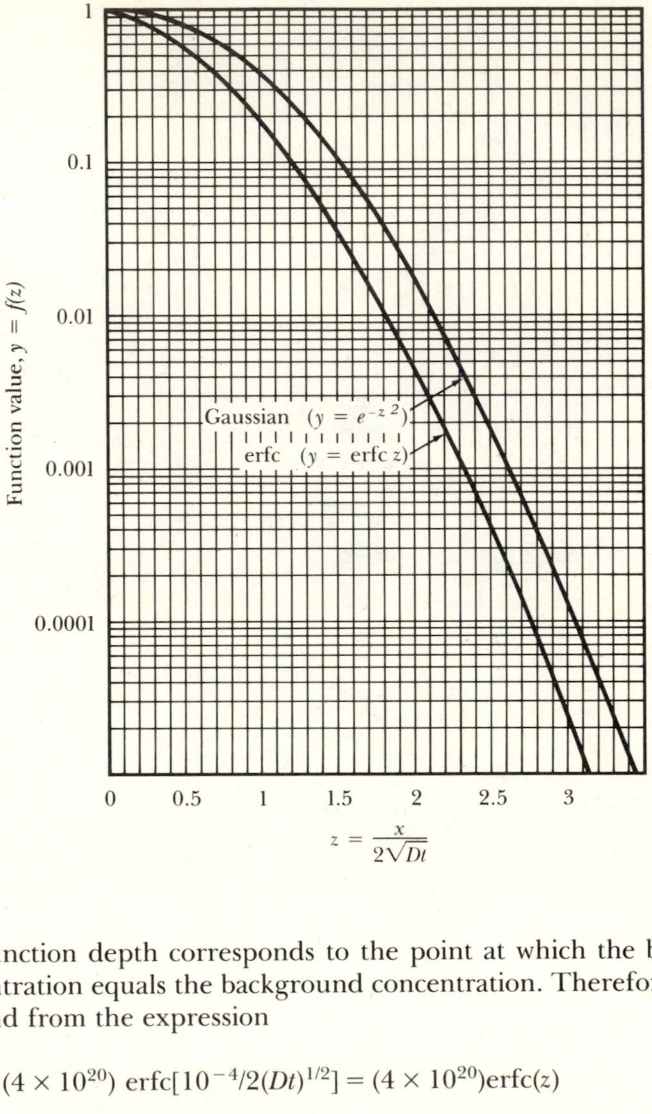

Solution The junction depth corresponds to the point at which the boron atom concentration equals the background concentration. Therefore, the time is found from the expression

$$10^{16} = (4 \times 10^{20})\ \mathrm{erfc}[10^{-4}/2(Dt)^{1/2}] = (4 \times 10^{20})\mathrm{erfc}(z)$$

Using Eq. (2.21) and the values of D_0 and E_A in Section 2.5.2, an approximation of D can be found:

$$D = 10.5\ \exp\left[\frac{-3.69}{(8.62 \times 10^{-5})(1050 + 273)}\right]$$

or

$D = 9.31 \times 10^{-14}\ \text{cm}^2/\text{s}$

Since $\text{erfc}(z) = 2.5 \times 10^{-5}$ in this example, z is found graphically from Figure 2.12 to equal 3.0. Therefore,

$t = (10^{-4}/6)^2\ D^{-1} = 2.98 \times 10^3\ \text{s} = 49.7\ \text{min}$

The second set of boundary conditions to be considered is the limited source, in which the diffusion is begun with Q/A impurities per unit area on the wafer surface. Furthermore, this initial condition is the sole source of diffusing impurities; there are no additional impurities in the diffusion atmosphere. Now the initial condition is

$$N(x,0^-) = (Q/A)\ \delta(x) \tag{2.39}$$

where $\delta(x)$ is the delta function. Again, the solution to Fick's second diffusion equation is Eq. (2.36). Solving for B yields

$$N(x,s) = \frac{Q}{A\sqrt{Ds}} \exp\left(-\sqrt{\frac{s}{D}}\,x\right) \tag{2.40}$$

which, after the inverse Laplace transform, becomes

$$N(x,t) = \frac{Q}{A\sqrt{\pi Dt}} \exp\left(\frac{-x^2}{4Dt}\right) \tag{2.41}$$

which is a Gaussian distribution.

The Gaussian profile is also illustrated in Figure 2.12, in which both the erfc and Gaussian distributions are normalized to unity at the surface. However, it should be noted that in contrast to constant-source diffusions, the integral of the limited-source diffusion profile is always Q/A. As the diffusion time increases, the surface-impurity concentration decreases for limited-source diffusions but remains constant for constant-source diffusions. This important difference is illustrated in Figure 2.13.

Integrated circuit diffusions are often done in two steps, the predeposition diffusion and the drive-in diffusion. The first is a constant-source diffusion, and the second is a variation on the limited-source diffusion. A typical predeposition temperature is 950°C and a typical time is 30 minutes. The relatively low temperatures and short times result in shallow diffusion profiles, which serve as the initial condition for the following drive-in diffusion. Since the drive-in is done in an atmosphere containing oxygen but not dopants, a limited-source condition applies. In this case the initial profile is not an ideal delta function but rather a shallow Gaussian profile. To a first approximation, however, the area under the predeposition profile may be in-

Figures 2.13 Gaussian and erfc diffusion profiles for increasing diffusion times. A. Limited source (Gaussian). B. Constant source (erfc).

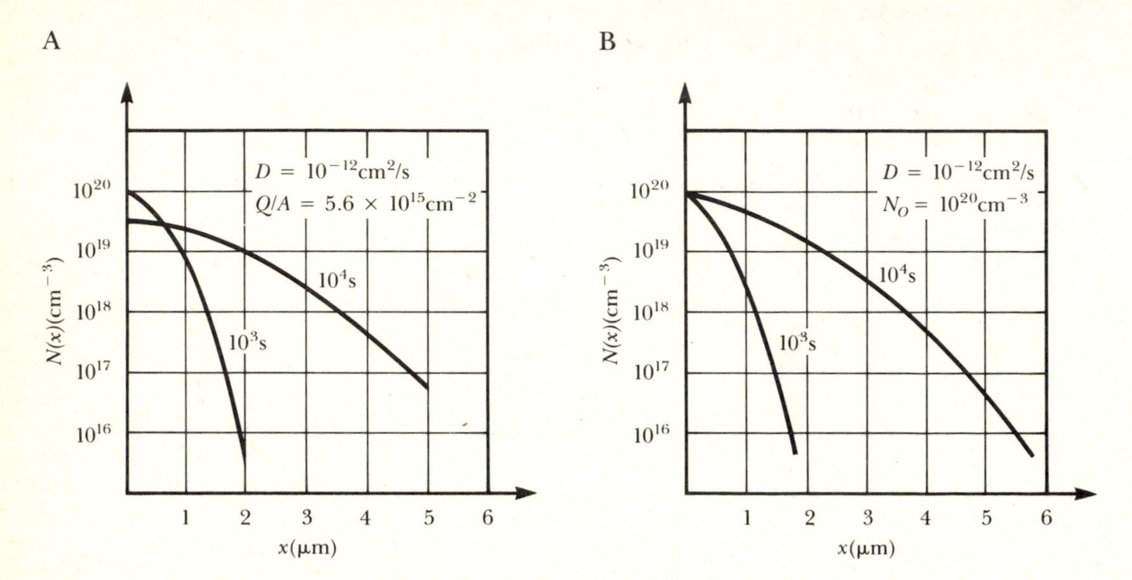

tegrated and treated as Q/A in Eq. (2.41). Following this approach, the result for the two-step diffusion process is

$$N(x) = \frac{2N_o}{\pi}\sqrt{\frac{D_1t_1}{D_2t_2}}\,\exp\left[\frac{-x^2}{4D_2t_2}\right] \tag{2.42}$$

where D_1 and t_1 refer to the predeposition, and D_2 and t_2 refer to the drive-in diffusion. N_o is the constant surface concentration for the predeposition.

There are multiple reasons for including the drive-in as part of the diffusion process. Often, a high dopant concentration near the wafer surface is not desirable. During the drive-in, the diffusion profile moves deeper into the wafer and the surface-impurity concentration drops, as shown in Figure 2.13. Furthermore, an oxide layer grows during the drive-in diffusion and is useful for subsequent masking steps. An oxide layer may also help to prevent diffusion of impurities out of the wafer.

That the wafer goes through several high-temperature processes and that impurities diffuse in each high-temperature step complicates the integrated circuit fabrication process. In a bipolar transistor, for example, the base diffusion is followed by an emitter diffusion and the base impurities will continue to diffuse during the emitter diffusion. Suppose, for example, that the base diffusion is characterized by D_1 and t_1 for the base predeposition and D_2 and t_2 for the base drive-in. Subsequently, an emitter diffusion is carried out with a predeposition time of t_3 and a drive-in time of t_4. Essen-

Figure 2.14 Phosphorus predeposition diffusion profiles.
SOURCE: From G. Masetti, D. Nobili, and S. Solmi, "Profiles of Phosphorus Predeposited in Silicon and Carrier Concentration in Equilibrium with SiP Precipitates," in *Semiconductor Silicon 1977*, eds. H. R. Huff and E. Sirtl (Princeton: Electrochemical Society, 1977), pp. 648–657. Figure presented first at the Spring, 1977, Electrochemical Society meeting, Philadelphia.

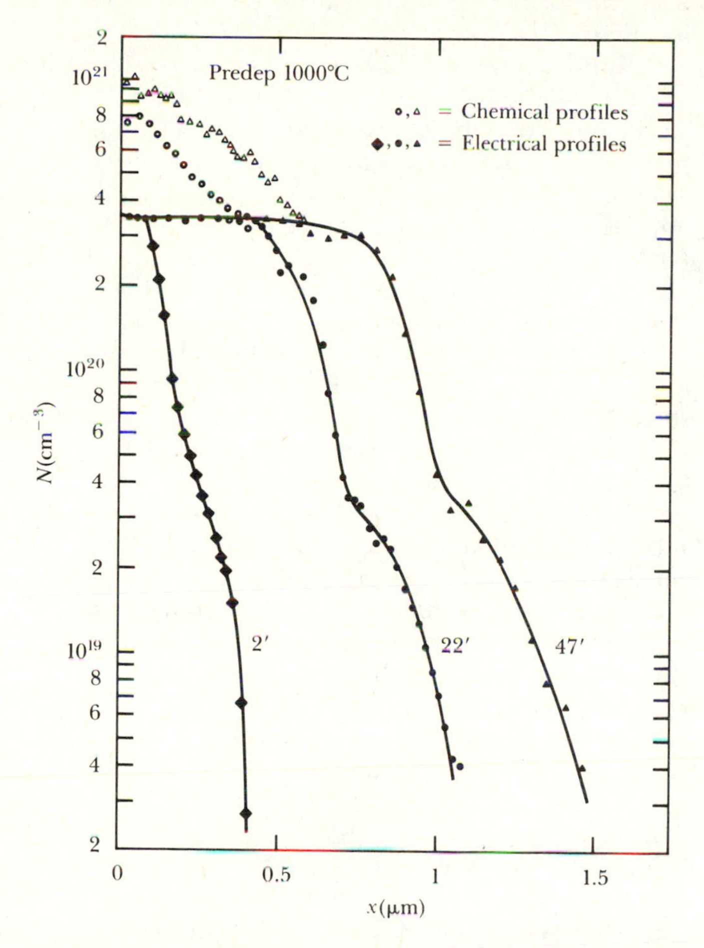

tially, the times t_3 and t_4 add to the drive-in of the base impurities. If D_3 and D_4 are the base-impurity diffusion coefficients corresponding to the temperatures of the emitter predeposition and drive-in, respectively, then in Eq. (2.42), D_2t_2 should be replaced by the sum $D_2t_2 + D_3t_3 + D_4t_4$ [10].

It should not be surprising, after the discussion in Section 2.5.3, to find that actual diffusion profiles may differ considerably from the Gaussian and erfc results that were obtained assuming a constant D. Differences between simple theory and experiment are particularly pronounced in the case of phosphorus, as shown in Figure 2.14 for a constant-source diffusion. Since phosphorus has a high solid solubility, field enhancement of diffusion near the surface is considerable. In addition, there is an obvious kink in the profile at concentrations somewhat below 10^{20} cm^{-3}. At around this concentration, for P, doubly charged vacancy pairs dissociate into singly charged vacancies, resulting in an increased diffusion coefficient [11]. Also, as shown in Figure 2.14, the peak electrically active phosphorus concentration is considerably less than N_o.

Although the diffusion equations were solved in one dimension, two- and three-dimensional effects come into play at the edge of a diffusion mask. As a rule of thumb, the extent of lateral diffusion at the edge of a diffusion mask is comparable to the vertical diffusion. Generally, this condition is considered a disadvantage, since it places a limit on how close to one another adjacent diffusion patterns can be placed. However, in double-diffused MOS (DMOS) the lateral diffusion effect is used to advantage to achieve submicron FET channel lengths. Edge effects for both a constant-source and limited-source diffusion are shown in Figure 2.15 [12]. In addition to edge effects, surface effects may also be appreciable. As discussed in Section 2.7, for example, an oxide growth results in a redistribution of impurities between the oxide layer and the silicon wafer. Furthermore, diffusion coefficients within the wafer are also affected by surface oxidation.

Another discrepancy between simple theory and experiment is due to energy band gap narrowing at very high impurity concentrations. This mechanism is thought to be responsible in part for the "emitter push" effect, for example, which is illustrated in Figure 2.16 [13]. In this case, as shown for an emitter/base region, the base impurities diffuse more rapidly in the presence of a high concentration of emitter impurities, resulting in a dip in the base profile under the emitter region. Effects such as these, involving concentration-dependent diffusion coefficients and redistribution of impurities, must be treated numerically, such as with the SUPREM program developed at Stanford University [4].

2.5.5 Diffusion Techniques

For silicon, the most often used acceptor dopant is boron. The reason for this is discovered by considering the alternatives. Considering other group III elements, indium has a relatively large ionization energy, about 0.16 eV, compared to 0.045 eV for boron. Aluminum is highly reactive and may combine with oxygen during heat treatment to give undesirable results. Gallium has a relatively large diffusion coefficient in SiO_2. Since SiO_2 is often used as a diffusion mask, Ga cannot be readily used in selective diffusions.

Phosphorus and arsenic are the donor impurities most often used in silicon. Phosphorus is used for general-purpose diffusions where a high diffusion coefficient is desirable. Arsenic's diffusion coefficient in silicon is an order of magnitude less than the diffusion coefficient for phosphorus. For early diffusion steps, the low D of arsenic may be advantageous in minimiz-

Figure 2.15 Edge diffusion effects. A. Constant source. B. Limited source.
SOURCE: From D. P. Kennedy and R. R. O'Brien, "Analysis of the Impurity Atom Distribution Near the Diffusion Mask for a Planar *p-n* Junction," *IBM Journal of Research and Development* 9[3]: 179–186, 1965. Copyright 1965 by International Business Machines Corporation; reprinted with permission.

A

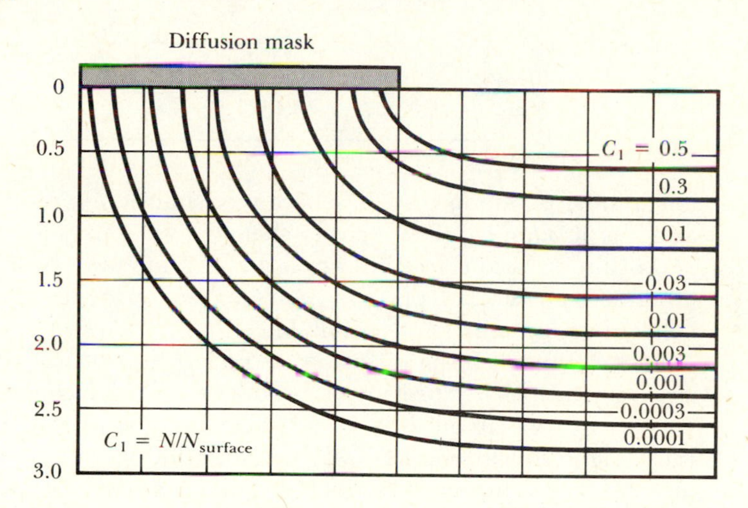

B

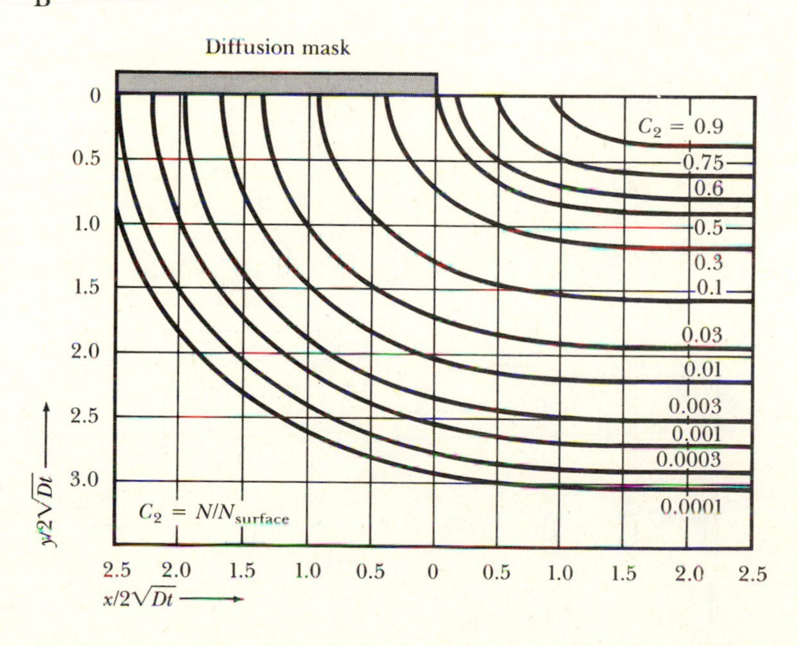

Figure 2.16 The emitter push effect.

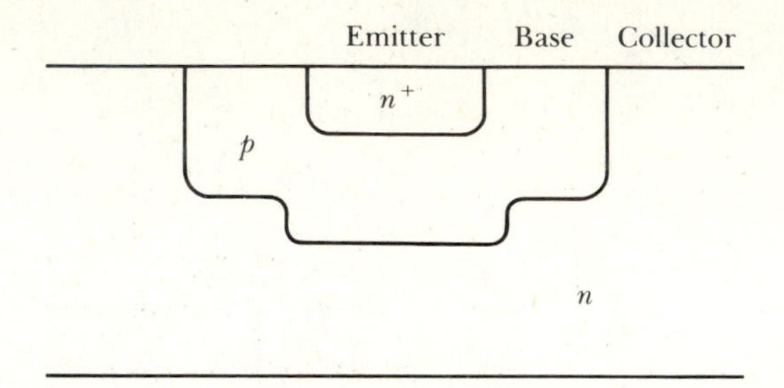

ing impurity motion during subsequent high-temperature processing steps, and it is also appropriate for very shallow junctions, such as MOSFET source and drain diffusions in VLSI applications and emitters of high-speed bipolar junction transistors (BJTs). Since arsenic also provides the best fit to the silicon lattice among donor possibilities, it produces little surface damage or strain. Antimony has a diffusion coefficient in silicon comparable to that of arsenic but does not provide as good a fit to the silicon lattice.

Integrated circuit diffusions are almost always done in an open-tube arrangement, in which a quartz processing tube containing the wafers is placed in a resistively heated diffusion furnace. A gaseous dopant source is shown for the boron diffusion system in Figure 2.17, although solid and liquid sources are also available. For the case illustrated, diborane reacts with oxygen via the reaction,

Figure 2.17 Schematic view of an open-tube boron diffusion system.

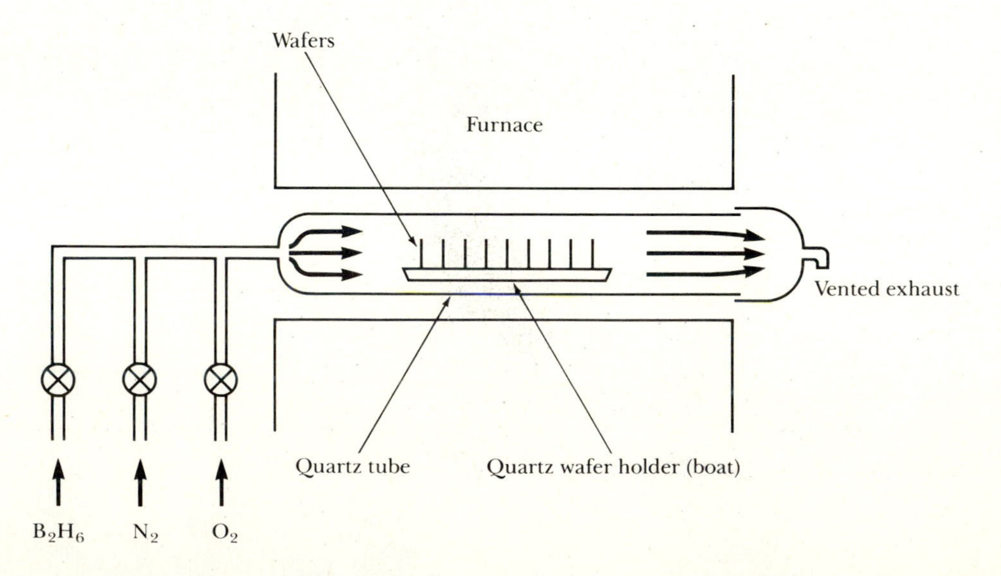

$$B_2H_6 + 3O_2 \rightarrow B_2O_3 + 3H_2O \quad (2.43)$$

Next, the silicon reacts with the boron trioxide as

$$2B_2O_3 + 3Si \rightarrow 4B + 3SiO_2 \quad (2.44)$$

forming a borosilicate "glass" that coats the wafer and provides a source of boron for the diffusion process. So long as the borosilicate glass is on the wafer, constant-source conditions prevail. For drive-in diffusions, the glass is removed to achieve a limited-source condition. A similar reaction occurs when phosphine is substituted for diborane. Now,

$$2PH_3 + 4O_2 \rightarrow P_2O_5 + 3H_2O \quad (2.45)$$

resulting in the formation of a phosphosilicate glass by the reaction,

$$2P_2O_5 + 5Si \rightarrow 4P + 5SiO_2 \quad (2.46)$$

Liquid dopant sources may also be used as feeds for open-tube diffusions. A common liquid source for n-type diffusions is $POCl_3$, via the reaction

$$4POCl_3 + 3O_2 \rightarrow 2P_2O_5 + 6Cl_2 \quad (2.47)$$

Again a phosphosilicate glass forms according to the reaction in Eq. (2.46). Both n- and p-type diffusions may also be carried out with solid sources. In one version of the solid source, wafers containing SiP_2O_7 in an inert substrate are interleaved with the silicon wafers in the diffusion furnace. Wafers containing B_2O_3 and boron nitride are used for p-type diffusions. Another variation on the use of solid diffusion sources involves coating the wafer with a dopant-laden "slurry."

2.6 Ion Implantation

Although diffusion is a widely used doping technique, it does have drawbacks. Principal among them are the following:

1. Donor and acceptor compensation must be used to define new regions of a device. Peak impurity concentrations are at, or near, the wafer surface. Therefore, subsequent diffusions must use increasingly large concentrations of impurities to convert a semiconductor region from one carrier type to another.
2. Little can be done to tailor the impurity doping profiles; the basic shapes are erfc or Gaussian.
3. Appreciable lateral diffusion takes place at the edges of diffusion masks.

Figure 2.18 Schematic drawing of ion implantation apparatus.
SOURCE: From A. Axmann, "Source Feed Materials in Ion Beam Technology," in F. F. Y. Wang, ed., *Impurity Doping Processes in Silicon,* Vol. 2, Amsterdam: North Holland Publishing Company, 1981, pp. 147–174.

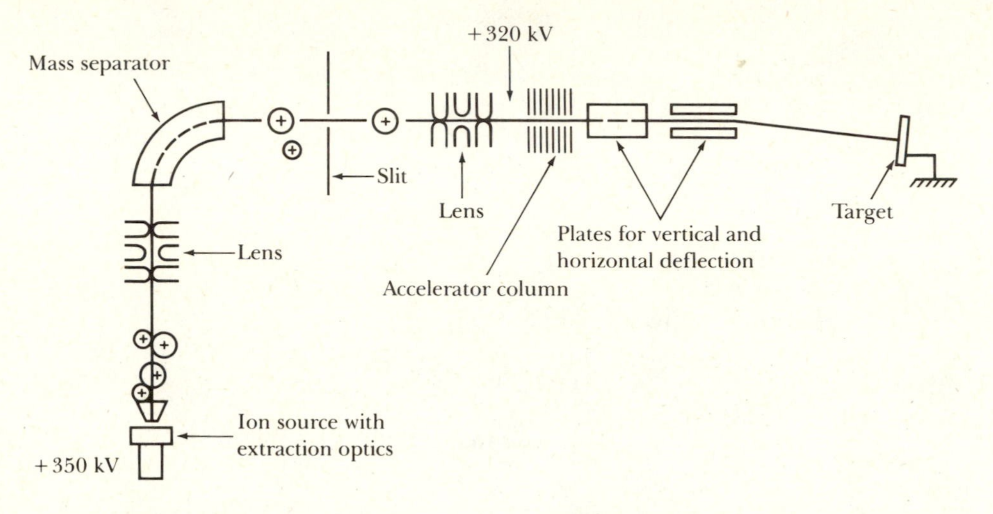

An alternative approach, which avoids or reduces the listed negative effects, is ion implantation.

The main features of an ion implant apparatus, are shown in Figure 2.18. Ions, typically generated in a plasma by an electric discharge, are extracted from the ion source, focused, and separated according to their atomic mass in a mass separator. Ions of the desired mass are selected by a slit aperature and then accelerated to the desired energy. A combination of electrical beam sweeping and wafer substrate rotation displays the ion beam across the wafer such that a uniform implant dose is achieved. Key parameters are the accelerating voltage, which determines the penetration of the ions, and the beam current-time product, which establishes the total number of impurities per unit area, or the dose. Figure 2.19 shows a photograph of a commercial ion implanter.

Ions are accelerated to energies in the approximate range of 10 keV to 1 MeV. Therefore, when a dopant ion collides with the wafer, it suffers many collisions with the silicon atoms before eventually stopping at some distance R beneath the surface. If a second identical ion is accelerated into the wafer at the same energy, it will stop at some distance generally not equal to R, since the stopping process involves a series of random events. For a monoenergetic-beam incident on an amorphous substrate, the distribution of implanted impurity atoms is Gaussian, and may be expressed as

$$N(x) = \frac{\Phi}{\sqrt{2\pi}\Delta R_P} \exp\left[-\frac{1}{2}\left(\frac{x - R_P}{\Delta R_P}\right)^2 \right] \tag{2.48}$$

Figure 2.19 Photograph of a commercial ion implanter.
SOURCE: Courtesy Veeco Instruments, Inc.

In Eq. (2.48), the mean penetration, R_P, is called the range and the standard deviation, ΔR_P, is called the straggle [16]. $N(x)$ is the implanted ion concentration per unit volume, and Φ is the dose,

$$\Phi \simeq \int_0^{\infty} N(x)\, dx \qquad (2.49)$$

where $x = 0$ is taken at the wafer surface and again the wafer is treated as being semi-infinite.

In crystals, the distribution is skewed from the ideal Gaussian profile to a greater or lesser degree, depending on how the angle of incidence lines up with the principal crystal directions. Preferential scattering along the directions formed by the dominant crystal planes is referred to as channeling. Channeling can be reduced by keeping the wafer slightly off normal with respect to the ion beam, and Eq. (2.48) is often used as a first approximation

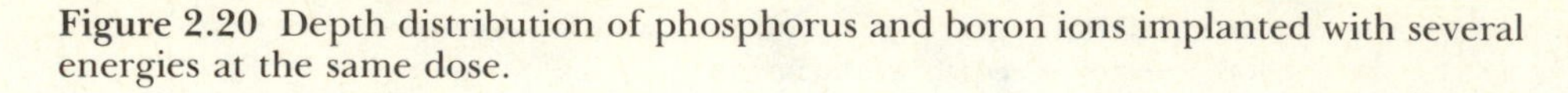

Figure 2.20 Depth distribution of phosphorus and boron ions implanted with several energies at the same dose.

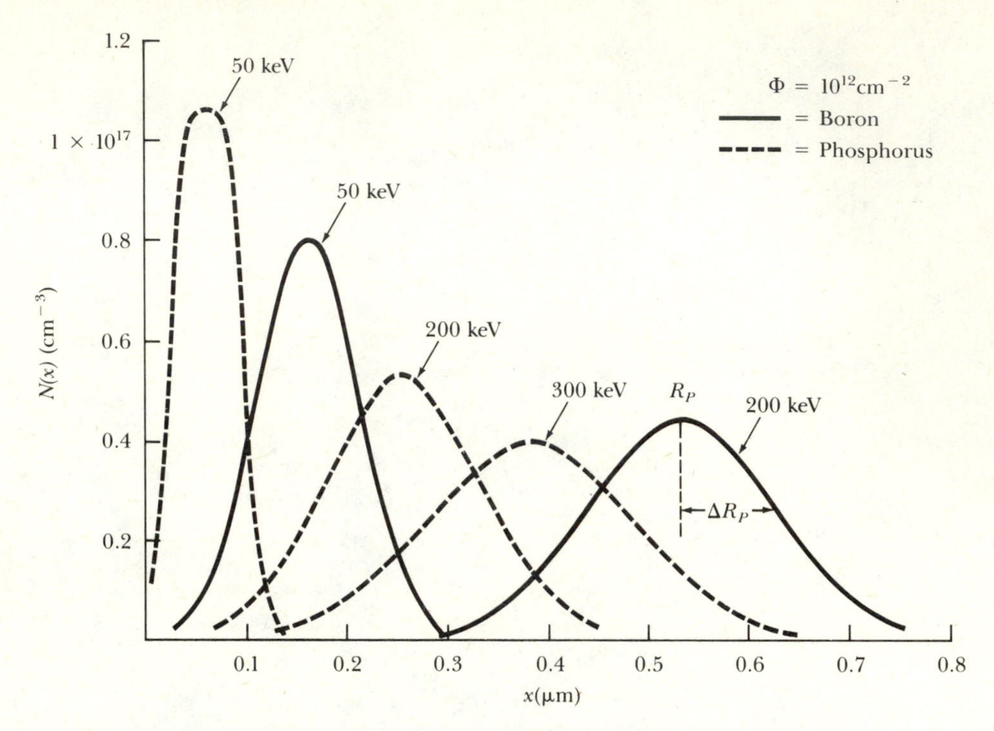

to the implant distribution in crystalline materials. For a given energy, the lighter the atom, the larger the range and the smaller the ratio $\Delta R_P/R_P$. This trend is illustrated in Figure 2.20 and in Table 2.3 for dopants in silicon.

Several advantages of ion implantation exist:

1. Doped regions may be buried beneath the surface, allowing low surface-impurity concentrations. This is particularly useful in MOS technologies where a high near-surface mobility is important.
2. Voltage, current, and time are more accurately controlled than high-temperature diffusion processes involving gas flow. Doses may be controlled well within 1%.
3. Ion beams are easily masked. Metal masks and even resist masks can be used, as well as oxide masks.
4. Unique doping profiles can be achieved by using multiple implants of different energy and merging them together.
5. The technique lends itself well to shallow junctions, which are required for small-dimension, high-speed devices.

However, some problems with ion implantation may also be anticipated.

Table 2.3 Calculated range and straggle for implantation of boron, phosphorus, and arsenic into amorphous silicon (in micrometers)

		Energy (keV)			
		20	50	100	200
B	Range	0.0662	0.1608	0.2994	0.5297
	Straggle	0.0283	0.0504	0.0710	0.0921
P	Range	0.0253	0.0607	0.1238	0.2539
	Straggle	0.0119	0.0256	0.0456	0.0775
As	Range	0.0159	0.0322	0.0582	0.1114
	Straggle	0.0059	0.0118	0.0207	0.0374

Source: Selected energies from J. F. Gibbons, W. S. Johnson, and S. W. Mylroie, *Projected Range Statistics, Semiconductors and Related Materials,* 2nd ed. (Stroudsburg, Penn.: Dowden, Hutchinson and Ross, 1975).

1. Heavy damage to the silicon lattice results.
2. The ion implant Gaussian profile is often highly peaked, which is usually not desirable.
3. Ion implant equipment is more expensive to buy and maintain than is diffusion equipment.

The first two drawbacks can be solved by a post-implant anneal in which sufficient energy is provided to allow the silicon atoms and dopant atoms to rearrange themselves back into the diamond crystal structure. The energy may be applied rapidly, such as with a few seconds exposure to an intense optical source. Alternatively, the wafers are annealed in a diffusion furnace. Higher implant doses usually require higher annealing temperatures, or longer annealing times. For example, a 150-keV boron implant with a dose of 2×10^{14} cm^{-2} requires 30 minutes at approximately 900°C to activate all the boron atoms. However, a dose of 2×10^{15} cm^{-2} requires a 30-minute anneal at 1000°C [18]. Since the ion implant profile has a large associated gradient, diffusion takes place during the anneal, such that a partial flattening of the profile results.

The diffusion theory of Section 2.5 applies to the post-implant anneal, with the added complication that D changes during the anneal because the vacancy concentration changes as annealing takes place. Provided a constant effective diffusion coefficient, D_A, can be assigned over the anneal time, t_A, the post-anneal profile is given by the Gaussian expression in Eq. (2.50), as follows [18]:

$$N(x) = \frac{\Phi}{(2\pi)^{1/2}[\Delta R_P{}^2 + 2D_At_A]^{1/2}} \exp\left[\frac{-(x - R_P)^2}{2(\Delta R_P{}^2 + 2D_At_A)}\right] \qquad (2.50)$$

In fact, because of spatial variations of D during the anneal, the resulting profiles are not as a rule exactly Gaussian.

EXAMPLE 2.8

An ion implantation is done on a uniformly doped n-type silicon wafer with background concentration $N_B = 10^{16}\ \mathrm{cm}^{-3}$. The implant is a 200-kV boron implant with a 10-μA/cm^2 beam current for 30 seconds. It is followed by a 1-hour anneal/diffusion process at 1000°C, for which the diffusion coefficient is $2.6 \times 10^{-14}\ \mathrm{cm}^2/\mathrm{s}$. Where is the junction?

Solution From Table 2.3, the range is approximately 0.53 μm and the straggle is 0.092 μm. Calculating the dose,

$$\Phi = It/q = (10^{-5}\mathrm{C/cm}^2\cdot\mathrm{s})\frac{30\ \mathrm{s}}{1.6 \times 10^{-19}\ \mathrm{C/ion}}$$

or

$$\Phi = 1.88 \times 10^{15}\ \mathrm{cm}^{-2}$$

The junction is positioned at the value of x where the boron concentration equals N_B, or in other words, when

$$N(x_J, 1\ \mathrm{h}) = 10^{16}\ \mathrm{cm}^{-3}$$

Using Eq. (2.50) for $N(x,t_A)$ results in

$$|x_J - R_P| = 0.68\ \mu\mathrm{m}$$

Therefore, the junction depth is at 1.21 μm below the wafer surface. For smaller values of x the material is p-type, and for larger values of x it is n-type.

An example of a measured implant profile before and after laser annealing is shown in Figure 2.21 [19].

As we see in later chapters, ion implantation plays a vital role in many integrated circuit designs, particularly at the LSI and VLSI levels.

2.7 Oxidation

Oxide layers play several important roles in integrated circuits, including the following:

1. Diffusion, or ion implant masks
2. MOSFET gate insulators

Figure 2.21 Dopant concentration of ion-implanted boron measured before and after laser annealing.
SOURCE: C. W. White, W. H. Christie, B. R. Appleton, S. R. Wilson, and P. P. Pronko, "Redistribution of Dopants in Ion Implanted Silicon by Pulsed Laser Annealing," *Applied Physics Letters*, 33 (1978), 654–656.

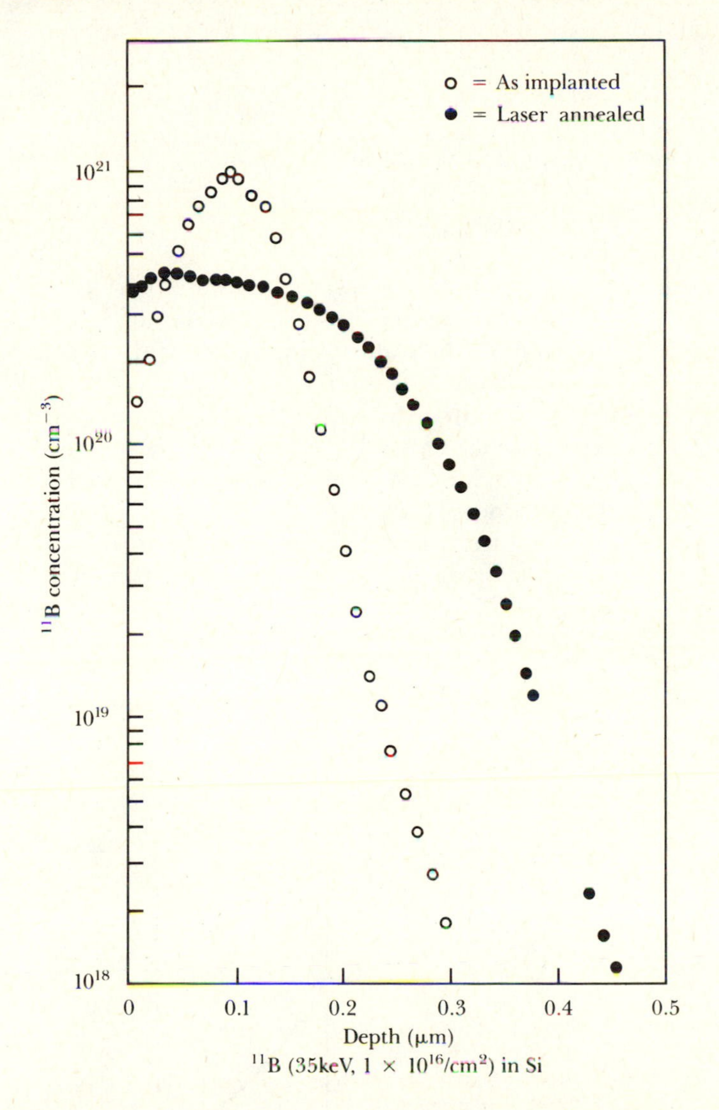

3. MOS capacitor dielectrics
4. Circuit passivation
5. Insulation for multiple metalization layers

The two basic processes for growing SiO_2 on silicon are referred to as wet and dry oxidation, and the respective reactions are,

$$Si + 2H_2O \rightarrow SiO_2 + 2H_2 \tag{2.51}$$

(neglecting some intermediate steps) and

$$Si + O_2 \rightarrow SiO_2 \tag{2.52}$$

In both cases, an oxide is grown on the wafer surface and silicon is consumed in the process. Alternatively, oxides may be deposited on the wafer, as opposed to grown on the wafer, using methods discussed in Section 2.10.

The reactions of Eqs. (2.51) and (2.52) take place at the wafer surface. This means that after the first monolayer of oxide, the oxidizing species, derived either from O_2 or H_2O, must diffuse to the wafer surface through the intervening oxide layer. Since diffusion is strongly temperature activated, the oxidation process is also strongly temperature dependent. A typical oxidation temperature is 1100°C. Furthermore, as the oxide thickness increases, the rate of new oxide growth decreases. This process is illustrated with the aid of Figure 2.22 for the case of dry oxidation where N_I is the oxidizing species concentration in the oxide at the Si-SiO_2 interface and N_{OX} is the concentration at the oxygen/SiO_2 interface.

A detailed theoretical treatment of oxidation is complex and, in fact,

Figure 2.22 The diffusion gradient for a dry oxidation.

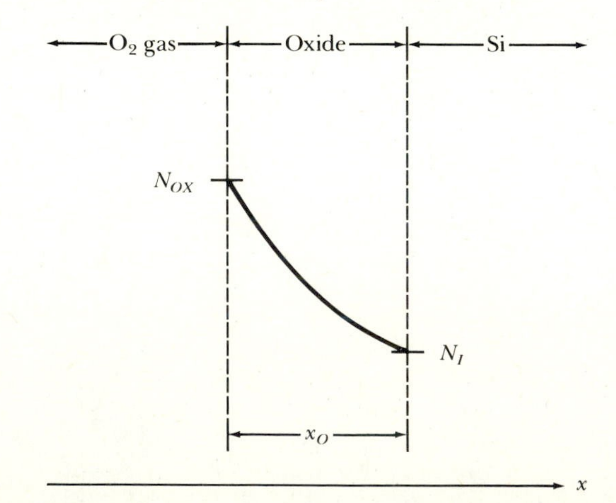

Figure 2.23 Wet oxidation growth rate.
SOURCE: From R. P. Donovan, "Oxidation," in R. M. Burger and R. P. Donovan, eds., *Fundamentals of Silicon Integrated Device Technology,* Englewood Cliffs, N.J.: Prentice-Hall. © 1967, pp. 41, 49. Reprinted by permission of Prentice-Hall.

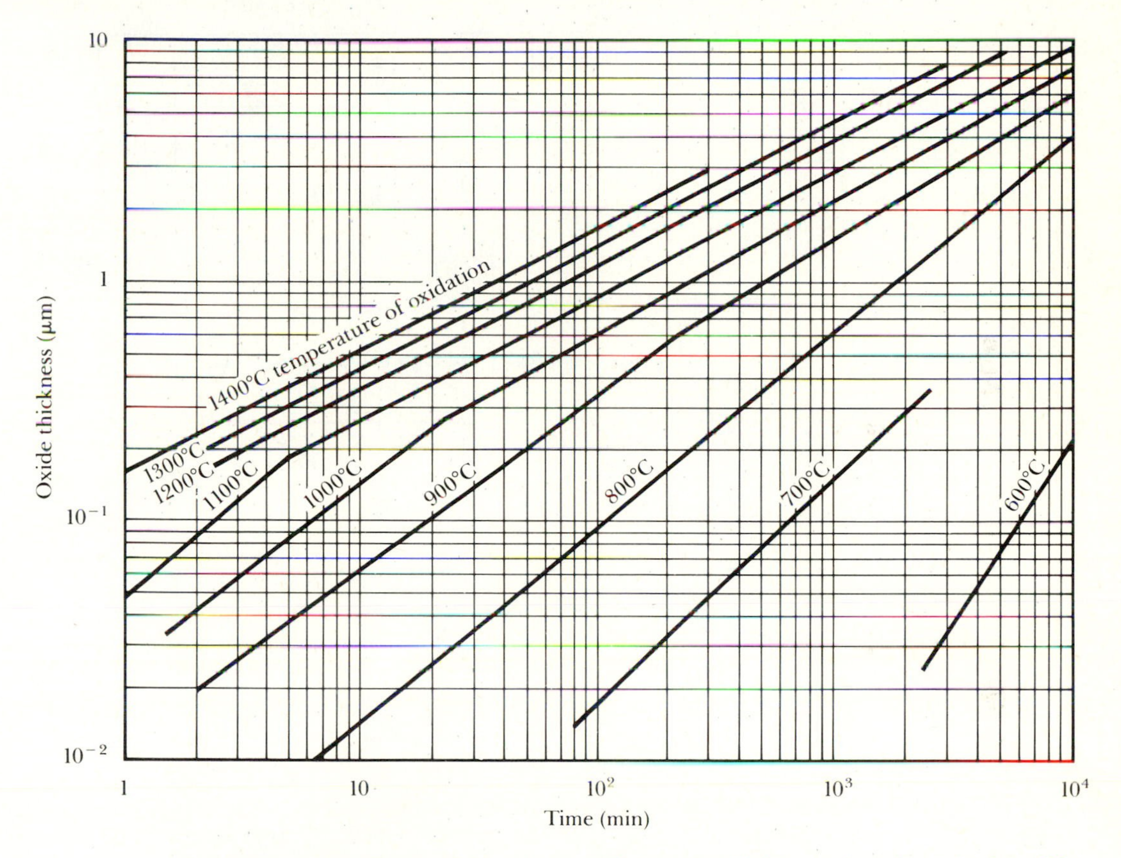

questions remain concerning the relative roles of charged and neutral oxidizing species and the roles of molecular and atomic species. However, applying the basic concepts of diffusion, it is noted that by Fick's first law, the flux of oxidizing species through the oxide is

$$j = -D(dN/dx) \approx D\left[\frac{(N_{OX} - N_I)}{x_O}\right] \tag{2.53}$$

where x_O is the oxide thickness. Clearly, as x_O increases, the flux of oxidizing species to the wafer surface decreases and the rate of the oxidation reaction shown in Eq. (2.52) decreases. In fact, it can be shown that, after an initial linear phase, the rate of growth is approximately proportional to the square root of the oxidation time.

Figures 2.23 and 2.24 show the oxide thickness on silicon as a function

Figure 2.24 Dry oxidation growth rate.
SOURCE: From R. P. Donovan, "Oxidation," in R. M. Burger and R. P. Donovan, eds., *Fundamentals of Silicon Integrated Device Technology,* Englewood Cliffs, N.J.: Prentice-Hall. © 1967, pp. 41, 49. Reprinted by permission of Prentice-Hall.

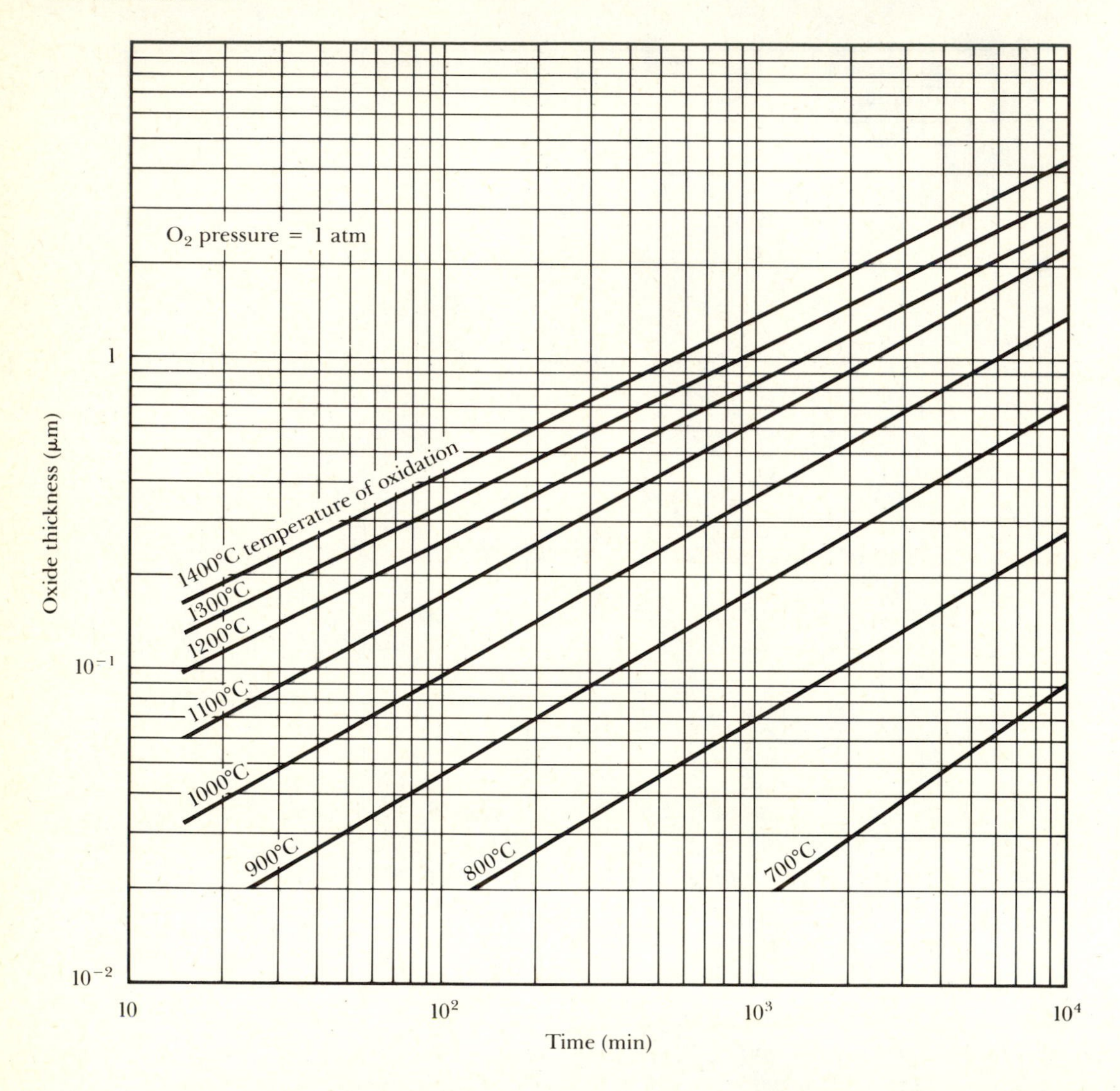

of time a~d temperature for wet and dry processes, respectively. It is apparent from these figures that wet oxidation proceeds at an appreciably faster rate than does dry oxidation. The rate is faster because the diffusion coefficient of H_2O in SiO_2 is considerably higher than is the case for O_2, owing in part to the polar nature of the water molecule.

The oxidation process is actually more complex than is indicated by Figures 2.23 and 2.24 since, in addition to depending on temperature and the oxidizing species, oxidation rates also depend on oxidant gas pressure, crystallographic orientation of the silicon substrate, and the substrate doping. Wafers with a (111) orientation oxidize somewhat more rapidly than do (100) wafers because the rate at which silicon atoms are incorporated into the growing oxide film depends on silicon surface atom concentration. At 1200°C, for example, a one hour dry oxidation on a (111) wafer produces an oxide thickness approximately 5% higher than the 0.2 μm shown in Figure 2.24 and, for a (100) wafer, an oxide thickness approximately 5% lower than 0.2 μm. At lower temperatures, the wafer orientation effect increases, and at 900°C the (111) oxidation rate is about 20% higher than the (100) rate.

Oxidation rates are higher for wafers heavily doped with boron and phosphorus since these dopants segregate into the oxide and weaken the oxide bond structure, which allows an increased incorporation and diffusion rate of oxidizing species in the oxide. Oxidation rates and oxide quality are also modified by the intentional introduction of impurities into the oxidizing atmosphere. Specifically, a small amount of HCl, in the range of 1 to 5%, is often added to improve the oxide quality. Chlorine reacts with certain impurities to form volatile chlorides and is particularly useful in reducing sodium contamination.

The consumption of silicon by the oxide growth may be calculated as follows. Let $x = 0$ represent the surface of the wafer before oxidation, x_O be the oxide thickness, and x_S be the amount of silicon consumed in the process, as shown in Figure 2.25. The number of Si atoms in the oxide per unit area is $(N_{\mathrm{OXIDE}})x_O$ where N_{OXIDE} is the concentration of SiO_2 molecules in the oxide. Each SiO_2 molecule contains one Si atom that came from the wafer. Therefore,

$$x_S(N_{\mathrm{Si}}) = x_O(N_{\mathrm{OXIDE}}) \tag{2.54}$$

where N_{Si} is the concentration of silicon atoms in the wafer. Since $N_{\mathrm{Si}} = 5 \times 10^{22}$ cm^{-3} and $N_{\mathrm{OXIDE}} = 2.3 \times 10^{22}$ cm^{-3},

$$x_S = 0.46\ x_O \tag{2.55}$$

Figure 2.25 Silicon consumption during oxidation.

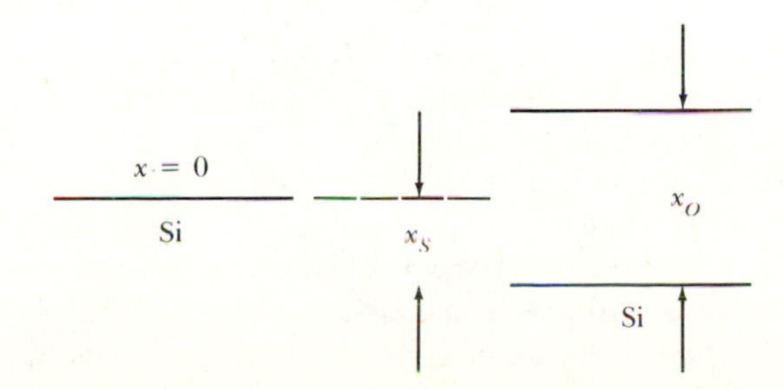

Figure 2.26 Wafer profile for Example 2.9.

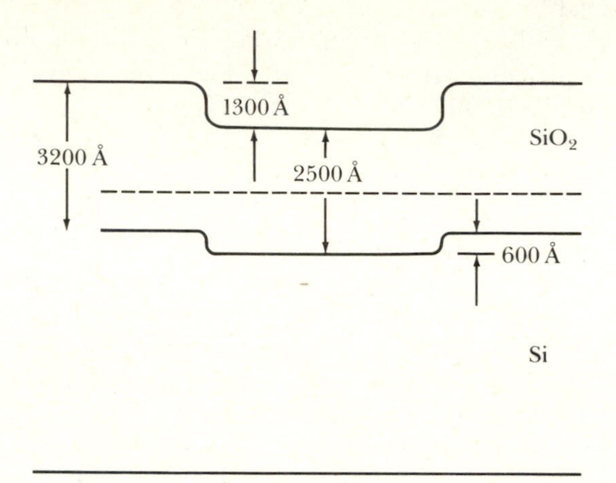

If, for example, 1 μm of silicon dioxide is grown in a localized region of the silicon wafer, it will be countersunk by about 0.46 μm into the silicon.

EXAMPLE 2.9

A silicon wafer is initially covered with a 2000-Å-thick SiO_2 layer. Then a window is opened in the oxide for purposes of defining a region to be doped. Finally, a drive-in diffusion is performed in a dry oxygen atmosphere for 3 hours at 1100°C. What is the resulting wafer profile?

Solution First, the amount of oxygen growing on the fresh silicon surface during the drive-in is seen from Figure 2.24 to be approximately 2500 Å. The amount of Si consumed in the process is $(2500)(0.46) \simeq 1150$ Å.

The same figure may be used to find the amount of oxide that grows on the silicon covered by 2000 Å. Starting at 2000 Å on the 1100°C curve, a 3-hour oxidation is seen to increase the thickness to about 3200 Å. The additional 1200 Å consumes approximately 550 Å of silicon. Therefore, as shown in Figure 2.26, the step in the silicon is 600 Å and the step in the oxide surface is 1300 Å. Since large steps in the wafer surface are a potential source of integrated circuit failure because of the development of discontinuities in overlying metal lines, information about the wafer profile is important.

Oxide layers may be used as effective diffusion masks for boron and phosphorous, for example, because of the relatively low diffusion coefficients of these elements in SiO_2. Of course, if a diffusion persists for a sufficiently long time, the dopants will eventually diffuse through the oxide layer, negating its usefulness as a mask. At 1100°C, the necessary oxide thickness for an effective mask against a 2-hour diffusion is approximately 7000 Å for a

Figure 2.27 Dopant diffusion coefficients in SiO_2.
SOURCE: Based on data from J. C. C. Tsai, "Diffusion," in *VLSI Technology,* ed. S. M. Sze (New York: McGraw-Hill, copyright © 1983), Chap. 5. Reprinted with permission of McGraw-Hill.

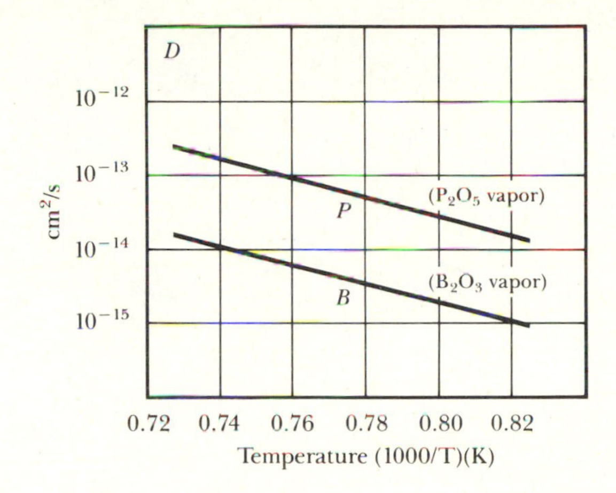

phosphorus diffusion and 2000 Å for a boron diffusion. Diffusion coefficients in SiO_2 for boron and phosphorus are plotted as a function of temperature in Figure 2.27.

Impurity concentrations in the wafer will change during high-temperature oxidation processes because of diffusion and impurity redistribution. At the silicon-oxide interface, the problem is complex and depends on the ratio C_S/C_O, where C_S and C_O are the solute concentrations by weight in silicon and the oxide, respectively. For boron, the ratio is less than 1 and, compared to the predictions of Eq. (2.42), there is a relative depletion of boron at the surface of the silicon wafer. For phosphorus, the opposite is true and there is a relative accumulation of phosphorus near the silicon surface, as shown in Figure 2.28.

Oxidation furnaces are similar to diffusion furnaces; an open quartz tube in a resistively heated furnace is the standard approach. For wet oxidations, oxygen is typically bubbled through a flask containing high-purity de-ionized water maintained at a temperature in the range of 95 to 98°C, such that a vapor enters the processing tube and flows over the wafers. For dry oxidations, the flask is bypassed. Often, small amounts of HCl are added to the gas stream to reduce impurities and improve oxide quality. Several variations on these basic schemes are possible, including both high- and low-pressure oxidations and plasma-assisted low-temperature growth of oxides.

2.8 Epitaxial Layers

Epitaxial layers are thin single-crystalline films deposited on crystalline substrates of the same or similar crystal structure. The word is derived from the

Figure 2.28 Impurity redistribution during oxidation.

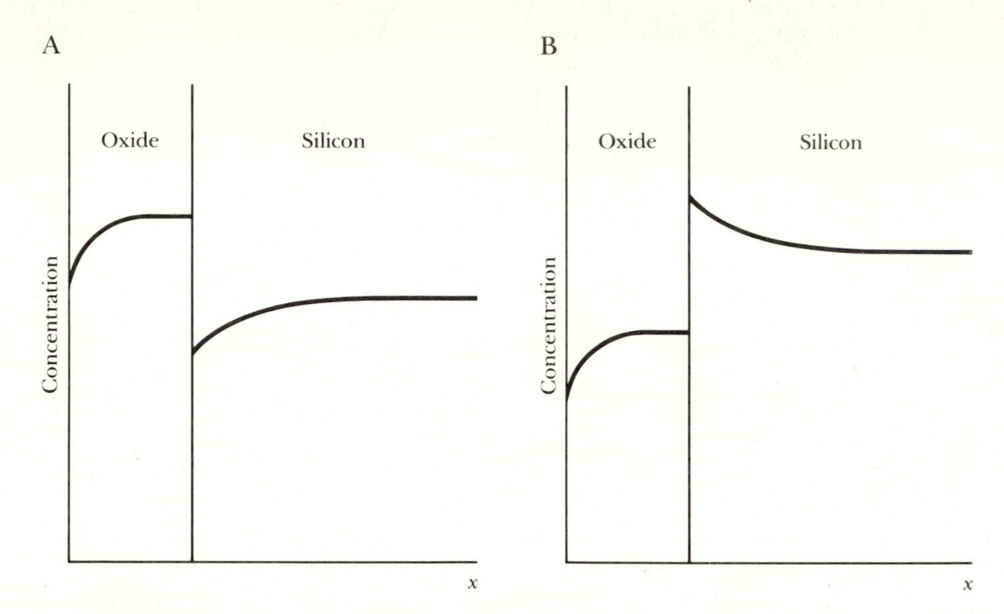

combination of two Greek words, *epi* meaning "upon," and *taxis* meaning "ordered." In silicon integrated circuits, an epitaxial layer refers to a thin film of single-crystalline silicon that is grown on the surface of the wafer. Typically, it is from 1 to 5 μm thick. Since it may be deposited in doped form, it offers another means of selectively doping regions of the integrated circuit, and is essentially an extension of the original wafer.

One example of the usefulness of the epitaxial layer is evident from the device-isolation problem associated with BJTs. Consider adjacent *n-p-n* BJTs formed by diffusing two *p*-type base regions and two *n*-type emitter regions in an *n*-type silicon wafer. If the *n*-type wafer is used as the transistor collectors, then the collectors are inherently joined. However, an *n*-type epitaxial layer overlying a *p*-type substrate provides a means of using reverse-biased junctions to isolate the devices, as shown in Figure 2.29. In principle, the

Figure 2.29 Use of an epitaxial layer for device isolation.

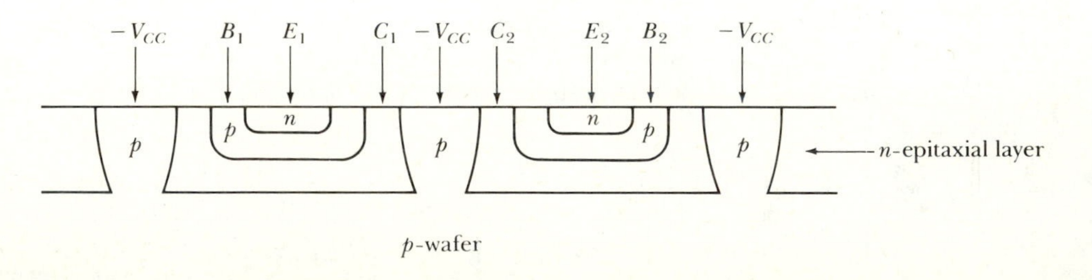

same isolation scheme could be achieved with three successive diffusions with increasing levels of impurity compensation; one diffusion for an n top layer in a p-type wafer, followed by the base and emitter diffusions. However, additional impurities reduce mobilities and represent, in general, a less desirable option. Indeed, a generally attractive feature of epitaxial layers is that they often have fewer undesirable impurities than the original wafer.

Silicon epitaxial layers are grown by employing chemical reactions in which silicon is a product and arranging for the silicon to deposit on a silicon wafer. With proper wafer temperatures and deposition rates, the resulting film is single-crystalline. One commonly used epitaxial silicon reaction is (with intermediate steps omitted),

$$SiCl_4 + 2H_2 \rightarrow Si + 4HCl \tag{2.56}$$

and another is

$$SiH_4 \rightarrow Si + 2H_2 \tag{2.57}$$

where Si is solid and the remaining are gasses.

Both of these reactions are vapor phase reactions, carried out in a high-temperature reactor such as the one sketched in Figure 2.30. In this figure—which shows a horizontal reactor, one of several possible approaches—the wafers are heated by radio frequency (rf) induction heating of the graphite susceptor holding the wafers. One advantage of heating only the wafers instead of the whole tube is that the epitaxial silicon deposits only on the wafer/susceptor and not on the whole tube. Another advantage of the selective heating is that the thermal gradient from the wafer to the tube tends to prevent impurities from being transported from the tube to the wafer.

For the apparatus shown in Figure 2.30, N_2 or another suitable purge gas is used under standby conditions to remove air from the open tube. The reactor is brought up to temperature in a hydrogen atmosphere, and HCl is added to the gas stream to etch the wafer just prior to the epitaxial growth, thus ensuring a surface free of oxides and impurities. Finally, the flows containing the source of silicon and the desired dopant are started. Growth rates are in the range of 0.1 to 3 μm/min.

2.9 Molecular Beam Epitaxy

In molecular beam epitaxy, epitaxial films are grown under ultra-high vacuum conditions. MBE systems are able to achieve pressures in the range of 10^{-10} to 10^{-11} torr, which is about 10^{-14} of atmospheric pressure and is lower than that obtained in space near a spacecraft. These systems are therefore well suited for high-purity processes. In further contrast to conventional epitaxial systems, the materials that make up an MBE layer are obtained

Figure 2.30 Simplified view of an epitaxial vapor phase reactor. A. Cross-sectional view of reactor. B. Gas flow system.

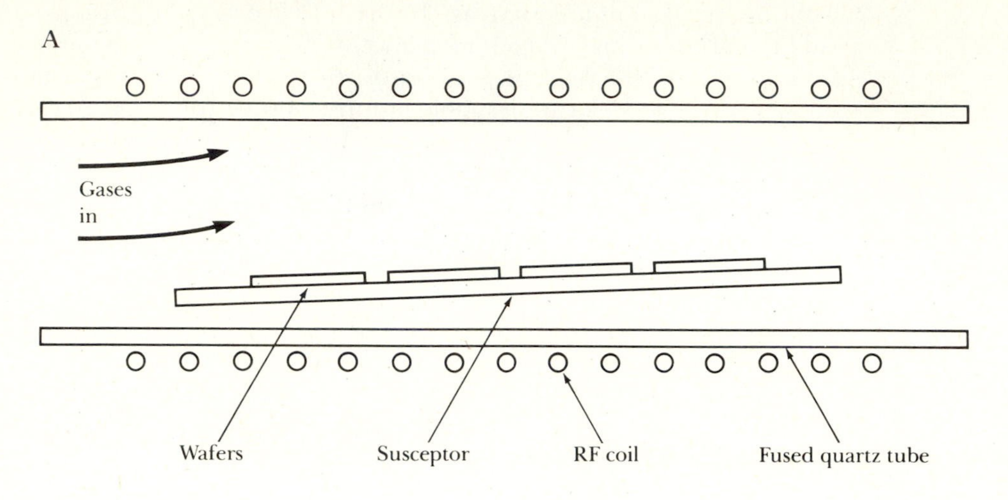

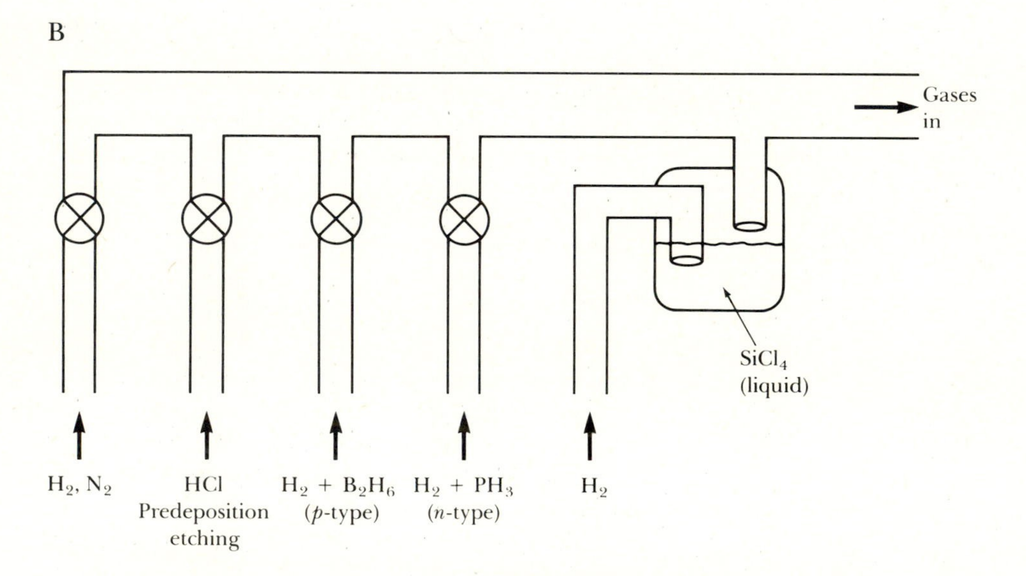

from effusive sources, rather than as by-products of chemical reactions. Most commonly, the film constituents are evaporated from ovens, or "effusion cells," in which a material is elevated to high temperatures either by resistive heating or by an incident electron beam.

The flux of evaporated atoms and molecules constitutes a beam controlled by the temperature of the source and by a mechanical shutter placed in front of the beam. Because collisions between atoms are unimportant un-

Figure 2.31 Schematic of an MBE system for silicon growth.
SOURCE: From Konig et al., "MBE Growth and Sb Doping," *Journal of Vacuum Science and Technology* 16: 985, © 1979.

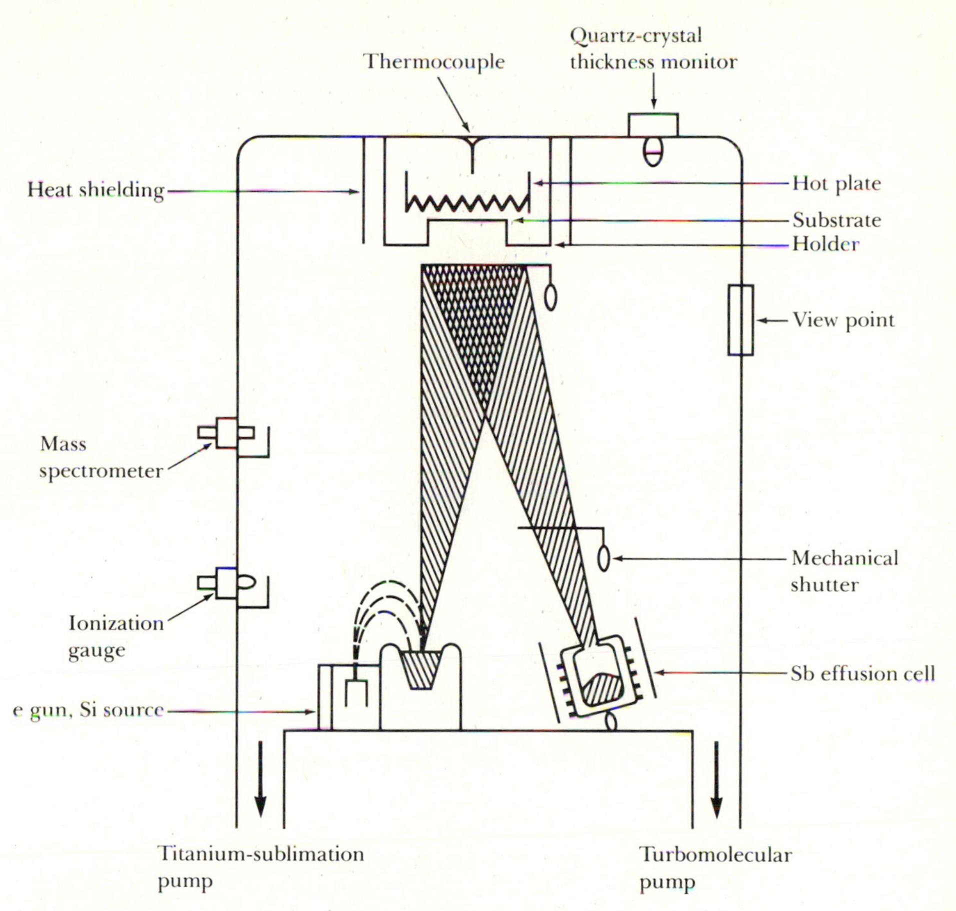

der ultra-high-vacuum conditions (the mean free path at 10^{-10} torr is 5×10^7 cm), the flux of the beam can be determined by the source characteristics. Furthermore, since thermal velocities are high and atoms and molecules do not interact until they reach the substrate, film properties change rapidly with source changes. This situation allows the growth of layered materials with very sharp interfaces. Energy is also supplied to the substrate, usually in the form of heat, so that the incident atoms or molecules have enough energy to find a proper lattice site.

A schematic drawing of a system for MBE growth of silicon is shown in Figure 2.31. In addition to the components already discussed, this figure shows a thickness monitor and a mass spectrometer. The thickness monitor in this drawing consists of an oscillating quartz crystal whose frequency changes as its exposed face is loaded by the deposited film. The mass spectrometer is useful for residual gas analysis. Depending on the use of the system, MBE systems often contain additional analytical apparatus.

An obviously attractive feature of MBE is that impurity levels in deposited films can be extremely low—in the parts-per-trillion range. However, a second important feature is that MBE is well suited for growing compound semiconductors, including new materials not found in nature, by using multiple beams incident on a common substrate. Consider, as an example of a simple compound, the MBE growth of GaAs. In the growing crystal, each Ga atom should be bonded to four As atoms, and each As atom should be bonded to four Ga atoms. For a crystal-plane-substrate orientation of (100), the film will consist of alternating layers of Ga atoms and As atoms with approximately a 2.8 Å center-to-center spacing between two Ga or two As planes.

Nearly all of the incident Ga atoms stick to the substrate and are incorporated into the film. However, some of the incident arsenic, supplied from the effusion cell as As_2 and As_4, does not form proper bonds with the growing film and is desorbed from the substrate. Therefore, separate effusive cells are used for gallium and for arsenic, which is shown in Figure 2.32. More arsenic than gallium is supplied to the substrate to grow a defect-free film.

By adding additional effusive sources and carefully controlling flux rates and substrate temperatures, high-quality MBE films of more complex compounds, such as GaAlAsSb and InGaAsP, may be grown. Since modern MBE systems may have eight effusion cells, considerable flexibility is possible. Furthermore, by shuttering cells on and off, compositional variation may actually be controlled monolayer by monolayer. Consequently, MBE plays a key role in the search for new electronic materials and devices.

2.10 Additional Film-Deposition Techniques

2.10.1 Introduction

Two methods have been discussed at this point for adding layers to the wafer as part of the integrated circuit fabrication process, namely oxidation and epitaxy. To provide device interconnects, it is also necessary to deposit layers, or films, of conductors such as metals and polysilicon. Furthermore, in many fabrication schemes, the deposition of insulators such as silicon nitride is required. In contrast to epitaxial deposition, in which the film's atomic structure is an extension of the wafer's crystalline structure, deposited conductors and insulators have atomic structures independent of the wafer. Usually, the films are amorphous or polycrystalline. The deposition techniques discussed in this section are evaporation, sputtering, chemical vapor deposition (CVD), plasma-assisted CVD, and optically assisted CVD.

Figure 2.32 Conceptual illustration of MBE growth of GaAs.

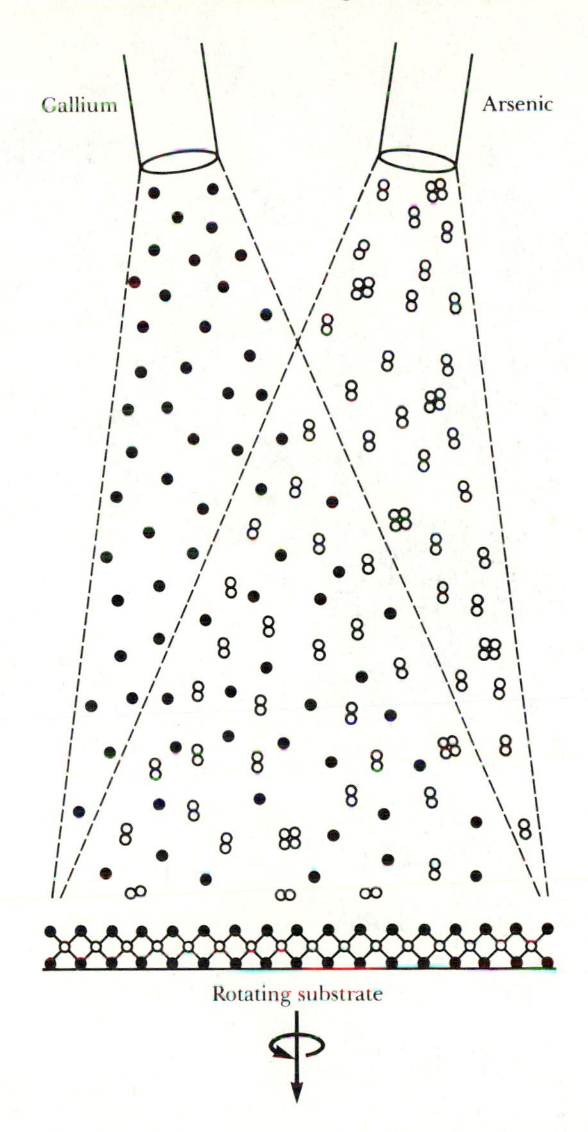

2.10.2 Evaporation

Evaporation is a common method used to metalize the wafer. The process is carried out in a vacuum and is illustrated schematically in Figure 2.33. Heat is provided by an electron beam, by rf induction heating, or by resistive heating in sufficient degree to vaporize a charge of material. A simple aluminum evaporation, for example, may be carried out by suspending aluminum clips from a tungsten wire coil and passing several amperes through the coil until it heats to the glowing point and melts the aluminum. The aluminum atoms

Figure 2.33 Film deposition by evaporation.

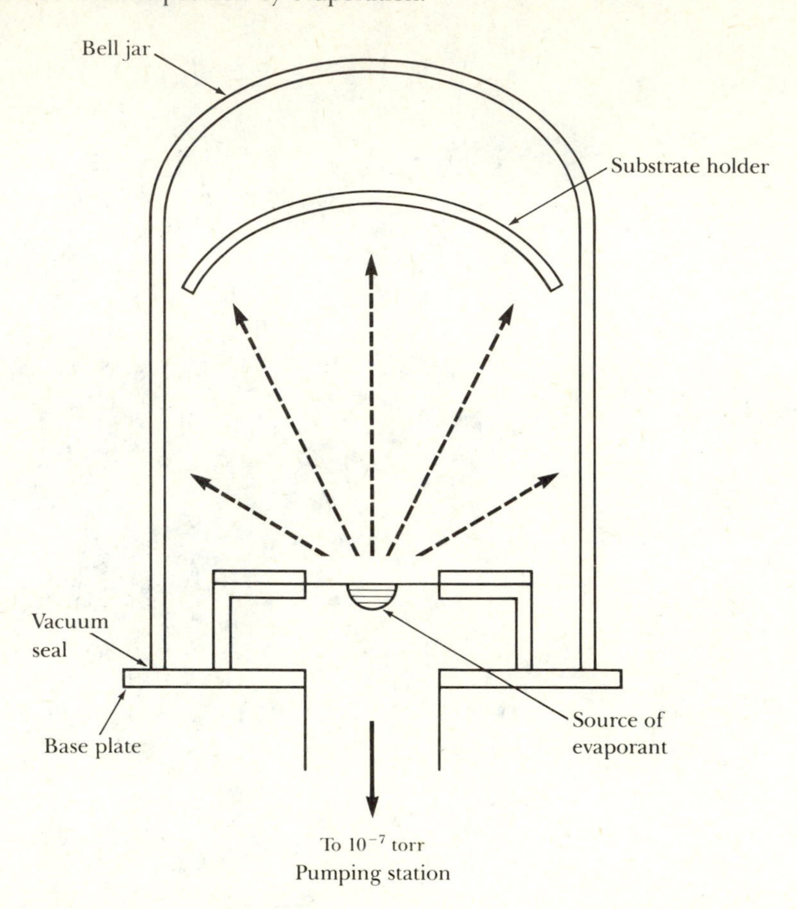

and atom clusters condense on cooler surfaces, including the wafers to be metalized.

EXAMPLE 2.10

The distance between an evaporation coil and the wafer holder is 14 cm. Seven aluminum clips, each $\frac{3}{4}$ in. long and $\frac{1}{16}$ in. in diameter, are evaporated. What is the thickness, d_F, of the resulting aluminum film on the wafer?

Solution Consider the evaporant source to be a point source and the evaporant stream to be isotropic. At a distance R from the coil, the film thickness is related to the mass of the evaporant, M, by

$$(4\pi R^2{}_{dF})\ \rho_D = M$$

where ρ_D is the density of the aluminum. It may be helpful in understanding this equation to consider the aluminum to be coating the interior of a sphere of radius R and surface area $4\pi R^2$. Therefore,

$$d_F = M/(4\pi R^2 \rho_D) = v/4\pi R^2$$

where v is the volume of the aluminum evaporated. In this example, the volume of each clip is 0.0377 cm^3, so

$$d_F = \frac{(7)(.0377)}{4\pi(14)^2} = 1.07 \times 10^{-4} \text{ cm}$$

or 1.07 μm.

If the wafer is centered above the evaporant source, this thickness is the value at the center of the wafer. Points away from the wafer center will have a smaller thickness because of the greater distance and oblique incidence of the evaporant stream. If the evaporant source can be treated as a point source, then the thickness of the film may be shown to fall off with distance y from the center, as follows:

$$d_F/d_F(0) = [1 + (y/R)^2]^{-3/2} \tag{2.58}$$

A potential problem in film deposition is step coverage, as shown in Figure 2.34. Shadowing of the evaporant stream by steps in the wafer surface may cause breaks as illustrated. Both the step coverage problem and thickness fall-off can be minimized by careful placement of the wafers around the source and by rotation of the wafer-holding assembly during evaporation.

Filament sources for evaporation, because of contaminants from the heated filament, are not desirable when ultra-high purity is required. For example, sodium is sometimes added to tungsten wire coils to improve the ductility of the wire. A superior evaporation method from a high-purity perspective is electron beam evaporation, in which an approximately 10-kV beam, at several kilowatts, melts the surface of the material to be evaporated.

Figure 2.34 Film-deposition step coverage. A. Side angle incidence. B. Multiangle incidence.

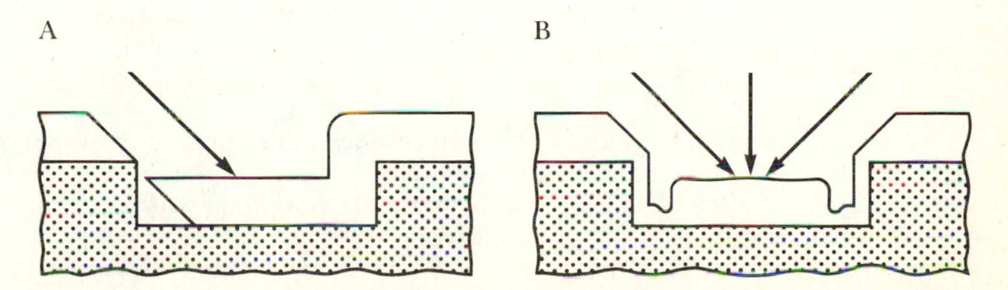

Figure 2.35 Schematic view of a dc sputtering system. A typical sputtering pressure is 10^{-2} torr.

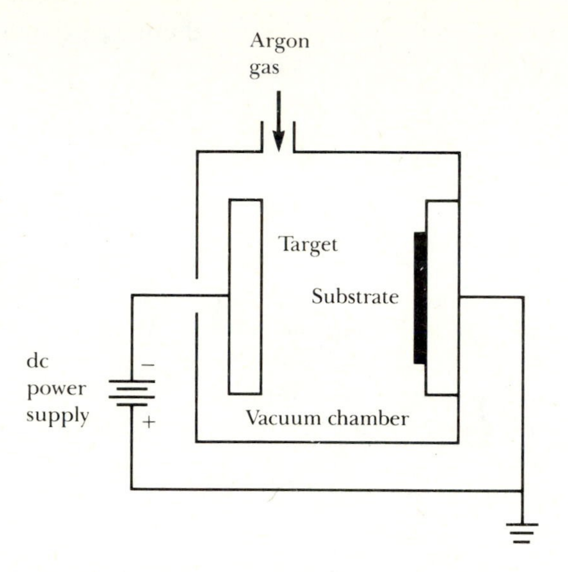

2.10.3 Sputtering

If an ion strikes the surface of a material with sufficient energy, atoms or clusters of atoms may be dislodged, or "sputtered" away from the surface. Under proper conditions, the sputtered material coats a nearby substrate. Sputter deposition systems most often employ two electrodes and an inert sputtering gas, usually argon. In dc sputtering systems, as shown in Figure 2.35, several hundred volts are applied to the two electrodes, which ionizes the argon, creating a glowing discharge [20]. The positively charged argon atoms sputter the negative "target" electrode, which contains the desired material, and deposition occurs on interior surfaces, including the opposite electrode which contains the substrates to be coated.

Sputtering takes place in a chamber that is first evacuated and then filled to a partial pressure, typically 10 mtorr, with a continuous flow of argon. The technique is useful for depositing materials with very high vaporization temperatures as well as alloy materials consisting of elements with greatly different vapor pressures, since the sputtering mechanism does not involve melting. If the dc voltage is replaced by an rf voltage, then insulators can be sputtered as well as conductors. Also, the addition of a magnetic field provides a more efficient discharge and higher sputtering rates. In some cases, sputtering provides superior step coverage, as compared to evaporation. By cosputtering from two targets simultaneously, compounds may be deposited. An example of cosputtering is silicide deposition, in which one target is silicon and the other is a refractory metal.

A related deposition method is ion beam coating. An ion beam, usually

Figure 2.36 Schematic view of ion beam coating. Typical pressures range from 10^{-4} to 10^{-5} torr.

SOURCE: From G. R. Thompson, "Ion Beam Coating: A New Deposition Method," *Solid State Technology* 21: 73–77, December 1978, Reprinted with permission of *Solid State Technology*, published by Technical Publishing, a company of Dun & Bradstreet.

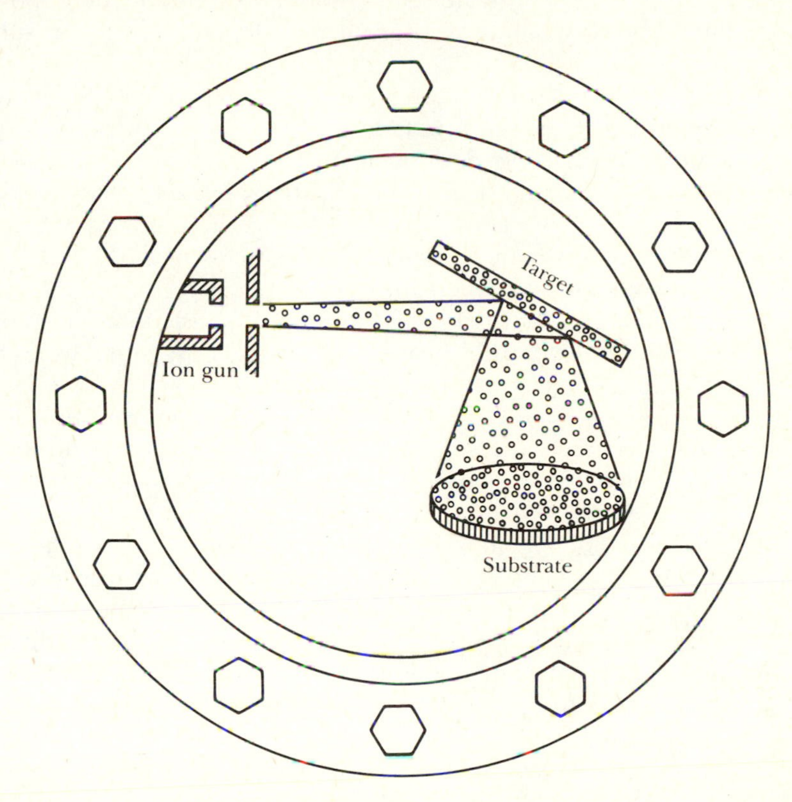

argon, sputters material from a target to a substrate located in a remote field-free region in which the vacuum is of higher quality than that in a sputtering chamber. Consequently, the substrate is exposed to a low thermal load and the degree of contamination from the environment is small, such that high-purity, high-quality films result. A schematic representation of the ion beam coating process is shown in Figure 2.36.

2.10.4 Chemical Vapor Deposition

Chemical vapor deposition is a deposition method in which energy is supplied to drive a gas phase reaction. In standard CVD, the energy is supplied in the form of heat. However, the reaction energy may also be supplied by plasma or optical excitation. Examples of CVD products are Si_3N_4, SiO_2, and silicon. The nature of the process is similar to that for epitaxial growth, as described in Section 2.8, but when the qualifying adjective *epitaxial* is not used, CVD films are generally assumed not to be single-crystalline. Consider, for example, the CVD of polysilicon. As indicated in Eq. (2.57), the decomposition of silane produces silicon and hydrogen. Under epitaxial conditions, the resulting silicon film is single-crystalline. However, if the substrate on which deposition takes place does not have the silicon diamond lattice structure, or if the substrate temperature is low, then the deposited film will be either polycrystalline or amorphous, depending on the substrate temperature and type.

Thermal CVD reactors are similar to epitaxial reactors in that a gas flow is provided over the wafers at a temperature sufficient to produce the desired reaction. "Hot wall" reactors, in which a quartz processing tube containing the wafers is placed in a resistively heated furnace similar to a diffusion apparatus, have been found to produce high-quality CVD films. However, "cold wall" systems, in which a susceptor is heated by rf induction heating, have a somewhat higher throughput.

In plasma-assisted CVD, most of the reaction energy is provided by a plasma, rather than by heat. Typically, rf energy establishes a glow discharge such that the gas constituents are in highly reactive states. A chief advantage of plasma-assisted CVD is that it is a low-temperature process. Therefore less wafer warpage and impurity motion occur, which is an appreciable advantage in VLSI processing. Bulk heating of the wafer is also avoided in laser-assisted CVD.

CVD polysilicon is often used as the gate electrode material in MOSFET integrated circuits and as a signal-conducting path where distances are not too great. Most commonly, decomposition of silane, as shown in Eq. (2.57), is the method of choice. Polysilicon may be doped after deposition, by either diffusion or ion implantation, or it may be doped during deposition by adding dopants to the gas flow. A typical thermal CVD deposition temperature is 650°C.

Silicon dioxide may be deposited, for example, by reacting silane in the presence of oxygen as

$$SiH_4 + O_2 \rightarrow SiO_2 + 2H_2 \tag{2.59}$$

This reaction may be carried out at fairly low temperatures, less than 500°C. Although the quality of the oxide is not as good as that of thermally grown oxides, this method provides a way of depositing an insulator over nonsilicon materials, such as over a metalization, which allows multiple metal paths separated by an insulating oxide layer.

Silicon nitride is a highly effective barrier to impurities and water vapor and therefore makes a good passivating layer over the surface of an integrated circuit to protect the device against aging effects. In addition, as discussed further in Chapter 3, it provides a means of selective oxidation in device fabrication schemes. Again, by reaction with ammonia, silane provides one chemical pathway to this CVD process, as follows:

$$3SiH_4 + 4NH_3 \rightarrow Si_3N_4 + 12H_2 \tag{2.60}$$

at temperatures in the range of 700 to 900°C.

The chemical vapor deposition of metals is also of interest as an alternative to evaporation and sputtering. For example, tungsten films may be deposited by

$$WF_6 \rightarrow W + 3F_2 \tag{2.61}$$

Again, the reaction energy may be provided thermally or, for low-temperature processing, by plasma or optical excitation. Supplying the optical energy in the form of a focused laser beam opens the possibility of writing metal patterns directly on wafers without a mask. With a narrow beam, CVD reactions can be localized to an area approximately 1 μm in diameter.

2.11 Etching

Much of the discussion so far has been concerned with the addition of materials and dopants to the wafer. However, it is also of vital importance to selectively remove materials by etching. Etching may be divided into two categories, wet etching and dry etching. In the former, resist-coated wafers are immersed in acid baths, and in the latter, the chemical processes are plasma driven. A chief advantage of the latter is the availability of anisotropic, or vertical wall etching. In addition, dry etching procedures offer safety advantages.

Consider, as a case in point, the etching of SiO_2 in an acid bath. The primary ingredient in the bath is hydrofluoric acid (HF), with ammonium fluoride (NH_4F) added for buffering purposes. Since the nature of this etch is isotropic, there will be lateral etching as well as vertical etching, as shown in Figure 2.37A. This limits line spacing and therefore density on the chip. However, SiO_2 may also be etched, with a higher degree of vertical etching, in a reactive plasma environment that contains F. Although plasma-etch chemistry is complex, involving a variety of excited species, the basic reason for the anisotropic nature of plasma etching is that the etch rate increases rapidly when energetic ions are incident on the wafer. If the ions are normally incident on the wafer, then the etching of the sidewalls proceeds at a much slower rate than that for the horizontal wafer surface, as shown in Figure 2.37B. On the other hand, if the plasma pressure is sufficiently high, or the voltage between the plasma and the wafer is sufficiently low, then the ions are randomly incident on the wafer and plasma etching becomes isotropic.

Before discussing different reactor designs, it is useful to consider the etching of silicon in a CF_4 plasma as an example of the chemical pathways.

Figure 2.37 Isotropic and anisotropic etching. A. Lateral undercutting with isotropic etching. B. Plasma anisotropic etching.

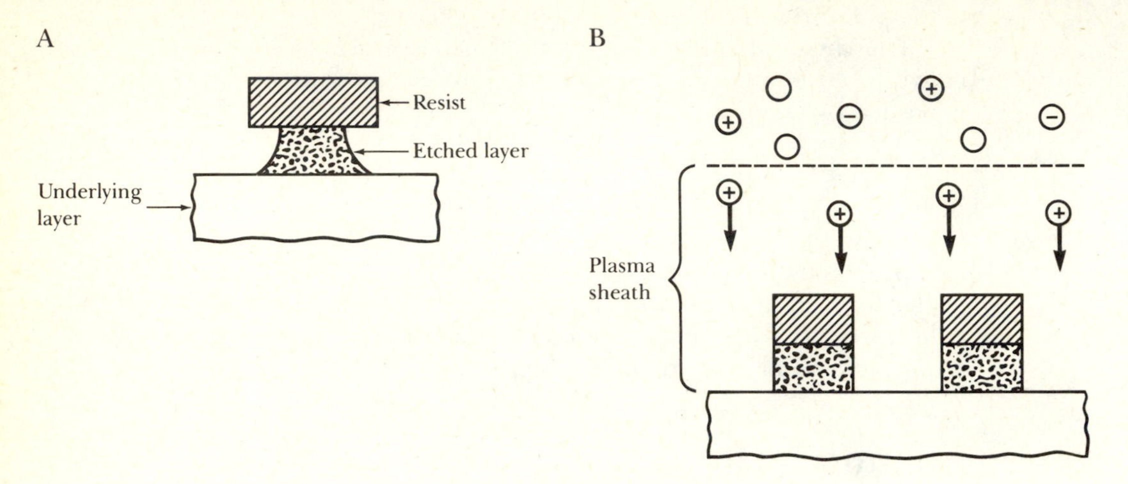

The chemically active species in the plasma is F, and the simplified steps in the process are as follows [21]:

Chemisorption	$(F_2)_{GAS} \rightarrow (F_2)_{ADS} \rightarrow 2(F)_{ADS}$	(2.62)
Reaction	$Si + 4F_{ADS} \rightarrow (SiF_4)_{ADS}$	(2.63)
Desorption	$(SiF_4)_{ADS} \rightarrow (SiF_4)_{GAS}$	(2.64)

where the ADS subscript refers to adsorbed. A possible explanation for ion-assisted etching here is that ions cause damage sites on the silicon surface, which enhances the dissociative chemisorption of F_2.

Several dry etch reactor types have been developed and are in use. Figure 2.38 shows a sampling of the various approaches. In barrel reactors, the wafers are not in direct contact with the rf discharge but are in a perforated metal tunnel, held at ground potential, that is surrounded by the discharge. The surface potential of the wafer is not much different than that of the surrounding glow, so ion bombardment is not appreciable and etching is due primarily to long-lived radicals. Operating pressures are typically high, on the order of 200 mtorr, and etching is generally isotropic. The approach is well suited for multiple wafer processing, but uniformity problems require overetching.

In parallel-plate reactors, also known as Reinberg reactors, the wafers are on the grounded plate of a diode plasma system and in direct contact with the plasma [22]. A sheath potential of approximately 75 to 250 V exists between the wafers and the plasma, and ion bombardment is significant. Also, short-lived radicals take part in the etching in this configuration. Both barrel

Figure 2.38 Dry etching techniques. A. Barrel reactor. B. Parallel plate reactor. C. Reactive ion etching. D. Ion beam milling. If reactive gases are used, this configuration is referred to as *reactive ion beam etching* (RIBE).
SOURCE: Figure 2.39D from C. J. Mogab, "Dry Etching," in S. M. Sze, *VLSI Technology*, New York: McGraw-Hill, 1983, Chapter 8. Reproduced with permission.

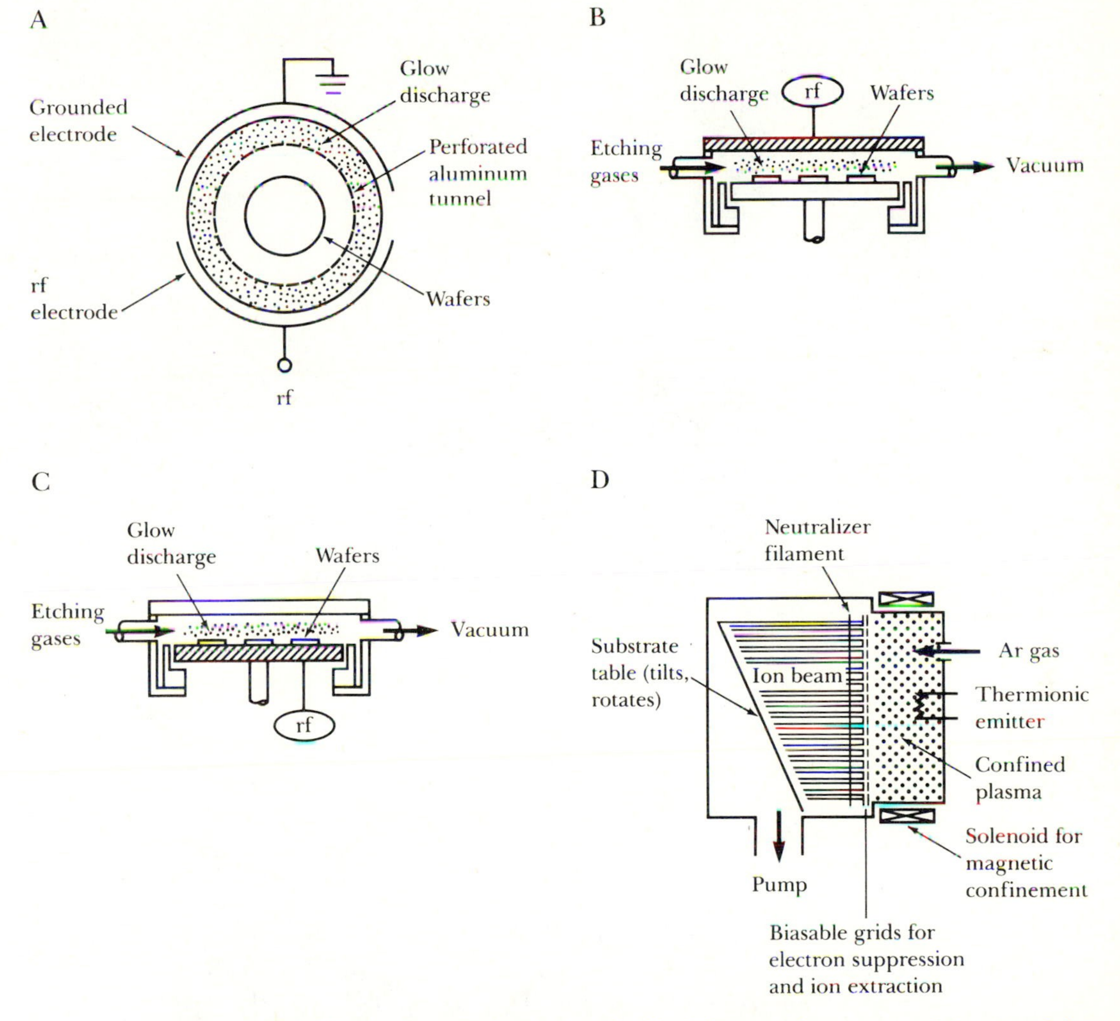

and parallel-plate reactors have received widespread commercial use, with the latter offering improved etch uniformity and the possibility of anisotropic etching.

Reactive ion etching (RIE), also known as reactive sputter etching, often employs apparatus similar to sputtering chambers, with the wafer taking the place of the target. In this mode, rf power is supplied to the wafer-holding substrate. Compared to the parallel-plate reactor, RIE systems are character-

Figure 2.39 Photograph of a dual ion source system. The source for ion beam milling is on the left and the source for ion beam coating is on the right.
SOURCE: Courtesy of Commonwealth Scientific Corporation, 500 Pendleton Street, Alexandria, Virginia 22314.

ized by generally lower pressures (10 mtorr to 100 mtorr) and higher substrate potentials. These traits combine to produce highly anisotropic etching.

Ion beam milling uses the ion beam sputtering process to remove material from a substrate. A photograph of a system that includes two ion beam sources, one for ion beam etching and one for ion beam coating, is shown in Figure 2.39. If the gas providing the ions is reactive, then the method is known as reactive ion beam etching (RIBE). RIBE offers the possibility of controlling the angle of ion incidence and operates at quite low-pressure levels of about 0.1 mtorr. Consequently a very high degree of etching anisotropy is achievable. The RIBE method also differs from RIE in that the wafer is not in actual exposure to the plasma discharge. Therefore, there is little interaction with atoms in short-lived excited states.

Another variation on ion beam etching is ion beam–assisted etching (IBAE), in which the chemically reactive species and incident energetic ions are independently controlled [23]. In this arrangement, a chemically active gas such as chlorine impinges on the sample from jets that direct a gas stream onto the wafer. Independently, the sample is bombarded by energetic ions,

Figure 2.40 Scanning electron microscopy view of a GaAs sample etched by ion beam–assisted etching. The grating is etched approximately 1.5 μm deep with an aspect ratio greater than 10 : 1.

SOURCE: From G. A. Lincoln, M. W. Geiss, S. Pang, and N. N. Efremow, "Large Area Ion Beam Assisted Etching of GaAs with High Etch Rate and Controlled Anisotropy," *Journal of Vacuum Science and Technology,* B1 (1983), 1043–1046. Reprinted with permission of Lincoln Laboratory, Massachusetts Institute of Technology, Lexington, MA.

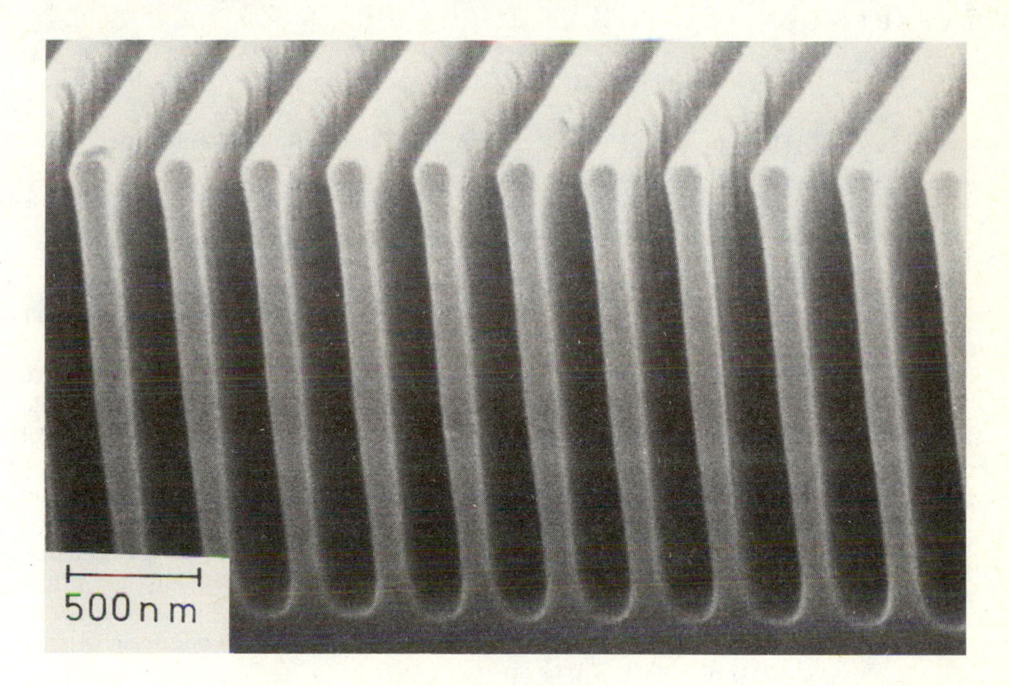

typically argon, from a broad-beam ion source. This method is capable of dramatic etching profiles, as shown in Figure 2.40.

Directional etching is also achieved with certain wet etches. For example, a mixture of hydrazine, potassium hydroxide, and ethylene diamine preferentially attacks the (100) plane of silicon. In etching (100)-oriented wafers, the etch exhibits a self-stop property along the (111) planes, such that the etch proceeds at an angle of 54.7 degrees. However, directional wet etches are rather an exception to the rule.

BIBLIOGRAPHY

The following sources contain further information on the physics and chemistry of processing, and offer advanced treatment of many of the topics in this chapter:

Elliott, D. J. *Integrated Circuit Fabrication Technology.* New York: McGraw-Hill, 1982.

Ghandhi, S. K. *VLSI Fabrication Principles—Silicon and Gallium Arsenide.* New York: Wiley, 1983.

Sze, S. M., ed. *VLSI Technology.* New York: McGraw-Hill, 1983.

The following reference contains a wealth of practical information concerning material properties and process data:

W. E. Beadle, J. C. C. Tsai, and R. D. Plummer, eds. *Quick Reference Manual For Silicon Integrated Circuit Technology.* New York: Wiley, 1985.

REFERENCES

1. H. Heiblum, E. E. Mendez, and F. Stern, "High Mobility Electron Gas in Selectively Doped *n*:AlGaAs/GaAs Heterojunctions," *Applied Physics Letters,* 44 (1984), 1064–1066.
2. H. Cramer, "Heterostructure Bipolar Transistors and Integrated Circuits," *IEEE Proceedings,* 70 (1982), 13–25.
3. C. D. Thurmond, "The Standard Thermodynamic Functions for the Formation of Electrons and Holes in Ge, Si, GaAs, and GaP," *Journal of the Electrochemical Society,* 122 (1975), 1133–1141.
4. C. P. Ho, J. D. Plummer, S. E. Hansen, and R. W. Dutton, "VLSI Process Modeling—SUPREM III," *IEEE Transactions on Electron Devices,* ED-30 (1983), 1438–1452.
5. S. K. Ghandhi, *VLSI Fabrication Principles* (New York: Wiley, 1983), Chap. 4.
6. S. K. Ghandhi, *The Theory and Practice of Microelectronics* (New York: Wiley, 1968).
7. R. A. Colclaser, *Microelectronics: Processing and Device Design* (New York: Wiley, 1980), Chap. 7.
8. R. B. Fair, "Recent Advances in Implantation and Diffusion Modeling for the Design and Process Control of Bipolar IC's," in *Semiconductor Silicon 1977,* eds. H. R. Huff and E. Sirtl (Princeton: Electrochemical Society, 1977), pp. 968–987.
9. F. A. Trumbore, "Solid Solubilities of Impurity Elements in Germanium and Silicon," *Bell System Technical Journal,* 39 (1960), 205–234.
10. D. J. Hamilton and W. G. Howard, *Basic Integrated Circuit Engineering* (New York: McGraw-Hill, 1975), Chap. 2.
11. G. Masetti, D. Nobili, and S. Solmi, "Profiles of Phosphorus Predeposited in Silicon and Carrier Concentration in Equilibrium with SiP Precipitates," in *Semiconductor Silicon 1977,* eds. H. R. Huff and E. Sirtl (Princeton: Electrochemical Society, 1977), pp. 648–657.
12. D. P. Kennedy and R. R. O'Brien, "Analysis of the Impurity Atom Distribution Near the Diffusion Mask for a Planar *p-n* Junction," *IBM Journal of Research and Development,* 9 (1965), 3.
13. R. W. Dutton, "Modeling of the Silicon Integrated Circuit Design and Manufacturing Process," *IEEE Transactions on Electron Devices,* ED-30 (1983), 968–985.
14. A. Axmann, "Source Feed Materials in Ion Beam Technology," in *Impurity Doping Processes in Silicon,* ed. F. F. Y. Wang, vol. 2 (Amsterdam: North Holland, 1981), pp. 147–174.
15. Veeco, Industrial Equipment Division, P. O. Box 17068, Austin, TX 78760.
16. A. B. Glaser and G. E. Subak-Sharpe, *Integrated Circuit Engineering* (Reading, Mass.: Addison-Wesley, 1977), Chap. 5.
17. J. F. Gibbons, W. S. Johnson, and S. W. Mylroie, *Projected Range Statistics, Semiconductors and Related Materials,* 2nd ed. (Stroudsburg, Penn.: Dowden, Hutchinson and Ross, 1975).

18. T. E. Seidel, "Ion Implantation," in *VLSI Technology*, ed. S. M. Sze (New York: McGraw-Hill, 1983), Chap. 6.
19. C. W. White, W. H. Christie, B. R. Appleton, S. R. Wilson, and P. P. Pronko, "Redistribution of Dopants in Ion Implanted Silicon by Pulsed Laser Annealing," *Applied Physics Letters*, 33 (1978), 654–656.
20. B. Chapman, *Glow Discharge Processes* (New York: Wiley, 1980), Chap. 6.
21. J. W. Colburn and H. F. Winters, "Plasma Etching—A Discussion of Mechanisms," *Journal of Vacuum Science Technology*, 16 (1979), 391–403.
22. A. R. Reinberg, "Radial Flow Reactor," U.S. Patent 3 757 733, 1973.
23. G. A. Lincoln, M. W. Geiss, S. Pang, and N. N. Efremow, "Large Area Ion Beam Assisted Etching of GaAs with High Etch Rate and Controlled Anisotropy," *Journal of Vacuum Science and Technology*, B1 (1983), 1043–1046.

PROBLEMS

1. One milligram of phosphorus is added to a 25-kg silicon melt in a Czochralski crystal grower. The distribution coefficient is 0.32 and phosphorus has an atomic weight of 30.97. Calculate the expected hole and electron concentrations in the resulting wafers at 27°C, 500°C, and 1000°C if the corresponding values of n_i are 1.45×10^{10} cm^{-3}, 8×10^{16} cm^{-3}, and 6.5×10^{18} cm^{-3}, respectively.

   Repeat the question if arsenic is added instead of phosphorus. The distribution coefficient for arsenic in silicon is 0.27 and the atomic weight of arsenic is 74.92.

2. A 30-min boron predeposition (constant-source) diffusion is carried out in a uniformly doped *n*-type silicon wafer at 960°C. The background doping level in the wafer is 3.5×10^{15} cm^{-3}, and the constant-source condition is determined by the solid solubility limit. Assuming that the diffusion coefficient can be approximated by Eq. (2.21), calculate the junction depth.

3. Processing on the wafer in Prob. 2 is continued with a drive-in diffusion at 1150°C for 2 hours.

   a. Find the boron concentration at the silicon surface, neglecting redistribution between the silicon and oxide.
   b. Calculate the new junction depth.

4. Assume the temperature accuracy on the diffusion furnace is ± 1°C.

   a. What is the resulting uncertainty in the predeposition junction depth of Prob. 2?
   b. What is the resulting uncertainty in the drive-in junction depth of Prob. 3? How many minutes uncertainty in the drive-in time would give the same uncertainty in the junction depth?

5. A constant-source boron isolation diffusion is performed in a 2-μm-thick epitaxial layer as shown in Figure 2.29. The epitaxial layer is n-type with $N_D = 10^{16}$ cm^{-3}, and the substrate is p-type with $N_A = 10^{15}$ cm^{-3}. The diffusion is carried out such that $D = 10^{-12}$ cm^2/s and the surface concentration is maintained at 10^{20} cm^{-3}. Your strategy for timing the diffusion is as follows: Find the time to just get through the epitaxial layer, then add 30 minutes for safety. What is the required diffusion time?

6. A uniformly doped n-epitaxial layer has 10^{15} donors per cm^3. A p-type base drive-in diffusion is carried out for 1 hour with $D = 10^{-12}$ cm^2/s. The wafers are removed from the furnace and measurements on a test wafer show the junction depth to be 1 μm. The wafers are returned to the same drive-in furnace for 1 more hour. Now where will the junction be?

7. Find the processing schedule for a silicon npn BJT with the following requirements for the finished device.

   $x_{JC} = 4.0\ \mu\text{m}$ $\quad$ $N(0)_{\text{BASE}} = 10^{19}\ \text{cm}^{-3}$

   $x_{JE} = 2.5\ \mu\text{m}$ $\quad$ $N(0)_{\text{EMITTER}} = 2 \times 10^{19}\ \text{cm}^{-3}$

   Here, x_{JE} represents the depth of the junction between the base and emitter and x_{JC} is the depth of the junction between the collector and base.

   The device is fabricated in an n-type epitaxial layer with a doping level of 10^{16} cm^{-3}. The drive-in temperatures are fixed at 1150°C and the predeposition temperatures are fixed at 960°C. The predeposition diffusions are solid solubility limited, and boron and phosphorus are the dopants. Specifically, find the predeposition times for the base and emitter, and also the drive-in times.

8. Show that Eqs. (2.30) and (2.31) result from Eq. (2.29).

9. Calculate diffusion coefficients for arsenic and boron at 1100°C for dopant concentrations as follows:

   a. $N_{\text{DOPANT}} << n_i$
   b. $N_{\text{DOPANT}} = 10^{18}$ cm^{-3}
   c. N_{DOPANT} = solid solubility limit

   The wafer background doping is 10^{16} cm^{-3}. Take into account field-assisted diffusion and multiple vacancy types. Take the intrinsic carrier concentration to be 10^{19} cm^{-3} for the boron diffusion and 6×10^{19} cm^{-3} for the arsenic diffusion. Although the temperature is the same for both cases, the high solid solubility of arsenic causes energy-gap narrowing and a resultant increase in intrinsic carrier concentration.

10. An ion implantation is carried out such that the peak impurity concentration after the implant is 10^{20} cm^{-3} and the peak is located 2 μm below the surface. Also, the concentration falls to 10^{19} cm^{-3} on either side of the peak at 1.8 μm and 2.2 μm. Find the range, straggle, and dose for this implant.

11. The implant in Prob. 10 is annealed for 30 minutes with a diffusion coefficient of 10^{-14} cm^2/s. If the background wafer doping is 10^{16} cm^{-3} and of a doping type opposite to the implant, where are the junctions?

12. One way to fabricate complementary MOS (CMOS) integrated circuits is to implant n-type wells into a lightly doped p-type substrate. Suppose a 200-kV phosphorus implant is carried out with a beam current density of 0.1 $\mu \text{Å}/cm^2$. The substrate doping is $N_A = 10^{15}$ cm^{-3}, and the desired depth of the n well is 5 μm. After the implant, an anneal is done for 5 hours with $D = 10^{-12}$ cm^2/s.

    a. What is the implant dose in atoms per square centimeter?
    b. What is the peak value of the phosphorus concentration after the anneal?
    c. How long should the ion beam be on to get the dose implied in this problem?

13. A silicon wafer is covered by an SiO_2 layer 0.3 μm thick.

    a. What is the time required to increase the thickness to 0.5 μm by a wet oxidation at 1200°C?
    b. Repeat the question for a dry oxidation at 1100°C.

14. An oxide layer 5000 Å thick is grown on a silicon wafer, and an opening is etched in it to the silicon surface. Then the wafer is oxidized again such that an additional 5000 Å of oxide are grown in the regions containing the old, unetched oxide. The second oxidation is a 1200°C wet oxidation. Find the resulting wafer steps.

15. Assume the distance between an aluminum evaporation coil and a wafer is 14 cm. Assuming the wafer is centered over the coil, the maximum film thickness will be at the center of the wafer. If a thickness fall-off of 10% at the wafer edge is tolerable, how large can the wafer be?

3 MOS Integrated Circuit Fundamentals

3.1 Introduction

Insulated-gate field-effect transistors (FETs) are well suited for high-density integrated circuits because of their small size and, with proper circuit design, low power consumption. Since the gate insulator is usually an oxide and since the first gate electrodes were metallic, such devices are usually referred to as metal-oxide-semiconductor devices, or MOSFETs. However, with current technology, the gate electrode is often polysilicon or a silicide.

Basic concepts behind MOSFET operation are illustrated in Figure 3.1 for an *n*-channel enhancement-mode device. The simplified cross-sectional view in this figure shows a three-terminal device with bilateral symmetry; that is, the source and drain appear identical and interchangeable. In most actual MOSFET realizations this symmetry is indeed present and, for *n*-channel devices, the drain is considered by convention to be the terminal at the more positive potential. The gate electrode may be considered the control electrode, in that it controls the conduction path between the source and drain.

On first inspection, a conducting path between the source and drain would appear unlikely because the two terminals are connected to back-to-back *p*-*n* junctions. However, the role of the gate is to selectively establish an *n*-type region of semiconductor, or a conducting "channel," beneath the oxide layer, such that varying degrees of conduction are achieved between the source and drain. This *n*-type channel is also referred to as an inversion region, and is illustrated from an energy band viewpoint in Figure 3.1B. The onset of the inversion layer is taken to be at a threshold gate-to-source voltage, V_T. As the band bending increases, the conduction path between the source and drain increases, and for a given drain-to-source voltage, the drain current increases as shown in the family of curves in Figure 3.1C.

The relationships between device properties such as V_T and fundamental processing variables such as wafer doping, wafer orientation, and oxide thickness are of considerable practical importance in integrated circuit fabrication and design. To appreciate the relationship between device per-

Figure 3.1 MOSFET fundamentals. A. Simplified n-channel MOSFET cross section. B. Energy band diagram for inversion. C. Family of curves: Drain current I_D, versus drain-to-source voltage, V_{DS}, and gate-to-source voltage, V_{GS}.

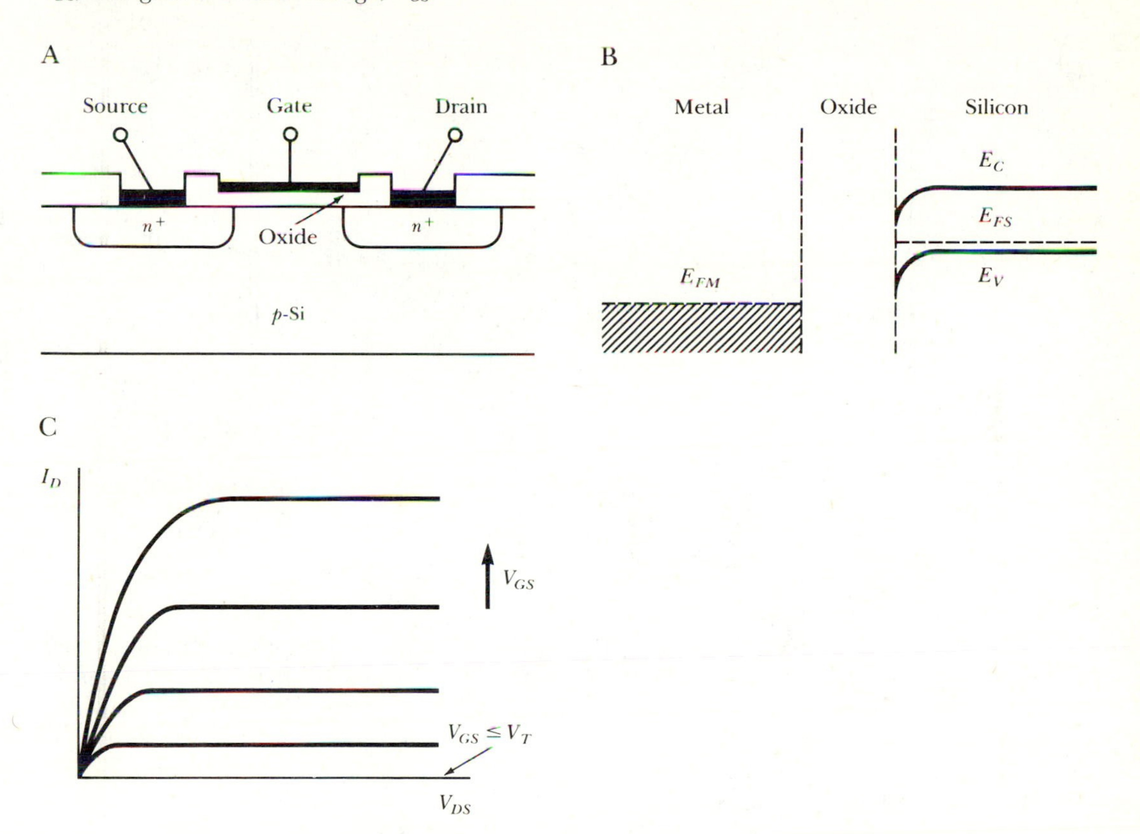

formance and wafer processing conditions, it is necessary to review the metal-oxide-semiconductor interface.

3.2 The Metal-Oxide-Semiconductor Interface

3.2.1 MOS Energy Bands

An energy band viewpoint offers the best insight into the MOS system. Figure 3.2A shows the energy band diagrams of three separated materials for the cases of an aluminum gate electrode, a silicon dioxide gate insulator, and a p-type silicon wafer doped at approximately 10^{16} cm^{-3}. The energy levels in this diagram are referenced to the zero energy of an electron in a vacuum, E_0, and the work function of the metal and of the semiconductor are labeled $q\Phi_M$ and $q\Phi_S$, respectively. Since the Fermi energy in the silicon lies in a

Figure 3.2 MOS energy bands. A. The separated materials. B. The MOS system in equilibrium.

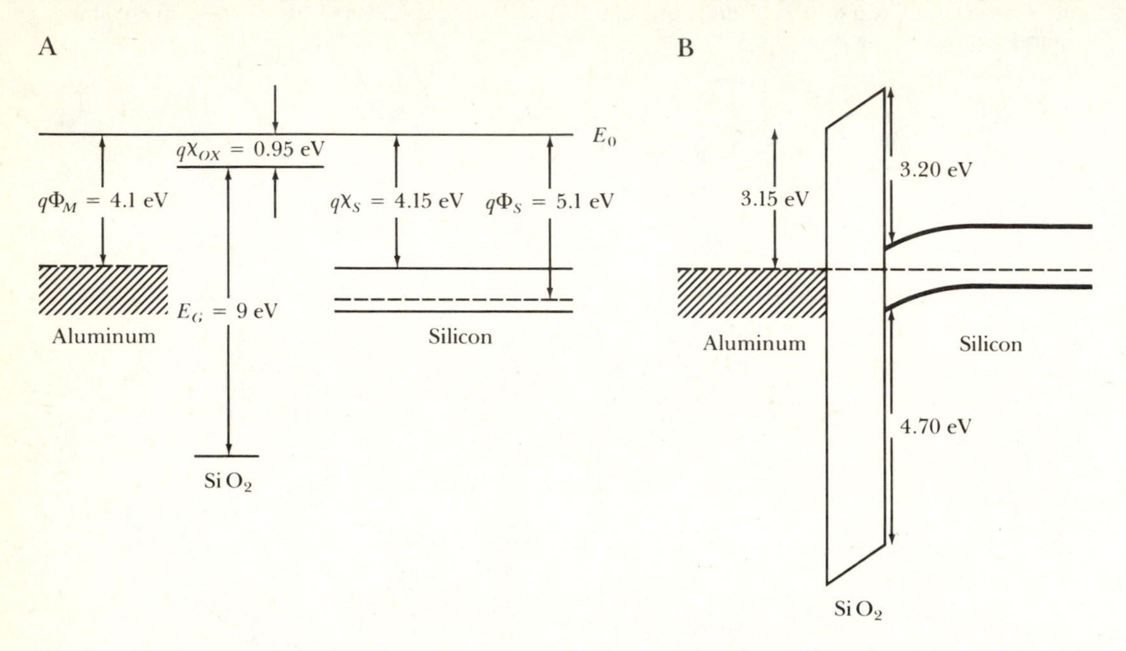

region of forbidden energy, the silicon work function cannot be measured directly. However, the electron affinity, $q\chi_S$, can be measured, which allows the work function value to be deduced. The energy gap of SiO_2 is shown to be 9 eV, although there is some experimental and theoretical ambiguity as to this value because of the amorphous atomic ordering in the oxide.

When the three materials are joined, the combined system in equilibrium must be characterized by a single Fermi energy, as shown in Figure 3.2B. The barrier heights at the metal-oxide and metal-semiconductor interfaces, of 3.15 and 3.20 eV respectively, are determined by the work-function and electron-affinity values. Given these barrier heights, the constant Fermi energy requirement leads to band bending in the oxide and semiconductor, as shown. More physically, since aluminum has a lower work function than silicon, aluminum electrons are in high energies relative to silicon electrons. Consequently there is negative charge transfer from the aluminum to the silicon. Band bending in the semiconductor results, since carrier concentrations have been changed near the interface. There is band bending in the oxide because of the charge stored on either side of it, and there is a corresponding oxide voltage drop, much as is the case for a capacitor dielectric.

For appreciable current flow between the source and drain, an inversion layer must be established in the semiconductor adjacent to the gate insulator; that is, band bending in the *p*-type silicon must be sufficient that the quasi-Fermi level should be close to the conduction band edge, E_C, near the interface. For the energy band diagram shown in Figure 3.2B, the band bending is not sufficient to achieve inversion. However, two factors related to band bending are yet to be considered. The first is a bias on the gate electrode relative to the silicon. The second concerns additional charges in the system, added intentionally by ion implantation or inadvertently due to contamination, or present because of defects in the MOS system.

Consider first the effect of bias on the energy band diagram and on the charge distribution in the semiconductor. The zero bias, or equilibrium condition, shown in Figure 3.2B indicates the existence of a depletion region in the semiconductor. Since the Fermi energy near the oxide interface is relatively far from both E_C and E_V, there are few holes and electrons near the interface under zero bias conditions. Near the interface, the net charge in the semiconductor is

$$\rho = q(p - n - N_A) \simeq -qN_A \tag{3.1}$$

and this charge extends the length of the depletion region, x_D. If a positive bias is applied to the gate, the voltage drop across the oxide is increased, as is the charge stored on either side of the oxide. Under bias, the system is no longer in equilibrium, so Fermi-Dirac statistics do not strictly apply and the Fermi level is replaced by the quasi-Fermi level. For low values of positive bias, the depletion layer simply extends from the equilibrium position more deeply into the semiconductor. Eventually, however, with further increase in bias and increased band bending, E_C near the interface is sufficiently close to the quasi-Fermi energy such that the electron concentration, n, becomes the dominant term in Eq. (3.1), and inversion is established. The effect of increasing positive gate bias is shown in Figure 3.3, where V_G refers to the voltage on the gate relative to the semiconductor. Note that once inversion is reached, further charge increases in the semiconductor are mostly due to increased electron concentration in the inversion layer, and x_D is pinned near a maximum value.

A negative gate bias opposes the built-in voltage across the oxide. When the bias is exactly equal to the work-function difference, there is no band bending in either the semiconductor or the oxide. The "flat band" voltage, neglecting oxide charges, may then be expressed as

$$V_{FB} = \Phi_M - \Phi_S \tag{3.2}$$

With gate biases more negative than V_{FB}, the voltage drop across the oxide becomes negative, and positive charge is stored in the semiconductor. From the energy band diagram shown in Figure 3.4, this corresponds to band bending such that E_V is near the quasi-Fermi level and the hole concentration, p, is the dominant term in Eq. (3.1). In this case there is an accumulation of holes near the interface and the MOS system is in "accumulation."

Figure 3.3 Application of positive bias to the MOS system. A. Equilibrium, $V_G = 0$. B. $V_G = V_{G1} > 0$. C. $V_G = V_{G2} > V_{G1}$ (inversion).

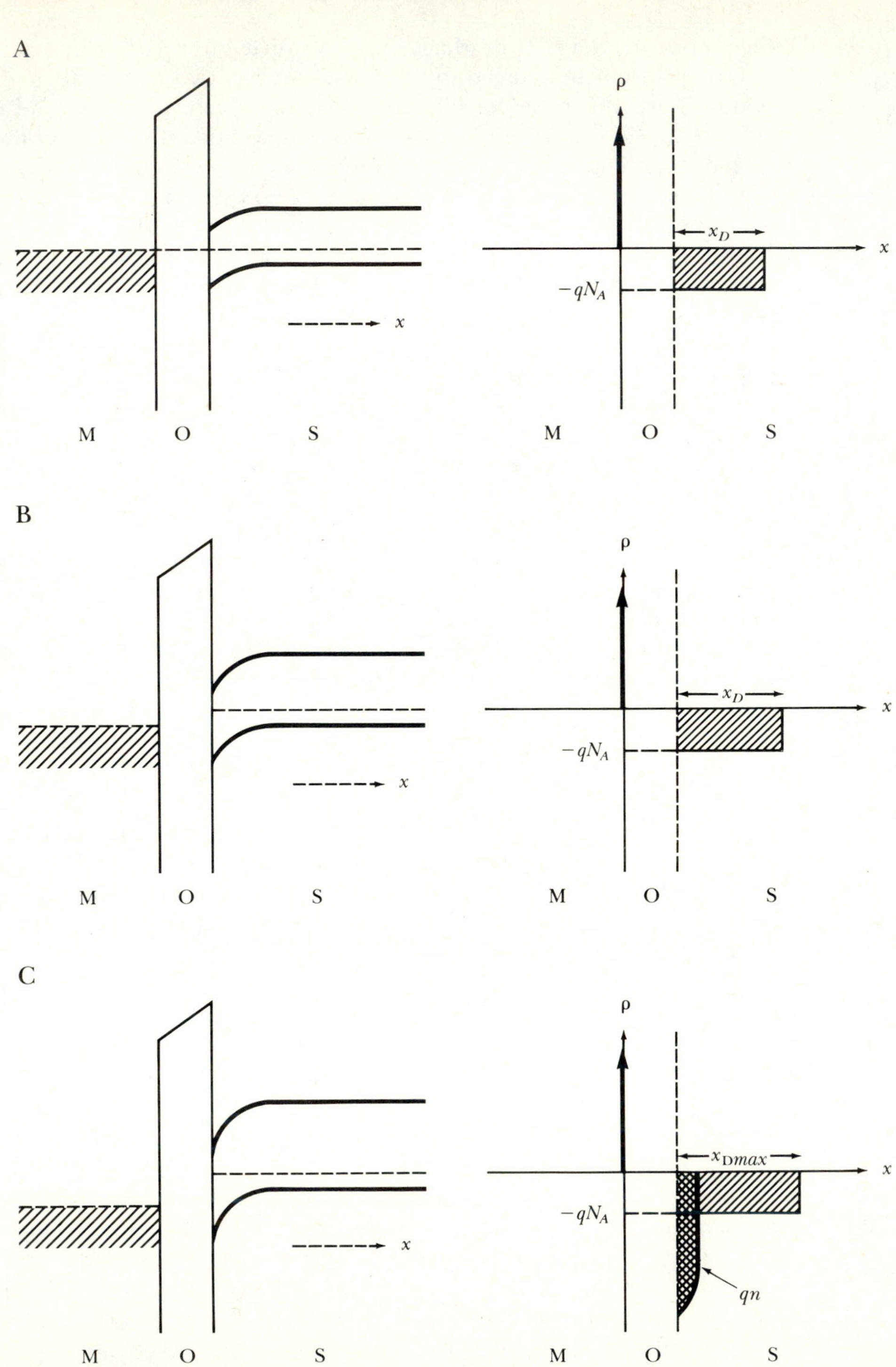

Figure 3.4 Application of negative bias to the MOS system. A. Equilibrium, $V_G = 0$. B. $V_G = V_{FB}$. C. $V_G < V_{FB}$.

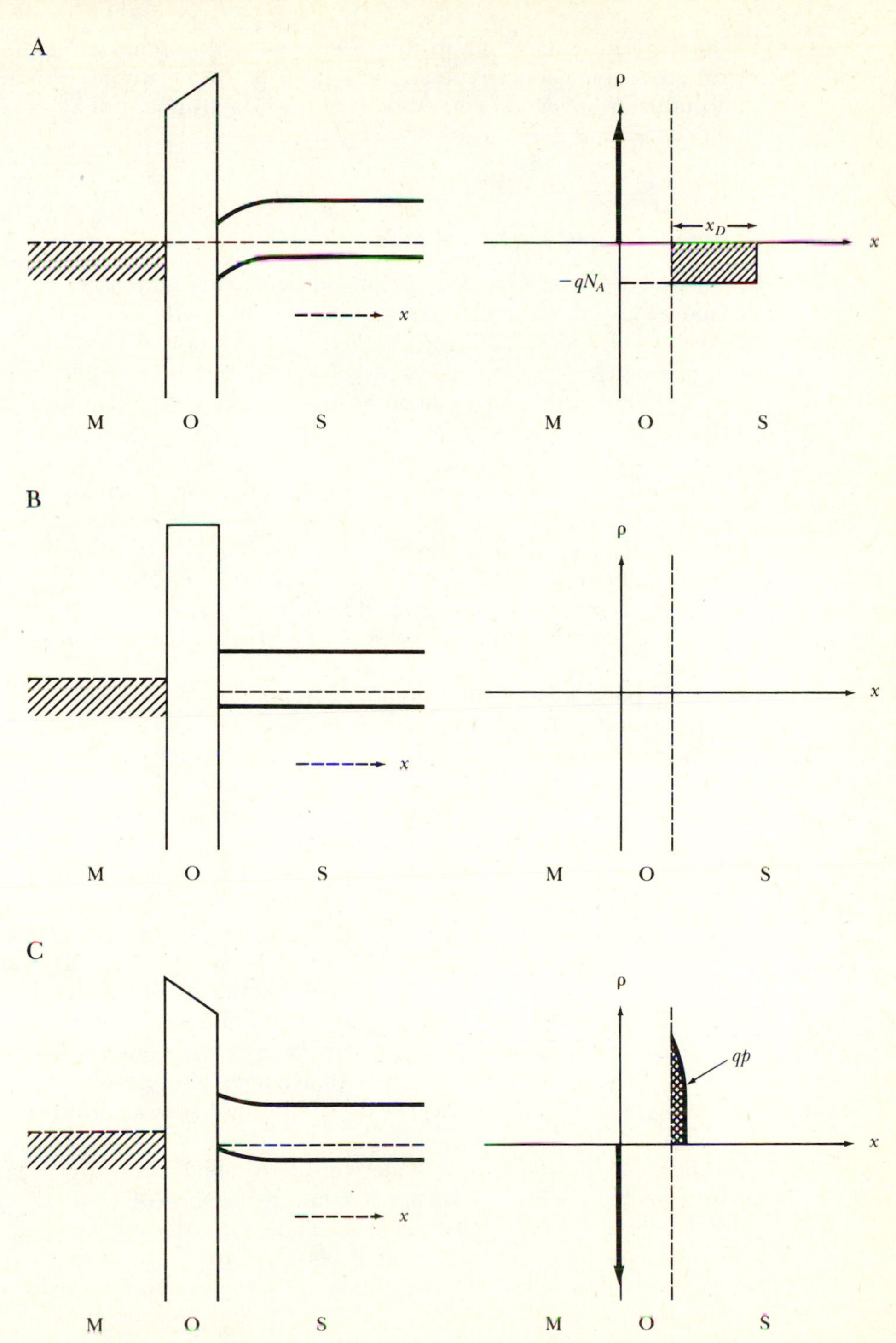

Similar concepts apply to the case of an *n*-type semiconductor, except for differences in carrier type and bias polarity. For example, with an *n*-type semiconductor, inversion corresponds to the establishment of a *p*-type region under the interface by means of a negative bias on the gate.

3.2.2 Oxide and Interface Charges

In practical MOS systems there is an additional charge, not accounted for in the above discussion, which can have appreciable effects on band bending and the establishment of inversion, accumulation, and flat band conditions. Four general types of charges in the SiO_2-Si system may be identified [1]. A charge Q_{it} is present at the silicon-oxide interface largely because of the interruption of the single-crystal silicon lattice. There are approximately 10^{15} surface atoms per square centimeter on a silicon wafer surface. When an SiO_2 layer is present, the first atomic layer of oxide is actually SiO_x where $1 < x < 2$. In other words, the silicon is incompletely oxidized. Most of the silicon surface atoms bond to the SiO_x layer, but the fit is not perfect and some unused or very weak silicon and oxygen bonds result. The energy levels associated with these dangling bonds lie in the silicon energy gap and may be charged. Furthermore, these states can readily exchange charge with the silicon crystal.

Other possible contributors to Q_{it} are bonded foreign atoms at the surface, structural defects in the wafer surface, and radiation-induced bond breaking at the silicon-oxide interface. The value of Q_{it} is highly dependent on oxide growth conditions and on subsequent annealing steps. If the wafer is annealed in hydrogen at temperatures of approximately 450°C, the interface charge density can be as small as 10^{10} cm^{-2} for (100)-oriented wafers [2, 3]. Silicon wafers with a (111) surface orientation have larger values of Q_{it}, other things being equal, because the (111) surface has more available silicon bonds per square centimeter. Wafers with (110) orientation have values of Q_{it} between the (100) and (111) cases. Since small Q_{it} values are desirable, MOS integrated circuits are routinely fabricated on (100) wafers.

Interface states giving rise to Q_{it} are called fast states because they are in intimate contact with the silicon. As bias changes, the occupancy of the interface states, and hence their charge state, change. Therefore, the value of Q_{it} is bias dependent. For historical reasons, interface states are also referred to as Tamm or Shockley states [2].

After the SiO_x layer, the oxide is primarily SiO_2. However, the first 30 to 40 Å of SiO_2 are strained and contain a number of charged sites with either excess silicon or excess oxygen. These sites do not exchange charge with the silicon over a wide range of bias and are denoted as Q_f, the oxide fixed charge. The value of Q_f depends on wafer orientation, oxidation conditions, and annealing steps. Typical values are in the range of 10^{10} to 10^{12}

positive charges per cm^{-2}. As is the case for Q_{it}, the smallest values of fixed charge result from wafers with the (100) orientation.

Several atomic layers away from the interface, the oxide becomes strain free and there are fewer defects. A certain number of defects are present throughout the remainder of the oxide, though, and give rise to energy levels deep within the oxide energy gap. Such energy levels are generally unoccupied and neutral in equilibrium conditions. However, they do act as traps for holes and electrons and can be charged positive or negative if they hold a trapped carrier. This condition gives rise to a third type of charge, an oxide-trapped charge, or Q_{ot}. Since the energy gap of SiO_2 is high, few holes and electrons are expected in the oxide under normal conditions. However, holes and electrons may be photogenerated within the oxide or photoexcited over either the metal-oxide barrier or the semiconductor-oxide barrier by photons with sufficient energy. In addition, hot electrons in the semiconductor, as might be produced by an FET operated in avalanche injection, can have sufficient energy to surmount the barrier.

The fourth contributor to oxide charge is Q_m, which refers to mobile ionic contaminant charges in the oxide. Na^+, in particular, is a potential oxide contaminant since it is a widely distributed trace chemical in metals and laboratory chemicals and is also easily picked up from human contact. Sodium ions in early MOS devices caused threshold voltages to drift with bias, especially under high temperatures, as the sodium ions were rearranged in the oxide layer. Positive ion contamination in the oxide causes the threshold voltage in an n-channel MOS (NMOS) device to become less positive. In fact, with a sufficiently high level of contamination, the equilibrium NMOS state is inversion, as shown in Figure 3.5, and the threshold voltage becomes negative. Other possible mobile oxide charges include Li^+ and K^+.

In addition to Q_{it}, Q_{ot}, Q_f, and Q_m, shallow ion implants at the silicon-oxide interface may be used to intentionally add charge and thereby tailor the threshold voltage. The ion implanted charge may be considered, to a first-order approximation, as adding to or subtracting from the fixed charge Q_f depending on whether the implanted ions are positive or negative.

The flat band voltage is no longer given by Eq. (3.2) when oxide charges are present. Rather, V_{FB} will be shifted by

$$\Delta V = \Delta V_f + \Delta V_m + \Delta V_{ot} \tag{3.3}$$

where the charge contributions are indicated by the subscripts. Since Q_f is located very close to the silicon interface, the amount that Q_f contributes to the voltage drop across the oxide, and therefore its contribution to the flat band voltage shift, is

$$\Delta V_f = \frac{-Q_f}{C_{OX}} \tag{3.4}$$

Figure 3.5 Effect of Na^+ ion contamination on the equilibrium MOS system.

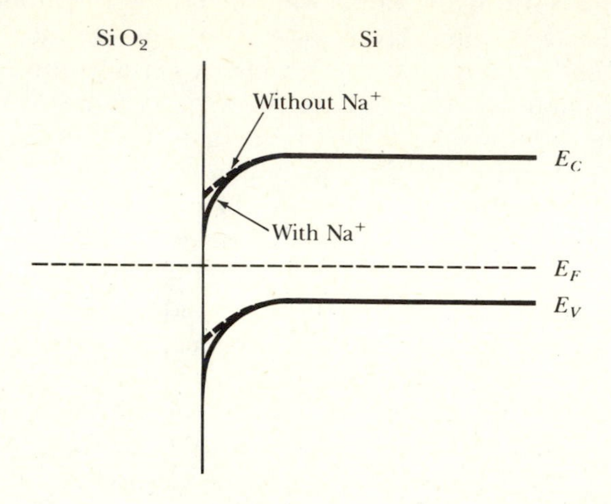

where C_{OX} is the oxide capacitance per unit area. Since Q_f is relatively fixed for a given process, the corresponding flat band voltage shift is well defined.

The mobile ions and oxide-trapped charge are distributed through the oxide, so their contribution to the flat band voltage requires a weighted integration [2],

$$\Delta V_m + \Delta V_{ot} = \frac{-1}{C_{OX}}\left\{\frac{1}{d_{OX}}\int_0^{d_{OX}} x[\rho_{ot}(x) + \rho_m(x)]dx\right\} \tag{3.5}$$

where the integration is over the oxide layer, beginning at the metal-oxide interface. In Eq. (3.5) the thickness of the oxide layer is taken to be d_{OX} and ρ is the volume charge density in the oxide.

EXAMPLE 3.1

Consider a silicon MOS structure with a 500-Å-thick oxide layer on a p-type wafer doped with 10^{16} cm^{-3} acceptors and characterized by a positive fixed charge density of 5×10^{10} cm^{-2}. If the gate electrode is aluminum, what is the room temperature value of the flat band voltage? Assume that Q_m, Q_{ot}, and Q_{it} are negligible. The work function of aluminum is 4.1 eV, the electron affinity of silicon is 4.15 eV, and the relative dielectric constant of SiO_2 is 3.9.

Solution The flat band voltage in this case is

$$V_{FB} = \Phi_M - \Phi_S - (Q_f/C_{OX})$$

where

$$C_{OX} = \epsilon_{OX}/d_{OX}$$

and

$$q\Phi_S = q\chi_S + (E_C - E_F)_{\text{SILICON}}$$

From Eq. (2.7),

$$E_F - E_V = kT[\ln(N_V/p_0)]$$

The room temperature value of the valence band density of states is 1.04×10^{19} cm^{-3} and, since $N_A >> n_i$, $p_0 \simeq 10^{16}$ cm^{-3}. Therefore, taking room temperature to be 300 K,

$$E_F - E_V = (.0259)\ln\left(\frac{1.04 \times 10^{19}}{10^{16}}\right) = 0.18 \text{ eV}$$

Since the energy gap of silicon is 1.12 eV,

$$E_C - E_F = 1.12 - 0.18 = 0.94 \text{ eV}$$

and

$$q\Phi_S = 4.15 + 0.94 = 5.09 \text{ eV}$$

Also,

$$C_{OX} = \frac{(3.9)(8.85 \times 10^{-14} \text{ F/cm})}{500 \times 10^{-8} \text{ cm}} = 6.9 \times 10^{-8} \text{ F/cm}^2$$

Expressing Q_f in coulombs,

$$Q_f = (1.6 \times 10^{-19})(5 \times 10^{10}) = 8.0 \times 10^{-9} \text{ C/cm}^2$$

Therefore,

$$V_{FB} = 4.1 - 5.09 - \frac{8.0 \times 10^{-9}}{6.9 \times 10^{-8}} = -1.11 \text{ V}$$

Work-function values for several other metals and silicides are given in Table 3.1 [2, 4].

EXAMPLE 3.2

Repeat Ex. 3.1 for the case of an n-type polysilicon gate. The polysilicon gate is doped with phosphorus at a concentration of approximately 10^{20} cm^{-3}.

Table 3.1 Work functions (determined by photoresponse)

Material	Φ (volts)
Al	4.10
Au	5.27
$MoSi_2$	4.73
WSi_2	4.62
$TiSi_2$	3.95
n-type polysilicon[a]	4.15
p-type polysilicon[a]	5.27

[a] Degenerately doped polysilicon with the Fermi energy at the band edges.

Source: From S. M. Sze, *Physics of Semiconductor Devices*, 2nd ed. (Copyright © 1981 by John Wiley & Sons; reprinted by permission of John Wiley & Sons), and T. P. Chow and A. J. Steckl, "Refractory Metal Silicides: Thin Film Properties and Processing Technology," *IEEE Transactions on Electron Devices*, ED-30 (1983), 1480–1497. © 1983 IEEE.

Solution Since the polysilicon is degenerately doped, the Fermi energy will be near the conduction band and the gate work function is essentially equal to the silicon electron affinity, or 4.15 eV. Therefore, using the results of Ex. 3.1,

$$V_{FB} = 4.15 - 5.09 - \frac{8.09 \times 10^{-9}}{6.9 \times 10^{-8}}$$

$$V_{FB} = -1.06 \text{ V}$$

3.3 MOS Field-Effect Transistors

3.3.1 Threshold-Voltage Considerations

The results of the previous section may be extended to a consideration of the current-voltage characteristics of MOSFETs. For example, the threshold voltage is commonly considered to be the gate bias at which conduction between the source and drain begins. As is clear from the earlier discussion, however, the establishment of an inversion layer is a gradual process. This fact is also evident in a laboratory investigation of V_T. If V_T is taken to be the gate bias at the onset of current flow from source to drain, then the more sensitive the measurement apparatus is for drain-current detection, the lower the apparent threshold voltage. In spite of this ambiguity, there is a standard

Figure 3.6 The onset of strong inversion, where $V_G = V_T$. ($\psi_S = 2\psi_F$; $n(0) = N_A$.)

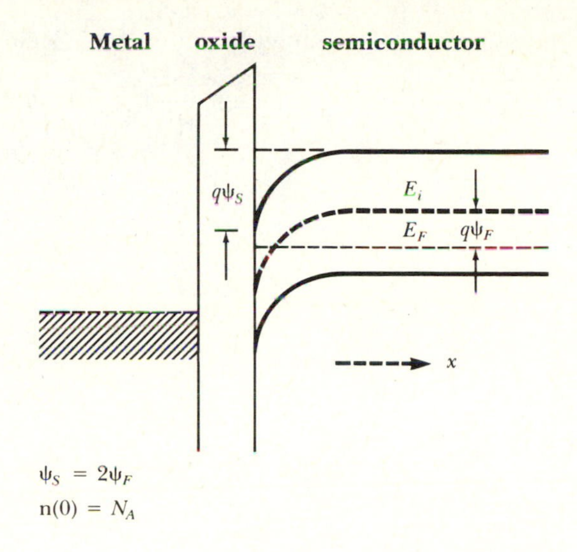

but somewhat arbitrary definition for V_T. Referring to Figure 3.6, the onset of threshold is taken to be when

$$(E_F - E_i)_{\text{INTERFACE}} = (E_i - E_F)_{\text{BULK}} \tag{3.6}$$

where E_i is the Fermi level for intrinsic material. This condition, which corresponds to the point when the electron concentration near the surface is equal to the acceptor doping level, is also known as the strong inversion condition. It marks the point at which n, in Eq. (3.1), becomes the largest term. Using the notation of Figure 3.6, Eq. (3.6) becomes

$$\psi_S = 2\psi_F \tag{3.7}$$

Equations (3.6) and (3.7) lead to an expression for V_T in terms of wafer doping, oxide thickness, and flat band voltage. This expression may be derived as follows for a p-type semiconductor and therefore an NMOS transistor.

Consider first the idealized case where zero gate bias corresponds to the flat band condition. When a gate-to-semiconductor voltage, V_G, is applied, a portion of the potential drop will be across the oxide and a portion across the semiconductor,

$$V_G = V_{OX} + \psi_S \tag{3.8}$$

where, at the onset of threshold,

$$V_G = V_{OX} + 2\psi_F \tag{3.9}$$

The quantity ψ_F is determined by wafer doping. Since

$$p = N_V \exp[-(E_F - E_V)/kT] = n_i \exp[(E_i - E_F)/kT] \tag{3.10}$$

and

$$p \simeq N_A$$

then

$$q\psi_F = (E_i - E_F) = kT[\ln(N_A/n_i)] \tag{3.11}$$

Therefore, at threshold,

$$\psi_S = 2(kT/q)[\ln(N_A/n_i)] \tag{3.12}$$

The oxide voltage drop may be found by considering the capacitance relationship between the charge stored on either side of the oxide and the voltage drop across it,

$$V_{OX} = Q/C_{OX} \tag{3.13}$$

If, referring to Figure 3.3, the approximation is made that at the onset of threshold the semiconductor charge density is still primarily $-qN_Ax_D$, but that x_D has reached its maximum value, then

$$Q = Q_B = -qN_Ax_{D\max} \tag{3.14}$$

where Q_B is the charge stored in the bulk of the semiconductor and $x_{D\max}$ is the depletion layer width when $\psi_S = 2\psi_F$. In the abrupt depletion approximation,

$$x_D = (2\epsilon_S\psi_S/qN_A)^{1/2} \tag{3.15}$$

so

$$Q_B = -(4\epsilon_S\psi_F qN_A)^{1/2} \tag{3.16}$$

From Eq. (3.8), the gate voltage at the onset of threshold is, for zero flat band voltage,

$$V_T = 2\psi_F - Q_B/C_{OX} \tag{3.17}$$

where ψ_F is given by Eq. (3.11) and Q_B is given by Eq. (3.16). The second term in equation (3.17) has a minus sign because Q_B is the charge on the negative side of the dielectric, taking positive polarity for V_G at the gate.

The previous discussion has assumed a zero flat band voltage. In the more general case of a nonzero V_{FB},

$$V_T = V_{FB} + 2\psi_F - Q_B/C_{OX} \tag{3.18}$$

where the several possible contributors to V_{FB} have been discussed in Section 3.2. Equation (3.18) may also be used for the case of an n-type semiconductor, but with changes in the expressions for ψ_F and Q_B. For an n-type semiconductor,

$$\psi_F = -(kT/q)[\ln(N_D/n_i)] \tag{3.19}$$

and

$$Q_B = (-4\epsilon_S\psi_F qN_D)^{1/2} \tag{3.20}$$

EXAMPLE 3.3

Calculate the threshold voltage for the polysilicon gate MOS system described in Ex. 3.2. The relative dielectric constant for silicon is 11.9.

Solution Since V_{FB} and C_{OX} have already been calculated in Exs. 3.1 and 3.2, it is only necessary to find ψ_F and Q_B. Applying Eq. (3.11),

$$\psi_F = (.0259)\ln\left(\frac{10^{16}}{1.45 \times 10^{10}}\right) = 0.348 \text{ V}$$

and, from Eq. (3.16),

$$\begin{aligned} Q_B &= -\sqrt{4(11.9)(8.85 \times 10^{-14})(0.348)(1.6 \times 10^{-19})(1 \times 10^{16})} \\ &= -4.84 \times 10^{-8} \text{ C/cm}^2 \end{aligned}$$

From Eq. (3.18), the threshold voltage is found to be

$$V_T = -1.06 + 2(0.348) + \frac{4.84 \times 10^{-8}}{6.9 \times 10^{-8}}$$

$$V_T = 0.34 \text{ V}$$

The threshold voltage discussion so far refers to a gate-to-semiconductor potential, V_G. However, the semiconductor is not at an equipotential for a MOSFET in circuit operation. The source, drain, and in some cases the substrate may all be at different voltages, as shown in Figure 3.7. Provided the

Figure 3.7 Four-terminal MOSFET operation.

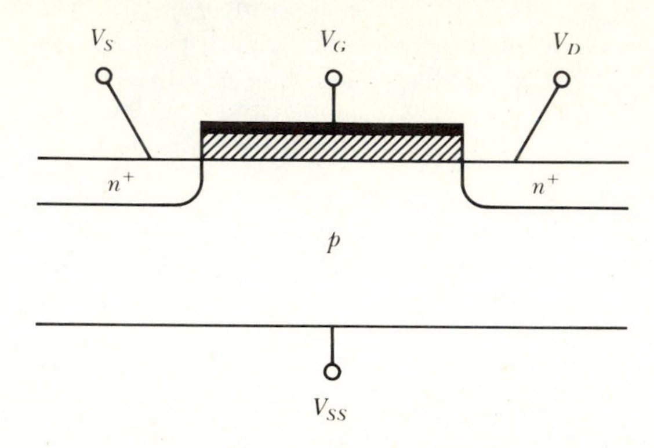

substrate and source are at the same potential, Eqs. (3.18), (3.16), and (3.11) may still be used to calculate the threshold condition for the gate-to-source voltage, V_{GS}. However, if the substrate is biased separately from the source, the expression for Q_B changes. For p-type material,

$$Q_B = -[2\epsilon_S q N_A[2\psi_F - (V_{SS} - V_S)]]^{1/2} \tag{3.21}$$

where the quantity $(V_{SS} - V_S)$ is negative, and for n-type material,

$$Q_B = [2\epsilon_S q N_D[-2\psi_F + (V_{SS} - V_S)]]^{1/2} \tag{3.22}$$

where the quantity $(V_{SS} - V_S)$ is positive. The variation of threshold voltage with the source-to-substrate voltage is sometimes referred to as the body effect, and the technique of applying a substrate bias to control the threshold voltage is referred to as body biasing.

To find the threshold voltage between the gate and channel at any point in the channel from the source to drain, Eq. (3.21) and (3.22) apply with V_S replaced by V_C, where V_C is the channel voltage at the point of interest.

EXAMPLE 3.4

An n-channel MOSFET with an n-type polysilicon gate has an oxide thickness of 500 Å and a substrate doping, N_A, of 10^{16} cm^{-3}. The substrate is biased at -5 V relative to the grounded source. Find the gate-to-source threshold voltage.

Solution V_{FB}, C_{OX}, and ψ_F have been calculated in Exs. 3.1, 3.2, and 3.3. However, Q_B must be recalculated as

$$Q_B = -[2\epsilon_S q N_A(2\psi_F + 5)]^{1/2} = -1.39 \times 10^{-7} \text{ C/cm}^2$$

From Eq. (3.18), the threshold voltage is

$$V_T = -1.06 + 2(0.348) + \frac{1.39 \times 10^{-7}}{6.9 \times 10^{-8}}$$

$$V_T = 1.64 \text{ V}$$

Although substrate biasing allows considerable control over threshold voltage values, it requires an extra power-supply voltage. Threshold voltages may also be altered by means of ion implantation. Provided the ion implantation is a shallow implant peaked near the silicon-oxide interface, the implant dose in atoms per square centimeter may be treated, to a first-order approximation, in the same fashion as Q_F. That is, the flat band voltage is shifted by an amount

$$\Delta V_{II} \simeq \frac{q\Phi_{\text{IMPLANT}}}{C_{OX}} \tag{3.23}$$

where Φ_{IMPLANT} is the ion implant dose as defined in Section 2.6.

EXAMPLE 3.5

It is often desirable to have both positive and negative threshold voltage transistors on NMOS integrated circuits to facilitate logic gate design. Assume that the basic device structure is a polysilicon (n-type) gate over a 500-Å-thick oxide, on a substrate doped with $N_A = 10^{16}$ cm^{-3}, as has been considered in Exs. 3.2, 3.3, and 3.4. Also, assume that a boron implant is available for positive threshold adjustment, and that phosphorus implants are available for negative threshold adjustment. The desired threshold voltages are +1 V and −2.5 V. Find the corresponding boron and phosphorus implant doses.

Solution From Ex. 3.3, it is known that the threshold voltage without an ion implant is 0.34 V. To establish a threshold voltage of −2.5 V, the phosphorus implant must result in $\Delta V_{II} = -2.5 - 0.34 = -2.84$. Using the previously calculated value of C_{OX},

$$q\Phi_{\text{PHOSPHORUS}} = (2.84 \text{ V})(6.9 \times 10^{-8} \text{ F/cm}^2)$$
$$= 1.96 \times 10^{-7} \text{ C/cm}^2$$

So $\Phi_{\text{PHOSPHORUS}} = 1.22 \times 10^{12}$ atoms/cm^2.

Likewise, for the boron implant, $\Delta V_{II} = 1 - 0.34 = 0.66$ V, and

$$q\Phi_{\text{BORON}} = (0.66 \text{ V})(6.9 \times 10^{-8} \text{ F/cm}^2)$$
$$= 4.55 \times 10^{-8} \text{ C/cm}^2$$

Therefore, $\Phi_{\text{BORON}} = 2.85 \times 10^{11}$ atoms/cm^2.

3.3.2 MOSFET Current-Voltage Characteristics

Thus far the discussion in this section has focused on the critical voltage, V_T, required to establish a conducting channel between source and drain. However, the magnitude of the drain current depends on both the gate-to-source voltage and the drain-to-source voltage. For example, if the semiconductor is in inversion, then at any point in the channel, referring to Figure 3.8, the current density is given by

$$\mathrm{J} = q\mu_n n\mathscr{E} + qD_n \nabla n \tag{3.24}$$

where μ_n is the electron mobility, D_n is the electron diffusion coefficient, n is the electron concentration, and $\mathscr{E}$ is the electric field vector. Since the current in the channel flows primarily in the x direction and is mainly due to drift, Eq. (3.24) simplifies to

$$\mathrm{J} \approx q\mu_n n\mathscr{E}_x \tag{3.25}$$

which may be written as

$$\mathrm{J} \approx -q\mu_n n dV/dx \tag{3.26}$$

where V is the channel voltage. For gate voltages greater than the threshold voltage, the expression Q_S for total charge in the semiconductor includes the contribution of the electrons in the inversion layer,

$$Q_S = Q_n + Q_B \tag{3.27}$$

where for a given value of x

$$Q_n = -\mathrm{q} \int_0^{y_c} n(x,y)dy \tag{3.28}$$

and

$$Q_B = -qN_A y_\mathrm{D} \tag{3.29}$$

Here, y_c corresponds to the channel depth and y_D is the depletion layer width at the value of x for which Q_S is evaluated.

Integrating Eq. (3.26) over y and z yields,

$$I_D = -\mu_n W Q_n (dV/dx) \tag{3.30}$$

Figure 3.8 Model for the MOSFET channel.

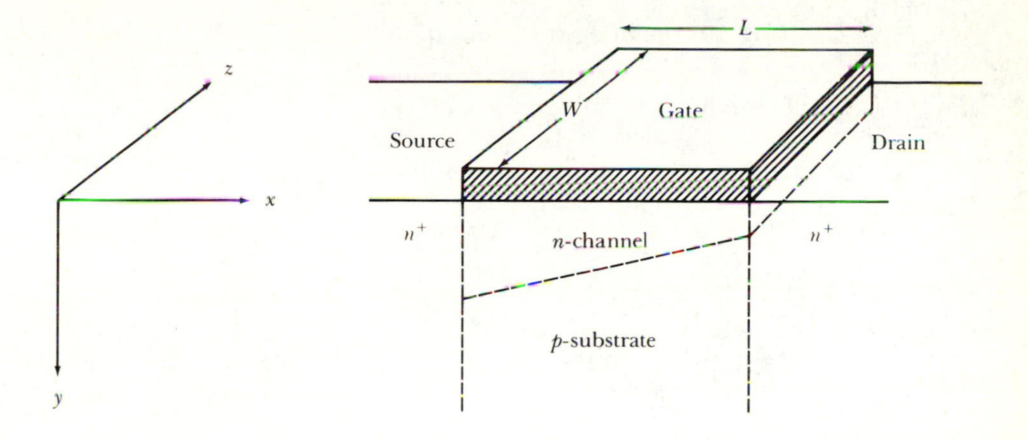

where the direction of positive drain current is into the drain. Further integration over x yields

$$I_D = -\frac{\mu_n W}{L}\int_0^{V_D} Q_n \, dV \tag{3.31}$$

where the voltage at the source is taken to be zero. From Eq. (3.8), V and Q_N are related via the expression

$$V_G = -(Q_n + Q_B)/C_{OX} + \psi_S \tag{3.32}$$

Assuming that the entire channel is in strong inversion,

$$\psi_S = 2\psi_F + V(x) \tag{3.33}$$

where V ranges from zero at the source to V_D at the drain. From Eqs. (3.15), (3.29), and (3.33), Q_B is found to be equal to

$$Q_B = -[2\epsilon_S[2\psi_F + V(x)]qN_A]^{1/2} \tag{3.34}$$

where y_D replaces x_D in Eq. (3.15). The depletion layer width y_D increases with increasing x since the channel voltage increases from source to drain. In the gradual channel approximation made here, it is assumed that the rate of change of potential and electric field in the x direction is small compared to the y direction such that the one-dimensional solution for Q_B in section 3.3.1 still holds.

If the further simplifying assumption is made that Q_B is constant throughout the channel, and equal to the value at the source, then substitution of Eqs. (3.33) and (3.34) into Eq. (3.32) yields [5]

$$Q_n(x) = -C_{OX}[V_G - V_T - V(x)] \tag{3.35}$$

Carrying out the integration in Eq. (3.31) then yields

$$I_D = \mu_n C_{OX}(W/L)[(V_G - V_T)V_D - (V_D)^2/2] \tag{3.36}$$

A more exact expression results if the voltage dependence of Q_B is included. Then the integration in Eq. (3.31) yields [6, 7]

$$I_D = \mu_n C_{OX}(W/L)[(V_G - 2\psi_F - V_D/2)V_D - (2/3)\frac{(2\epsilon_S q N_A)^{1/2}}{C_{OX}}[(2\psi_F + V_D)^{3/2} - (2\psi_F)^{3/2}]] \tag{3.37}$$

Eq. (3.37) assumes a grounded source and substrate, and a zero flat band voltage. If V_{FB} is not zero, then Eq. (3.8) is replaced by $V_G - V_{FB} = V_{OX} + \psi_S$ and, in Eq. (3.37), V_G is replaced by $V_G - V_{FB}$.

Clearly, MOSFET current-voltage characteristics depend strongly on processing variables, such as the wafer doping and gate oxide thickness. Device properties are also seen to be determined by inherent material properties, such as carrier mobilities, and by purely geometric factors, such as the ratio of gate width to length, W/L.

For gate voltages less than V_T, the drain current is not zero even though the channel is not in strong inversion. An n-type channel begins to form between the source and drain when $\psi_S = \psi_F$ and the semiconductor surface is in weak inversion. This weak inversion condition holds until the gate bias is large enough to cause $\psi_S = 2\psi_F$, which is the condition for strong inversion and the definition of threshold. Consequently, there is a smooth decrease in current as the gate voltage decreases from values above V_T to values below V_T.

Under weak inversion conditions, for V_G less than V_T, the channel current is called subthreshold current. This current may be shown to be largely independent of V_D and to increase exponentially with V_G. In the subthreshold region, the gate voltage swing required to reduce the current by a factor of 10 is on the order of 100 mV or less [2].

Derivation of Eqs. (3.36) and (3.37) assumes strong inversion throughout the channel. This condition corresponds to the "triode" region of MOSFET operation, as shown in Figure 3.9A. However, if the drain voltage is increased sufficiently, the strong inversion layer disappears at the drain end of the channel. This condition is shown in Figure 3.9B and corresponds to the saturation region of MOSFET operation. The saturation condition may be expressed approximately as

$$V_{GD} < V_T \tag{3.38}$$

and the onset of saturation is therefore given approximately by

$$V_{D\text{sat}} \simeq V_G - V_T \tag{3.39}$$

More exactly, the onset of saturation occurs when $Q_n(L) = 0$. This results in Eq. (3.40) [7]:

$$V_{D\text{sat}} = V_G - V_{FB} - 2\psi_F - \frac{\epsilon_S q N_A}{(C_{OX})^2}\left[[1 + \frac{2(C_{OX})^2}{q\epsilon_S N_A}(V_G - V_{FB})]^{1/2} - 1\right] \tag{3.40}$$

The drain current in saturation may be found approximately by inserting Eq. (3.39) in Eq. (3.36), yielding

$$I_{D\text{sat}} \simeq (\mu_n W C_{OX}/2L)(V_G - V_T)^2 \tag{3.41}$$

More rigorously, the drain current in saturation is found by inserting Eq. (3.40) in Eq. (3.37).

Two comments are in order at this point. First, Eq. (3.41) shows that the drain current does not go to zero at the point of saturation, even though the strong inversion layer no longer exists near the drain end of the device. In saturation, carriers traveling from the source are injected across the pinched-off region into the drain, and current continues to flow. That current continues to flow after the onset of pinch-off may seem counterintuitive. Note, however, that a more exact analysis shows that the carrier concentration in the pinched-off region at the channel end is relatively small compared to device doping levels, but is not zero. There are fewer carriers in this part of the device than elsewhere, but they move faster because there is a large electric field in the pinched-off region. (It may also be helpful to recall that currents flow through depleted regions of other solid state devices as well. For example, a large current flows through the depletion layer of a forward biased diode, even though the number of carriers in the depletion region is small compared to other parts of the diode.)

Figure 3.9 MOSFET regions of operation. A. Triode. B. Saturation.

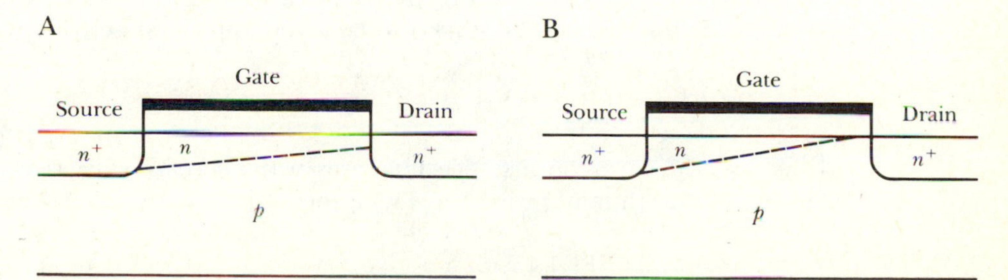

Second, as the drain voltage is increased beyond the onset of saturation, the additional voltage drop is mostly across the pinched-off region. Since the rest of the channel is largely unaffected by the increased drain voltage, the carrier-flux incident on the pinched-off region is, to a first approximation, unchanged by the increase in drain voltage. Consequently, the drain current in saturation is, to a first approximation, independent of drain-to-source voltage. In fact, however, as the drain voltage is increased beyond the onset of saturation, the length of the pinched-off region does increase approximately as the square root of $(V_D - V_{D\text{sat}})$ such that the effective length of the channel is decreased by an amount ΔL. As a result, the quantity L in Eq. (3.41) is a function of drain voltage, and the slope of I_D versus V_D in the saturation region is not zero, but rather I_D increases somewhat with increasing V_D. This effect is referred to as channel-length modulation and is of particular importance in short-channel devices where ΔL is not small compared to L. Channel length modulation is sometimes accounted for by using an empirical approximation for the saturated drain current in which the current in Eq. (3.41) is multiplied by the quantity $(1 + \lambda V_{DS})$. In SPICE (Simulation Program with Integrated Circuit Emphasis), λ is called the channel-length modulation parameter. Typical values of λ are in the range of 0.1 to 0.01 V^{-1}.

In addition to channel-length modulation, other short-channel effects come into play when the depletion-layer widths of the source and drain are comparable to the channel length. In that case, the gradual channel approximation used in developing current-voltage characteristics in this section no longer applies, and two dimensional effects become important. Furthermore, the expressions for threshold voltage are also modified because two-dimensional effects must be considered in calculating the bulk charge distribution under the gate [2].

Another useful MOSFET device parameter related to the current-voltage characteristics of the FET is the transconductance, g_m, defined as

$$g_m \simeq \left.\frac{\partial I_D}{\partial V_G}\right| V_D = \text{constant} \qquad (3.42)$$

The transconductance, as seen from Eq. (3.42), is a gain parameter. In saturation, g_m is found from Eq. (3.41) to be given approximately by

$$g_m \simeq (\mu_n W C_{OX}/L)(V_G - V_T) \qquad (3.43)$$

Therefore, the transconductance increases with increasing carrier mobility and channel width, and it decreases with increasing channel length and gate oxide thickness.

n-channel MOSFETs with positive threshold voltages are traditionally referred to as enhancement-mode MOSFETs and are distinguished from depletion-mode *n*-channel MOSFETs, which have negative threshold voltages. Although different circuit symbols are used for the two devices, the current-voltage expressions in Eqs. (3.36) and (3.37) apply to both. For *p*-channel MOSFETs, enhancement-mode devices have negative threshold voltages and depletion-mode devices have positive threshold voltages. Equations (3.36) and (3.37) may also be applied to *p*-channel MOSFETs; however, the resulting drain current has a polarity opposite to the *n*-channel case, as shown in Figure 3.10. The circuit symbols shown in Figures 3.10A and 3.10B are not unique; several variations are found in the literature. Figure 3.10C shows one set of alternate symbols that are also often used.

3.4 MOS Capacitor Diagnostics

When the MOS structure is connected as a two-terminal device, with one electrode attached to the gate and the other attached to the semiconductor, a voltage-dependent capacitance results. The nature of the voltage dependence offers considerable diagnostic information that is useful both in evaluating the MOS fabrication process and in predicting MOSFET characteristics. For these reasons, MOS capacitors are often included on chip test sites.

It is convenient to consider the structure as consisting of two capacitors in series, C_{OX} and C_S, as shown in Figure 3.11. C_{OX} is determined by the thickness and dielectric constant of the dielectric slab that constitutes the gate insulator; C_S corresponds to the space-charge layer capacitance in the semiconductor. The values per unit area are given by

$$C_{OX} = \epsilon_{OX}/d_{OX} \tag{3.44}$$

and

$$C_S = \epsilon_S/x_D \tag{3.45}$$

The combined capacitance is therefore

$$C = \frac{1}{(C_{OX})^{-1} + (C_S)^{-1}} \tag{3.46}$$

A plot of C as a function of applied voltage shows three main regions. If, for a *p*-type semiconductor, the gate voltage is more negative than the flat band voltage, then the semiconductor is in accumulation, and provided the charge in the accumulation region can be considered as a sheet or delta function, the MOS capacitance is constant and given by

Figure 3.10 MOSFET circuit symbols. A. n-channel MOSFETs. B. p-channel MOSFETs. C. IEEE standard symbols.

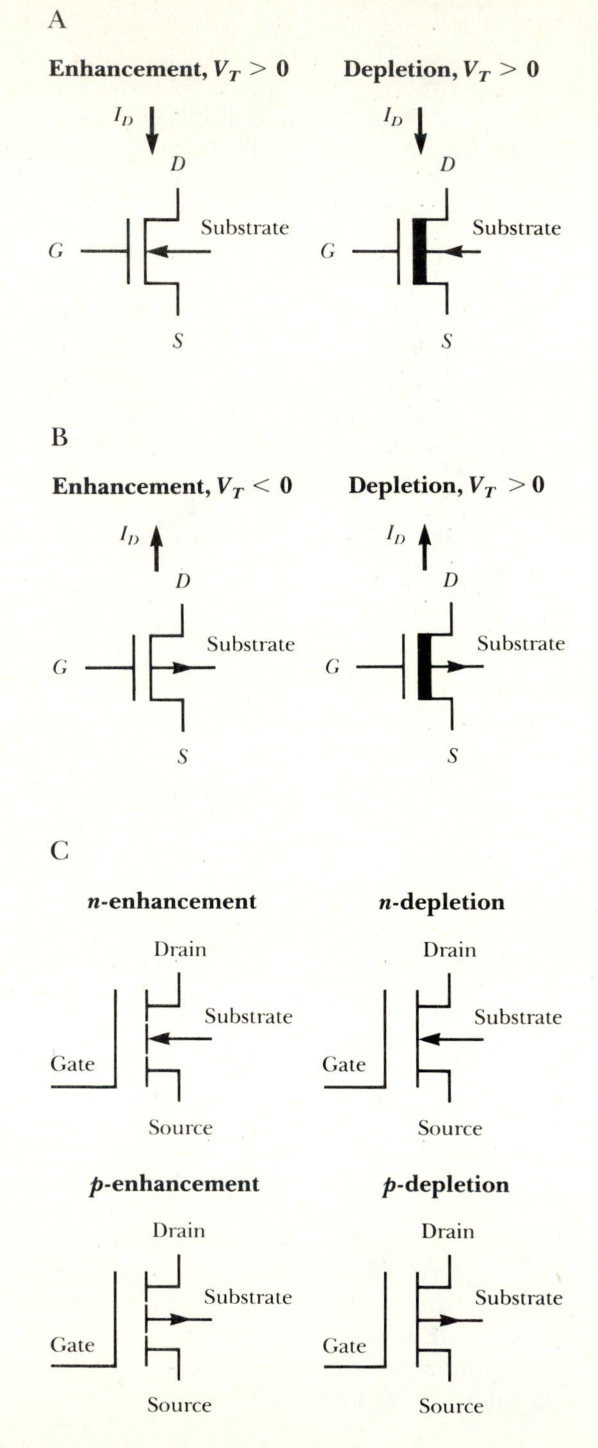

$$C \simeq C_{OX} \tag{3.47}$$

For voltages more positive than the threshold voltage, the semiconductor is in strong inversion and the depletion-layer width is pinned at $x_{D\max}$. Under this condition, the capacitance is constant and equal to

$$C = \frac{1}{(d_{OX}/\epsilon_{OX}) + (x_{D\max}/\epsilon_S)} \tag{3.48}$$

where $x_{D\max}$ is given by

$$x_{D\max} = (4\epsilon_S \psi_F / qN_A)^{1/2} \tag{3.49}$$

However, for voltages between V_{FB} and V_T, the capacitance varies with voltage, since x_D is voltage dependent. In the depletion-layer approximation, x_D is given by Eq. (3.15) and the MOS capacitance is

$$C = \frac{1}{d_{OX}/\epsilon_{OX} + x_D/\epsilon_S} \tag{3.50}$$

Combining Eqs. (3.15) and (3.50) results in a relationship between C and V, given by

$$C = C_{OX}\left[1 + \frac{2V(\epsilon_{OX})^2}{qN_A\epsilon_S(d_{OX})^2}\right]^{-1/2} \tag{3.51}$$

The derivation of Eq. (3.51) is a chapter exercise.

A variety of experimental methods have been established for measuring C as a function of voltage [3]. In a typical arrangement, a capacitance bridge is used with a biasing circuit that automatically sweeps the desired voltage range. The ac signal for the bridge, which is superimposed on the dc bias, is small, approximately 10 mV, and a typical frequency for the bridge signal is in the range of 100 kHz to 10 MHz.

Figure 3.11 The MOS capacitor.

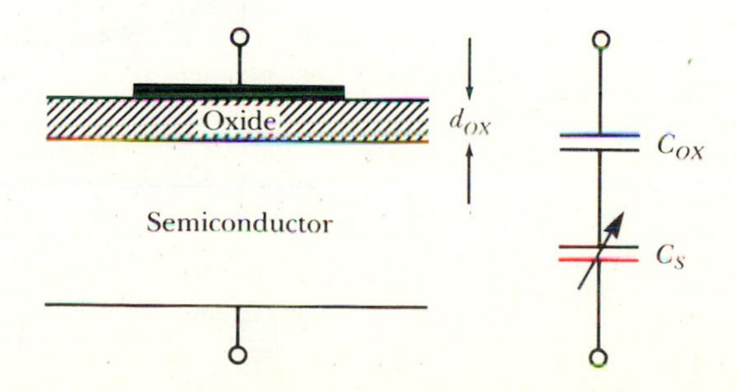

The solid line in Figure 3.12 shows the capacitance voltage plot resulting from Eqs. (3.47), (3.48), and (3.51), the dotted line shows the results of a more exact analysis [8]. Differences between the two plots are particularly apparent near the transition regions between accumulation and depletion and between depletion and inversion. For example, at the flat band voltage, a more rigorous analysis yields [7]

$$C_{FB} = \frac{1}{d_{OX}/\epsilon_{OX} + L_D/\epsilon_S} \tag{3.52}$$

where L_D is the extrinsic Debye length given by

$$L_D = (\epsilon_S kT/q^2 N_A)^{1/2} \tag{3.53}$$

As a further practical note, the capacitance in the inversion region is frequency dependent. Since the inversion-layer charge results from generation of minority carriers, the population in the inversion region can change only as rapidly as carriers can be generated within the depletion region near the surface. At sufficiently low bridge frequencies, of approximately 100 Hz or less, the capacitance in the inversion region is therefore approximately equal to C_{OX}, as for the accumulation region. However, at high frequencies, Eq. (3.48) applies. This frequency dependency is illustrated in Figure 3.13.

MOS capacitance-voltage plots therefore allow determination of V_{FB} and,

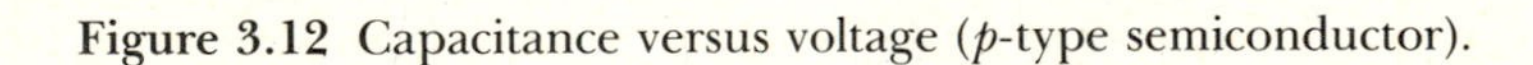

Figure 3.12 Capacitance versus voltage (*p*-type semiconductor).

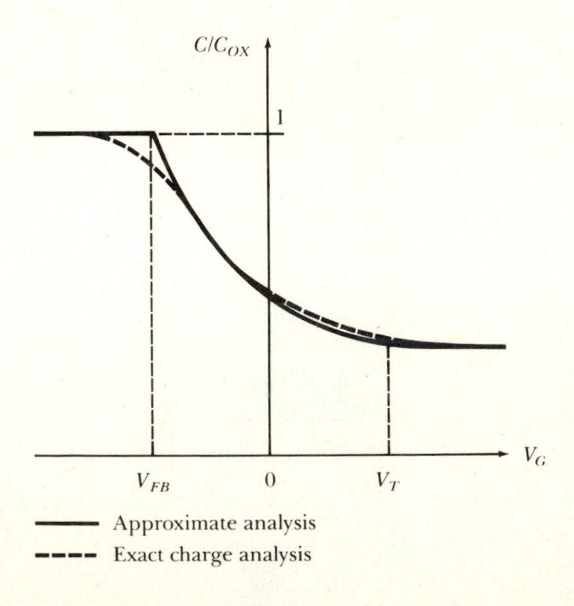

Figure 3.13 Frequency dependence of the MOS capacitor.

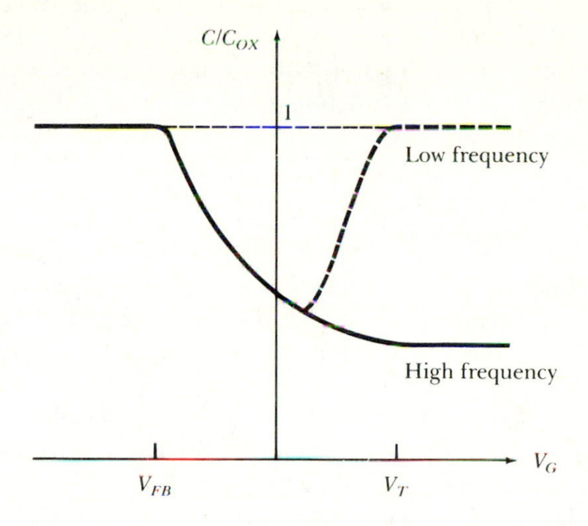

consequently, facilitate the documentation of intended and nonintended contributors to the flat-band-voltage shift, as discussed in Section 3.2. The effect of slow states, such as Q_f, is to produce a voltage shift in the *C-V* curve, as shown in Figure 3.14A; but the shape of the curve is unchanged, since the slow states do not change charge state with bias. The Q_{it} fast states, on the other hand, cause a "stretch-out" in the *C-V* plot, as shown in Figure 3.14B.

A measurement and analysis sequence for characterizing features of an MOS fabrication process could proceed for example, as follows:

Figure 3.14 Contribution of charged states to C-V plots. A. Q_f. B. Q_{it}.

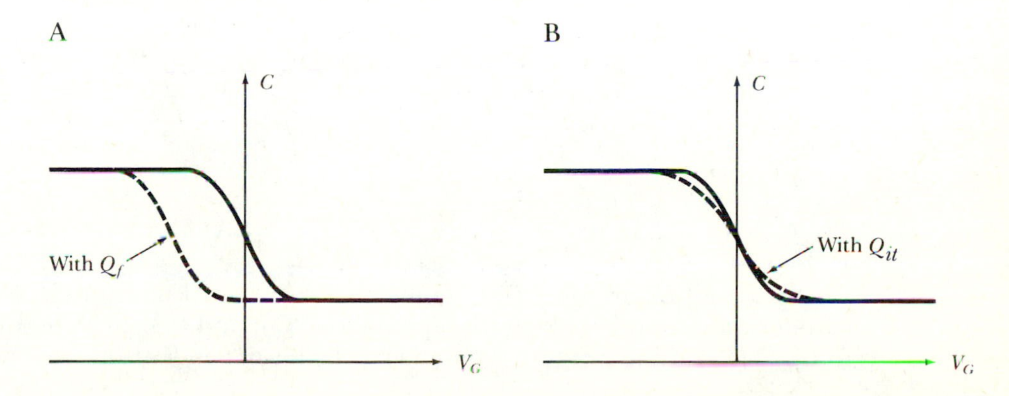

1. Measure C in the accumulation region. This provides an experimental value for C_{OX}, and therefore for the oxide thickness, d_{OX}.
2. Calculate L_D using Eq. (3.53).
3. Use the result of step 2 and Eq. (3.52) to calculate the ratio C_{FB}/C_{OX}.
4. Measure C as a function of voltage through the depletion and inversion regions.
5. Using the result of step 3, determine an experimental value for V_{FB}.
6. Calculate the expected value of V_{FB}, as in Ex. 3.1 and compare with the experimental value.

A comparison between the calculated and measured values of the flat band voltage provides quantitative information about the combined contribution of Q_f, Q_m, and Q_{ot} to the flat-band-voltage shift. Bias and temperature stress indicate the relative role of Q_m.

For increased accuracy, the C-V plot should be corrected for interface-trap stretch-out before the flat band voltage is determined. Additional information on Q_{it} may be obtained by measuring the MOS conductance, as well as capacitance, as a function of voltage and frequency. A variety of methods for determining the relative roles of Q_{it}, Q_f, Q_{ot}, and Q_m have been developed and are reviewed in reference 3.

3.5 MOSFET Parasitics

Device performance is often compromised, and in some cases severely limited, by parasitic elements. The cross-sectional view of a MOSFET shown in Figure 3.15 indicates the physical origin of several parasitic capacitors and resistors. Some of these parasitics are unavoidable, such as the gate-to-drain and gate-to-source capacitances. Others, particularly the overlap capacitances C_{o1} and C_{o2}, may be minimized by choice of fabrication technology, as for example by the use of self-aligned gates.

A corresponding large-signal model for the transistor is shown in Figure 3.16. Note that many of the elements in this model are nonlinear, so computer-aided analysis is of obvious advantage. The voltage dependence of the current source is described, for the triode region, by Eq. (3.37) and, for the saturation region, by the result of inserting Eq. (3.40) into Eq. (3.37).

The overlap capacitances may be expressed as

$$C_{o1} = C_{OX}WL_S \tag{3.54}$$

$$C_{o2} = C_{OX}WL_D \tag{3.55}$$

where L_S and L_D are the overlap dimensions indicated in Figure 3.15. The gate-to-source and gate-to-drain capacitances, C_{GS} and C_{GD}, are proportional to $C_{OX}WL$ with the proportionality factor being dependent on the drain-to-

Figure 3.15 Origin of MOSFET parasitics. A. Top view. B. Cross-sectional view.

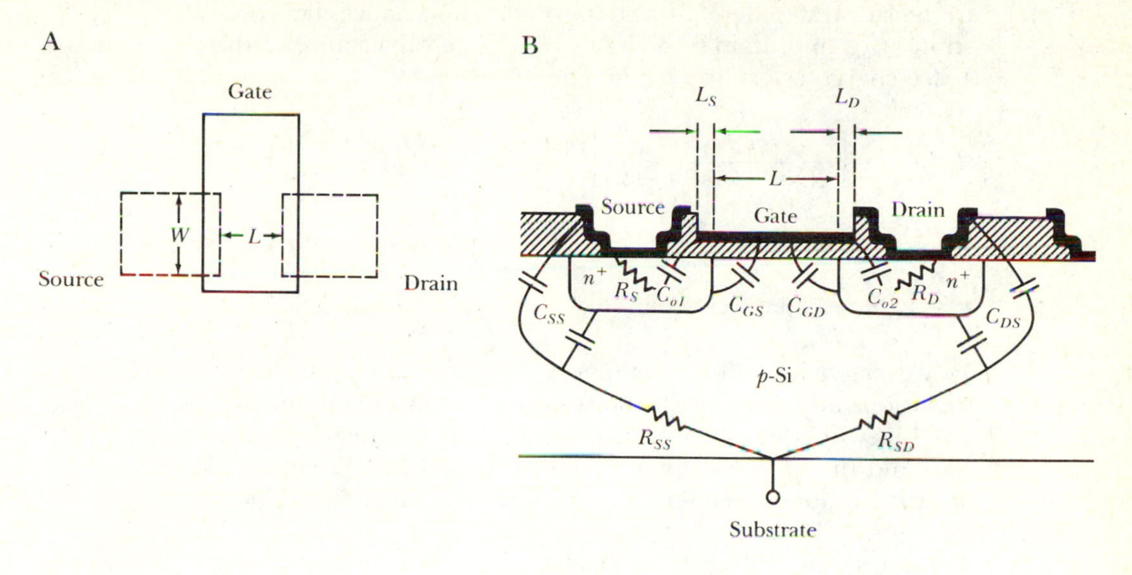

Figure 3.16 Large-signal MOSFET model.

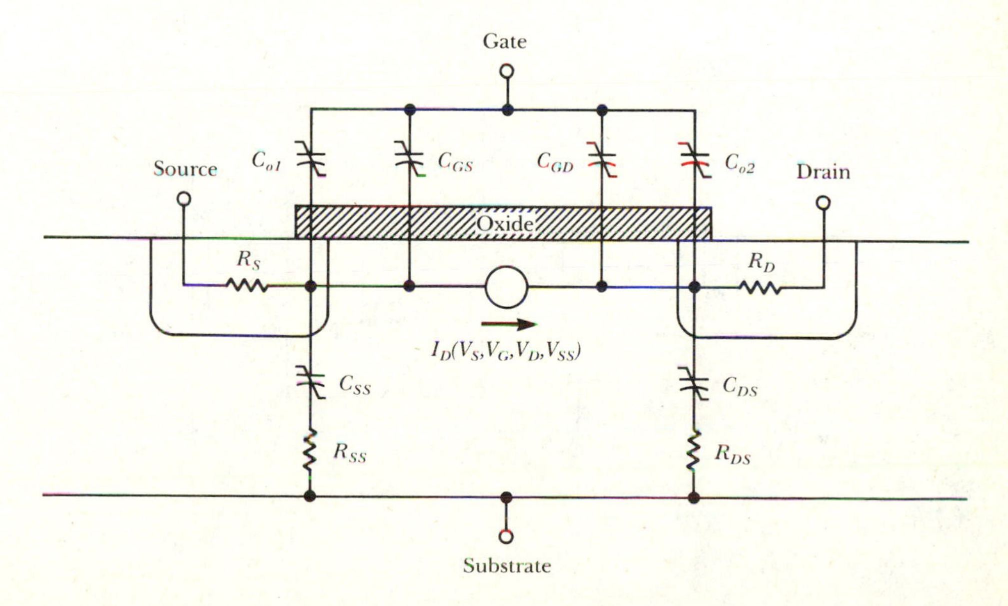

source voltage [9]. Since the source and drain regions are of opposite doping to the substrate, there are also inevitable capacitances between source to substrate, C_{SS}, and drain to substrate, C_{DS}. The capacitance of the depletion, or space-charge layer for a *p-n* junction is given by

$$C_{scl} = \left[\frac{q\epsilon_S N_A N_D}{2(N_A + N_D)(V_{bi} - V)} \right]^{1/2} \tag{3.56}$$

where V is the applied voltage and V_{bi} is the built-in voltage given by

$$V_{bi} = (kT/q)\ln[N_A N_D/(n_i)^2] \tag{3.57}$$

However, the situation is complicated because doping profiles resulting from diffusions and ion implantations do not result in constant values of N_A and N_D. Likewise, the values of the parasitic resistors depend on the doping profiles and the geometric dimensions of the MOSFET. For small-signal analysis, the situation is simplified because a linear model may be used, as shown in Figure 3.17.

These models treat nonideal FET phenomena by using circuit models with lumped elements. An alternative approach is to use numerical analysis to solve the basic semiconductor transport equations. This approach is generally more accurate than the lumped-parameter circuit model, and allows

Figure 3.17 Small-signal MOSFET model for the saturation region. The dependence of i_d on v_{ds} is modeled by the conductance, g_D.

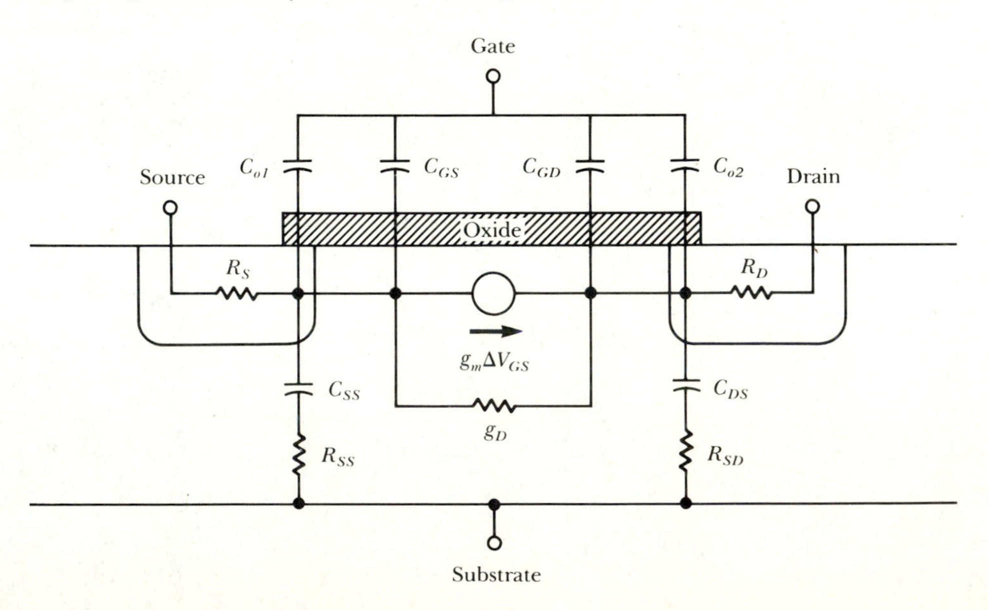

the direct incorporation of effects such as field- and temperature-dependent mobilities. An example of the numerical analysis approach is the FIELDAY program developed at IBM, in which the current-transport equations, the current-continuity equations, and Poisson's equation are solved by the finite element method [10]. The obvious trade-off between numerical analysis and lumped parameter circuit models is computational complexity.

3.6 MOSFET Fabrication Alternatives

3.6.1 Metal-Gate Process

Historically, the first commercial MOSFET integrated circuits were fabricated by the metal-gate, thick-oxide process discussed in Chapter 1 and illustrated in Figure 1.2. Although the figure shows an *n*-channel device, early MOS integrated circuits were *p*-channel because of practical difficulties with *n*-channel threshold voltages in the early 1970s. Large values of Q_f led to negative threshold voltages and therefore to "normally on" *n*-channel devices. With the development of annealing techniques to reduce Q_f and the use of ion implantation to tailor threshold voltages, this problem has been solved.

The thick-oxide process demonstrates an inherent advantage of MOS integrated circuits, namely self-isolation of the transistors. Since the threshold voltage increases with increasing oxide thickness, oxide thicknesses can be chosen such that only the semiconductor material under the gate is ever in inversion. The thick oxide, or field oxide is sufficiently thick that the corresponding threshold voltage is greater than any voltage encountered in the circuit. A channel stop implant, described in Section 3.6.2, is often used to further raise the field threshold voltage. Consequently a conducting path does not exist between the source and drain of one device and another, unless it is intentionally established such as by a metalization or diffusion path.

However, the process shown in Figure 1.2 also suffers from an inherent drawback. Since the pattern definition for the gate oxide region and the source and drain diffusions are on separate masks, the gate oxide must necessarily overlap the source and drain to allow for mask-alignment errors. In addition to limiting the scaling of the device, this overlap also leads to appreciable overlap parasitic capacitance. Several MOS fabrication alternatives have been developed to improve packing density or to reduce parasitics or both.

3.6.2 Self-Aligned Gates

Self-aligned gate fabrication processes represent the dominant MOS integrated circuit fabrication technique. If the gate oxide and electrode are defined before the source and drain, then the gate structure can act as the mask for the diffusion or, more commonly, for the ion implantation step, either of which establishes the source and drain. This places a practical constraint on the choice of gate electrode material, since it must be able to withstand the high temperatures of a diffusion furnace or an ion implant annealing step. Aluminum, for example, would not be appropriate for the gate electrode since it melts at 660°C. The general classes of materials suitable for self-aligned gates are represented by refractory metals, such as molybdenum, tungsten, tantalum, titanium, and platinum; silicides, such as $MoSi_2$, $TiSi_2$, PtSi, and WSi_2; and polysilicon. Several manufacturers use composites of polysilicon and silicides called polycides, in which the first gate layer is polysilicon. For polysilicon layers greater than 1500 Å, an overlying silicide layer has negligible effect on threshold voltages but enhances the gate conductivity.

Figure 3.18 shows an example of a fabrication sequence for a polysilicon self-aligned transistor. This sequence includes an ion implantation to tailor the threshold voltage. The first mask defines an opening in the field oxide, which is about 5,000 to 10,000 Å thick, where the transistor is to be placed. A widely used method to pattern the field oxide with smoothly tapered steps is the localized oxidation of silicon (LOCOS) technique. In this approach, the wafer is coated with a triple insulator layer: first an oxide layer; then a chemical vapor deposition (CVD) silicon nitride layer, as described in Section 2.10.4; and finally a CVD silicon dioxide layer. The top oxide is used to pattern the nitride layer that covers the active device region. This method allows selective growth of the thick-field oxide since the nitride oxidation rate is very small. After the nitride definition and before the field oxidation, the wafer is exposed to a low-energy p-type implant. This exposure establishes p^+ regions in the silicon that is not protected by the nitride. These p^+ regions—called channel stops—are in the silicon that will underly the field oxide, and serve to raise the field threshold voltage sufficiently to prevent parasitic MOSFET action. Following the channel-stop implant, the field oxide is grown. An etch is then used to remove the nitride layer, and a subsequent oxide etch exposes bare silicon where the source, drain, and channel of the FET are to be established.

The next step is the gate oxidation. This step is very critical, since the quality of the gate oxide has a large effect on the final device performance. Following the gate oxidation, ion implants are used to tailor the threshold voltage. If there is more than one device type in the integrated circuit, multiple threshold adjusting implants may be used. For example, a boron implant may be used to adjust the threshold voltage of enhancement mode devices and a phosphorus implant may be used to adjust the threshold voltage of depletion mode devices. In Figure 3.18B, a patterning procedure for a threshold tailoring implant is shown. A layer of photoresist, approximately 10,000 Å thick, is spun onto the wafer to provide a patterning medium for the ion implantation. The second mask opens a region that encompasses the gate area for the ion implantation. After the ion implanta-

Figure 3.18 A fabrication sequence for a polysilicon self-aligned gate MOSFET. A. Opening the field oxide. B. Ion implantation. C. Polysilicon etching. D. Diffusion or implant. E. Contact cuts. F. Metal definition.

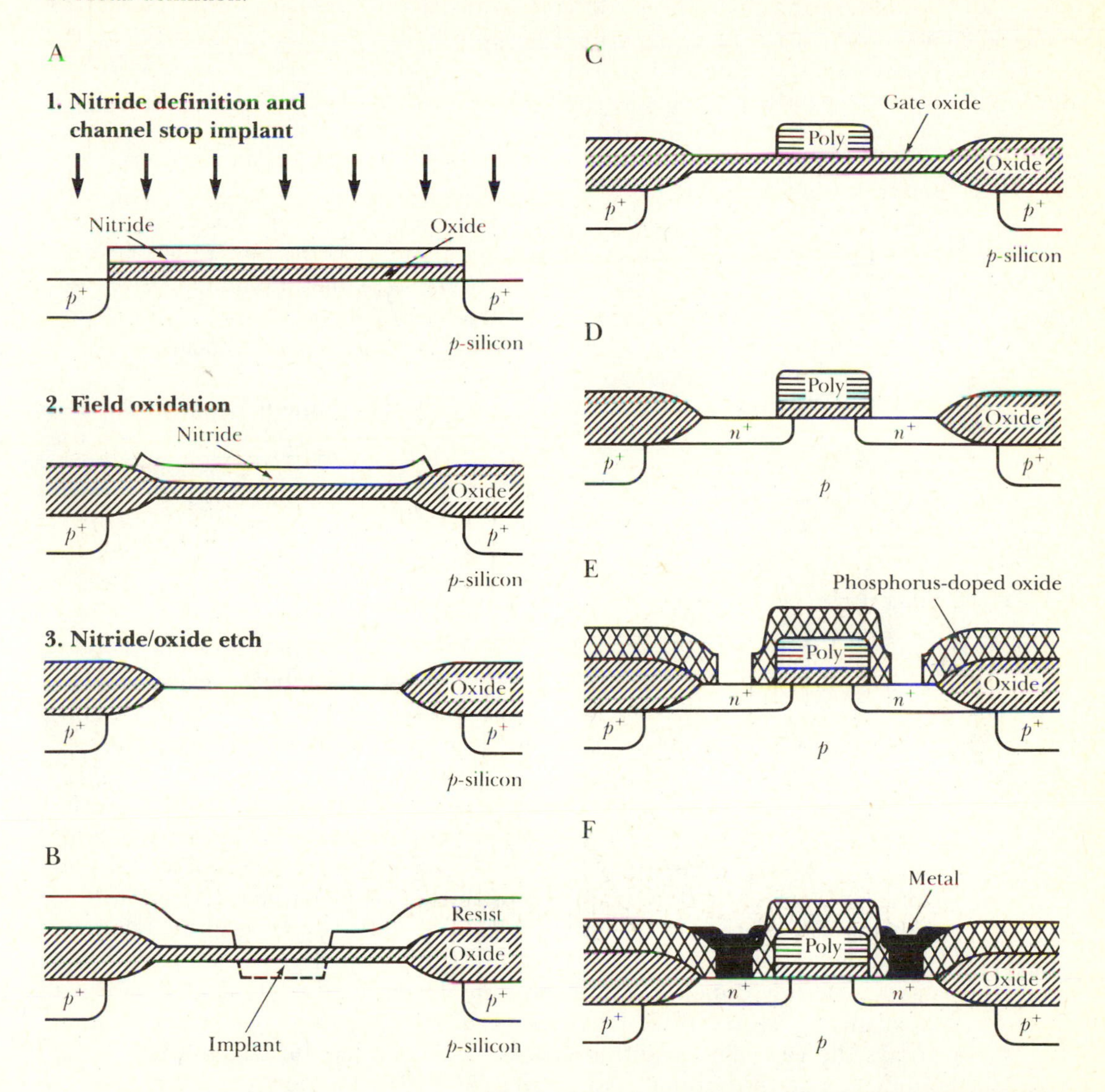

tion, the photoresist is stripped. This step is followed by chemical vapor deposition of polysilicon, which is doped n-type, and the third mask is used to pattern the polysilicon. The wafer is now ready for either diffusion or ion implantation to establish the source and drain regions. The patterning for this step is provided by the field oxide and polysilicon gate. Next an insulating CVD oxide layer is deposited over the wafer to insulate the polysilicon layer from the metal layer. This oxide is phosphorus doped and is sometimes referred to as P-glass. Multiple benefits arise from adding the phosphorus. It reduces the viscosity of the glass at high temperatures, such that the glass flows over the polysilicon edges and smooths the surface steps, tending to planarize the surface. Also, the phosphorus in the glass protects the active portion of the wafer from mobile-ion contamination by reacting with the impurities and entrapping them, a process known as gettering. A fourth mask defines the contact cuts. Then, after the contact cut etch, metal is deposited over the entire wafer and a fifth mask defines the metalization pattern. The wafer is annealed to form good ohmic contacts and, to protect the chip from hostile ambient conditions, a passivation layer is deposited. A final mask, not shown in Figure 3.18, is used to provide openings in the passivation layer to the chip bonding pads. Additional masking steps may be added to this sequence to introduce other features. For example, in Chapter 5, a buried contact mask is added to facilitate gate layout by providing a means for electrical contact between the polysilicon and diffusion layers.

Note that the polysilicon gate acts as a self-alignment structure twice—once to align the gate electrode with the gate oxide, and once to align the gate structure with the source and drain regions. Several variations on this basic sequence are available.

3.6.3 DMOS

The self-aligned process described in the previous section was aimed at reducing the overlap capacitances. Other processes have been developed with the motive of reducing the channel length. A short channel length is generally desirable because device speed increases as the channel length decreases. The most obvious way to achieve short channels is to shrink device dimensions by using high-resolution lithography. However, techniques have also been developed to achieve very small channel lengths with standard lithography. One such method is double-diffused MOS (DMOS), which is illustrated in Figure 3.19, in simplified form, for an n-channel device. Starting with an oxide coated n^--type substrate, a p-type diffusion is performed through a window in the oxide. The n^- refers to lightly doped n-type material. Alternatively, lightly doped p^- substrates can also be used in the DMOS process. Next an n^+ source diffusion is performed using the same window. As is always the case, lateral diffusion takes place under the edges of the oxide window. Then the wafer is reoxidized, and the subsequent mask establishes an adjacent window for an n^+ diffusion that establishes the drain regions.

Note that the channel length, L, is the difference between the lateral diffusions of the p and n^+ diffusions. Consequently, L can be in the submicron region without the need for submicron lithography. DMOS has found only limited commercial use in MOS integrated circuits. However the DMOS technique, and variations of it, have been an important part of power MOS-

Figure 3.19 The DMOS process. A. Boron diffusion. B. Phosphorus diffusions.

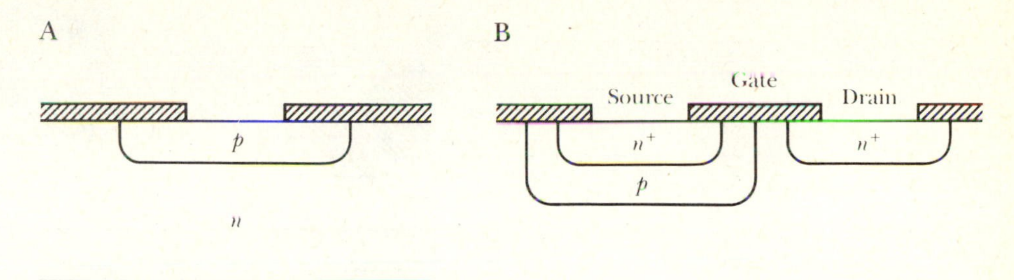

FET technology because small channel lengths correspond to small "on" resistances. Also, the n^- or p^- region, between the active channel and the drain, provides a high punch-through voltage even though the active channel is short.

3.6.4 VMOS

V-groove MOS (VMOS) also establishes a short channel by a rather indirect route. One of several versions of this process is shown in Figure 3.20. The starting point here is a heavily doped n^+ wafer; the preferred dopant is arsenic or antimony because each has a low diffusion coefficient that reduces impurity motion during subsequent fabrication steps. A light boron ion implantation precedes the growth of a lightly doped epitaxial layer. The epitaxial layer may be n^- or p^-. In this example, a p^- epitaxial layer is shown. During the epitaxial growth, out-diffusion of the boron atoms, whose diffusion coefficient is approximately 10 times that of arsenic or antimony, creates a thin p-type region in the epitaxial layer adjacent to the n^+ substrate. As will be seen, the vertical dimension of this p-type layer determines the channel length. Next, a field oxide is grown using the LOCOS method described in Section 3.6.2. The remaining nitride is then patterned to open areas for the n^+ drain region.

For access to the buried p-type layer where the channel is to be induced, the silicon must be etched. By using an anisotropic etch, such as KOH plus hydrazine as described in Section 2.11, a V-groove is etched through the structure, reaching the underlying n^+ wafer. The gate oxide is grown, and metalization is carried out as shown. In the example shown, the source corresponds to the bottom n^+ layer, which is common to all transistor sources. This would be well suited for memory applications, but not for circuit applications where separate sources are required. However, it is possible to alter the process so that the source and drain are on opposite sides of the V-groove.

Figure 3.20 A VMOS processing sequence. A Boron implantation. B. Growth of epitaxial layer, and out-diffusion of boron. C. Field oxidation and channel-stop implant. D. Ion implantation of the n^+ drain region. E. Second oxidation. F. Cutaway view of the V-groove etch. G. Gate oxidation, contact cuts, and metalization.

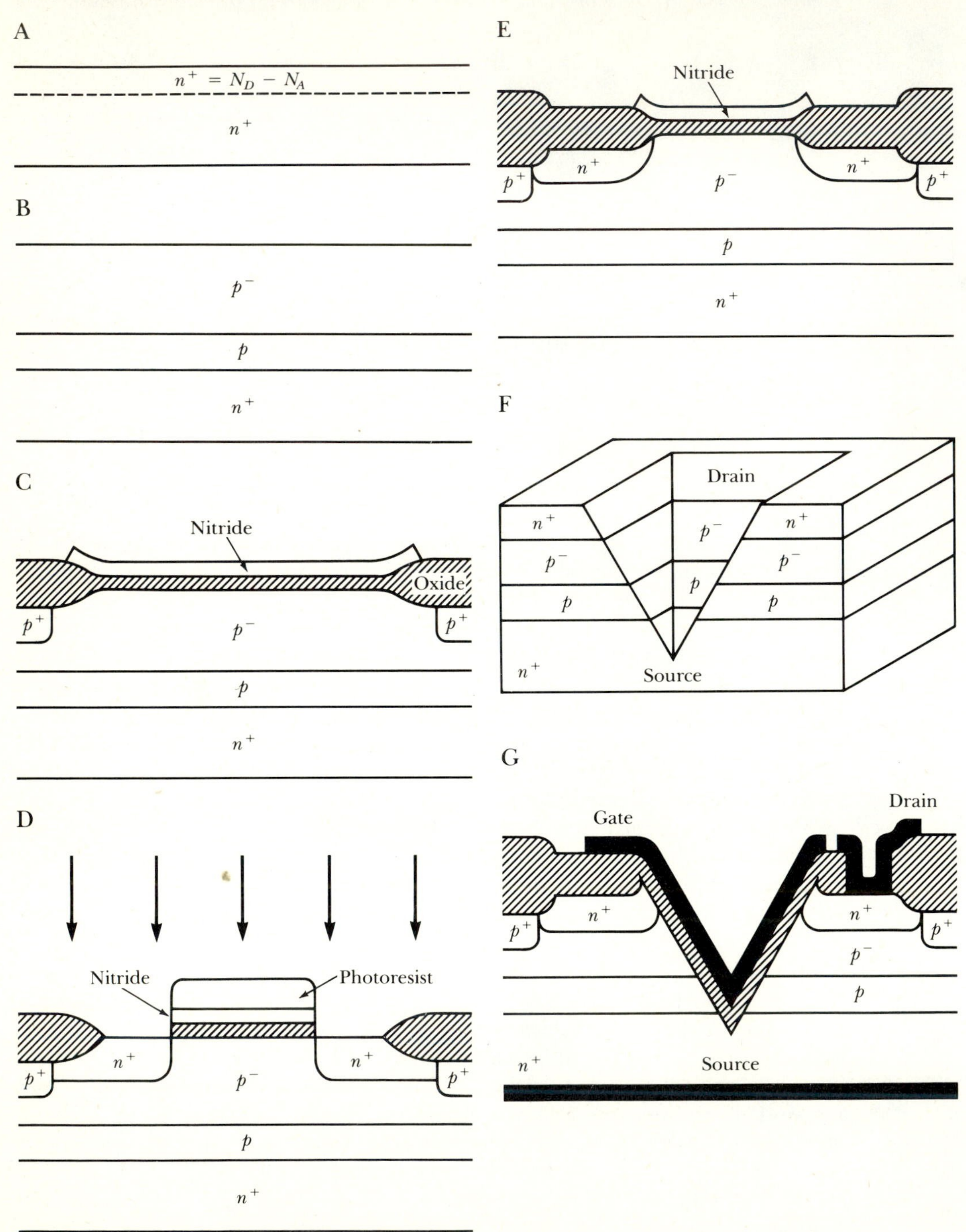

Note that the n-channel transistor is formed along the sides of the V-groove. The structure may be thought of as two transistors in series, corresponding to the p and p^- regions, but with different threshold voltages so that the overall transistor is dependent on the higher threshold. Since the p region has the larger V_T, the effective channel length is determined by the degree of out-diffusion of boron into the lightly doped epitaxial layer. Also, note that the VMOS structure leads to large W/L ratios since the channel wraps around the inverted pyramid formed by the anisotropic etch. Although originally developed for integrated circuit applications, VMOS has had limited impact on commercial MOS integrated circuits. However, as is the case for DMOS, VMOS concepts have been useful in power MOSFET design.

3.6.5 CMOS

Complementary MOS (CMOS) circuitry requires the fabrication of adjacent n- and p-channel devices. If the starting wafer is n-type, then p-type "wells" or "tubs" must be formed for the n-channel devices. Alternatively, if the starting wafer is p-type, then n-type wells are required. A third option is the twin-tub method, in which both n and p wells are established in a lightly doped, nearly intrinsic epitaxial layer. All three techniques are used. The choice of processing method sometimes depends on circuit considerations. For example, when the circuitry contains a larger number of n-channel devices than p-channel ones, as may be the case for analog circuits, then the n-well approach leads to a more efficient layout. Mobility arguments can also be made for n-well layouts in such cases, since the high-mobility and therefore high-speed properties of NMOS transistors are most fully realized when the n-channel devices are formed in the substrate. Impurity scattering in p-wells can place an appreciable cost on electron mobility.

CMOS processing is more complex than NMOS processing. Consider the modifications that would be necessary to accommodate an adjacent p-channel device in the polysilicon self-aligned gate NMOS process described in Section 3.6.2. A comparable n-well CMOS sequence is shown in Figure 3.21. For simplicity of illustration, any threshold-voltage-tailoring ion implant steps or channel-stop implants are not shown here. The first mask establishes the n-well in which p-channel transistors will reside. A typical depth for the well, after drive-in, is in the range of 2 to 4 μm. After the well is established, the next step is to pattern the field oxide. As for NMOS, a thin gate oxide is grown, followed by polysilicon deposition. The third mask patterns the polysilicon. Note that the gate resist is retained for the next step, in what is referred to as a double-resist process.

Separate implants are required for the source and drain regions of the n- and p-channel devices. Mask D establishes the p-type source and drain

Figure 3.21 A CMOS fabrication sequence. A. n-well definition. B. Field oxide patterning. C. Polysilicon definition. D. p-type implant for PMOS source and drain. E. n-type implant for NMOS source and drain (negative of step D). F. Contact cuts. (Not shown: Threshold voltage implant, channel stop implant, guard ring implant, substrate contacts, metalization, and passivation.)

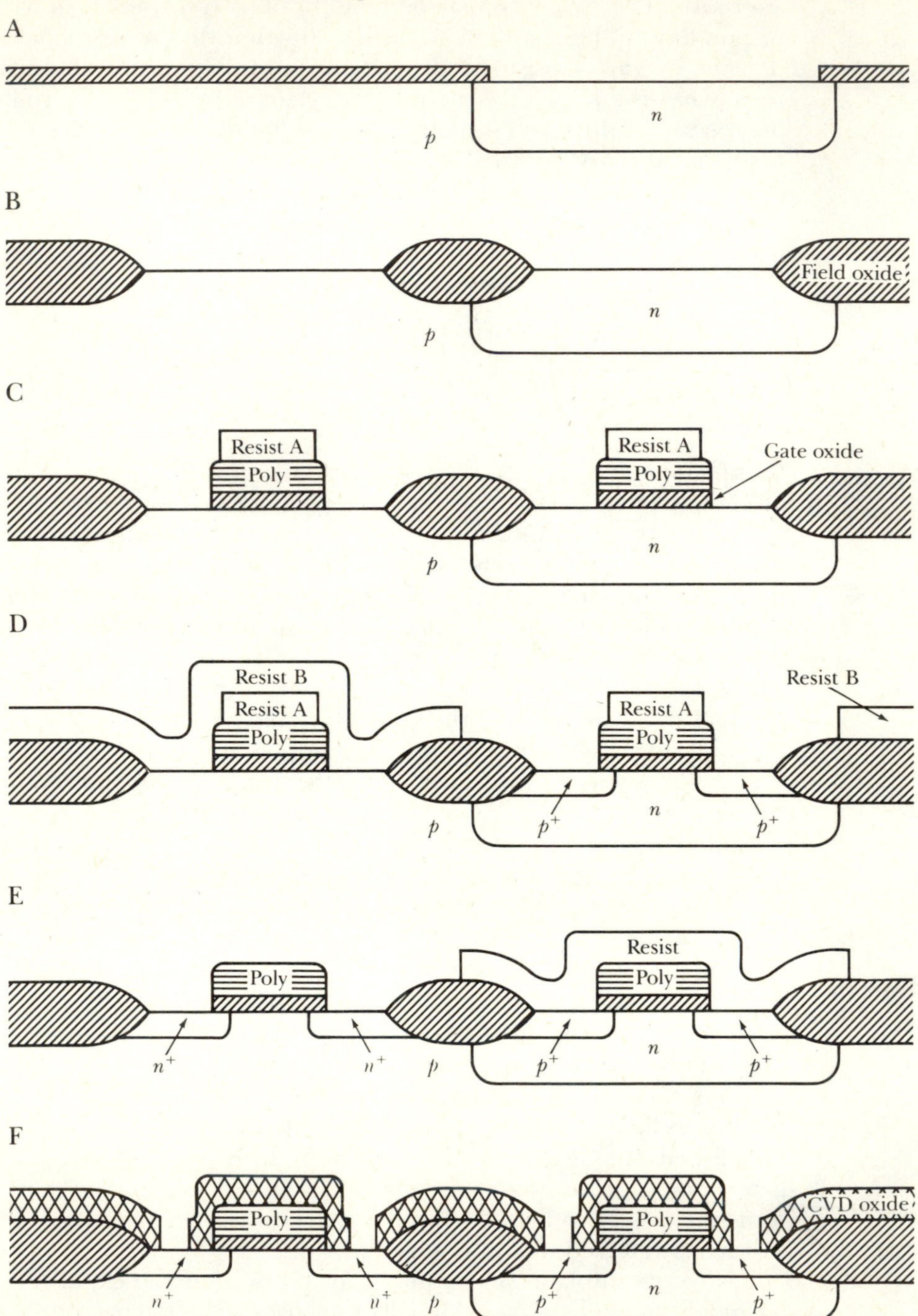

regions for the *n*-channel device, with resist A protecting the *n*-poly gate of the PMOS device. Mask E is used for the *n*-type implant for the source and drain of the *p*-channel device. Alternatively, mask D can be used again with positive resist. The remaining masks are similar to the NMOS case, namely mask F for the contact cuts and the final two masks (not shown) for the metalization and for openings in the passivation layer.

CMOS circuitry is prone to a latch-up problem owing to active parasitics. This problem can be circumvented by additional processing steps, as discussed in Section 3.9. With the further inclusion of threshold-voltage implants and channel-stop implants, CMOS processing is one of the more complex integrated circuit processes. For example, a silicon gate CMOS process reported by Texas Instruments uses 12 mask levels [11].

3.6.6 SOS/SOI

OMIT

MOS circuitry can also be created on sapphire substrates, for silicon-on-sapphire (SOS) integrated circuits. Sapphire, or Al_2O_3, has the useful property of being both a good electrical insulator and a relatively good thermal conductor. Furthermore, its crystalline structure is close to silicon, so that silicon epitaxial layers can be grown on sapphire substrates. In conventional silicon integrated circuits, device isolation is achieved by virtue of the small currents associated with *p-n* junctions in the off state. Although this is adequate for most low-frequency applications, the *p-n* junction capacitance causes isolation to fail at high frequencies.

With SOS, superior high-frequency device isolation can be achieved by physically separating adjacent devices using a silicon etch. The combination of CMOS and SOS is of particular interest because it combines the high-frequency performance of SOS with the low power consumption of CMOS.

A disadvantage of SOS is high substrate cost; the cost per finished wafer is 4 to 5 times that of a comparable bulk silicon wafer. Also, there are interface problems at the sapphire-silicon junction because the crystal matches are not perfect. An alternative to SOS is silicon-on-insulator (SOI) processing, in which single-crystalline silicon pads are isolated on an insulator such as SiO_2.

One approach to SOI uses a deep oxygen ion implant, followed by an 1150°C anneal, to form a buried oxide layer under the surface of the silicon wafer, as shown in Figure 3.22 [12]. Another method is to deposit polycrystalline or amorphous layers on an oxide and then to supply annealing energy from a laser beam, electron beam, or graphite heater to crystallize the deposited silicon. As shown in the laser recrystallization scheme in Figure 3.23, crystallization proceeds laterally by using the silicon wafer as a seed crystal [13]. An additional attraction of SOI is that multiple-layer integrated circuits open the possibility of three-dimensional integrated circuits.

Figure 3.22 Buried-oxide-layer approach to SOI.
SOURCE: T. R. Lineback, "Buried Oxide Marks Route to SOI Chips," *Electronics Week,* Oct. 1, 1984, pp. 11–12.

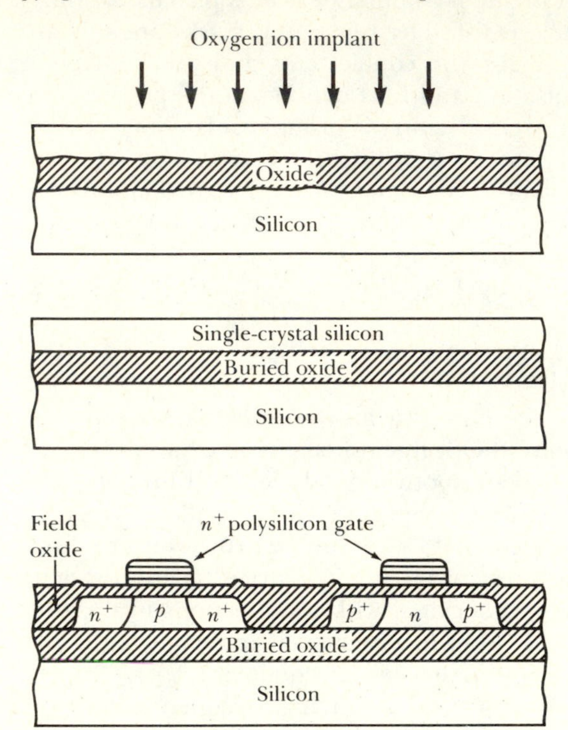

3.7 MOSFET Scaling

3.7.1 Constant-Channel-Field Scaling

The reduction in size, or scaling of a MOSFET requires changes in processing parameters such as wafer doping and oxide thickness, as well as changes in the length and width of the device. Consider, for example, the basic MOSFET structure shown in Figure 3.24 and assume that the gate length is to be reduced by a scaling factor S such that the new gate length is L/S. What are the implications for the other design parameters of the device? What follows is one rationale for obtaining the broad features of MOSFET scaling.

If L is scaled by a factor S, then it is reasonable to scale the gate width by the same factor, since transistor current-voltage characteristics depend on the ratio W/L, as seen from Eq. (3.36). Therefore the new gate width should be W/S. One eventual limitation on L concerns the width of the space-charge layers, since punch-through results if space-charge layers overlap. For the space-charge layers to occupy the same percentage of the channel length,

Figure 3.23 Laser beam recrystallization of polysilicon for SOI.
SOURCE: From R. T. Gallagher, "SOI Method Can Be Used in Submicrometer IC's," *Electronics Week*, Nov. 19, 1984, pp. 28–29.

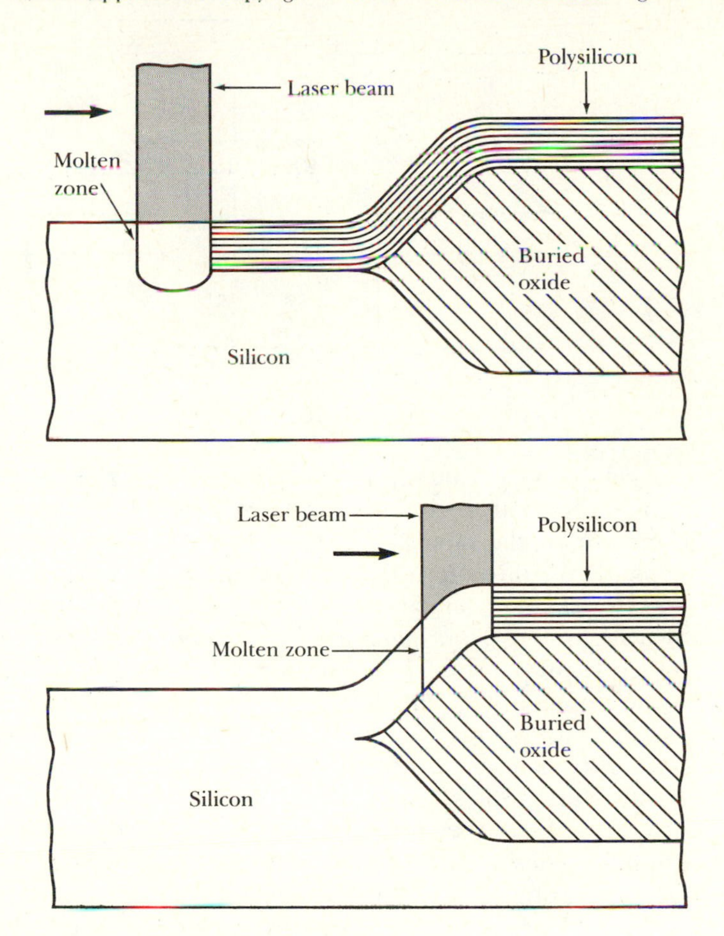

they should be scaled such that the new width is W_{scl}/S. In the depletion approximation, the largest value of W_{scl} is

$$W_{scl} = \left[\frac{2\epsilon_S(V_{bi} + V_{DD})}{qN_A}\right]^{1/2} \tag{3.58}$$

where V_{DD} is the power supply voltage. One way to approximately achieve the desired effect is to increase the wafer doping to a value SN_A and decrease V_{DD} to V_{DD}/S. Furthermore, by reducing the power supply voltage by the factor $1/S$, the electric field in the MOSFET channel is held approximately constant, which eliminates the problem of excessive semiconductor electric fields in the scaled-down devices.

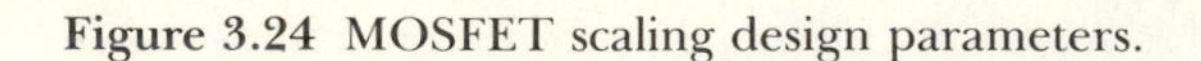

Figure 3.24 MOSFET scaling design parameters.

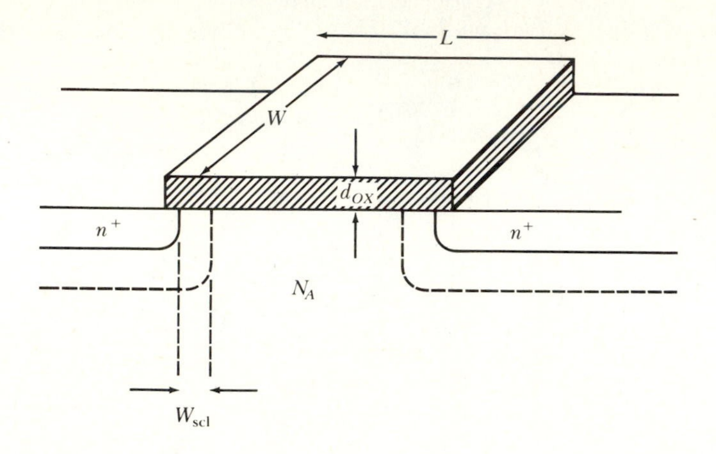

Since the power supply voltage has been reduced by a factor $1/S$, it is also appropriate to reduce the threshold voltage required to turn on the transistor by the same factor. Referring to Eq. (3.18), the threshold voltage depends on three terms, V_{FB}, ψ_F, and Q_B/C_{OX}. For a given gate electrode, wafer orientation, and annealing protocol, V_{FB} is primarily controlled by means of a low energy ion implant as illustrated in Ex. 3.5. The quantity C_{OX} is inversely proportional to the oxide thickness, and ψ_F and Q_B depend on substrate doping. Since the doping has already been fixed at SN_A, the primary variables for scaling V_T are d_{OX} and the implant dose. Furthermore, it is desirable to scale d_{OX} by $1/S$ so that the oxide electric field does not change greatly during the scaling process. (The relationship between the electric field in the oxide and scaling parameters such as circuit voltages and substrate doping is considered in more detail in Section 3.7.2.)

These scaling rules, first set forth by Dennard et al. [14], provide a constant electric field in the channel as the device is scaled. Results of this scaling approach are summarized in Table 3.2, along with the effects of scaling on several features of interest for the device and chip. The current per device, as can be seen from Eqs. (3.36) and (3.41), is proportional to the product of C_{OX}, W/L, and V^2. Since the oxide capacitance increases by S, while W/L is constant and V^2 decreases by $1/S^2$, the current per device decreases by a factor of $1/S$. Therefore the power dissipated per device, which is proportional to the IV product, decreases as $1/S^2$. The packing density, or number of devices per unit area on the chip, increases by a factor of S^2, so the power dissipation per chip is independent of scaling using this set of scaling laws.

In addition to packing density, scaling offers substantial speed advantages. In practical MOS digital integrated circuits, gate delay is primarily determined by the ability of a transistor to charge or discharge a load capacitor. Assuming that the load is approximately determined by the input capaci-

Table 3.2 Scaling for constant channel field

Parameter	Scaling factor
Channel length	$1/S$
Channel width	$1/S$
Oxide thickness	$1/S$
Substrate doping	S
Power supply voltage	$1/S$
Device-packing density	S^2
Power dissipation per device	$1/S^2$
Power dissipation per unit area on chip	1
Speed	S
Power-gate delay product	$1/S^3$

tance presented by a transistor on a chip, the load capacitance scales as $1/S$, since

$$C_{\mathrm{LOAD}} \propto C_{OX}WL \tag{3.59}$$

The current-driving ability of a transistor is decreased by $1/S$, but the voltage through which the capacitive load must be charged or discharged has also been decreased by $1/S$, so that the net effect is to reduce the gate delay by a factor of $1/S$. More dramatically, the power-delay product, which is a common figure of merit for digital integrated circuits, is reduced by a factor of $1/S^3$. However, secondary effects cause the actual improvement in power-delay product to be somewhat less.

3.7.2 Maximum-Oxide-Field Considerations

The electric field in the oxide, $\mathscr{E}_{OX}$, may be expressed as

$$\mathscr{E}_{OX} = \frac{V_G - V_T}{d_{OX}} - \frac{Q_B}{\epsilon_{OX}} \tag{3.60}$$

Note that as the substrate doping increases, the oxide field required to establish a strong inversion layer, $\mathscr{E}_{OX} = -Q_B/\epsilon_{OX}$, also increases. The oxide field must be less than the insulator dielectric strength, which has a maximum value of about 10^7 V/cm for SiO_2. For a given circuit and for a given limit on the oxide electric field, a relationship between minimum oxide thickness and substrate doping may be developed as follows [15]. Consider the case of an FET with the source and substrate at zero potential. Referring to Eq. (3.8), the applied gate voltage is equal to the sum of the voltage drops across the oxide, V_{OX}, and across the semiconductor, ψ_S where, under conditions of strong inversion,

$$V_G - V_{FB} = V_{OX} + 2\psi_F \tag{3.61}$$

The oxide voltage can therefore be expressed as

$$V_{OX} = V_G - V_{FB} - 2\psi_F \tag{3.62}$$

For a given circuit configuration, with a maximum gate voltage, $V_{G\max}$, and a specified maximum allowed oxide field, $\mathcal{E}_{\max}$, the minimum oxide thickness is then

$$d_{\min} = \frac{V_{G\max} - V_{FB} - 2\psi_F}{\mathcal{E}_{\max}} \tag{3.63}$$

Equation (3.63) leads to an expression for maximum substrate doping as follows. Recall that there is a relationship between oxide thickness and threshold voltage. From Eq. (3.18),

$$d_{OX} = \frac{\epsilon_{OX}(V_T - V_{FB} - 2\psi_F)}{-Q_B} \tag{3.64}$$

or

$$d_{OX} = \frac{\epsilon_{OX}(V_T - V_{FB} - 2\psi_F)}{(4\epsilon_S\psi_F q N_A)^{1/2}} \tag{3.65}$$

Equating Eq. (3.65) to Eq. (3.63) results in an expression for the maximum substrate doping concentration,

$$N_{\max} = \left[\frac{V_T - V_{FB} - 2\psi_F}{V_{G\max} - V_{FB} - 2\psi_F}\right]^2 \frac{(\epsilon_{OX}\mathcal{E}_{\max})^2}{4q\epsilon_S\psi_F} \tag{3.66}$$

Since ψ_F is also a function of substrate doping. Eq. (3.66) represents a transcendental equation that may be solved for N_A.

EXAMPLE 3.6

Consider the maximum gate voltage in an *n*-channel MOSFET circuit to be equal to the power supply voltage, V_{DD}, and assume that the threshold voltage of an enhancement-mode transistor is chosen to be $0.2(V_{DD})$. The FET source is grounded. Assume the maximum oxide field is 3×10^6 V/cm and that the flat band voltage is -1 V. What is the maximum allowed substrate doping and corresponding minimum oxide thickness if the power supply is 5 V?

Solution Equation (3.66) becomes

$$N_A = \left[\frac{2 - 2\psi_F}{6 - 2\psi_F}\right]^2 (1.59 \times 10^{18}/\psi_F)$$

where, from Eq. (3.11),

$$\psi_F = .0259 1n(N_A/1.45 \times 10^{10})$$

Solving for N_A and then ψ_F yields,

$$N_A = 1.87 \times 10^{17} \text{ cm}^{-3}$$
$$\psi_F = 0.424 \text{ V}$$

The corresponding oxide thickness is found from Eq. (3.63) to be

$$d_{OX} = \frac{(5 + 1) - 2(0.424)}{3 \times 10^6}$$
$$d_{OX} = 172 \text{ Å}$$

Example 3.6 demonstrates a method for calculating a minimum oxide thickness and a maximum substrate doping given the transistor's biasing environment and a maximum prescribed oxide field. However, it is certainly possible to choose a lower substrate doping and therefore a lower oxide field and thicker gate oxide. Indeed, the latter, more conservative approach offers certain advantages, since very thin oxides, such as in Ex. 3.6, lead to yield problems, and high substrate doping degrades carrier mobilities. The trade-off is related in part to channel-length considerations, as now shown in Ex. 3.7.

EXAMPLE 3.7

Consider the MOSFET in Ex. 3.6. Assume that the width of the space-charge layer, with maximum reverse bias, is to equal $L/5$ to reduce space-charge-layer width modulation effects on the transistor characteristics. What is the channel length L? Assume that the source and drain are degenerately doped.

Solution Equation (3.58) can be used to calculate the width of the space-charge layer once the built-in potential is known. Treating the source and drain junctions as one-sided,

$$V_{bi} \simeq E_G - (E_F - E_V)_{\text{substrate}}$$

where

$$(E_F - E_V) = kT\ln(N_V/N_A)$$

$$(E_F - E_V) = (0.0259)\ln\left(\frac{1.04 \times 10^{19}}{1.87 \times 10^{17}}\right)$$

$$(E_F - E_V) = 0.10 \text{ eV}$$

Therefore,

$$V_{bi} = 1.12 - 0.10 = 1.02 \text{ V}$$

Applying equation (3.58),

$$W_{scl} = \left[\frac{2(11.9)\,(8.85 \times 10^{-14})(5 + 1.02)}{(1.6 \times 10^{-19})(1.87 \times 10^{17})}\right]^{1/2}$$

$$W_{scl} = 0.206\ \mu\text{m}$$

The resulting channel length is

$$L = 1.03\ \mu\text{m}$$

Note that a smaller substrate doping would require a larger channel length. Quantum mechanical tunneling places a fundamental lower limit on oxide thickness of about 50 Å. This, in turn, places a fundamental limit on MOSFET scaling, a matter that is considered further in the chapter exercises.

3.7.3 Practical Scaling Considerations

In practice, scaling efforts have not strictly followed the protocol outlined in Table 3.2. Voltage supplies, in particular, have tended to stay at the transistor-transistor-logic (TTL) standard of 5 V, even though other device properties were scaled. This situation has led to a number of effects not considered in the simple scaling theory. Most obviously, the power dissipation per unit of chip area increases. Also, since the device's internal electric fields increase, a number of high-field device phenomena come into play if L is sufficiently small. These phenomena include carrier-velocity saturation and hot-electron effects [16, 2].

In addition, to the extent that scaling laws have been followed, the performance gains have been less than are predicted in Table 3.2 for a number of reasons. For example, as devices are scaled, source and drain junctions become shallower by a factor of approximately $1/S$ to reduce parasitic capacitances. Also contact cut openings become smaller in area by $1/S^2$. Consequently, the parasitic resistances associated with the source and drain, shown in Figure 3.15, become larger. In addition, decreasing line widths mean larger

Figure 3.25 Lightly doped drain for field spreading in short-channel devices. The n^- source is not necessary unless the device is to be operated bidirectionally, but to eliminate it would require an extra masking stop.

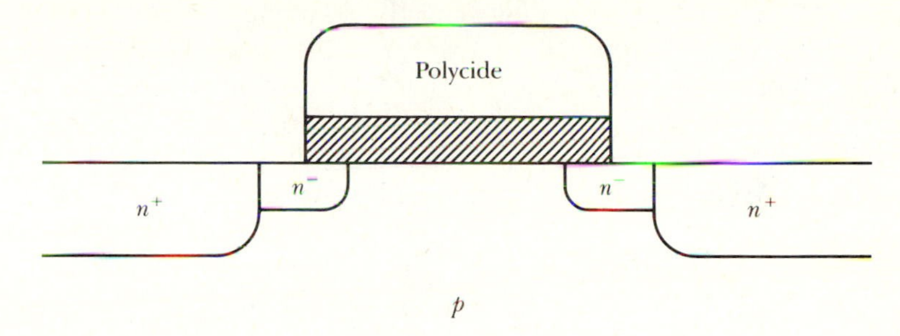

interconnect-line resistances. This concern applies particularly for polysilicon interconnect lines, because of the relatively low conductivity of polysilicon.

Another class of scaling-related problems concerns short-channel effects in MOSFETs. The basic device theory developed in Section 3.3 is appropriate for long-channel devices, but when L approaches 2 μm or less, additional phenomena come into play. Subthreshold currents increase in importance, and edge effects cause a shift in the threshold voltage [2]. Also, if the width of the space-charge layer is an appreciable portion of the channel length, then channel-length modulation causes the drain current to depend markedly on drain voltage in the saturation region.

Finally, scaling may lead to problems unique to a particular processing sequence. A classic example is the "bird's beak" effect in nitride-defined field oxide growth [17]. At the edge of the nitride layer, oxygen diffuses laterally into the active device region as shown in Figure 3.18A. This physical encroachment is approximately 1 to 2 μm and reduces packing density when device dimensions are in the 1-μm range.

On the other hand, processing methods also offer solutions to scaling-related problems. For example, lightly doped drain structures are often used in short-channel devices to reduce high field effects at the drain end of the channel. As shown in Figure 3.25, n^- regions between the heavily doped source and drain regions and the channel region are used to spread out and reduce the maximum value of the electric field near the end of the channel.

3.8 The MESFET Alternative

The metal-semiconductor field-effect transistor, or MESFET, presents an FET alternative to the MOSFET in cases where high-quality gate oxides are difficult to achieve, such as is the case for GaAs. Furthermore, since oxide integrity becomes a concern in silicon as scaling laws reduce the oxide thickness, the MESFET is also of potential interest as a high-density silicon device.

The MESFET, illustrated in Figure 3.26 [18], is a variation on the junction field-effect transistor (JFET), in which the gate structure is a Schottky barrier. As is the case for the MOSFET, the gate voltage modulates the conductivity of a channel between the source and drain. However, in the case of the MESFET, the channel width is determined by modulation of the space-charge-layer width associated with the metal-semiconductor junction at the gate. Both normally on and normally off devices are available, depending on the zero-gate-bias width of the Schottky space-charge layer.

MESFET current-voltage characteristics are given by the JFET relation-

Figure 3.26 The MESFET: An alternative to the MOSFET. A. Schematic drawing. B. Planar implanted depletion mode gallium arsenide MESFET. The lightly doped n-region under the gate is contacted by the ion implanted source and drain regions. SOURCE: B from R. C. Eden, A. R. Livingston, and B. M. Welch, "Integrated Circuits: The Case for Gallium Arsenide," *IEEE Spectrum,* 20 (Dec. 1983), 30–37. © 1983 IEEE.

A

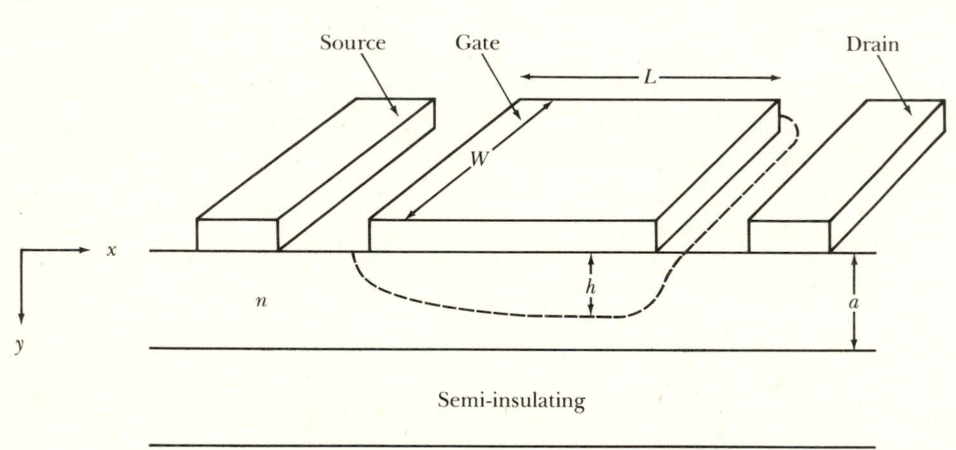

B

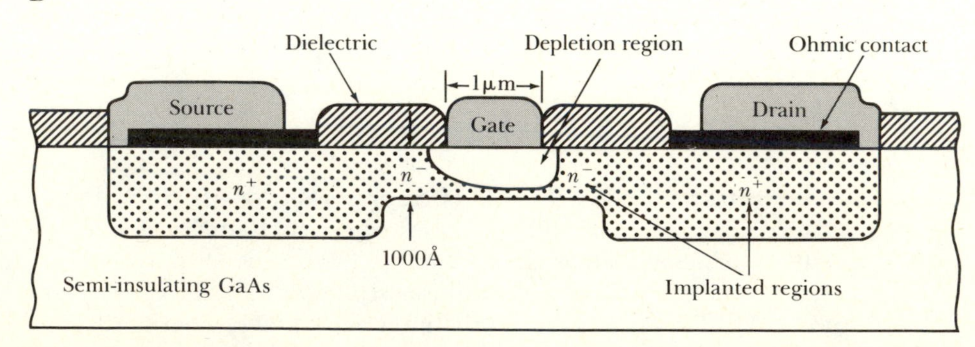

ships [2]. For an n-type semiconductor, the current density at any point in the channel is given by Eq. (3.24), repeated here for convenience, $J = q\mu_n n\mathcal{E} + qD_n\nabla n$. For constant doping and assuming no variation in the channel voltage in the y direction, the current is mainly due to drift in the x direction and the current density is

$$J = -qN_D\mu_n(dV/dx) \tag{3.67}$$

and the drain current is, referring to Figure 3.26,

$$I_D = qN_D\mu_n(a - h)W(dV/dx) \tag{3.68}$$

For the triode region, where $h(L) < a$, it may be shown that Eq. (3.68) leads to [2]

$$I_D = I_P\left[3\left(\frac{V_D}{V_P}\right) - 2\left[\frac{(V_D - V_G + V_{bi})^{3/2} - (V_{bi} - V_G)^{3/2}}{(V_P)^{3/2}}\right]\right] \tag{3.69}$$

where is V_P is given by the expression

$$V_P = qN_Da^2/2\epsilon_S \tag{3.70}$$

and is equal to the voltage required for channel pinch-off, including V_{bi}. Note that V_G is negative for the n-channel device. I_P is given by the expression,

$$I_P = \frac{\mu_n a^3(qN_D)^2}{6\epsilon_S}(W/L) \tag{3.71}$$

In the saturation region, where $h(L) = a$, the drain current is independent of drain voltage and given by the expression,

$$I_{D\text{sat}} = I_P\left[1 - 3\left(\frac{V_{bi} - V_G}{V_P}\right) + 2\left(\frac{V_{bi} - V_G}{V_P}\right)^{3/2}\right] \tag{3.72}$$

A useful approximation of the saturation current, not obvious from equation (3.72), is the square law

$$I_{D\text{sat}} \simeq I_{D0}\left[I - \frac{|V_{GS}|}{V_P'}\right]^2 \tag{3.73}$$

where I_{D0} is the saturation current for zero gate-to-source voltage and V_P' is the gate-to-source pinch-off voltage. V_P and V_P' differ in that the former includes the built-in junction voltage. A qualitative family of curves demon-

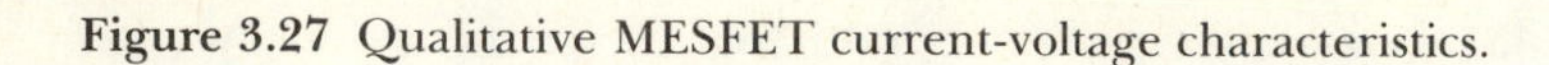
Figure 3.27 Qualitative MESFET current-voltage characteristics.

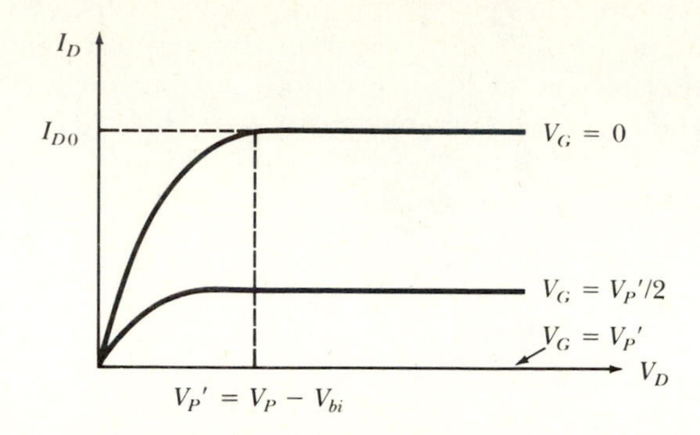

strating the current-voltage characteristics of a MESFET are shown in Figure 3.27.

For normally off MESFETs, the space-charge layer occludes the channel for zero gate voltage. A forward bias is applied to the gate to reduce W_{scl} and allow current flow. Positive gate biases are generally limited to less than 0.5 V to avoid large gate currents. Normally off MESFETs approximately follow a square-law current-voltage characteristic similar to enhancement-mode MOSFETs, with

$$I_{D\text{sat}} = \frac{\mu \epsilon_S W}{2aL}(V_G - V_T)^2 \tag{3.74}$$

where the threshold voltage is given by [2]

$$V_T \simeq V_{bi} - V_P \tag{3.75}$$

MESFET-based integrated circuits represent the dominant GaAs integrated circuit technology, since high-quality gate oxides on GaAs are difficult to achieve. However, the MESFET is not widely used in silicon-based integrated circuits. MESFETs demonstrate a high radiation resistance, which is an advantage for specialized applications.

3.9 MOSFET Gate Configurations

3.9.1 NMOS

The inverter, with a driving transistor and a load, is a basic building block in digital circuits. Four different inverter circuits, all of which use an n-channel enhancement-mode MOSFET as the driving transistor, are shown in Figure 3.28. By placing driving transistors in parallel, the circuit is expanded to a

Figure 3.28 Variations on the NMOS inverter. A. Resistor load. B. Saturated enhancement load. C. Nonsaturated enhancement load. D. Depletion mode load.

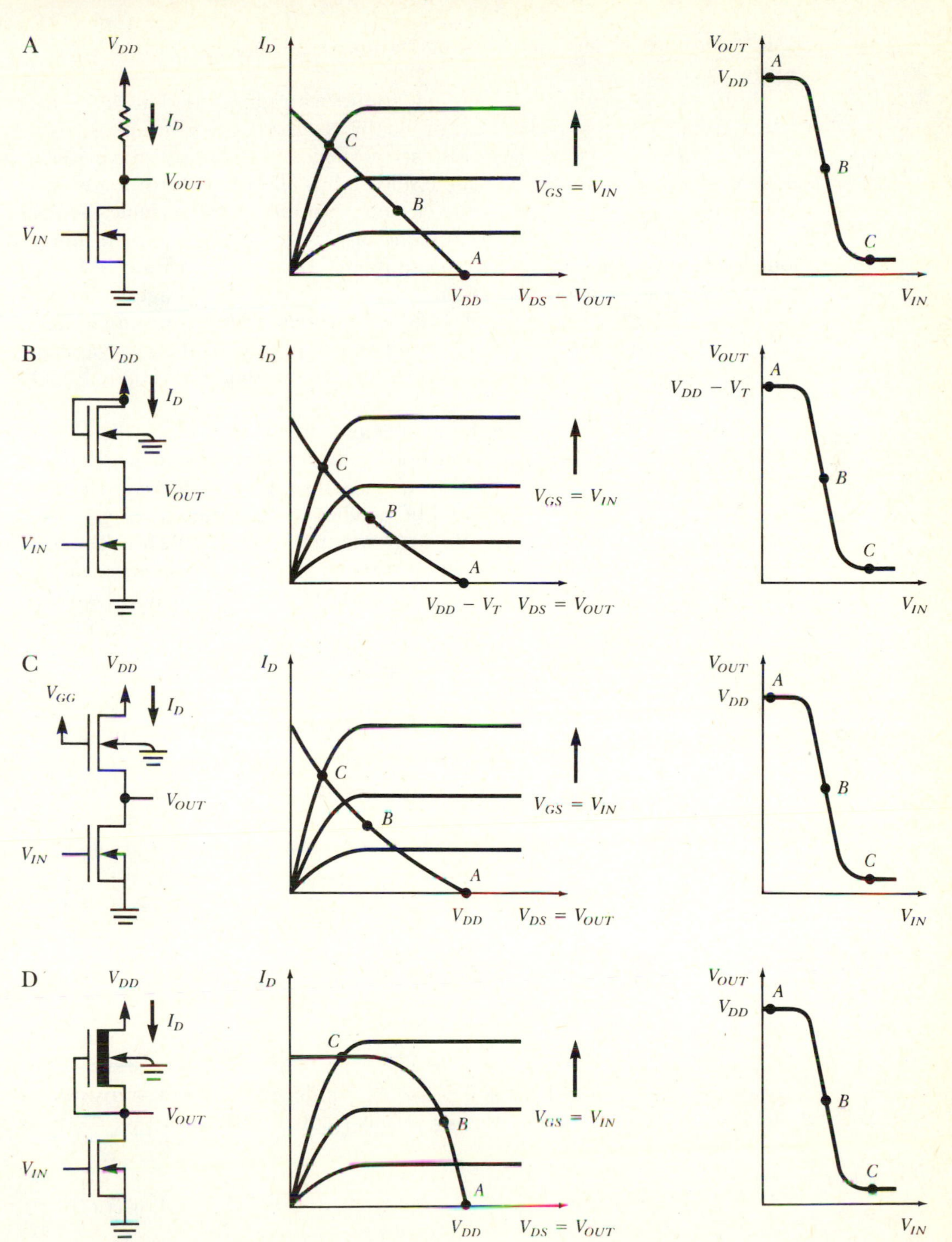

NOR gate, and by series connecting the driving transistors, a NAND gate results.

Resistive-load logic gates have not found widespread use in MOS integrated circuits because resistors traditionally require large chip areas. However, in a few cases, using high-resistivity polysilicon for the resistor material, rather than diffused single-crystalline silicon, has allowed resistor loads to be used in large-scale MOS integrated-circuits without an appreciable sacrifice in chip density. Considering the options of transistor loads, enhancement-mode loads have the advantage of being relatively easy to fabricate, although the logic swing does not fully utilize the power supply voltage range if a single supply is used. If ion implantation is available, depletion-mode loads can be fabricated and, all other things being equal, provide a larger logic swing than do enhancement-mode loads. At present, depletion-mode NMOS technology is widely used.

An analysis of the transfer characteristics of the depletion-mode inverter also indicates how the other inverter options could be analyzed as well. Note that the depletion-mode MOSFET is always on, since its gate is connected to the source and a negative V_{GS} would be needed to turn the device off. Considering the graphic construct of the load line in Figure 3.29B, four separate regions are observed. For $V_{\mathrm{IN}} < V_T$ the driver is off. For somewhat larger values of V_{IN}, the driver transistor is saturated but the load is not. For large values of V_{IN}, the load is saturated but the driver is not. At intermediate three points, both devices may be saturated. On the transfer characteristic, the regions are as shown in Figure 3.29A.

The transfer characteristics may be generated by computer-aided analysis, based on solutions of Eqs. (3.37) and (3.40). However, insight into the circuit's features are obtained by carrying out an approximate analytical solution based on Eqs. (3.36) and (3.41). This treatment makes the further simplifying assumption that each transistor has the source connected to the substrate, so that the threshold voltage is constant with circuit operation. In fact, the load transistor source is not connected to the substrate, so the threshold voltage of the depletion-mode device is not constant as V_{IN} of the inverter varies. The implications of this fact will be discussed later.

When the load is saturated, then from Eq. (3.41),

$$I_D = (\mu_n/2)\ (W_{\mathrm{PU}}/L_{\mathrm{PU}})(\epsilon_{OX}/d_{OX})\ (V_{\mathrm{DEP}})^2 \tag{3.76}$$

where the threshold voltage of the load is taken to be V_{DEP} and the subscripts PU refer to "pull-up," reflecting the role of the depletion mode transistor in this circuit. In the triode region, the current-voltage relationship for the load results in

$$I_D = \mu_n(W_{\mathrm{PU}}/L_{\mathrm{PU}})(\epsilon_{OX}/d_{OX})[(-V_{\mathrm{DEP}})\ (V_{DD} - V_{\mathrm{OUT}}) - (V_{DD} - V_{\mathrm{OUT}})^2/2] \tag{3.77}$$

Figure 3.29 Qualitative transfer characteristic of the depletion-mode load MOS inverter. A. V_{OUT} versus V_{IN}. B. Corresponding load line points.

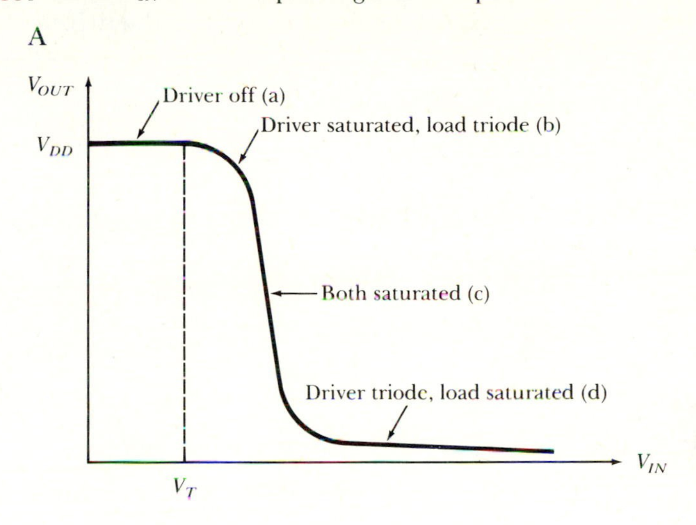

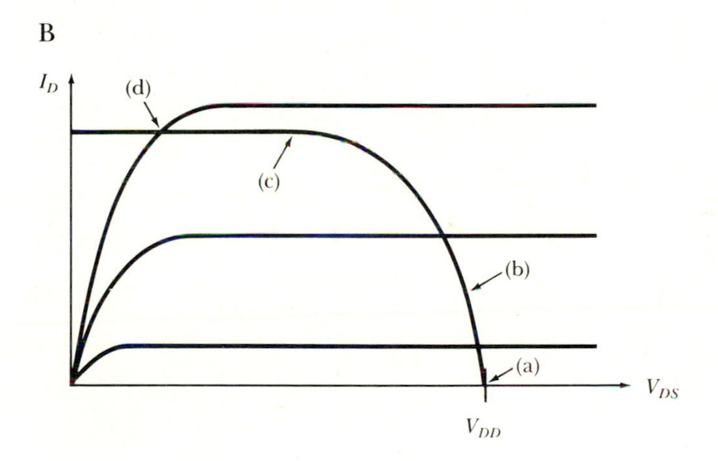

The load is saturated, from Eq. (3.38), when

$$V_{OUT} < V_{DD} + V_{DEP} \tag{3.78}$$

When the driving transistor is saturated,

$$I_D = (\mu_n/2)(W_{PD}/L_{PD})\ (\epsilon_{OX}/d_{OX})(V_{IN} - V_T)^2 \tag{3.79}$$

where the threshold voltage of the driver is taken to be V_T and the subscripts PD again refer to the circuit function of the transistor, in this case "pull down."

For simplicity, it is assumed that both load and driver have the same channel mobility and the same oxide thickness, although this is not always the case. In the triode region, the drain current of the driver is given by

$$I_D = \mu_n(W_{PD}/L_{PD})(\epsilon_{OX}/d_{OX})[(V_{IN} - V_T)V_{OUT} - (V_{OUT})^2/2] \quad (3.80)$$

The driver is in saturation when

$$V_{OUT} > V_{IN} - V_T \quad (3.81)$$

The transfer characteristics are found by equating the drain currents for the load and driver transistors, and solving for V_{OUT} as a function of V_{IN}. For example, consider the case when the load is saturated and the driver is in the triode region. Also, let

$$Z_{PU} = L_{PU}/W_{PU} \quad (3.82)$$

and

$$Z_{PD} = L_{PD}/W_{PD} \quad (3.83)$$

Then, equating Eqs. (3.80) and (3.76) results in the following quadratic expression:

$$(Z_{PU}/Z_{PD})(V_{OUT})^2/2 - (Z_{PU}/Z_{PD})(V_{IN} - V_T)V_{OUT} + (V_{DEP})^2/2 = 0 \quad (3.84)$$

which, for a given V_{IN}, may be solved for V_{OUT}.

When the driver is saturated and the load is in the triode region, a similar quadratic equation results. However, when both transistors are saturated, equating Eq. (3.76) and (3.79) results in

$$Z_{PU}(V_{IN} - V_T)^2 = Z_{PD}(V_{DEP})^2 \quad (3.85)$$

which implies that V_{OUT} is independent of the input, or that the slope of V_{OUT} versus V_{IN} is infinite. In fact, the slope is finite because I_D does change somewhat with V_{DS}; MOSFETs are not ideal current sources in the saturation region. Furthermore, because of the body effect, V_{DEP} is not constant.

EXAMPLE 3.8

Consider a depletion-mode load NMOS inverter with $V_{DD} = 5$ V, $V_T = 1$ V, and $V_{DEP} = -3$ V. Assume for purposes of this example that both transistors are of equal length and width, that is, that $Z_{PU} = Z_{PD}$. Calculate the output logic swing if the input voltage varies between 0 and 5 V.

Solution When $V_{IN} = 0$, the driver is off, so the output is equal to 5 V. When $V_{IN} = 5$ V, the driver is in the triode region and the load is saturated. Applying Eq. (3.84) results in the quadratic expression

$$(V_{OUT})^2 - 8V_{OUT} + 9 = 0$$

which yields $V_{OUT} = 1.35$ V. Therefore the output logic swing is approximately 3.65 V. A SPICE simulation provides a somewhat more accurate analysis that accounts for the varying threshold of the load resulting from the body effect on threshold voltage, which was described in Section 3.3.1. For a 400-Å oxide thickness and a substrate doping of 10^{15} cm^{-3}, the output swing calculated by SPICE for this example is 3.78 V. In any case, since the output logic swing is considerably less than the input swing, the transfer characteristic of this particular gate is poor.

EXAMPLE 3.9

Consider the gate of Ex. 3.8, except that the transistor geometries are such that $Z_{PU}/Z_{PD} = 8$. Calculate the output logic swing if V_{IN} varies from 0 to 5 V.

Solution Again, when V_{IN} is 0 V, the output is 5 V. When V_{IN} is 5 V, the quadratic equation becomes

$$4(V_{OUT})^2 - 64V_{OUT} + 9 = 0$$

which yields $V_{OUT} = 0.14$ V. Now the output logic swing has increased sharply to 4.86 V. In this example, a SPICE simulation that accounts for the body-bias effect on the load yields an essentially identical answer for the output swing. Since V_{OUT} is driven near ground in this circuit, the substrate and source of the load are near the same potential and the body effect on threshold voltage plays a negligible role on output-voltage swing.

Examples 3.8 and 3.9 demonstrate that the transfer characteristics of depletion-mode NMOS logic are quite geometry dependent. To obtain good transfer characteristics, values of Z_{PU} are required to be considerably larger than Z_{PD}, with typical ratios of Z_{PU}/Z_{PD} being in the range of 4 to 8. Because the driving and load transistors have different ratios of length to width, this type of logic circuit is sometimes referred to as ratioed logic. Computer-generated transfer characteristics for ratioed NMOS depletion-mode inverters, with several values of Z_{PU}/Z_{PD} are shown in Figure 3.30.

3.9.2 CMOS

The CMOS inverter uses an enhancement-mode NMOS transistor as the driver and an enhancement-mode PMOS transistor as the load, arranged as shown

Figure 3.30 SPICE-generated NMOS depletion-mode ratioed inverter characteristics. The threshold voltage is 1 V for the driver and −3 V for the load, the substrate doping is 10^{15} cm^{-3}, and the oxide thickness is 400 Å. Z_{PU}/Z_{PD} ratios are as shown.

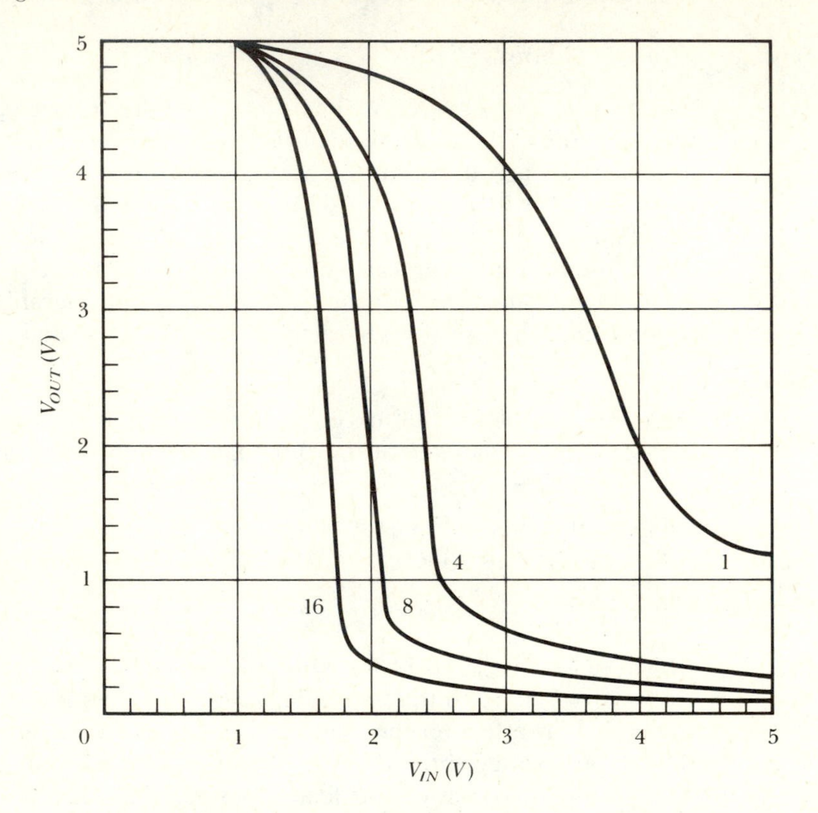

in Figure 3.31A. CMOS is characterized by low power consumption in the quiescent states, since, as shown in Figure 3.31B, one of the two series transistors is always off except during a switching transition from one logic level to the other. Since power dissipation is a concern for VLSI NMOS chips, because of practical scaling considerations discussed in Section 3.7, CMOS is an attractive alternative for VLSI applications.

The analysis of the CMOS inverter proceeds as follows. Consider a 5-V power supply and a threshold voltage of 1 V for the NMOS transistor and −1 V for the PMOS transistor. If the input voltage is less than 1 V, the driver is off and the load is in the triode region, causing the output to be 5 V. When the input voltage is greater than 4 V, the load is off and the driver is in the triode region, causing the output to be 0 V. Therefore, the output logic swing is 5 V regardless of the Z ratios of the two transistors. Since ratioed logic is not necessary for good transfer characteristics, CMOS logic is sometimes referred to as "ratioless" logic.

Figure 3.31 The CMOS inverter. A. Circuit. B. Voltage and current characteristics.

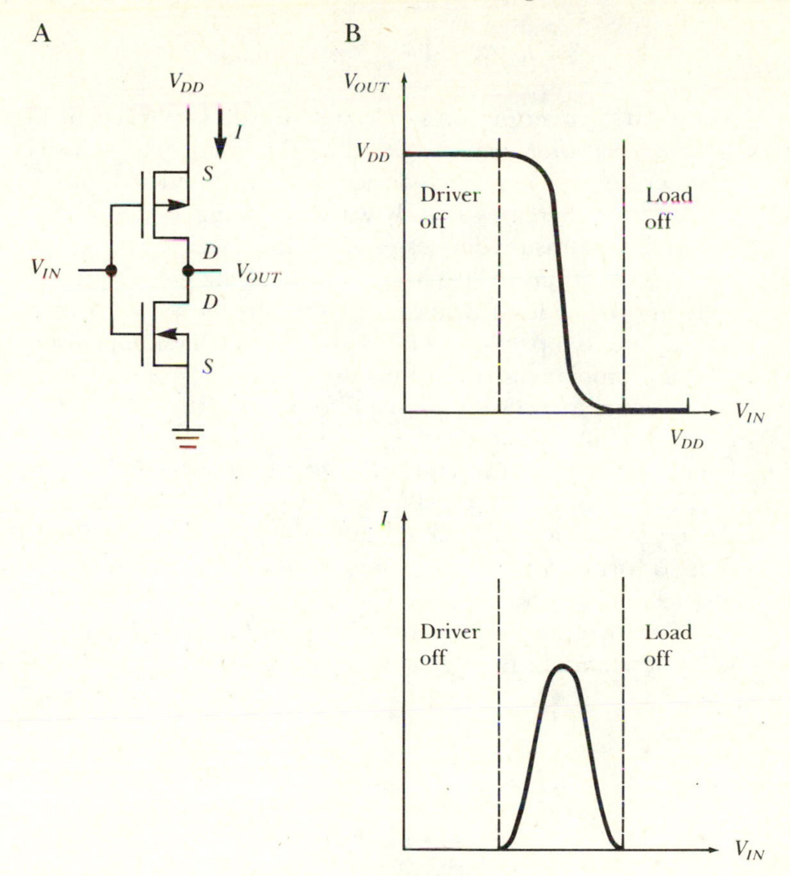

When the input voltage becomes greater than 1 V the driver enters the saturated region and the load is still in the triode region for a limited range of V_{IN}. Equating Eqs. (3.36) and (3.41) yields a quadratic expression for V_{OUT}:

$$(V_{\mathrm{OUT}})^2 - [2(V_{\mathrm{IN}} + 1)]V_{\mathrm{OUT}} - [15 - 10V_{\mathrm{IN}} - K(V_{\mathrm{IN}} - 1)^2] = 0 \tag{3.86}$$

where

$$K = (\mu_n Z_{\mathrm{PU}} / \mu_p Z_{\mathrm{PD}}) \tag{3.87}$$

Likewise, when the input voltage is slightly less than 4 V, the load enters the saturated region and the driver is in the triode region. Under these conditions, the quadratic expression for V_{OUT} is

$$(V_{\mathrm{OUT}})^2 - 2(V_{\mathrm{IN}} - 1)V_{\mathrm{OUT}} + (1/K)(V_{\mathrm{IN}} - 4)^2 = 0 \tag{3.88}$$

Finally, when both transistors are saturated,

$$(V_{\mathrm{IN}} - 4)^2 = K(V_{\mathrm{IN}} - 1)^2 \tag{3.89}$$

The CMOS inverter transfer characteristic is plotted in Figure 3.32 for the specific case of $K = 1$. As for the NMOS inverter, the actual slope of the transfer curve when both devices are saturated is finite, since in reality the saturated I_D increases slightly with increasing V_{DS}.

Active parasitic devices exist in the CMOS inverter, and can lead under certain conditions to "latch-up," in which the gate locks into a low impedance state and provides a near short circuit to the power supply. This phenomenon is due to a parasitic *p-n-p-n* structure, which may also be modeled as two coupled bipolar junction transistors (BJTs). In *p*-well technology, for example, there is a vertical *n-p-n* BJT formed by the n^+ drain and source regions of the NMOS transistor, the *p*-well, and the *n*-type substrate. Also, there is a lateral *p-n-p* structure consisting of the *p*-well, the *n*-substrate, and the p^+ drain and source of the PMOS device. As shown in Figure 3.33, and discussed in the next two paragraphs, the *n-p-n* and *p-n-p* parasitic devices combine to form a two-transistor model of a four-region structure that is capable of regenerative switching action.

Referring to the currents labeled in Figure 3.33, it is instructive, in terms of understanding the qualitative nature of latch-up, to consider the standard

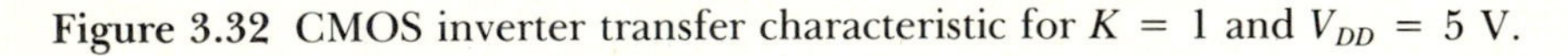

Figure 3.32 CMOS inverter transfer characteristic for $K = 1$ and $V_{DD} = 5$ V.

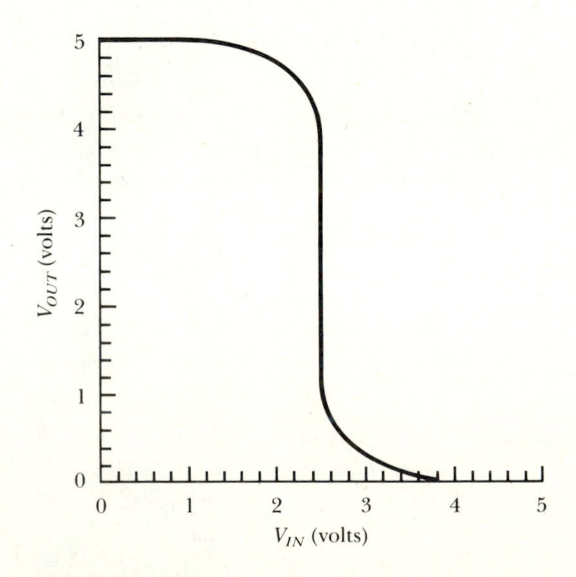

Figure 3.33 CMOS latch-up. A. Origin of active parasitic *p-n-p-n* structure, shown as interconnected BJTs. B. Regenerative switching.

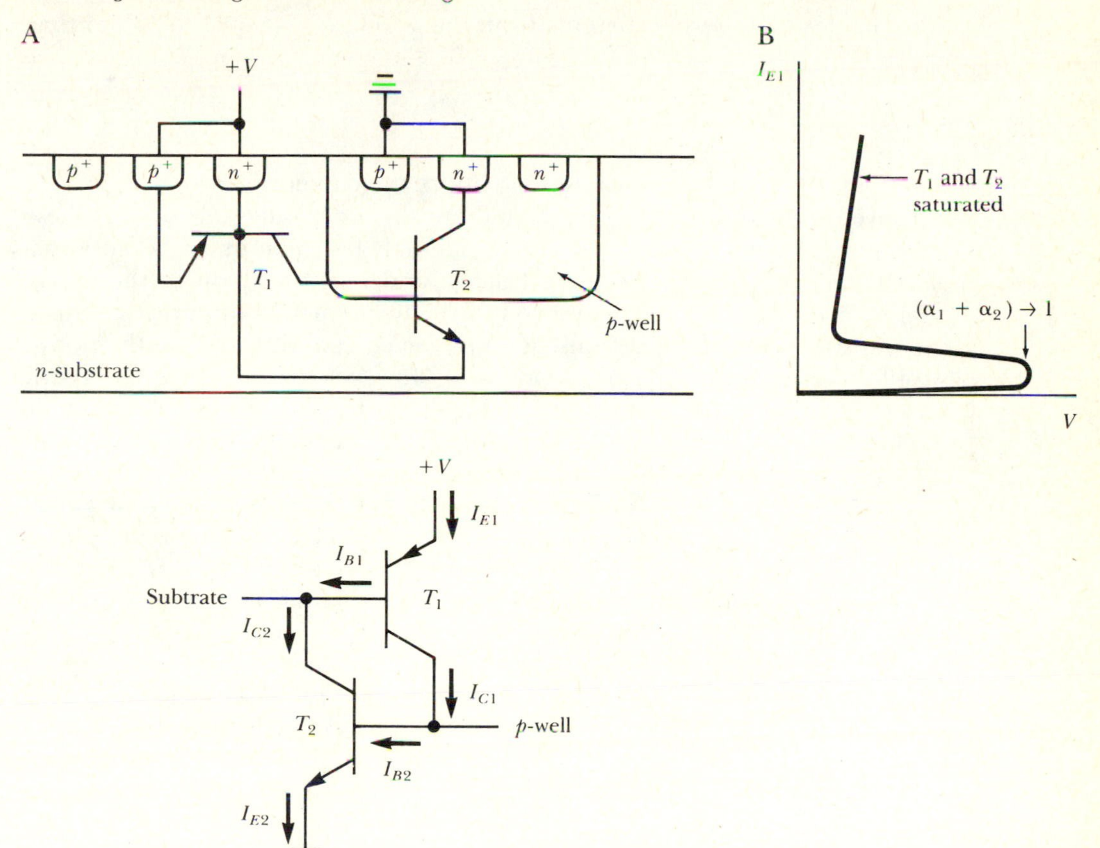

two-transistor analysis of forward blocking currents in 4-layer *p-n-p-n* structures [2]. The base current of T_1 may be expressed as

$$I_{B1} = (1 - \alpha_1)I_{E1} - I_{CO1} \tag{3.90}$$

where α is the common-base current gain and I_{CO} is the collector-base reverse saturation current. The collector current of the *n-p-n* transistor, T_2, is given by

$$I_{C2} = \alpha_2 I_{E2} + I_{CO2} \tag{3.91}$$

Neglecting internal currents from the substrate, I_{B1} and I_{C2} may be equated to yield

$$(1 - \alpha_1)\, I_{E1} - I_{CO1} = \alpha_2 I_{E2} + I_{CO2} \tag{3.92}$$

Also, neglecting internal currents from the p-well, I_{E1} is equal to I_{E2}, so that

$$I_{E1} = \frac{I_{CO1} + I_{CO2}}{1 - (\alpha_1 + \alpha_2)} \tag{3.93}$$

Under normal CMOS circuit operation, the currents through T_1 and T_2 are reverse-biased junction currents and are therefore quite low. Under these conditions, the quantity $(\alpha_1 + \alpha_2)$ is considerably less than unity, because carrier losses resulting from surface leakage and recombination in the space-charge layer are a significant fraction of the total current and greatly reduce the transistor current gain. Consequently, the current drawn from the power supply is small under normal conditions. However, if the structure experiences a sufficiently large transient voltage because of noise or electrostatic discharge that causes increased junction-leakage currents, or if sufficient internal substrate or well current occurs due to avalanche, then I_E increases. And as I_E increases, the current gain α increases, since the transistor enters a more efficient region of operation where recombination currents play a minor role. As $(\alpha_1 + \alpha_2)$ approaches unity, the denominator of Eq. (3.93) approaches zero and forward breakdown, or switching occurs, as shown in Figure 3.33B. The switching action is regenerative, since the high switching current keeps the α values high. As a result, the gate is locked in a state that draws a large current from the supply and latch-up has taken place. It should be noted that after switching, Eq. (3.93) no longer holds, because all junctions are forward biased such that both T_1 and T_2 are saturated. If the current is limited to a safe value, turning the power supply off and then back on will eliminate the latch-up. Without protective circuits that limit the current, physical damage results.

More elaborate models, including parasitic resistances, are required for predictive analysis of latch-up for a given layout. A rigorous analysis requires a two-dimensional numerical analysis. Several methods, using both layout techniques and process techniques, have been developed to prevent and control latch-up. For example, the spacing between the p-well and the source of the PMOS transistor can be increased to decrease the gain of the lateral p-n-p parasitic BJT, thereby increasing the voltage required for latch-up. However, this approach is not conducive to high packing densities. Another method for reducing the gain of active parasitics is to place heavily doped guard rings and buried layers in strategic places to reduce minority-carrier lifetime and therefore the parasitic BJT current gains. For example, n^+ guard rings and buried layers may be placed with the n-wells. Also, the SOS/SOI process described in Section 3.6.6 eliminates the problem by breaking the four-region structure.

3.10 Speed Considerations

The transit time of a carrier between source and drain places a fundamental limit on the switching speed and high-frequency performance of a MOSFET. Assuming that the electric field in the channel is simply the drain-to-source voltage divided by the channel length, the transit time is

$$\tau = (L)^2/\mu V_{DS} \tag{3.94}$$

In practical circuits, however, switching speed and frequency response are limited not by transit time but by capacitance charge and discharge times.

For small-signal analog circuits, the approach is to use the small-signal MOSFET model shown in Figure 3.17 and to carry out the circuit analysis for the particular circuit in which the FET is used. For a digital MOS circuit, the main capacitance of interest is the effective load capacitance of the gate, C_L, where C_L is due to the input capacitance of the next gate, or gates, plus any parasitic capacitance. Speed, therefore, is related to the current-driving capability of the transistors.

For the gates discussed in Section 3.9, pull-up time is limited by the load capacitance and the charging current provided by the pull-up, or load, transistor. Pull-down switching times are also determined by the load capacitance and the primary discharge current path is through the driving, or pull-down, transistor. Since the current-driving ability of a transistor is inversely proportional to the Z value, NMOS ratioed-logic is characterized by asymmetric rise and fall times. For CMOS logic, however, rise and fall times may be made nearly equal.

Since the load capacitance is to some degree a function of voltage, an exact solution for the rise and fall times of a gate is not simple. However, insight into the speed limitations is obtained by an approximate method that provides a rough estimate of switching speeds. The pull-down time, t_{PD}, may be estimated as [5]

$$t_{\text{PD}} = C_L V_H / I_{\text{PD}} \tag{3.95}$$

where I_{PD} is an averaged pull-down current and V_H is the output voltage when the output state is high. Implicit in Eq. (3.95) is the assumption that the low-state output voltage is near zero. Likewise, for the pull-up time,

$$t_{\text{PU}} = C_L V_H / I_{\text{PU}} \tag{3.96}$$

where I_{PU} is the average pull-up current. The average currents through the load capacitor may be calculated by

$$I_{\text{average}} = \frac{\int_0^{V_H} I(V)dV}{\int_0^{V_H} dV} \tag{3.97}$$

As an example of this method, consider its application to the CMOS inverter. Assume the input voltage goes in a step-wise fashion from 0 V to V_{DD}. The p-channel load transistor is cut off, and the driving transistor is in saturation until the output voltage reaches $V_{DD} - V_{TD}$, where V_{TD} is the threshold voltage for the driver. Thereafter, the driver is in the triode region. Applying Eq. (3.97) yields

$$I_{\text{PD}} = \frac{\mu_n C_{OX}}{2Z_{\text{PD}}V_{DD}}\left[\int_{V_{DD}-V_{TD}}^{V_{DD}} (V_{DD} - V_{TD})^2\, dV + \int_0^{V_{DD}-V_{TD}} [2(V_{DD} - V_{TD})V - V^2]dV\right] \tag{3.98}$$

which yields

$$I_{PD} = \frac{\mu_n C_{OX}}{6Z_{PD}V_{DD}}(V_{DD} - V_{TD})^2\,(2V_{DD} + V_{TD}) \tag{3.99}$$

Consequently the pull-down time is approximated by the expression,

$$t_{\text{PD}} \simeq \frac{6Z_{PD}(V_{DD})^2 C_L}{\mu_n C_{OX}(V_{DD} - V_{TD})^2\,(2V_{DD} + V_{TD})} \tag{3.100}$$

The CMOS inverter pull-up response to a step input, found by the same technique, is

$$t_{\text{PU}} \simeq \frac{6Z_{PU}(V_{DD})^2 C_L}{\mu_p C_{OX}(V_{DD} + V_{TL})^2(2V_{DD} - V_{TL})} \tag{3.101}$$

where V_{TL} is the threshold voltage for the load, which is a negative number. Provided the load transistor has the same value of $\mu C_{OX}/Z$ and the threshold voltages for the two transistors are of equal absolute magnitude, then the pull-up time will be the same as the pull-down time for the CMOS inverter.

For the NMOS inverter with a depletion mode load, assuming a step input from 0 V to V_{DD}, the driving transistor is saturated until the output voltage is $V_{DD} - V_{TD}$, after which it enters the triode region. In contrast to the CMOS inverter, the load transistor also conducts current during the pull-down event. To a first approximation, however, the contribution of the load transistor to I_{PD} may be ignored because of the large L/W ratio for this device. In that case, the pull-down time for the NMOS load is also given approximately by Eq. (3.100). On the pull-up portion of the switching cycle, the load is saturated until the output voltage is $V_{DD} + V_{\mathrm{DEP}}$. Recall that V_{DEP} is negative. Applying equation (3.97) results in

$$t_{PU} = \frac{2Z_{PU}(V_{DD})^2 C_L}{\mu_n C_{OX}[(-V_{\mathrm{DEP}})^2 (V_{DD} + V_{\mathrm{DEP}}/3)]} \tag{3.102}$$

provided V_{DD} is greater than or equal to $-V_{\mathrm{DEP}}$, as is generally the case.

These equations have been developed assuming a constant value of C_L. Usually, however, the gate is driving another gate and C_L is not constant. In fact, referring to the discussion in Section 3.5 on MOSFET parasitics and the capacitances shown in Figure 3.15, the input capacitance for even a single transistor is a combination of several capacitors including C_{GD}, C_{GS}, and the junction capacitances. Furthermore, the values of these capacitances change with voltage. However, it is possible to develop an approximate value of C_L when the load is another gate.

Consider the case when the load is a simple two-transistor NMOS inverter. What is the capacitive load presented by this circuit? Since the input is to the gate of an NMOS transistor, the primary capacitances that will contribute to C_L are C_{GS} and C_{GD}. As a first approximation, it may be estimated that C_{GD} and C_{GS} are each half the value of C_{OX}A, where A is the gate area $A = WL$ [19]. Therefore, the approximate values are

$$C_{GD} \simeq C_{GS} \simeq (1/2)(\epsilon_{OX}/d_{OX})(WL) \tag{3.103}$$

The gate-to-source capacitance may be directly accounted for in determining an effective value of C_L. However, the gate-to-drain capacitance will be charged in one direction for one polarity of input voltage and in the opposite direction for the opposite polarity input. Including a factor of 2 for the Miller effect, the input capacitance associated with a single inverter, neglecting stray capacitance owing to path interconnects, is roughly estimated as

$$C_L \simeq C_{GS} + 2C_{GD} \simeq (1.5)(\epsilon_{OX}/d_{OX})(WL) \tag{3.104}$$

For a CMOS inverter with two transistors connected to the gate input, the value of C_L is approximately

$$C_L \simeq (1.5)(\epsilon_{OX}/d_{OX})[(WL)_n + (WL)_p] \tag{3.105}$$

where the subscripts n and p refer to the n- and p-channel devices respectively.

Note that the factor of 1.5 in Eqs. (3.104) and (3.105) will tend to overestimate C_L, and the resulting values for t_{PU} and t_{PD} will be conservative estimates. This is because of the rough nature of the approximation in Eq. (3.103). In the linear portion of the triode region, $C_{GS} \simeq C_{GD} \simeq (1/2)(WL)C_{OX}$ is a good approximation. But in the saturation region, $C_{GD} \simeq 0$ because the channel vanishes at the drain end. And it may be shown that $C_{GS} \simeq (2/3)(WL)C_{OX}$ in saturation [9]. Under this condition, C_{GS} and C_{GD} combine to give C_L somewhat smaller than $(WL)C_{OX}$.

The discussion in this section provides methods for estimating rise and fall times in basic inverters. Clearly, computer simulations based on more sophisticated models are required for precise transient analysis.

EXAMPLE 3.10

Consider a CMOS inverter, with V_{DD} equal to 5 V, whose transistors have identical geometry and oxide thickness, with $W = 12\ \mu m$, $L = 4\ \mu m$, and $d_{OX} = 400$ Å. Assume the electron mobility for the NMOS device is 1000 $cm^2/V \cdot sec$ and the hole mobility for the PMOS device is 400 $cm^2/V \cdot sec$. The threshold voltages are +1 V for the NMOS transistor and −1 V for the PMOS transistor. Estimate the pull-up and pull-down times if the load is an identical inverter. Compare the results with the carrier transit times.

Solution Based on Eq. (3.105), the load capacitance can be considered to be approximately,

$$C_L \simeq 3WLC_{OX}$$

where

$$C_{OX} = \frac{(3.9)(8.85 \times 10^{-14})}{(400 \times 10^{-8})}$$

$$C_{OX} = 8.63 \times 10^{-8}\ \text{F/cm}^2$$

Therefore,

$$C_L \simeq 0.12\ \text{pF}$$

Applying Eq. (3.100),

$$t_{PD} \simeq 0.4\ \text{nsec}$$

From Eq. (3.101),

$$t_{PU} \simeq 1\ \text{nsec}$$

The transit times, calculated from Eq. (3.94), are only 32 psec and 80 psec for the NMOS and PMOS devices, respectively, when the drain-to-source voltage is 5 V.

Clearly high-speed gates should use transistors having short channels and low operating voltages and should be based on high-mobility materials. However, the load capacitance presented by the circuit in which the transistor is embedded also plays a critical role.

BIBLIOGRAPHY

Further reading on the MOS system and field-effect devices can be found in the following:

Nicollian, E. H., and J. R. Brews. *MOS (Metal Oxide Semiconductor) Physics and Technology*. New York: Wiley, 1982.

Pierret, R. F. *Field Effect Devices.* Modular Series on Solid State Devices. Edited by R. F. Pierret and G. W. Neudeck, vol. 4. Reading, Mass.: Addison-Wesley, 1983.

Sze, S. M. *Physics of Semiconductor Devices.* 2nd ed. New York: Wiley, 1981.

For further reading on MOS integrated circuit techniques, the following are recommended.

Botchek, C. M. *VLSI: Basic MOS Engineering.* Saratoga, Calif.: Pacific Technical Group, 1983.

Ong, D. G. *Modern MOS Technology.* New York: McGraw-Hill, 1984.

REFERENCES

1. B. E. Deal, "Standardized Terminology for Oxide Charges Associated with Thermally Oxidized Silicon," *IEEE Transactions on Electron Devices,* ED-27 (1980), 606–608.
2. S. M. Sze, *Physics of Semiconductor Devices,* 2nd ed. (New York: Wiley, 1981).
3. E. H. Nicollian and J. R. Brews, *MOS (Metal Oxide Semiconductor) Physics and Technology* (New York: Wiley, 1982).
4. T. P. Chow and A. J. Steckl, "Refractory Metal Silicides: Thin Film Properties and Processing Technology," *IEEE Transactions on Electron Devices,* ED-30 (1983), 1480–1497.
5. D. J. Hamilton and W. G. Howard, *Basic Integrated Circuit Engineering* (New York: McGraw-Hill, 1975).
6. A. S. Grove, *Physics and Technology of Semiconductor Devices* (New York: Wiley, 1967).
7. R. A. Colclaser, *Microelectronics Processing and Device Design* (New York: Wiley, 1980).
8. R. F. Pierret, *Field Effect Devices,* Modular Series on Solid State Devices, eds. R. F. Pierret and G. W. Neudeck, vol. 4 (Reading, Mass.: Addison-Wesley, 1983).
9. A. B. Glaser and G. E. Subak-Sharpe, *Integrated Circuit Engineering* (Reading, Mass.: Addison-Wesley, 1977).
10. K. A. Salsburg, P. E. Cottrell, and E. M. Buturla, "FIELDAY—Finite Element Device Analysis," *Process and Device Simulation for MOS-VLSI Circuits,* eds. P.

Antognetti, D. A. Antoniadis, R. W. Dutton, W. G. Oldham (Boston: Martinus Nijhoff, 1983), pp. 582–619.
11. L. Wafford, "Silicon Gates Spur Linear CMOS to Bipolar Speeds, Hold Offsets Stable," *Electronics,* March 24, 1982, pp. 137–140.
12. J. R. Lineback, "Buried Oxide Marks Route to SOI Chips," *Electronics Week,* Oct. 1, 1984, pp. 11–12.
13. R. T. Gallagher, "SOI Method Can Be Used in Submicrometer IC's," *Electronics Week,* Nov. 19, 1984, pp. 28–29.
14. R. H. Dennard, F. H. Gaensslen, H. Yu, V. L. Rideout, E. Bassons, and A. R. LeBlanc, "Design of Ion-Implanted MOSFET's with Very Small Physical Dimensions," *IEEE Journal of Solid State Circuits,* SC-19 (1974), 256.
15. B. Hoeneisen and C. A. Mead, "Fundamental Limitations in Microelectronics—I. MOS Technology," *Solid State Electronics,* 15 (1972), 819–829.
16. Y. El-Mansy, "MOS Device and Technology Constraints in VLSI," *IEEE Transactions on Electron Devices,* ED-29 (1982), 567–573.
17. Texas Instruments, Inc., VLSI Laboratory, "Technology and Design Challenges of MOS VLSI," *IEEE Journal of Solid-State Circuits,* SC-17 (1982), 442–448.
18. R. C. Eden, A. R. Livingston, and B. M. Welch, "Integrated Circuits: The Case for Gallium Arsenide," *IEEE Spectrum,* 20 (Dec. 1983), 30–37.
19. A. Barna, *VHSIC Technologies and Tradeoffs* (New York: Wiley, 1981), Chap. 3.

PROBLEMS

1. Typically, 256K silicon MOSFET-based random access memories (RAMs) use gate oxide thicknesses in the range of 200–300 Å. Calculate the flat band voltage and threshold voltage, in the absence of ion implantation, associated with each of the gate materials listed in Table 3.1. Do so for a gate oxide thickness of 250 Å and for both *n*- and *p*-channel devices. Assume that the oxide charges combine to give a positive effective Q_f of 5×10^{10} cm^{-2} and that the substrate doping is 2×10^{16} cm^{-3}.

2. An *n*-channel silicon MOSFET has an oxide thickness, d_{OX}, equal to 800 Å and a threshold voltage, V_T, equal to 2 V. Using identical processes, another MOSFET is made with $d_{OX} = 400$ Å. For the second device, $V_T = 1$ V. What is the difference between the gate-metal and semiconductor work functions?

3. Consider a silicon *n*-channel MOSFET with a gate oxide thickness of 750 Å, a substrate doping of 10^{15} cm^{-3}, and a flat band voltage of -0.44 V. The source is grounded to the substrate, and the channel length and width are both 5 μm. Assume that the channel mobility is 1000 cm^2/V·s.

   a. Calculate the threshold voltage, V_T.
   b. Find the transconductance, g_m, in saturation if the gate and drain are tied to $+5$ V.
   c. If the gate is aluminum, what is the implied total effective oxide charge? Is there evidence for an ion implantation?

4. Show that Eq. (3.51), the MOS capacitor-versus-voltage equation, follows from Eq. (3.50).

5. Consider the scaling of an NMOS inverter with a depletion-mode load, as shown in Figure 3.28D. Assume that the enhancement-mode driver has a threshold voltage $V_T = 0.2V_{DD}$, and that the depletion-mode load has a threshold voltage $V_{DEP} = -0.6V_{DD}$. The substrate is characterized by a flat band voltage of -1 V. Scaling is to be carried out such that the oxide electric field is maintained at 3×10^6 V/cm. The minimum oxide thickness, as limited by quantum mechanical tunneling, is 50 Å. What is the corresponding mimimum channel length, L, for each transistor, if L is twice the maximum width of the space-charge layer? Your answer will indicate a scaling limit for silicon MOSFETs.

6. Consider the depletion-mode load NMOS inverter of Ex. 3.9. Calculate and plot the power dissipated in the gate as a function of V_{IN}. Assume the oxide thickness is 400 Å, and take $Z_{PU} = 4$, $Z_{PD} = \frac{1}{2}$, and $\mu_n = 1000$ cm^2/V·s.

7. Consider the CMOS inverter of Ex. 3.10. Calculate and plot the power dissipated in the gate as a function of V_{IN}.

8. Consider an NMOS inverter with a saturated enhancement load, as shown in Figure 3.28B. Let $V_{DD} = 5$ V and $V_T = 1$ V.

   a. What is the relationship between output and input voltage when the input voltage is small—less than the threshold voltage?
   b. Calculate the relationship between output and input voltages when the output voltage is somewhat larger, such that both devices are saturated.
   c. For even larger values of V_{IN}, the driver is in the triode region. Again solve for V_{OUT} versus the input voltage.
   d. Obtain a plot of V_{OUT} versus V_{IN} for the cases of $Z_{PU}/Z_{PD} = 2$ and $Z_{PU}/Z_{PD} = 8$.

9. The p-channel loads in Figure 3P.1 have $V_T = -1$ V, and the n-channel drivers have $V_T = +1$ V.

   a. What logic function does the CMOS gate perform?
   b. Calculate Y, in volts, if $X_1 = X_2 = 1.5$ V. All transistors have equal values of $\mu(W/L)C_{OX}$.
   c. Calculate Y if both inputs are at 3 V.

10. The five-stage ring oscillator shown in Figure 3P.2 is built of silicon CMOS inverters. The transistors are characterized by $|V_T| = 1$ V, $L = 4$ μm,

Figure 3P.1 A CMOS logic gate.

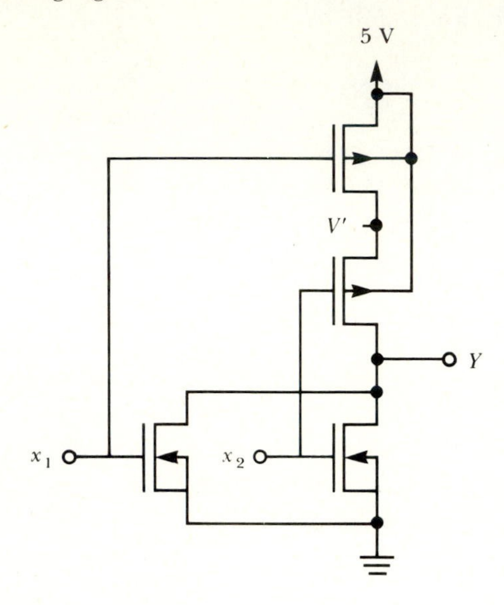

Figure 3P.2 A CMOS ring oscillator.

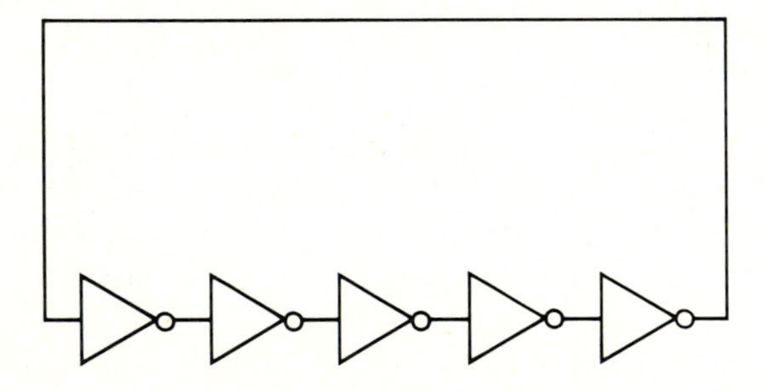

$W = 12\mu$m, and oxide thickness $d = 400$ Å. The power supply is 5 V. Assume electron and hole mobilities of 1000 and 400 $cm^2/v \cdot s$, respectively.

a. Calculate the pull-up and pull-down times of each gate.
b. Estimate the oscillation frequency.

11. Use Eqs. (3.95) and (3.96) to derive approximate expressions for pull-up and pull-down times for an NMOS inverter with a saturated enhancement-mode load.

4 Bipolar Integrated Circuit Fundamentals

4.1 Introduction

The first generation of integrated circuits relied predominantly on the bipolar junction transistor (BJT), and bipolar circuits are still responsible for the majority of small-scale integrated standard logic families. However, the standard BJT process proved to be less suitable than the metal-oxide-semiconductor (MOS) process for large-scale integration because MOS technology offered higher packing densities and lower power dissipation. Consequently, the larger market share has shifted from bipolar integrated circuits to MOS integrated circuits. For example, in 1975 the U.S. market for static random access memory (RAM) chips was 57% bipolar, 37% *n*-channel MOS (NMOS), and 6% complementary MOS (CMOS) [1]. By 1984, the market percentages reversed to approximately 11% bipolar, 36% NMOS, and 52% CMOS [2]. Furthermore, many of the modern developments in digital integrated circuits have been MOS based. For the past several years MOS has accounted for nearly 90% of the microprocessor market, and for VLSI dynamic RAM chips—64K and larger—the MOS percentage is even higher.

Nevertheless, the BJT remains an important design option for integrated circuitry for several reasons. Compared to the MOS field-effect transistor (MOSFET), the BJT has a much larger transconductance. In other words, a larger change in output current results for a given change in input voltage. This has important consequences for circuit performance. Also, advances in bipolar fabrication techniques, including trench-isolation and self-alignment methods, have combined to make bipolar technology more amenable to high packing densities. Furthermore, high-density bipolar layouts have been developed that use merged-transistor structures such as integrated-injection-logic (I^2L) and integrated Schottky logic (ISL). Bipolar VLSI options, which use advanced fabrication and layout methods, have been developed by several companies.

Bipolar circuits offer superior analog performance because of lower noise and higher voltage gains, and bipolar digital circuits provide large output currents, which may be used to rapidly charge and discharge capacitive loads.

In digital applications where speed is of primary importance, and where high packing density and low power dissipation may be sacrificed for the sake of speed, high speed bipolar circuits such as emitter-coupled-logic (ECL) have been a traditional option. Moreover, mixed-process chips, which combine NMOS with bipolar (BiMOS), and CMOS with bipolar (BiCMOS), allow designers to take advantage of positive features of both technologies. Consequently, bipolar integrated circuits continue to add an important dimension to design decisions regarding density/performance and power/performance trade-offs.

There are several physical distinctions between BJTs and MOSFETs [3]. Current in the MOSFET results from majority carrier flow from source to drain, but current in the BJT results from minority carrier flow through the base from emitter to collector. At moderate current levels, the FET current is primarily a drift current, whereas the BJT minority carrier current results from diffusion, although for BJTs with graded base doping, drift current is also present. The MOSFET gate electrode is capacitively coupled to the control circuit, but the BJT is resistively coupled. Also, the MOSFET current is controlled by the gate electric field, whereas the BJT current is controlled by the base current.

A comparison of speed performance between the two device types cannot be entirely separated from circuit considerations. The MOSFET has the inherent advantage of being a majority carrier device, so that minority carrier storage delay times and diffusion capacitances are absent. However, because the BJT has a larger transconductance, it can supply a large amount of charge with a small change in input voltage [3]. In ring oscillator circuits, where the switching devices are driving identical devices, BJT speeds and MOSFET speeds are comparable. However, the BJT is better for driving large capacitive loads because of the high transconductance. To fully realize that speed advantage, the bipolar circuit must be nonsaturating.

This chapter also discusses the fabrication of circuit elements other than bipolar transistors, including resistors and diodes. Although the use of such elements is not limited to bipolar circuits, they tend to be utilized more in bipolar integrated circuits than in MOS, and so have been included in this chapter.

4.2 The BJT: Basic Properties and Models

4.2.1 Structural and Bias Considerations

A cross section of one of several versions of the integrated circuit BJT is shown in Figure 4.1. Unlike the MOSFET, the BJT is not self-isolating. In the layout shown, electrical isolation is provided at the sidewalls by a collar of SiO_2 and at the bottom by a reverse-biased junction between the p-substrate and the n-type collector layer.

Figure 4.1 An example of an integrated circuit BJT structure (simplified cross section).

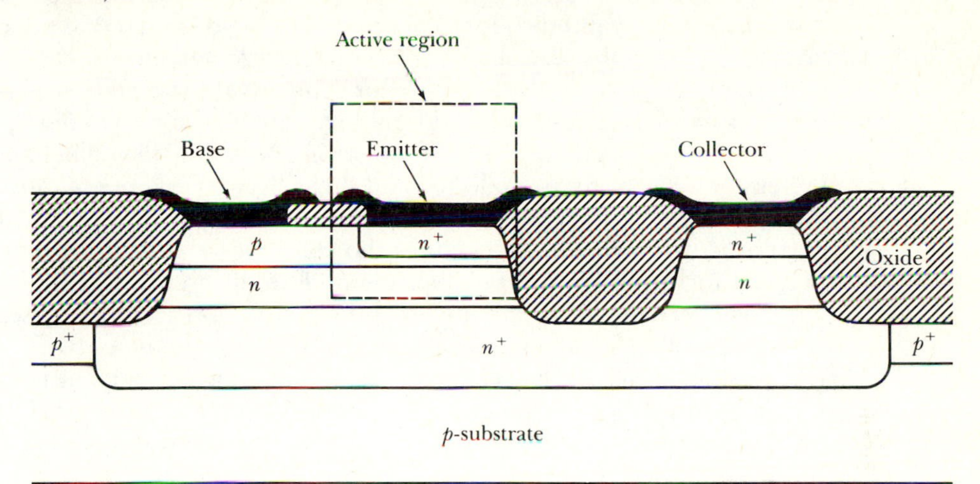

The p^+ region under the oxide is a channel-stop that raises the doping level of the substrate. By preventing surface inversion of the p-type substrate, which would connect adjacent n^+ layers, this serves a role similar to the channel-stop implant shown in the MOSFET fabrication sequence in Figure 3.18. However, alternative isolation techniques exist, as discussed in Section 4.4. The active region of the device—that is, the portion where the transistor action takes place—is shown inside the dotted box in Figure 4.1, and the remaining structure is either for the purpose of contacting the active region or for device isolation.

Under normal operation, the collector of an n-p-n BJT is biased positively relative to the emitter. These two terminals are connected to back-to-back p-n junctions, as are the source and drain terminals of the MOSFET. In further similarity to the n-channel MOSFET, a third control terminal enables conduction from one n-type region to the other through the intervening p-type material. In contrast to the gate of a MOSFET, however, the base terminal of a BJT has a direct electrical connection to the p-type semiconductor.

Consider the case when the base-emitter junction is forward biased and the base-collector junction is reverse biased with, for example, 5 V on the collector, 0.7 V on the base, and a grounded emitter. Current flow in the base-emitter junction consists mainly of electrons that are emitted into the p-type base from the heavily doped n-type emitter. Most of these injected electrons impinge on the space-charge layer of the collector-base junction because of a special geometric design feature of the BJT, namely that the collector and emitter are physically separated by a broad, very thin sheet of p-type material. The concentration gradient of electrons in the base causes an electron diffusion current across the base toward the collector junction. From the reverse-biased collector-base junction's view, these electrons are minority carriers, and the space-charge electric field is such that the electrons are swept across the junction. As a result, a large current flows through the reverse-biased junction because the junction is supplied with a large supply of minority carriers from a nearby source, namely the n^+ emitter.

In a well-designed *n-p-n* BJT operating with this bias scheme, most of the emitted electrons are collected and the collector current is almost as large as the emitter current. Consequently, by Kirchhoff's current law, the base current is much smaller than either the collector or emitter current, since it is equal to the difference between them. The difference in emitter current and collector current is due mainly to back injection from the base into the emitter and to a small amount of recombination in the base.

Furthermore, a relatively large change in collector current is associated with a relatively small change in base current, such that the ratio $\Delta I_C/\Delta I_B$ is much greater than unity. Therefore, the BJT is often thought of as a current-gain device. It is emphasized that this gain is due directly to the geometric structure of the device, namely the thin base. If the base width were not small compared to the carrier diffusion length, the emitted electrons would recombine before reaching the collector. As a result, the structure would consist simply of two back-to-back diodes and the active transistor action would be lost.

Hole and electron distributions for the four possible bias states of the BJT, corresponding to forward- or reverse-biased collector-base and emitter-base junctions, are shown in Figure 4.2. In this figure, the notation n_p refers to the electron concentration in *p*-type material and the notation n_{p0} refers to the equilibrium concentration. Although the majority concentrations are not specifically shown, they may be deduced from the doping profile if it is also assumed that the excess hole and electron concentrations at a given point are nearly equal, that is, if

$$\delta p \simeq \delta n \tag{4.1}$$

The quasineutrality condition expressed in Eq. (4.1) is good provided the times of interest are small compared to the dielectric relaxation time, τ_{dr}, where

$$\tau_{\mathrm{dr}} \equiv \epsilon/\sigma \tag{4.2}$$

For silicon with typical doping concentrations, the dielectric relaxation time is picoseconds or less and the quasineutrality approximation is sound. However, for nearly intrinsic or semiinsulating materials and very short time frames, Eq. (4.1) does not hold.

All four biasing conditions shown in Figure 4.2—forward active, inverse

Figure 4.2 Bias states for a BJT (not to scale). A. Forward active. B. Inverse (or reverse) active. C. Cut off. D. Saturation.

A

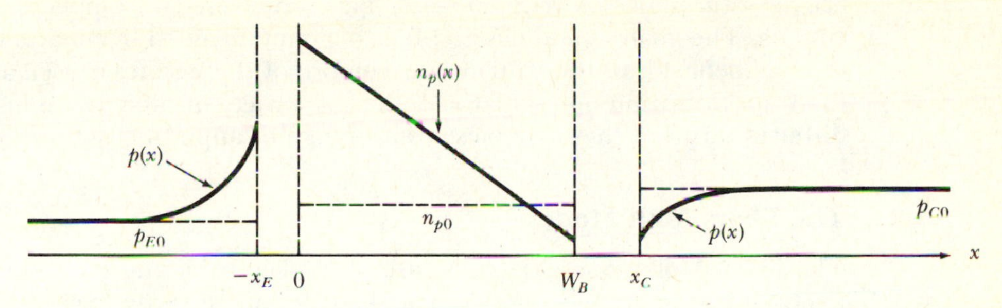

B

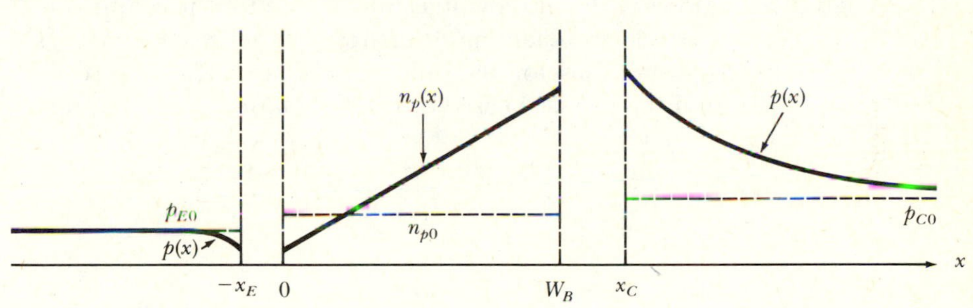

C

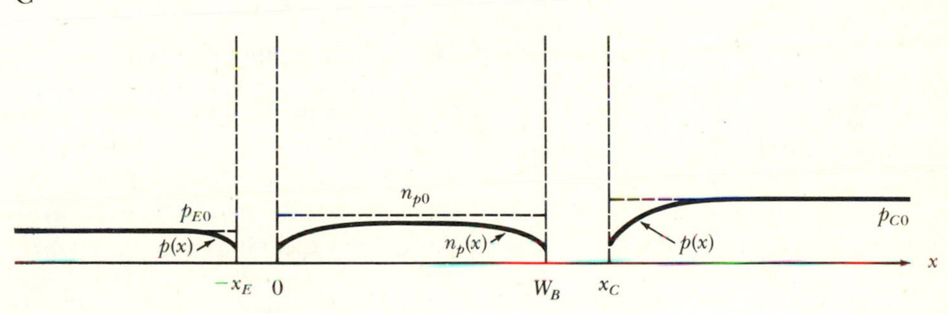

D

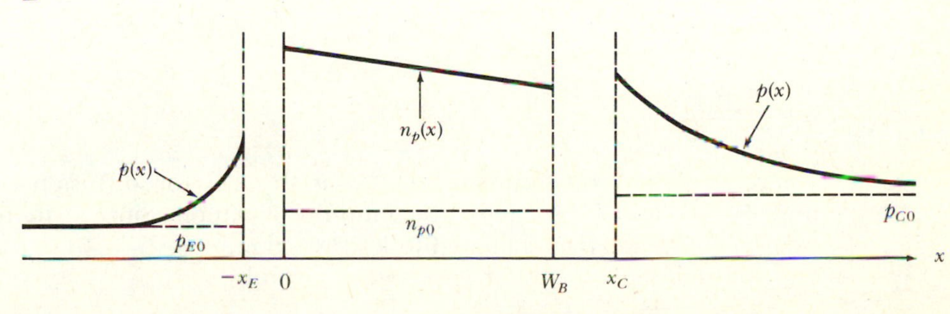

active, saturation, and cut off—may be encountered in bipolar integrated circuits. The analysis and design of such circuits is aided by models that predict the behavior of the BJT in one or more of these states as well as during the transition from one state to another. A variety of models are in use that differ in terms of their complexity and range of applicability.

4.2.2 The Ebers-Moll Model

The Ebers-Moll model [4] can be used for all four biasing states and is relatively simple to apply to circuit analysis. It follows directly from the solution to the electron- and hole-continuity equations and is quite useful, subject to the limitations of dc and low injection. Therefore it is appropriate for first-order dc analysis in obtaining estimates for the three transistor currents I_E, I_C, and I_B, as well as for the voltages V_{BE} and V_{BC}. There follows a brief discussion of the physical basis for the Ebers-Moll model, using an *n-p-n* BJT as a case in point.

The emitter and collector currents may be expressed as

$$-I_E = I_n(0) + I_p(0) \simeq I_n(0) + I_p(-x_E) \tag{4.3}$$

and

$$-I_C = I_n(W_B) + I_p(W_B) \simeq I_n(W_B) + I_p(x_C) \tag{4.4}$$

where $I_n(x)$ and $I_p(x)$ are the hole and electron currents, respectively, and the x-axis is labeled as in Figure 4.2. The minus signs in Eqs. (4.3) and (4.4) result from the convention of taking the direction of positive current flow in the *n-p-n* transistor to be from collector to emitter, such that in forward active bias both I_E and I_C are positive quantities. Neglecting any electric field in the base, the electron current in the base results from diffusion and, in the one-dimensional approximation, is expressed as

$$I_n(0) \simeq qD_nA \left.\frac{d(\delta n)}{dx}\right|_{x=0} \tag{4.5}$$

and

$$I_n(W_B) \simeq qD_nA \left.\frac{d(\delta n)}{dx}\right|_{x=W_B} \tag{4.6}$$

where A is the cross-sectional area, D_n is the electron diffusion coefficient and δn is found by solving the continuity equation. Since drift terms are neglected, the analysis is limited to low injection.

Assuming the recombination rate is given by

$$R = \delta n/\tau_n \tag{4.7}$$

where τ_n is the electron lifetime, then the steady-state electron continuity equation in the base is

$$0 = D_n \frac{d^2\delta n}{dx^2} - \frac{\delta n}{\tau_n} \tag{4.8}$$

The general solution is

$$\delta n(x) = C_1 \exp(x/L_n) + C_2 \exp(-x/L_n) \tag{4.9}$$

where L_n is the electron diffusion length given by

$$L_n = (D_n \tau_n)^{1/2} \tag{4.10}$$

and the constants C_1 and C_2 are determined by the boundary conditions for δn, namely,

$$\delta n(0) = n_{p0}\left[\exp\left(\frac{qV_{BE}}{kT}\right) - 1\right] \tag{4.11}$$

and

$$\delta n(W_B) = n_{p0}\left[\exp\left(\frac{qV_{BC}}{kT}\right) - 1\right] \tag{4.12}$$

The electron contributions to the emitter and collector currents are found from Eqs. (4.5) and (4.6) to be

$$I_n(0) = qA(D_n/L_n)(C_1 - C_2) \tag{4.13}$$

for the contribution to the emitter current and

$$I_n(W_B) = qA(D_n/L_n)[C_1 \exp(W_B/L_n) - C_2 \exp(-W_B/L_n)] \tag{4.14}$$

for the contribution to the collector current. Solving for C_1 and C_2 yields

$$C_1 = \frac{n_{p0}[\exp(qV_{BC}/kT) - 1] - n_{p0}\exp(-W_B/L_n)[\exp(qV_{BE}/kT) - 1]}{\exp(W_B/L_n) - \exp(-W_B/L_n)} \tag{4.15}$$

and

$$C_2 = \frac{n_{p0}\exp(W_B/L_n)[\exp(qV_{BE}/kT) - 1] - n_{p0}[\exp(qV_{BC}/kT) - 1]}{\exp(W_B/L_n) - \exp(-W_B/L_n)} \tag{4.16}$$

The hole contributions to the emitter and collector currents are also found by solving the continuity equation and calculating the diffusion current. Solving for the excess hole concentration $\delta p(x)$ in the emitter and collector results in a sum of exponentials as in Eq. (4.9). In this case, the junction boundary conditions are

$$\delta p(-x_E) = p_{E0}\left[\exp\left(\frac{qV_{BE}}{kT}\right) - 1\right] \tag{4.17}$$

and

$$\delta p(x_C) = p_{C0}\left[\exp\left(\frac{qV_{BC}}{kT}\right) - 1\right] \tag{4.18}$$

where p_{E0} and p_{C0} are the equilibrium minority carrier concentrations in the emitter and collector, respectively. The boundary conditions at the emitter and collector ohmic contacts require,

$$\delta p(\text{contact}) \simeq 0 \tag{4.19}$$

The resulting currents are given by the expressions,

$$I_p(-x_E) = -qAD_p \left.\frac{d(\delta p)}{dx}\right|_{x=-x_E} \tag{4.20}$$

and

$$I_p(x_C) = -qAD_p \left.\frac{d(\delta p)}{dx}\right|_{x=x_C} \tag{4.21}$$

A general solution for I_p yields hyperbolic functions as in Eqs. (4.13) through (4.16). However, it is convenient to write the approximate expressions,

$$I_p(-x_E) \simeq -qAD_{pE}\left(\frac{p_{E0}}{{}^*L_{pE}}\right)\left[\exp\left(\frac{qV_{BE}}{kT}\right) - 1\right] \tag{4.22}$$

and

$$I_p(x_C) \simeq qAD_{pC}\left(\frac{p_{C0}}{{}^*L_{pC}}\right)\left[\exp\left(\frac{qV_{BC}}{kT}\right) - 1\right] \tag{4.23}$$

where $^*L_{pE}$ and $^*L_{pC}$ are effective diffusion lengths in the emitter and collector, respectively. For cases where the ohmic contacts are far from the junctions, the effective diffusion lengths are the diffusion lengths as usually defined.

When the contacts are quite close to the junction, as is more likely the case in bipolar integrated circuits, the effective diffusion lengths are approximately the physical distances from the contacts to the junctions.

Adding the contributions to the currents yields the Ebers-Moll equations, which can be written as

$$I_E = I_{ES}\left[\exp\left(\frac{qV_{BE}}{kT}\right) - 1\right] - \alpha_R I_{CS}\left[\exp\left(\frac{qV_{BC}}{kT}\right) - 1\right] \tag{4.24}$$

and

$$I_C = \alpha_F I_{ES}\left[\exp\left(\frac{qV_{BE}}{kT}\right) - 1\right] - I_{CS}\left[\exp\left(\frac{qV_{BC}}{kT}\right) - 1\right] \tag{4.25}$$

where the parameters I_{ES} and I_{CS} are given by the expressions

$$I_{ES} = qA\left[\frac{D_{pE}p_{E0}}{*L_{PE}} + \frac{D_n n_{p0}}{L_n}\operatorname{ctnh}(W_B/L_n)\right] \tag{4.26}$$

$$I_{CS} = qA\left[\frac{D_{pC}p_{C0}}{*L_{pC}} + \frac{D_n n_{P0}}{L_n}\operatorname{ctnh}(W_B/L_n)\right] \tag{4.27}$$

where ctnh is the hyperbolic cotangent function. Also,

$$\alpha_R I_{CS} = \alpha_F I_{ES} = qA\left[\frac{D_n n_{p0}}{L_n \sinh(W_B/L_n)}\right] \tag{4.28}$$

where sinh is the hyperbolic sine function.

In most cases, the base width is considerably smaller than the diffusion length and the expressions are simplified by keeping only the first term in the power-series expansion of the hyperbolic functions, that is,

$$\sinh(W_B/L_n) \simeq W_B/L_n \tag{4.29}$$

and

$$\operatorname{ctnh}(W_B/L_n) \simeq L_n/W_B \tag{4.30}$$

The Ebers-Moll equivalent circuit corresponding to Eqs. (4.24) and (4.25) is shown in Figure 4.3. The model predicts most of the basic features of the transistor family of curves, such as the common emitter family of curves shown in Figure 4.4. An attraction of the model is its intuitive appeal. The

Figure 4.3 Ebers-Moll circuit model for an *n-p-n* transistor.

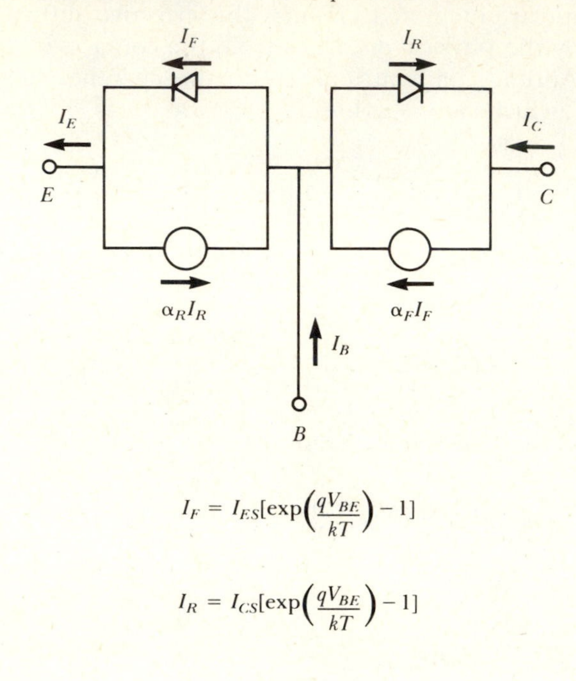

Figure 4.4 Common emitter family of curves for an *n-p-n* BJT.

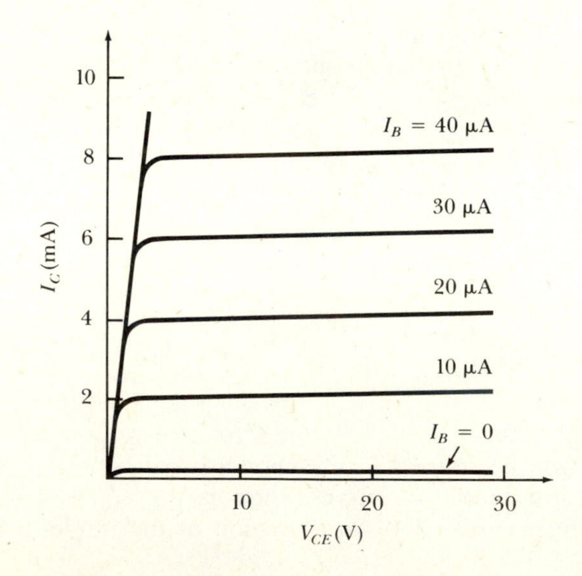

diodes in the circuit model represent the base-emitter junction and the collector-base junction; and the currents through them, I_F and I_R, are given by the usual diode equations, with I_{ES} and I_{CS} as the effective saturation currents. However, a transistor is more than back-to-back diodes, and the current sources represent the coupling between the two junctions that occurs because of the very thin base. Indeed, if the base width is much larger than the diffusion length, L_n, the Ebers-Moll model leads simply to two back-to-back diodes.

EXAMPLE 4.1

Consider the transistor biased with a floating collector as shown in Figure 4.5. Use the Ebers-Moll model to derive an expression for I. Will V_{BC} be zero, positive, or negative?

Solution From Eqs. (4.24) and (4.25),

$$I = I_E = I_{ES}\left[\exp\left(\frac{qV_{BE}}{kT}\right) - 1\right] - \alpha_R I_{CS}\left[\exp\left(\frac{qV_{BC}}{kT}\right) - 1\right]$$

and

$$0 = I_C = \alpha_F I_{ES}\left[\exp\left(\frac{qV_{BE}}{kT}\right) - 1\right] - I_{CS}\left[\exp\left(\frac{qV_{BC}}{kT}\right) - 1\right]$$

Note that for the last equation to equal zero, V_{BC} must be positive. Solving for I_{CS} from the second expression and inserting that into the first expression, yields

$$I = I_{EO}\left[\exp\left(\frac{qV_{BE}}{kT}\right) - 1\right]$$

where $I_{EO} = I_{ES}(1 - \alpha_F\alpha_R)$.

Figure 4.5 Illustration for Example 4.1.

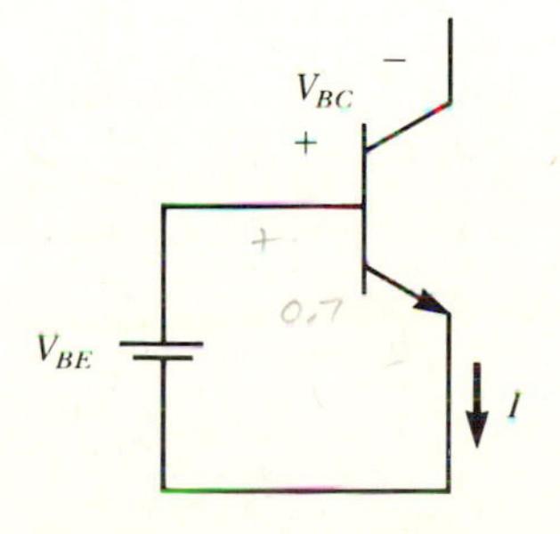

The development of the Ebers-Moll model also provides insight into the relationship between device performance and transistor doping and geometry. For example, the forward-active bias base transport factor, α_T, is defined as

$$\alpha_T = I_n(W_B)/I_n(0) \tag{4.31}$$

From Eqs. (4.13) and (4.14), this ratio may be expressed as

$$\alpha_T \simeq \text{sech}(W_B/L_n) \approx 1 - 1/2(W_B/L_n)^2 \tag{4.32}$$

where sech is the hyperbolic secant. The terms involving V_{BC} have been dropped since V_{BC} is negative for forward-active bias conditions. Also, the emitter-injection efficiency is defined in forward-active bias as

$$\gamma = I_n(0)/I_E \tag{4.33}$$

It is left as an exercise to show from Eqs. (4.3), (4.13), and (4.22) that

$$\gamma \simeq \left[1 + \frac{D_{pE}p_{E0}W_B}{D_n n_{p0}{}^* L_{pE}}\right]^{-1} \tag{4.34}$$

When the base transport factor and the emitter-injection efficiency have been found, then the dc values of the common-base current gain, α, and the common emitter-current gain, β, can be found from their basic definitions,

$$\alpha = I_C/I_E = \alpha_T\gamma \tag{4.35}$$

and

$$\beta = I_C/I_B = \alpha/(1 - \alpha) \tag{4.36}$$

EXAMPLE 4.2

Consider an *n-p-n* BJT doped such that the base minority-carrier concentration is 20 times larger than the emitter minority-carrier concentration, and the minority-carrier mobility in the base is 5 times larger than the minority-carrier mobility in the emitter. The base diffusion length is 5 times the effective diffusion length in the emitter and 25 times the base width. Calculate the β of the transistor.

Solution Applying Eq. (4.34), the emitter-injection efficiency is

$$\gamma = [1 + (1/5)(1/20)(5/25)]^{-1} = 0.9980$$

The base-transport factor, from Eq. (4.32), is

$$\alpha_T = [(1/2)[\exp(.04) + \exp(-0.04)]]^{-1} = 0.9992$$

Therefore,

$$\alpha = (0.9980)(0.9992) = 0.9972$$

and

$$\beta = (0.9972)/(1 - 0.9972) \simeq 356$$

It is clear from Ex. 4.2 that a high-gain transistor should have a thin base and a heavily doped emitter relative to the base. However, the expressions in Ex. 4.2 provide only semiquantitative information about the expected current gains of the transistor, since the effect of doping gradients and base electric fields, for example, have not been included. Also several parasitic effects are not included. Figure 4.6 illustrates a modification of the Ebers-Moll model in which capacitors and resistors have been added to model the effects of parasitic resistances and stored charges. Another current source also has been added to model the Early effect arising from base-width modulation [5].

4.2.3 A Charge-Control Model

Since the Ebers-Moll model, as presented in the previous section and shown in Figure 4.3, was based on the solution of the steady-state continuity equations, it does not directly apply to time varying conditions, although the ca-

Figure 4.6 Modified Ebers-Moll model.

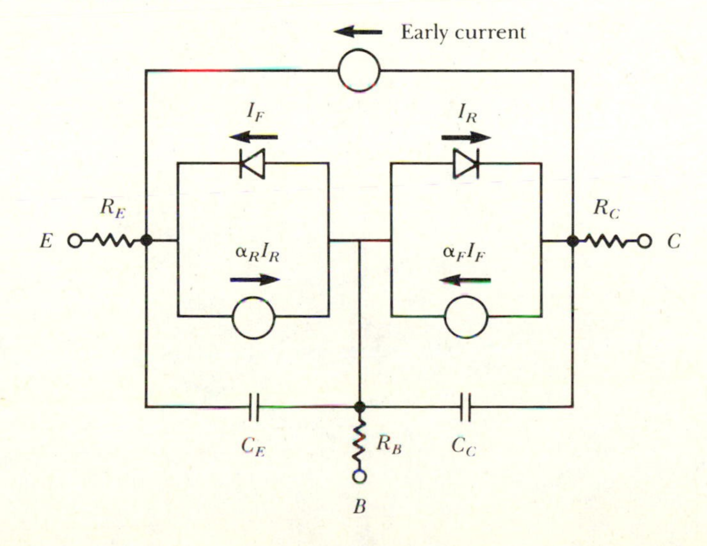

pacitors added in Figure 4.6 do model a frequency-dependent response. The charge-control model is an alternative useful approach for ac and transient analysis in which the terminal currents are related to charge stored in the BJT.

Consider the charge associated with excess minority-carriers stored in the base for a given bias situation. The base region is actually quasi-neutral since the number of excess electrons is approximately equal to the number of excess holes. However, the excess minority carriers represent charge that must be added to, or removed from, the base as the bias of the transistor changes. Let Q_N be the minority-carrier charge stored when the transistor is "normally" biased, that is, when it is in the forward-active bias state. Also let Q_I be the stored charge in the base when the transistor is in the inverse-active bias state. Then for the general bias case, the total base charge, Q_T, shown in Figure 4.7 may be written as

$$Q_T = Q_N + Q_I \tag{4.37}$$

The base current supports both recombination in the base and changes in the amount of base charge. For forward-active bias, therefore, $i_B(t)$ is related to $Q_N(t)$ by the expression,

$$i_B = Q_N/\tau_n + dQ_N/dt \tag{4.38}$$

The collector current, $i_C(t)$, is the charge Q_N divided by the mean time required to collect it, τ_t, which is also the average time it takes a carrier to cross the base region. Therefore, i_C is expressed as

$$i_C = Q_N/\tau_t \tag{4.39}$$

By Kirchhoff's current law, the emitter current is then

Figure 4.7 Charge storage in the base of a BJT.

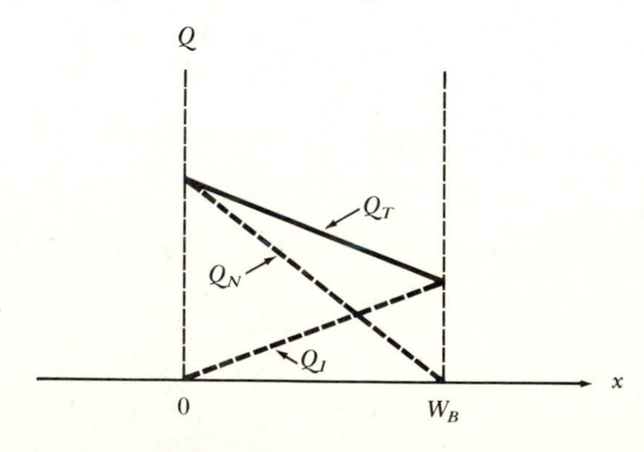

$$i_E = Q_N/\tau_t + Q_N/\tau_n + dQ_N/dt \tag{4.40}$$

For the case of inverse bias, the equations are similar but Q_I replaces Q_N and the collector plays the role of the emitter. For the general bias case, where both Q_N and Q_I are present, the terminal currents are expressed as

$$i_B = (Q_N + Q_I)/\tau_n + dQ_N/dt + dQ_I/dt \tag{4.41}$$

$$i_C = Q_N/\tau_t - Q_I(1/\tau_n + 1/\tau_t) - dQ_I/dt \tag{4.42}$$

$$i_E = Q_N(1/\tau_n + 1/\tau_t) - Q_I/\tau_t + dQ_N/dt \tag{4.43}$$

EXAMPLE 4.3

Using the charge-control expressions, calculate the turn-on time of a common emitter *n-p-n* transistor driven by a base-current step from 0 to I_B, where I_B is sufficient to drive the transistor to saturation.

Solution Consider the transition through the forward-bias region until the collector current reaches its maximum value, I_{Csat}, at the verge of saturation. Taking the Laplace transform of Eq. (4.38),

$$I_B(s)/s = Q_N(s)/\tau_n + sQ_N(s)$$

Solving for Q_N,

$$Q_N(s) = \frac{I_B}{s(s + 1/\tau_n)}$$

and

$$Q_N(t) = I_B\tau_n[1 - \exp(-t/\tau_n)]$$

The collector current follows Q_N according to Eq. (4.39) until it reaches its maximum value. Therefore the turn-on time, t_R, can be calculated from the relationship

$$I_{Csat} = Q_N(t_R)/\tau_t$$

or

$$I_{Csat} = (I_B\tau_n/\tau_t)[1 - \exp(-t_R/\tau_n)]$$

Consequently,

$$t_R = \tau_n \ln\left[\frac{1}{1 - (I_{Csat}\tau_t/I_B\tau_n)}\right]$$

Since it may be shown that [6]

$$\beta = \tau_n/\tau_t$$

the turn-on time may also be written as

$$t_R = \tau_n \ln\left[\frac{1}{1 - (I_{C\mathrm{sat}}/\beta I_B)}\right]$$

Faster turn-on times may be achieved by using materials with small values of τ_n or by making βI_B large compared to $I_{C\mathrm{sat}}$. However, a price is paid in either case. Small values of τ_n lead to smaller values of the base-transport factor and hence to smaller values of β, as well as to larger junction leakage currents. And large values of βI_B compared to $I_{C\mathrm{sat}}$ lead to overdriving or heavy saturation and consequently to long turn-off times.

The turn-on conditions for i_C and Q_t resulting from a step increase in base current from 0 to I_B are illustrated in Figure 4.8. When the base current is turned off, the base charge decreases exponentially with time constant τ_n from its maximum value of $I_B\tau_n$. However, the transistor remains saturated until the base charge reaches Q_S, which is the value of stored charge at the onset of saturation. During this time, which is referred to as the storage delay time, the collector current remains approximately constant. The quantity Q_S may be expressed as

$$Q_S = I_{C\mathrm{sat}}\tau_t \tag{4.44}$$

Therefore, the storage delay time, t_{sd}, is found from the expression

$$I_{C\mathrm{sat}}\tau_t = I_B\tau_n \exp(-t_{\mathrm{sd}}/\tau_n) \tag{4.45}$$

Solving for t_{sd} yields the expression

$$t_{\mathrm{sd}} = \tau_n \ln(\beta I_B/I_{C\mathrm{sat}}) \tag{4.46}$$

After the storage delay period, the base-stored charge and the collector current continue to decrease exponentially with time constant τ_n. Turn-off times may be decreased by providing a negative base current during the turn-off to aid in sweeping out stored charge from the base.

More sophisticated charge-control models also include the effect of the charging and discharging of junction capacitances, of minority carriers injected into the emitter and collector, and of parasitic resistances.

Figure 4.8 BJT switching: Currents and stored charge.

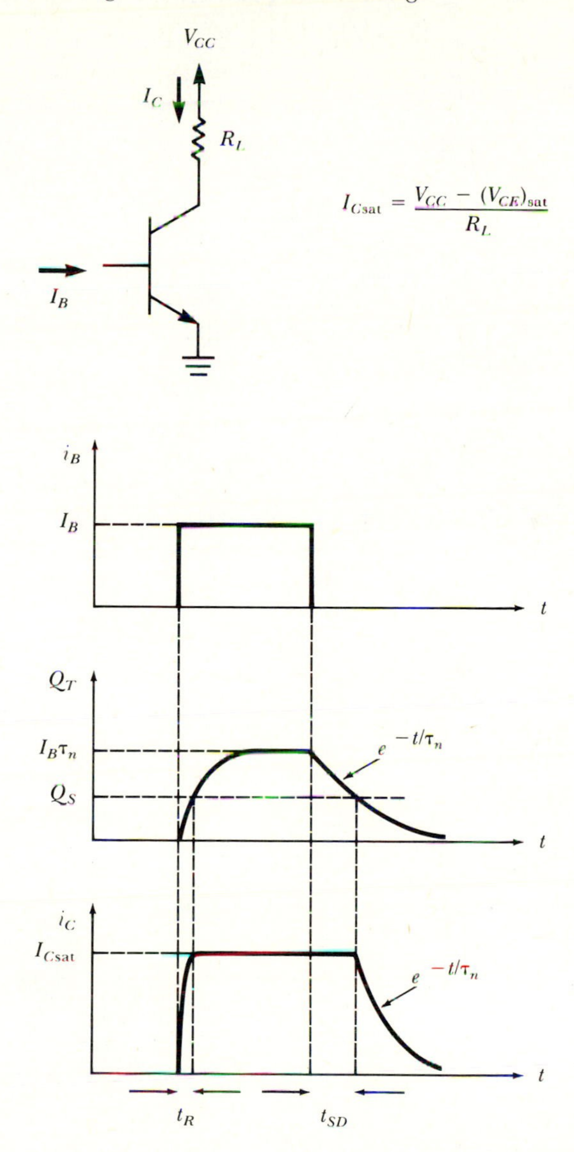

4.2.4 The Small-Signal Hybrid-Pi Model

For small-signal modeling, the circuit model may be linearized. A commonly used circuit for the forward-active bias region is the hybrid-pi, or Giacoletto, model shown in Figure 4.9 [7]. In this model, the resistors r_{CC}, r_{EE}, and r_{BB} represent the parasitic resistances associated with the collector, emitter, and base, respectively. The capacitance C_{BE} is due in part to the junction capacitance at the base-emitter junction and in part to the diffusion capacitance associated with injected minority carriers. The quantity C_{BC} represents the junction capacitance associated with the base-collector junction, and C_{CS} represents the capacitance between collector and substrate. The resistance r_{BE} represents the linearized small-signal resistance of the base-emitter input junction and is given by

$$r_{BE} = (kT/qI_C)\beta \tag{4.47}$$

where I_C is the collector current at the bias point. The dependence of i_c on the input signal is modeled by the voltage-controlled current source, which has a value of $g_m v'_{be}$, where the transconductance is given by

$$g_m = qI_C/kT \tag{4.48}$$

and v'_{be} is the small-signal voltage across r_{BE}. The effects of base width modulation are accounted for by resistors r_{BC} and r_{CE}.

The physical nature of the various parasitics in the context of the integrated circuit BJT is discussed further in Section 4.3.

Figure 4.9 The hybrid-pi (Giacoletto) small-signal BJT model.

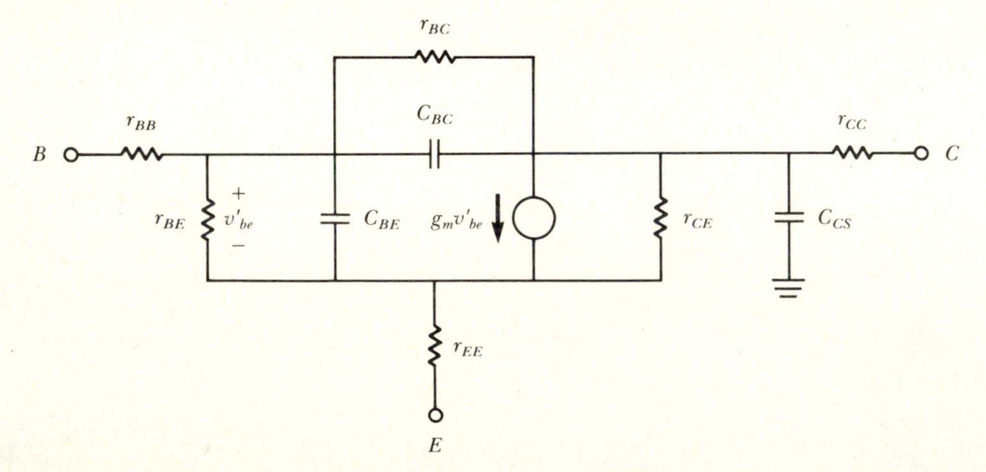

4.2.5 Gummel-Poon Model

The Gummel-Poon model is a sophisticated charge-control model suitable for computer simulation. It relates device-terminal characteristics to the base charge via charge-integral relationships [8]. The model, as available in modified form on the SPICE simulation program, requires up to 40 input parameters [9]. It accounts for base-width modulation, charge storage, lifetime dependence on current, β dependence on current, and some high-injection effects. With a simplified set of input parameters, the SPICE simulation reduces to the Ebers-Moll model [10].

4.2.6 Emitter Crowding

Perhaps the major deficiency of the models discussed in this section is that they are based on one-dimensional solutions to the semiconductor equations. Therefore, two- and three-dimensional effects like emitter crowding are not directly taken into account.

The base-current flow, referring to Figure 4.1, is transverse and consequently a transverse voltage drop exists across the base. Therefore, V_{BE} is less at the far end of the base than at the near end, so that emitter injection at the far end is curtailed. This situation is referred to as emitter crowding, and results in β degradation.

Another source of β degradation at large currents is the high-injection effect, which occurs when the excess minority-carrier concentration exceeds the equilibrium majority-carrier concentration. Under these conditions, the emitter injection efficiency decreases and consequently β is reduced. Whether emitter crowding or high injection occurs first depends on the transistor design. When the device is designed such that the two effects occur at about the same current level, then high-current β degradation begins when the collector current reaches the value [11],

$$I_C = (kT/q)(h\sigma_B)(2\beta)^{1/2} \tag{4.49}$$

where h is the emitter periphery and σ_B is the base conductivity. To reduce emitter crowding effects, emitters are elongated, rather than square, to achieve a higher value of h for a given area. Emitter crowding effects may be included in circuit models by allowing r_{BB} to increase with current.

4.2.7 Breakdown

With sufficiently high V_{CE}, the current begins to increase sharply either because of junction breakdown or punch-through. Junction breakdown may be due to either tunneling or avalanche breakdown; the former dominates for breakdown voltages of less than 5 V, and the latter dominates for breakdown voltages of more than about 8 V. For the collector-base junction, avalanche breakdown is usually the mechanism. Alternatively, BJT breakdown may be due to punch-through that occurs when the space-charge layers of the two junctions meet. Whether junction breakdown or punch-through occurs first depends on the doping and on the base width.

Collector-base breakdown effects can be included in the circuit model by adding, between the collector and the base, a current source with a value of $(M - 1)I_C$, where M is the avalanche multiplication factor given by [12]

$$M = [1 - (V_{CB}/V_{CBO})^n]^{-1} \tag{4.50}$$

where V_{CBO} is the breakdown voltage of the collector-base junction with the emitter terminal open, and n is between 2 and 4 for silicon.

4.3 BJT Parasitics

Several parasitic elements are an inherent part of the integrated circuit bipolar transistor, as shown in Figure 4.10. These parasitic resistors and capacitors all have deleterious effects on performance in that they may reduce small-signal gain, frequency response, and switching speed. Steps can be taken to reduce the parasitics, but interesting trade-offs are involved:

1. Increasing the base doping can reduce r_{BB}. However, this action reduces the emitter efficiency and therefore reduces β.
2. Increasing the collector doping can reduce r_{CC}. But this action causes the reverse breakdown voltage of the *c-b* junction, V_{CBO}, to decrease. Also base-modulation effects would increase since more of the space-charge layer at the *c-b* junction would extend into the base. The buried n^+ region shown in Figure 4.10 is quite an effective shunt and is vital to BJT design to reduce r_{CC} to acceptable levels. However, it does give rise to an increase in C_{CS}.
3. Often, r_{EE} is small owing to the heavy doping in the emitter. However, in the case of extremely small contacts, contact resistance may be nonnegligible at high currents.
4. The quantity C_{BE} arises because of two sources of charge storage, that stored in the base and that stored in the *b-e* space-charge layer. In other words, C_{BE} is expressed as

$$C_{BE} = (C_{BE})_{\text{diff}} + (C_{BE})_{\text{scl}} \tag{4.51}$$

The space-charge layer's contribution is given by Eq. (3.56), repeated here for convenience,

Figure 4.10 BJT parasitics.

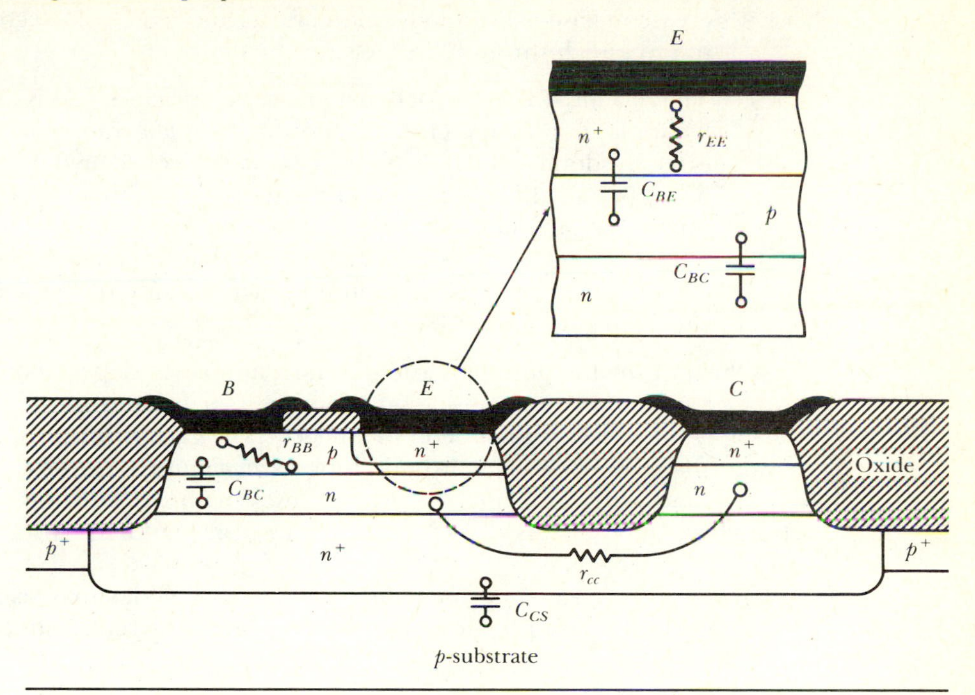

$$C_{\text{scl}} = \left[\frac{q\epsilon_S N_A N_D}{2(N_A + N_D)(V_{bi} - V)}\right]^{1/2} \tag{4.52}$$

where, for an n-p-n transistor, N_D is the effective donor concentration in the emitter, accounting for compensation, N_A is the effective acceptor concentration in the base, and C_{scl} is the capacitance per unit area. Because the doping profiles in the emitter and base are not constant with depth in the wafer, a quantitative assessment of the sidewall contribution to the capacitance requires either a weighted average or, more exactly, three-dimensional modeling. In active bias regions, however, Eq. (4.51) is often dominated by the first term because of the charge stored in the base. That term can be expressed as

$$(C_{BE})_{\text{diff}} = \frac{g_m(W_B)^2}{2D_B} \tag{4.53}$$

where D_B is the minority carrier's diffusion length in the base. Therefore, C_{BE} can be reduced by decreasing the width of the base.

The negative factors associated with base-width reduction are an increase in base-width-modulation effects and a possible decrease, owing to punch-through, in breakdown voltage.

5. The quantity C_{BC} is a space-charge layer capacitance and is also found by application of Eq. (4.52), where for an *n-p-n* transistor, N_A is the effective doping in the base and N_D is the collector doping. If the base doping and collector doping are decreased, then C_{BC} decreases, but an increase in r_{BB} and r_{CC} results.
6. The quantity C_{CS} is also a space-charge layer capacitance, which for a given layout, cannot be reduced much because of the necessity of the n^+ region that reduces r_{CC}.

Which of the foregoing parasitics is most important depends on the performance criterion being considered. Consider, for instance, the cut-off frequency, f_T, at which the common emitter-current gain is unity. Specifically, consider the common emitter hybrid-pi model in Figure 4.11, which is driven by a current source of value i_{in}. For purposes of this calculation, the base-width-modulation effects caused by r_{BC} and r_{CE} are ignored. Also, r_{EE} is assumed to be negligible.

Defining the frequency-dependent β as $i_{\text{out}}/i_{\text{in}}$, a nodal analysis results in the following equation for the frequency-dependent current gain, $\beta(s)$:

$$\beta = \frac{\beta_0(1 - sC_{BC}/g_m)}{1 + s\beta_0[C_1/g_m + r_{CC}(C_{BC}(1 - sC_{BC}/g_m) + C_2/\beta_0 + sC_1C_2/g_m)]} \tag{4.54}$$

where β_0 is the low-frequency value of β, equal to $g_m r_{BE}$, and C_1 and C_2 are given by

$$C_1 = C_{BE} + C_{BC} \tag{4.55}$$

and

$$C_2 = C_{BC} + C_{CS} \tag{4.56}$$

Figure 4.11 Hybrid-pi model for the calculation of cut-off frequency.

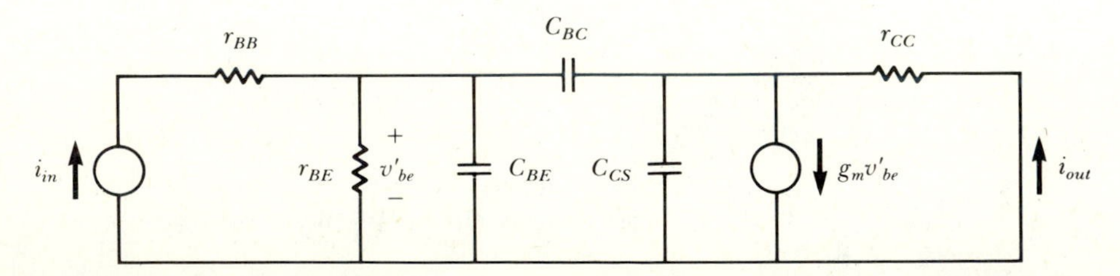

The unity-gain cut-off frequency, f_T, is found by setting the complex frequency, s, equal to $j\omega$ or $j2\pi f$ and solving for $|\beta| = 1$.

Under two conditions, which are often but not always met, Eq. (4.54) is simplified. Provided

$$sC_{BC} << g_m \tag{4.57}$$

and

$$C_{BC} >> C_2/\beta_0 + sC_1C_2/g_m \tag{4.58}$$

then the expression for $\beta(s)$ is

$$\beta(s) \simeq \frac{\beta_0}{1 + s\beta_0(C_1/g_m + r_{CC}C_{BC})} \tag{4.59}$$

Based on Eq. (4.59), the value of β falls 3 dB below the low frequency β_0 at frequency ω_β, where

$$(\omega_\beta)^{-1} = \beta_0(C_1/g_m + r_{CC}C_{BC}) \tag{4.60}$$

The unity-gain cut-off frequency for this one-pole approximation is

$$(\omega_T)^{-1} \simeq C_1/g_m + r_{CC}C_{BC} \tag{4.61}$$

Figure 4.12 illustrates the β frequency response and the relative positions of ω_β and ω_T.

It is instructive to expand Eq. (4.61) using Eqs. (4.51), (4.53), and (4.55), yielding the expression

$$(\omega_T)^{-1} \simeq \frac{W_B{}^2}{2D_B} + \frac{(C_{BE})_{\text{scl}} + (C_{BC})_{\text{scl}}}{g_m} + r_{CC}C_{BC} \tag{4.62}$$

Figure 4.12 BJT common-emitter short-circuit current gain as a function of frequency, for the hybrid-pi model one-pole approximation.

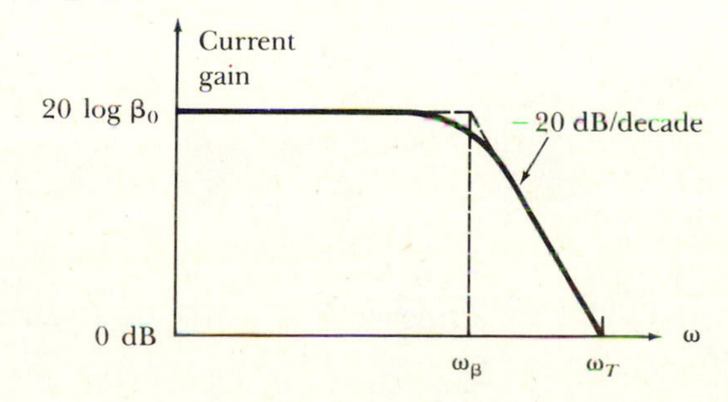

The first term in Eq. (4.62) represents the base transit time, τ_t, which is defined as

$$\tau_t = Q_N/I_C \tag{4.63}$$

under forward-active bias conditions where Q_N is the charge stored in the base. In the same sense that the source-to-drain transit time places a fundamental limit on the frequency response of an FET, the base transit time places a fundamental limit on the frequency response of a BJT. The second term shows the combined effect of the space-charge layer capacitors and the last term is due to the parasitic resistance, r_{CC}. Note that r_{BB} does not enter the expression for cut-off frequency because the input is driven by a current source. However, for the frequency response of the small-signal voltage gain, r_{BB} does become a factor.

EXAMPLE 4.4

Consider a bipolar transistor biased in the active region with $I_C = 0.25$ mA. The base width is 1 μm, and the parameters of the small-signal model are $\beta_0 = 150$, $r_{CC} = 100\ \Omega$, $r_{BB} = 200\ \Omega$, $C_{BC} = 0.5$ pF, $C_{CS} = 1.5$ pF, and $(C_{BE})_{\text{scl}} =$ 1 pF. The base is doped at 3×10^{17} cm^{-3}; consequently, the electron mobility in the p-type base is only 500 V/cm$^2\cdot$s. Find the cut-off frequency, f_T. In comparison, what would be the cut-off frequency if all parasitics were absent and performance was only limited by the base transit time?

Solution As a first approximation, Eq. (4.62) may be applied. Calculating D_B and g_m,

$$D_B = 0.0259(500) = 13 \text{ cm}^2/\text{s}$$

and

$$g_m = 0.25 \text{ mA}/25.9 \text{ mV} = 9.65 \times 10^{-3}\ \Omega^{-1}$$

Therefore,

$$(\omega_T)^{-1} \simeq \frac{(10^{-4})^2}{2(13)} + \frac{1.5 \times 10^{-12}}{9.65 \times 10^{-3}} + (100)(0.5 \times 10^{-12})$$

yielding

$$(\omega_T)^{-1} \simeq 3.85 \times 10^{-10} + 1.55 \times 10^{-10} + 0.5 \times 10^{-10}$$

So the cut-off frequency is

$$f_T \simeq 270 \text{ MHz}$$

For the numerical results of this example, however, the condition expressed in Eq. (4.58) is not satisfied and the cut-off frequency is more properly found from Eq. (4.54). Solving for the cut-off frequency from Eq. (4.54) yields 259 MHz.

If only the base-transit time limited f_T, the new value of the cut-off frequency would be

$$f_{T\text{max}} \simeq [(2\pi)(3.85 \times 10^{-10})]^{-1} = 413 \text{ MHz}$$

For this example, the base-transit time is the dominant term in the f_T expression and the r_{CC} term is the smallest term.

EXAMPLE 4.5

If the buried n^+ layer were not present, r_{CC} would be about an order of magnitude higher. Find the cut-off frequency for the bipolar transistor of Ex. 4.4 if r_{CC} is increased to 1000 Ω.

Solution Equation (4.62) yields

$$(\omega_T)^{-1} \simeq 3.85 \times 10^{-10} + 1.55 \times 10^{-10} + 5.0 \times 10^{-10}$$

and the r_{CC} term is now the largest. The corresponding f_T is

$$f_T \simeq [2\pi(1.04 \times 10^{-9})]^{-1} = 153 \text{ MHz}$$

The importance of the buried n^+ layer is clear.

In addition to the parasitic resistors and capacitors just discussed, there is a parasitic transistor associated with the integrated circuit BJT, as shown in Figure 4.13. Note that if the desired transistor, labeled T_1, is cut off or active, then the parasitic transistor, labeled T_2, is cut off. In this case, T_2 acts only as a passive parasitic through its junction capacitances. That effect has been considered in the previous discussion through the terms C_{CS} and C_{BC}. But when T_1 is saturated, T_2 can turn on and be an active parasitic. Under this condition, T_2 may provide a large current path to the substrate. One way of reducing this effect is to add a gold diffusion to the processing schedule. Gold impurities introduce recombination centers that reduce the minority carrier lifetime and lower the β of the *p-n-p* parasitic, which has a fairly large base width. The buried n^+ layer also has the effect of lowering the current gain of the *p-n-p* parasitic.

Figure 4.13 The parasitic *p-n-p* transistor. A. Cross-sectional view. B. Effective circuit.

A

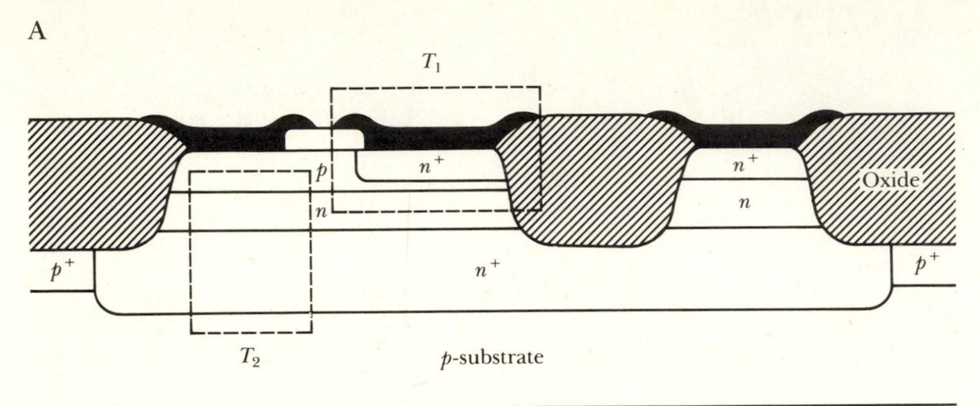

B

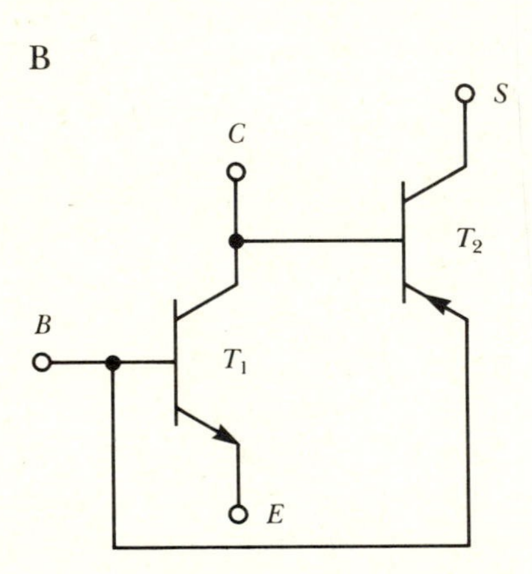

4.4 Examples of Isolation Techniques

4.4.1 Junction Isolation

Several bipolar integrated circuit processing methods have been developed to achieve device isolation. A relatively simple isolation scheme, and one that was widely used in the earlier years of bipolar integrated circuits, is the junction-isolated method. Cross-sectional and top views for a junction-isolated device are shown in Figure 4.14. In this figure, m is taken to be the minimum size of a lateral feature of the device, such as a window opening, and also the minimum clearance between adjacent features. The emitter elongation factor is k, consequently a minimum-area transistor corresponds to $k = 1$ [13].

In the sequence shown in Figure 4.15, the first processing step is to

Figure 4.14 The junction-isolated transistor. A. Top view. B. Cross-sectional view. Horizontal dimensions are compressed relative to vertical dimensions.

A

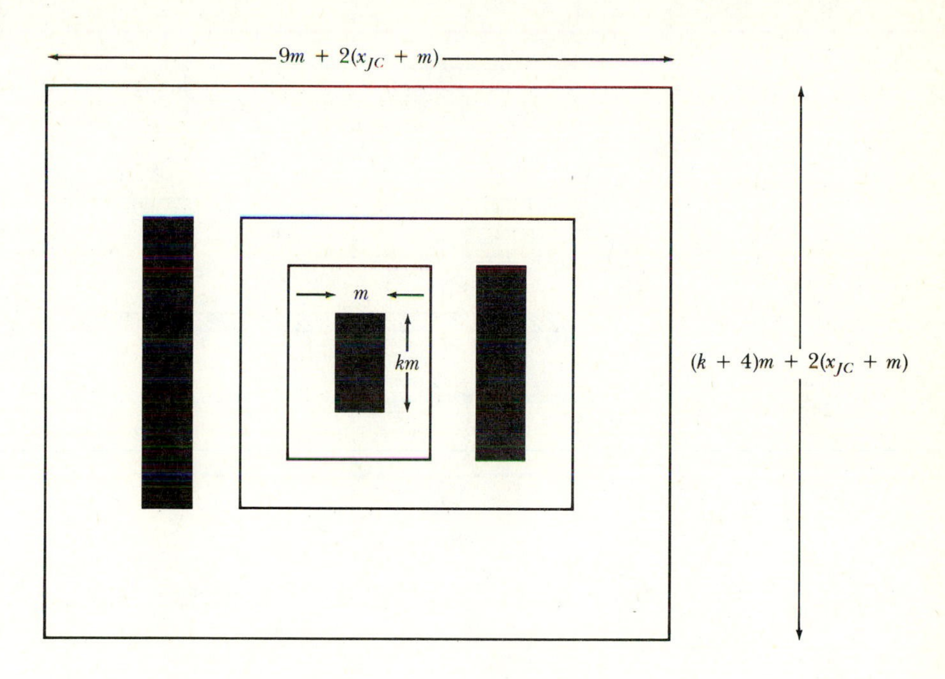

B

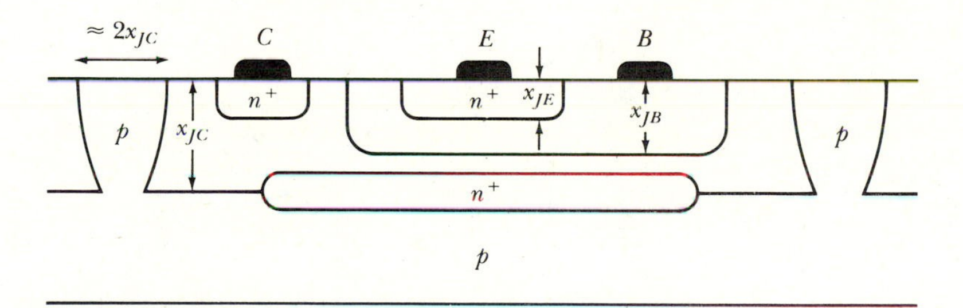

establish the n^+ buried layer in the p-type wafer prior to growth of an n-type epitaxial layer. Arsenic and antimony are good candidates as donors for the buried layer, since they have a high solid solubility in Si, which facilitates heavy doping, and because they have relatively small diffusion coefficients, which reduces subsequent impurity motion. Arsenic offers the advantage of a better fit to the silicon lattice with less resulting surface strain, yielding a better-quality epitaxial layer. Next, a deep boron isolation diffusion establishes the p-type isolation walls around the device. Note that device isolation

Figure 4.15 The junction-isolated BJT fabrication sequence. A. Buried layer. B. Epitaxial growth. C. Isolation diffusion. D. Base diffusion. Not shown: n^+ emitter diffusion, contact cuts, and metalization.

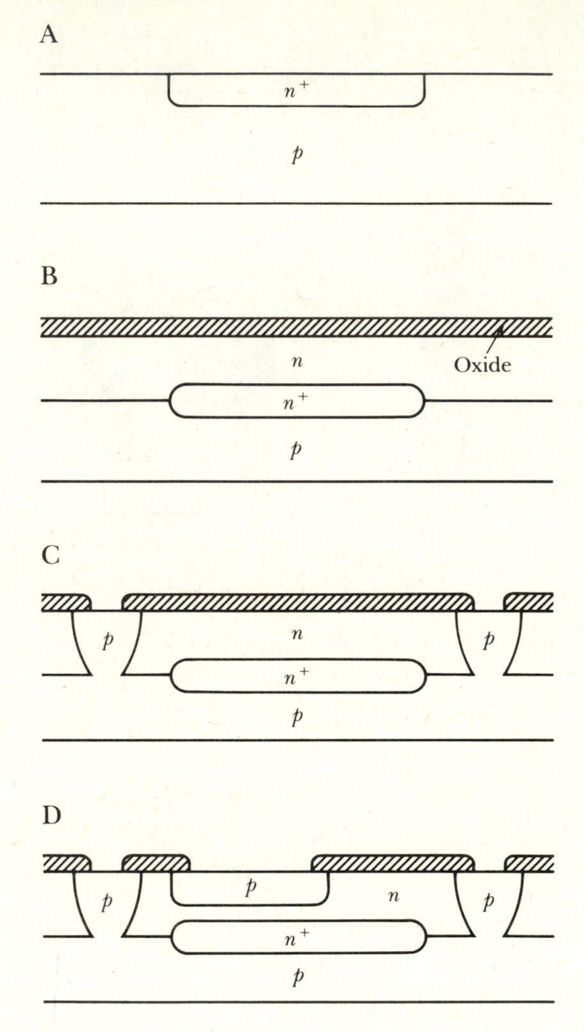

requires that the p-type substrate and isolation walls always be connected to the most negative voltage in the circuit. A second, shorter boron diffusion is used to establish the base. The base mask may also include the isolation regions to increase the surface concentration of the isolation walls. Recall that impurity redistribution during subsequent high-temperature processing steps causes a relative depletion of boron at the oxide-silicon interface, as shown in Figure 2.28.

An n-type diffusion establishes the n^+ emitter region and also provides an ohmic contact to the collector. In the standard process, aluminum is used as the metalization. Although a low-temperature sintering process provides an ohmic contact between the p-type base and aluminum, a nonohmic Schottky barrier results for lightly doped n-type material. However, with heavily doped n-type silicon, the barrier width is sufficiently narrow to allow tunneling, and an ohmic contact results. Phosphorus is most often used as the emitter dopant in the standard junction-isolated BJT process, but for very shallow junction devices, arsenic is used. Note that the base width, referring to Figure 4.14, is

$$W_B = x_{JB} - x_{JE} - W_{\text{scl}} \tag{4.64}$$

where W_{scl} is the total extension of the space-charge layers into the p-type base region.

EXAMPLE 4.6

If the minimum dimension, m, is 3 μm, and the emitter elongation factor, k, is 3, calculate the total area and the active area of the junction-isolated BJT, assuming an epitaxial-layer thickness of 5 μm.

Solution From Figure 4.14, the total area is

$$A_t \simeq [9m + 2(x_{JC} + m)][(k + 4)m + 2(x_{JC} + m)]$$

where x_{JC} is the thickness of the epitaxial layer. Taking the active area of the transistor to be the emitter area, the active area is

$$A_a \simeq [(k + 2)m][3m]$$

For this example, A_t is found to be 1591 μm^2 and A_a is 135 μm^2. Therefore the active area is only 8.5% of the total device area.

Example 4.6 shows that the junction-isolated BJT is not a compact structure. Furthermore, large device areas give rise to large parasitic capacitances that degrade speed performance. Alternative isolation schemes are aimed at reducing both area and capacitance.

4.4.2 Oxide Isolation

In Figure 4.1, the sidewall isolation is provided by a collar of silicon dioxide, which essentially eliminates the sidewall contribution to the collector-to-substrate capacitance, C_{CS}. The oxide isolation regions may be grown by the localized oxidation of silicon (LOCOS) process described in Chapter 3. When LOCOS is combined with trench etching in the epitaxial layer, as shown in

Figure 4.16, a highly planar structure results. A thermal oxidation converts the silicon in the grooves into SiO_2, with a resulting increase in volume as described by Eq. (2.55). The isolation oxide is grown such that the top of the oxide and the top of the silicon surface are in the same plane. This technique is sometimes referred to as isoplanar oxide-isolation [13]. The cross-sectional view shown in Figure 4.16 is in simplified form, in that the effects of lateral oxide growth (the bird's beak effect) are not shown.

The oxide isolation method offers an area reduction because the collector-metal stripe and the base stripe can overlap the sidewall oxide. Furthermore, the collector n^+ contact and the p-type base can extend to the isolation region. The result is a reduction in total device area of somewhat greater than 50%. In addition to an increased packing density, lower parasitic capacitances result from the smaller area. The emitter may be walled by the oxide on two sides, as in Figure 4.16, or on three sides, as in Figure 4.1.

EXAMPLE 4.7

Consider an oxide isolated BJT, as shown in Figure 4.1, that has an area of 700 μm^2. If the substrate doping is 10^{16} cm^{-3}, calculate C_{CS} at 5 V reverse bias between collector and substrate.

Solution Neglecting the sidewall contributions to C_{CS}, the collector-to-substrate capacitance is due to the space-charge layer capacitance at that junction. The collector-substrate junction in Figure 4.1 is one sided because of the n^+ region. Under these conditions, Eq. (4.52) becomes

$$C_{scl} \simeq \left[\frac{q\epsilon_S N_S}{2(V_{bi} + 5)}\right]^{1/2}$$

where N_S is the substrate doping. Calculating the built-in voltage, as in Ex. 3.7,

Figure 4.16 Isoplanar oxide isolation. A. Trench etching. B. Planarized isolation (idealized).

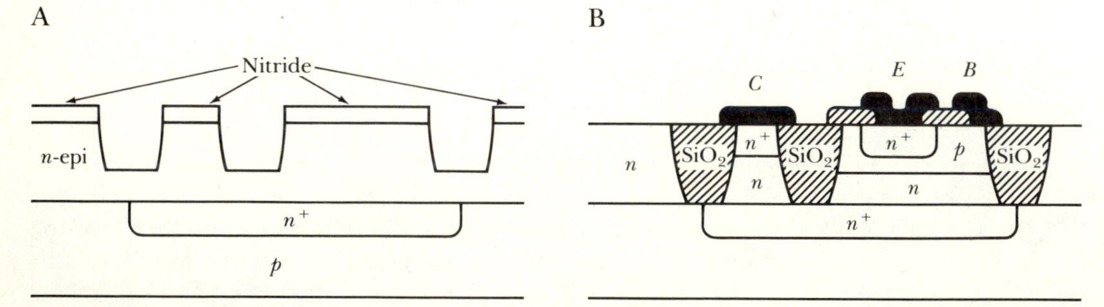

$$qV_{bi} \simeq E_G - (E_F - E_V)$$

where

$$(E_F - E_V) = kT\ln(N_V/N_A)$$

$$(E_F - E_V) = (0.0259)\ln\frac{1.04 \times 10^{19}}{10^{16}}$$

$$(E_F - E_V) = 0.18 \text{ eV}$$

Therefore,

$$V_{bi} = 1.12 - 0.18 = 0.94 \text{ V}$$

The space-charge layer's capacitance is,

$$C_{scl} = \left[\frac{(1.6 \times 10^{-19})(11.9)(8.85 \times 10^{-14})(10^{16})}{2(5.94)}\right]^{1/2}$$

$$C_{scl} = 1.2 \times 10^{-8} \text{ F/cm}^2$$

Therefore,

$$C_{CS} = (1.2 \times 10^{-8})(700 \times 10^{-8}) = 0.08 \text{ pF}$$

4.4.3 Collector Isolation

The collector-diffusion isolation (CDI) method has been available for many years but has not received wide acceptance. However, the advent of VLSI has sparked renewed interest in the CDI approach for certain applica-ions because of its relatively compact structure and simple processing. It is included here as another example of a solution to the bipolar isolation problem.

A cross-sectional view of one version of a CDI transistor is shown in Figure 4.17. The first patterning step establishes an n^+ buried layer in a p-type substrate. The next step is to grow a p-type epitaxial layer, which will serve as the base material. The diffusion of the n^+ collector contact reaches through the epitaxial layer to the buried layer and provides an n-type collar around the device. A separate n^+ diffusion or implant establishes the emitter. In all, only five masks are required for the CDI process, with a single metalization layer, and the transistor is appreciably smaller than the junction-isolated BJT, although not as small as oxide isolated devices. Since the collector is heavily doped, the inverse β is large. A negative feature of the

Figure 4.17 The collector-diffusion-isolated BJT.

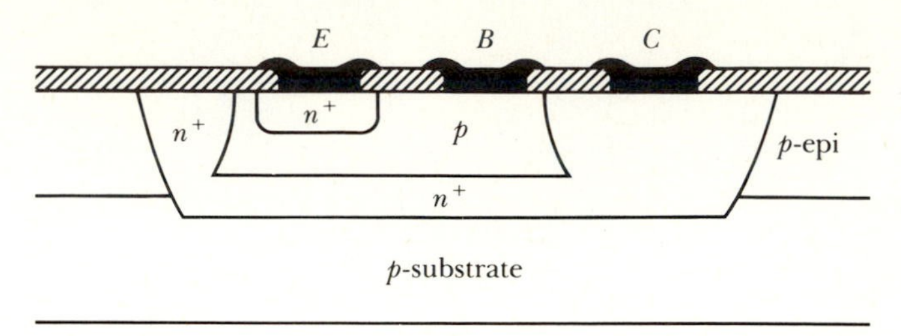

CDI approach is that the process is less versatile in terms of incorporating other device types such as *p-n-p* transistors and Schottky barrier diodes.

Another version of collector isolation omits the buried layer and uses ion implantation to implant an *n*-type collector region directly into the *p*-type substrate. The base region is formed by subsequent *p*-type implantation that is surrounded on all sides by the collector implant. Then the emitter and the collector contact are formed by an n^+ implant. Again isolation is achieved by reverse biasing the *p*-type substrate relative to the *n*-type collector region in which the BJT resides.

4.4.4 Deep-Groove Isolation

Deep-groove, or trench, dielectric isolation techniques have been developed to achieve VLSI densities by further reducing the area associated with the isolation regions. An example of this technique, which represents a rich blend of the processing methods discussed in Chapter 2, is the U-groove isolation technology developed at Hitachi and shown in Figure 4.18 [14]. The initial wafer processing is similar to the standard process, with n^+ buried layers being established in a *p*-substrate prior to the growth of an *n*-type epitaxial layer. Processing of the isolation regions begins with the formation of an SiO_2 layer and an Si_3N_4 overlayer.

Openings in the silicon nitride layer, which will define the isolation regions, are created by nitride etching. This step is followed by a wet isotropic oxide etch, which provides an Si_3N_4 overhang, as shown in Figure 4.18A. An anisotropic wet silicon etch, as described in Section 2.11, forms a well with tapered sides, as shown by the dotted lines in Figure 4.18A. These tapered sides reduce the surface-step problem. The silicon etch is continued with an anisotropic dry silicon etch that forms vertical sidewalls, as shown in Figure 4.18B. A boron ion implant provides a channel-stop that assures isolation between adjacent n^+ layers.

Next the grooves are oxidized and covered with a silicon nitride film to provide dielectric isolation between neighboring devices. At this point isola-

Figure 4.18 A deep-groove isolation technique. A. Nitride etch, oxide etch, and anisotropic wet silicon etch. B. Anisotropic dry silicon etch and channel-stop implant. C. Dielectric isolation and CVD polysilicon. D. Planarization and passivation.
SOURCE: From A. Hayasaka, Y. Tamaki, M. Kawamura, K. Ogiue, and S. Ohwaki, "U-Groove Isolation Technique for High Speed Bipolar VLSI's," *IEEE IEDM Technical Digest* (1982), 62–65. © 1982 IEEE.

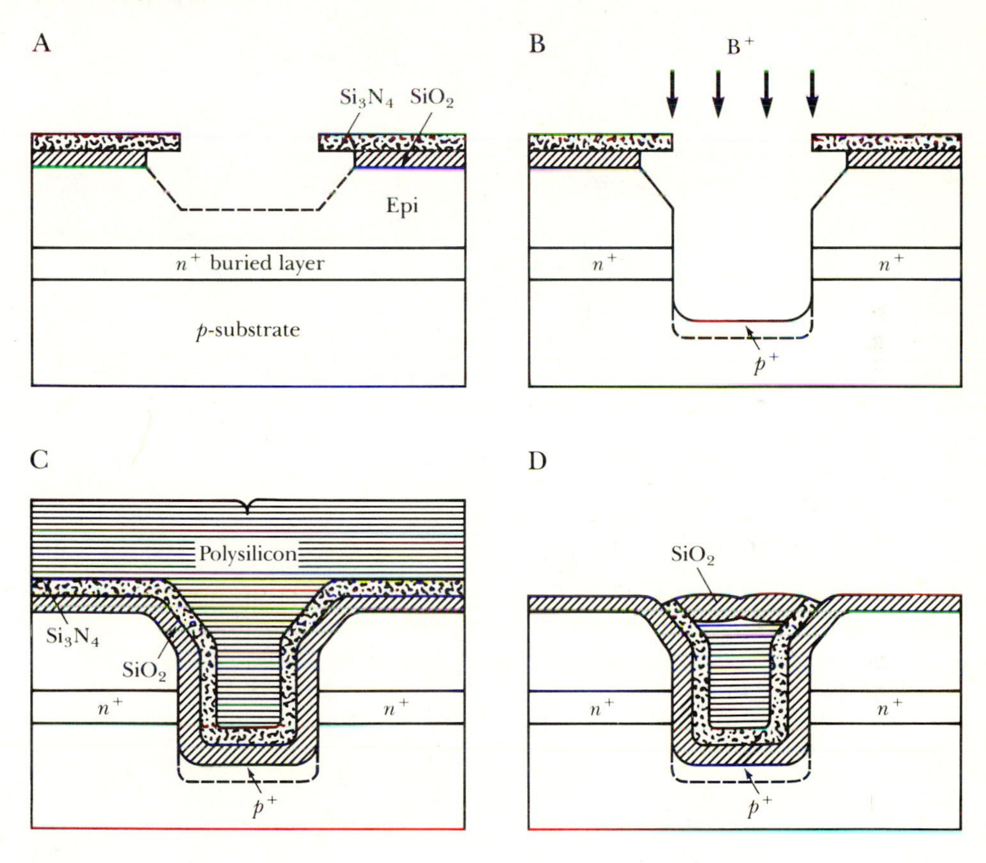

tion has been achieved, but the structure is highly nonplanar. Planarization is achieved by depositing and partially etching a chemical vapor deposition (CVD) polysilicon layer, as shown in parts C and D of Figure 4.18. Finally, the polysilicon surface is passivated with an oxide layer.

Using lithography of 1-μm resolution, this deep groove method achieves a 2.5-μm-wide isolation region. With the LOCOS method, the same lithographic resolution results in a 7-μm-wide isolation region, partly because of the bird's beak effect. A cross-sectional comparison of the two isolation methods is shown in Figure 4.19.

Figure 4.19 A comparison of deep-groove (A) and LOCOS (B) isolation methods. For 1 μm lithography, $x = 1$ μm, and $x + 2y + 2z = 7$ μm.
SOURCE: Adapted from A. Hayasaka, Y. Tamaki, M. Kawamura, K. Ogiue, and S. Ohwaki, "U-Groove Isolation Technique for High Speed Bipolar VLSI's," *IEEE IEDM Technical Digest* (1982), 62–65. © 1982 IEEE.

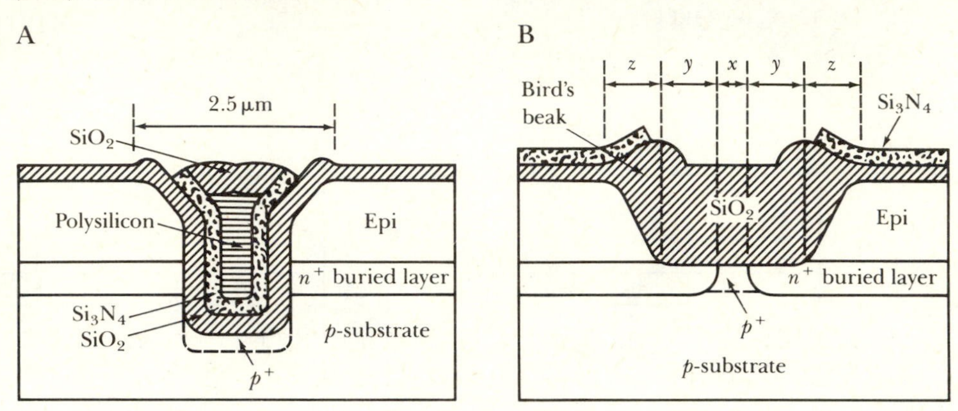

4.5 Variations on the Integrated Circuit BJT

4.5.1 The Schottky Clamp

High-speed BJT switching circuits require some means of limiting charge storage in the base, and a commonly used technique is a Schottky barrier diode (SBD) clamp between the BJT base and collector as shown in Figure 4.20. The current-voltage characteristics of the SBD typically show a forward voltage drop of approximately 0.4 V, which is lower than that of the base-collector *p-n* junction under forward-bias conditions with appreciable current flow. Therefore, if the transistor is turned on sufficiently hard to drive it toward saturation, the SBD's shunting action clamps the collector-base voltage at approximately 0.4 V. As a result, the charge stored in the base is not increased much beyond the case for normal active bias. Furthermore, the SBD is predominantly a majority carrier device, so there is negligible charge storage in the diode. Consequently, the storage delay time associated with switching the BJT is appreciably reduced. For example, the gate delay is about 9 ns for a non-clamped TTL circuit but approximately 3 ns for a Schottky-clamped TTL gate [15].

The SBD structure is merged with that of the BJT in the processing sequence, as shown in Figure 4.20. The contact window to the base overlaps the collector such that the overlying metal contacts both the base and the collector. The contact to the *p*-type base is essentially ohmic, but to the *n*-type collector it is rectifying. Often the SBD is formed in two steps. The initial metalization is a platinum deposition that is sintered to form a plati-

Figure 4.20 A Schottky-clamped BJT. A. Circuit symbol. B. Layout cross section.

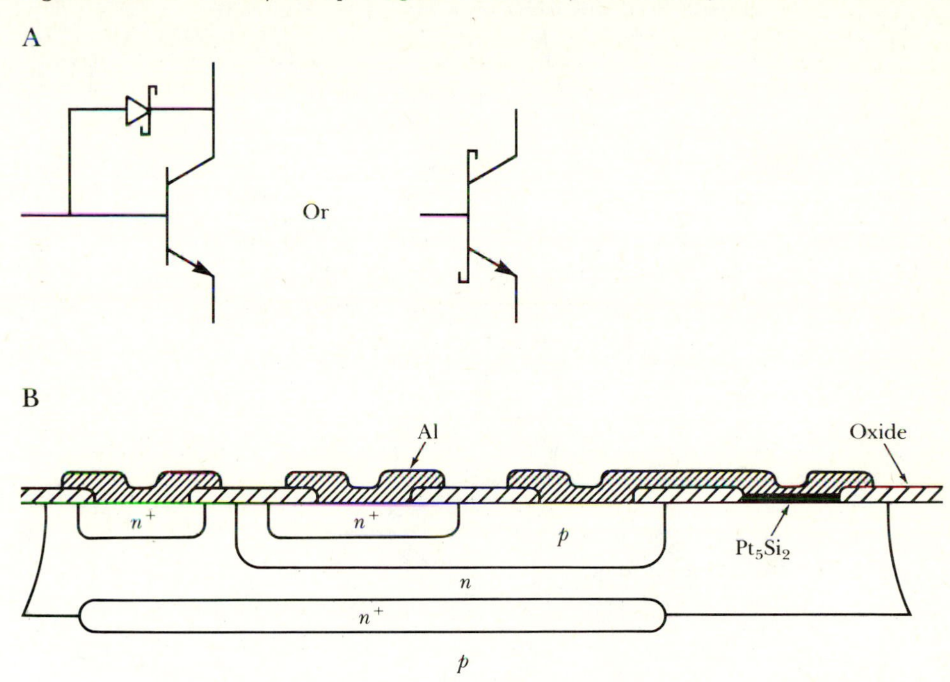

num silicide characterized by a barrier height of about 0.84 eV on n-Si. The Pt_5Si_2 anode is contacted by aluminum to complete the structure.

4.5.2 Polysilicon Self-Aligned Structures

As demonstrated in Ex. 4.6, a negative feature of many BJT structures is the large total device area compared to the active emitter area. This drawback is reduced by polysilicon self-aligned (PSA) schemes, in which heavily doped polysilicon is used both to define and to contact base and emitter regions.

One version of the PSA BJT approach is illustrated in Figure 4.21, which begins with a polysilicon layer deposited on single-crystalline n-type silicon [16]. The polysilicon is heavily doped with boron. The overlying oxide layer and a portion of the polysilicon are patterned by anisotropic reactive-ion etching (RIE) to define the emitter. The next step takes advantage of the doping dependence of certain etches. A wet etch consisting of $HF{:}HNO_3{:}CH_3COOH$, in ratios of 1:3:8, preferentially etches the remaining heavily doped polysilicon but not the more lightly doped n-type material. Then a thin oxide layer is thermally grown, followed by CVD deposition of a thicker oxide layer. Next, the p^+ polysilicon layer is used as a boron diffu-

Figure 4.21 A polysilicon self-aligned (PSA) BJT scheme. The processing steps associated with each cross-sectional view are listed in order.
SOURCE: From T. H. Ning, R. D. Isaac, P. M. Solomon, D. Duan-Lee Tang, Hwa-Nien Yu, G. C. Feth, and S. K. Wiedmann, "Self-Aligned Bipolar Transistors for High-Performance and Low-Power-Delay VLSI," *IEEE Transactions on Electron Devices,* ED-28 (1981), 1010–1013. © 1981 IEEE.

A

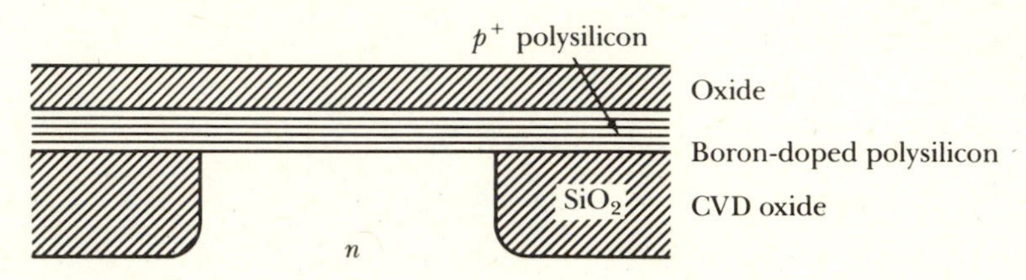

B

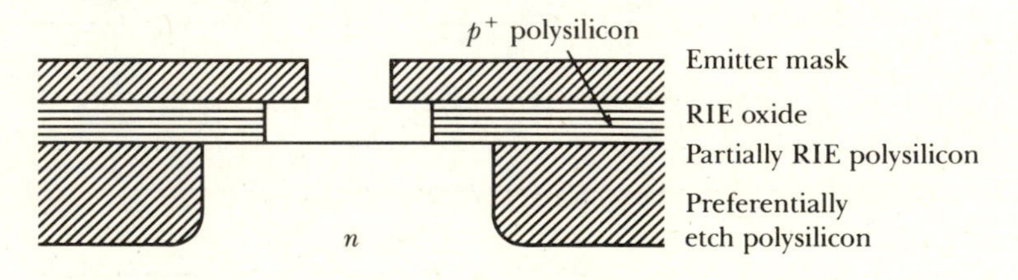

C

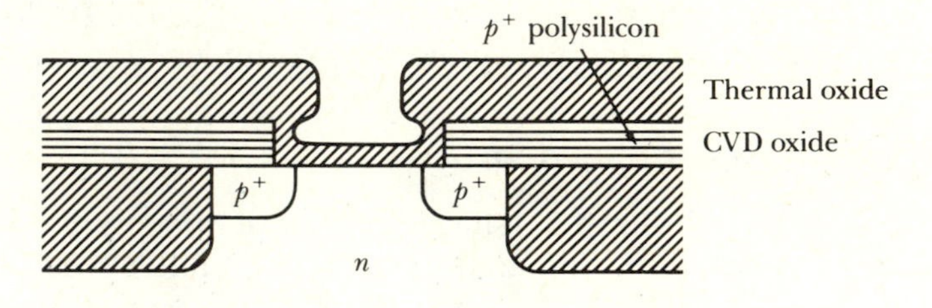

D

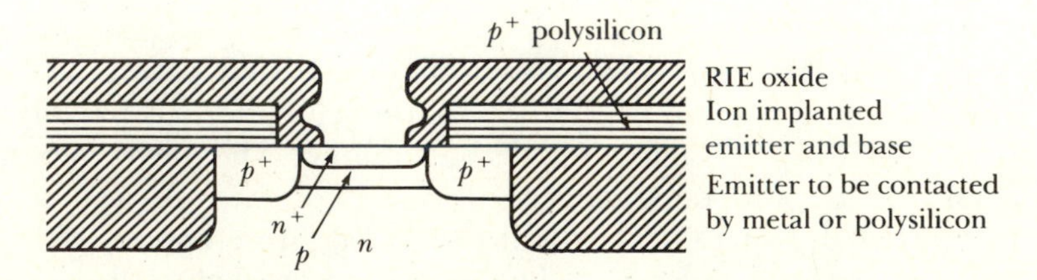

sion source to provide p^+ regions to contact the active portion of the lightly doped p-type base, which is formed by ion implantation. Prior to the implantation, the oxide is etched in a RIE plasma, consisting of CF_4 and H_2, which has a low etch rate for silicon. Since the RIE etch rate is very slow for sidewalls, as described in Section 2.11, an oxide coating remains on the polysilicon sidewalls. Often the lightly doped p-region under the emitter is referred to as the intrinsic base, and the p^+ contacts that are lateral to the emitter are referred to as the extrinsic base.

The polysilicon opening serves as the alignment for both the p-type base implant and the subsequent n-type emitter implant. Consequently, the base contact provides self-alignment for the emitter formation. In this compact device design, the emitter contact and base polysilicon contact are separated only by the thickness of the sidewall oxide on the p^+ polysilicon. The emitter contact may be metal, metal-silicide, or polysilicon. It has been found that n^+ polysilicon-contacted emitters provide a substantial increase in the BJT current gain [17]. This represents an important advantage since the effect of higher gain is to reduce the amount of power needed to drive a gate, thereby reducing the power dissipation of the chip.

In addition to providing high packing densities, the small transistor areas associated with PSA schemes lead to low values for parasitics. For example, the PSA transistor with an emitter of 2 μm × 2 μm, shown in cross section in Figure 4.22, has parasitic capacitances with values of several femtofarads and a base transit frequency of 5 GHz [18] (Tables 4.1, 4.2). Also of interest for this device is the four-level metalization. The device shown in Figure 4.22 was developed for a macrocell array, in which each macrocell consisted of 10 unconnected transistors and 12 resistors. The first metal layer is actually a

Figure 4.22 A high-speed PSA BJT example.
SOURCE: From M. Suzuki and S. Horiguchi, "A 333 ps/800 MHz 7 K-Gate Bipolar Macrocell Array Employing 4 level Metalization," *IEEE Journal of Solid State Circuits,* SC-19 (1984), 474–479. © 1984 IEEE.

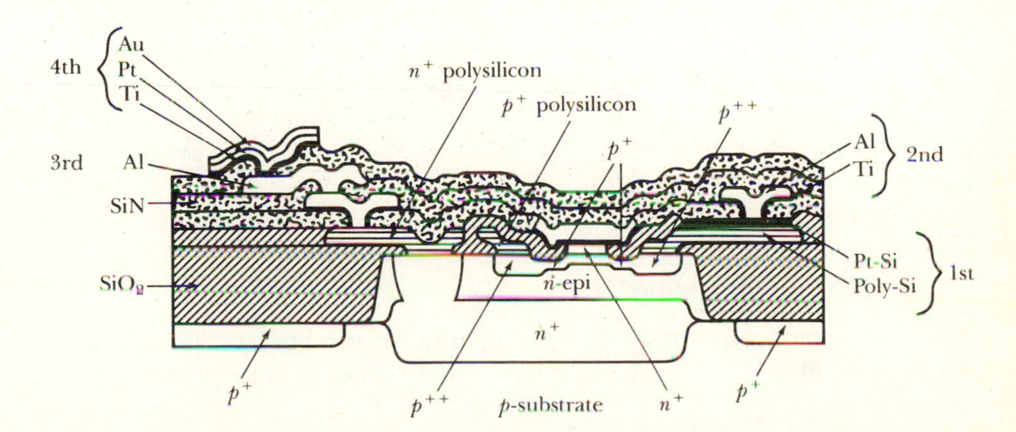

Table 4.1 Process main features of a high-speed PSA BJT example

Epitaxial layer	n-type, 1 $\Omega \cdot$cm, 1.5 μm
Isolation	High-pressure oxidation
Junction depth	$x_{jE} = 0.1$ μm, $x_{jB} = 0.25$ μm
Emitter	As, minimum 2×2 μm^2
Resistor	B-doped polysilicon
First metal	Pt-silicide, 7 μm pitch
Second metal	Ti and Al, 6 μm pitch
Third metal	Al, 7 μm pitch
Fourth metal	Ti, Pt, and Au, 20 μm pitch

Source: From M. Suzuki and S. Horiguchi, "A 333 ps/800 MHz 7 K-Gate Bipolar Macrocell Array Employing 4 level Metalization," *IEEE Journal of Solid State Circuits,* SC-19 (1984), 474–479. © 1984 IEEE.

Table 4.2 Transistor characteristics associated with the process in Table 4.1

Base series resistance	R_B	500 Ω
Collector series resistance	R_C	200 Ω
Emitter series resistance	R_E	50 Ω
Base-emitter junction capacitance	$(C_{BE})_{scl}$	0.010 pF
Base-collector junction capacitance	C_{BC}	0.016 pF
Collector-isolation capacitance	C_{CS}	0.043 pF
Transit frequency	f_t	5 GHz

Source: From M. Suzuki and S. Horiguchi, "A 333 ps/800 MHz 7 K-Gate Bipolar Macrocell Array Employing 4 level Metalization," *IEEE Journal of Solid State Circuits,* SC-19 (1984), 474–479. © 1984 IEEE.

platinum-silicide and is used only in the macrocell formation. The second metal layer consists of Al and Ti, and is used for macrocell formation and also internal wiring. The third metal layer is Al and is used for internal wiring, and finally a layer of Ti, Pt, and Au is used for power lines and reference voltage lines.

4.5.3 Multiple-Stripe Transistors

The transistor layouts shown so far in this chapter have mostly used single contacts for the base, emitter, and collector. However, device performance at high currents can be improved at the expense of device size by using multiple contacts, or stripes. Double base contacts on either side of the emitter, for example, reduce the parasitic base resistance by approximately half and also reduce emitter crowding effects. Likewise, multiple collector contacts can be made to the *n*-type epitaxial layer. A high-current BJT device with multiple emitters and interleaved base stripes is shown in Figure 4.23 [11].

Figure 4.23 A high-current BJT layout. Dimensions are in mils.
SOURCE: Adapted from J. R. Hauser, "Integrated Circuit Transistors," in R. M. Burger and R. P. Donovan, eds. *Fundamentals of Silicon Integrated Device Technology,* vol. *2, Bipolar and Unipolar Transistors.* © 1968, p. 253. Reprinted by permission of Prentice-Hall, Englewood Cliffs, N.J.

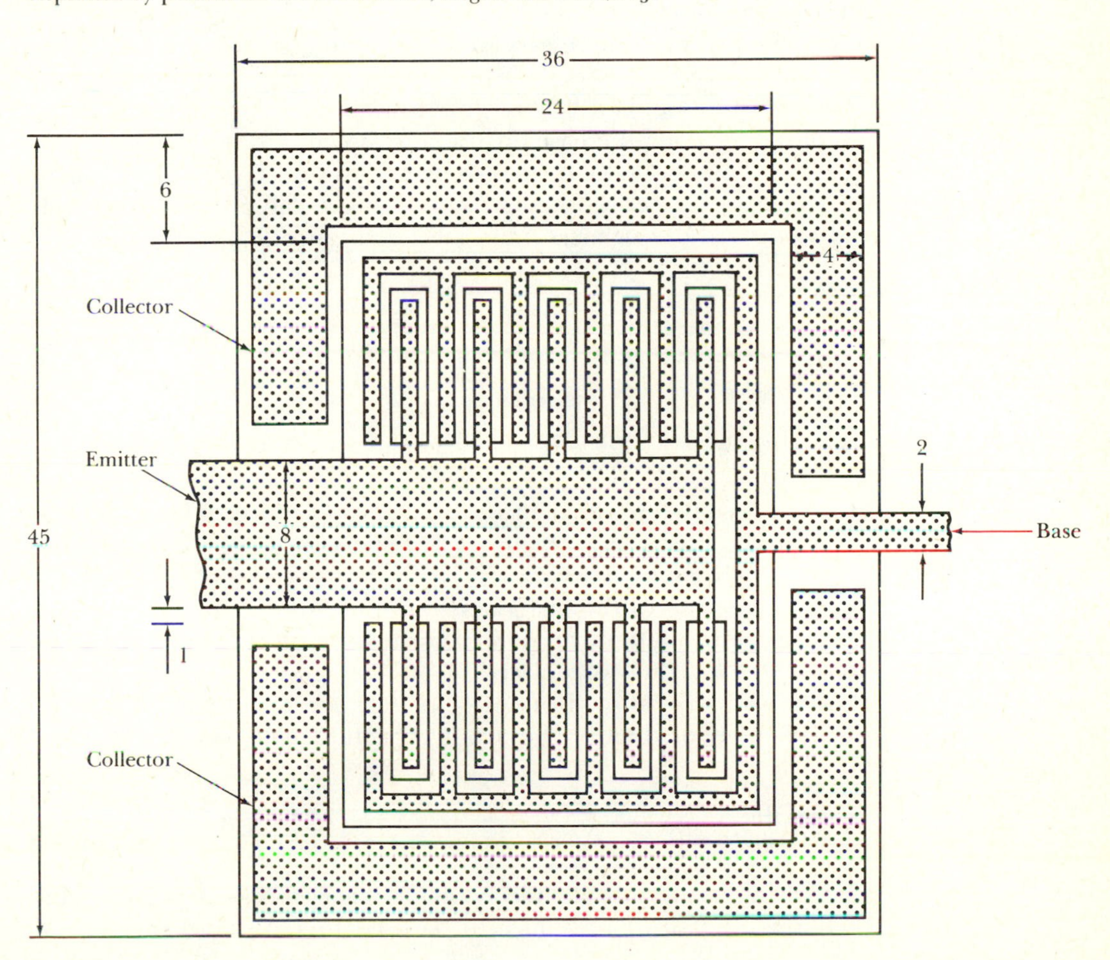

4.6 Additional Integrated Circuit Devices

4.6.1 Introduction

Both BJT and MOS integrated circuits often use devices other than transistors. This section describes how a variety of devices such as resistors, capacitors, and diodes can be incorporated into the integrated circuit. For the most part, these devices are discussed in the specific context of a bipolar integrated circuit fabrication process. However, many of the basic concepts are also applicable to MOS circuits.

4.6.2 Resistors

Although *n-p-n* transistors are the main devices used by bipolar integrated circuit designers, it is often necessary to incorporate resistors, diodes, capacitors, and *p-n-p* transistors as well. Several methods are available for incorporating resistors into standard BJT processing sequences, but most commonly they are realized by using the base diffusion to also fabricate resistors.

Discussion of integrated circuit resistor design and layout often makes use of the concept of sheet resistance. Considering a slab of material of uniform resistivity ρ and of thickness t, the sheet resistance, $R_\square$, is given as

$$R_\square = \rho/\mathrm{t} \tag{4.65}$$

Therefore the resistance of a bar of such material of length L and width W may be expressed as

$$R = R_\square(L/W) \tag{4.66}$$

In words, the resistance is equal to the sheet resistance multiplied by the number of squares, as shown in Figure 4.24. The units of $R_\square$, are often descriptively described as ohms per square, or $\Omega/\square$.

When the resistivity of the material is not constant with depth, as is the case for regions with nonuniform doping formed by diffusions and ion implantations, then Eq. (4.65) is replaced by an averaged expression. For example, consider a region of semiconductor of thickness t in which the doping profile is a function of depth. The averaged conductivity is found by the integration

$$\sigma_{\text{ave}} = (1/t) \int_0^t q\mu(x)[N(x) - N_B]dx \tag{4.67}$$

where $N(x)$ is the doping profile and N_B is the background doping. Under these conditions, the sheet resistance to be used in Eq. (4.66) is obtained from

Figure 4.24 The sheet-resistance concept.

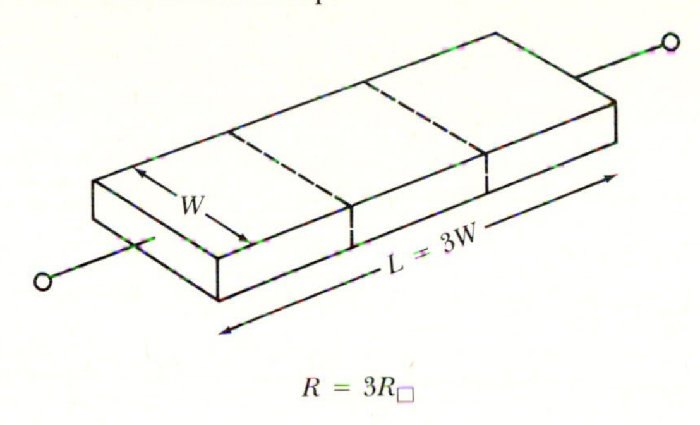

$$R_{\square} = 1/\sigma_{\mathrm{ave}}t \tag{4.68}$$

Standard graphs of sheet resistance as a function of background doping and surface concentration have been prepared for erfc and Gaussian profiles by Irvin [19]. The curves are useful for obtaining approximate values of sheet resistances resulting from diffusions and are discussed further in Chapter 7. However, since the ideal erfc and Gaussian curves do not include several effects such as impurity redistribution and concentration-dependent diffusion coefficients, as described in Chapter 2, measured values of sheet resistance may differ considerably from Irvin's curve predictions.

Resistors formed in the integrated circuit bipolar process must be isolated, and the isolation scheme is usually the same used for the transistors. That is, if the BJT isolation method is by junction isolation, the resistors are also isolated by junction isolation. Likewise, if the transistors are isolated by deep-groove isolation, the resistors are also isolated by deep-groove isolation. A cross-sectional view of a junction-isolated resistor formed by a base-diffusion step is shown in Figure 4.25. Note that a parasitic junction capacitance

Figure 4.25 Integrated circuit resistor formed by the base diffusion.

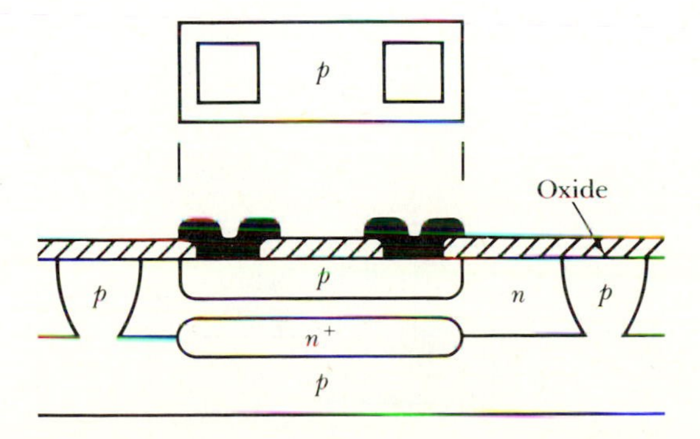

is associated with the resistor and the n-type epitaxial region, as is a p-n-p parasitic transistor between the resistor and substrate. The n^+ buried layer reduces the gain of the parasitic transistor.

It is possible and desirable to put base resistors in a common isolation region, since they are self-isolated from each other provided the surrounding n-type material is at the positive supply voltage, as shown in Figure 4.26.

EXAMPLE 4.8

A 50-kΩ resistor is to be formed in a bipolar integrated circuit using the base diffusion that results in a sheet resistance of 200 Ω/□. If the minimum diffusion width is 5 μm, as constrained by the particular lithographic process used, what are the minimum dimensions of the resistor? Also, if the sheet resistance has a tolerance variation of ±10% and the width has a tolerance variation of ±5%, what is the resulting variation in resistance?

Figure 4.26 Self-isolation of base resistors in a common isolation region. A. Cross-sectional view. B. Top view.

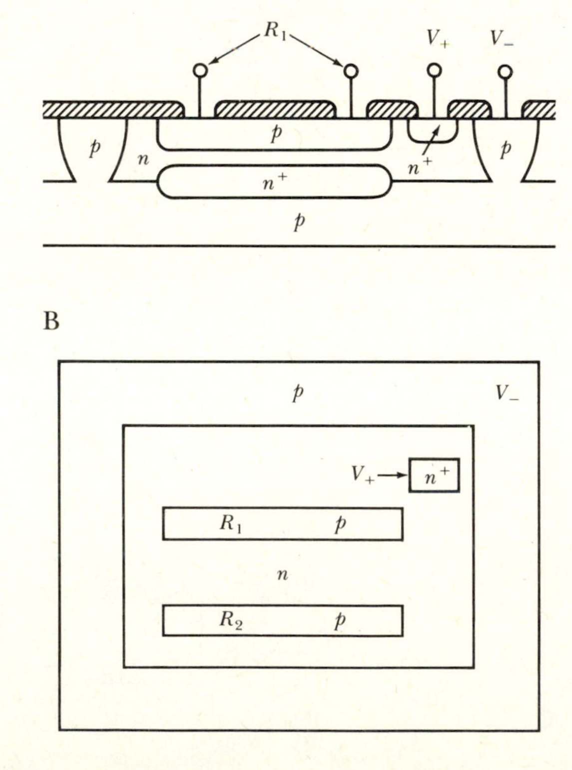

Solution The ratio of L to W, or the aspect ratio, is

$$L/W = (50 \times 10^3)/200 = 250$$

Therefore, if the minimum width is 5 μm, then the corresponding length is 1250 μm. The worst-case resistor values are

$$R_{\min} = \frac{(0.9)(200)(1250)}{(1.05)(5)} = 42.9 \text{ k}\Omega$$

and

$$R_{\max} = \frac{(1.1)(200)(1250)}{(0.95)(5)} = 57.9 \text{ k}\Omega$$

Although integrated circuit resistors may not show outstanding precision, it is easy to obtain matched resistors on a chip because of the common processing conditions. Resistor precision can be improved by using larger widths, since the percent variation in width due to overetching, and so on, decreases as the width increases. It may also be necessary to use larger widths because of power considerations, as will be seen by example in Chapter 5.

The previous example shows that base-diffused resistors are not very compact devices. To achieve more efficient layouts, large resistors are laid out in a meander design, as shown in Figure 4.27. To a first approximation, corner squares are weighted as 0.65 to account for the two-dimensional nature of the current flow around the resistor corners. Two dimensional considerations also enter into the picture at the contact points. Again, the terminations can be approximately accounted for by assigning a fraction of a square to the end, as shown in Figure 4.27 [20].

Resistors formed by the base diffusion are usually limited to a range from approximately 50 Ω to 100 kΩ; small values are subject to large variations in resistance because of their short lengths, and large resistor values require formidable amounts of chip real estate. Alternatively, the emitter diffusion, with its much smaller sheet resistance, can also be used for low-value resistors. An alternate method for larger resistors is the "pinch resistor," which uses a base diffusion that has an overlying emitter diffusion, such that the height of the p-type diffused region is reduced, thereby increasing the sheet resistance [20].

Resistors may also be achieved by thin-film deposition, in which the sheet resistance is determined by the deposited material. Examples of thin-film resistor materials are nichrome (80% nickel and 20% chromium), tantalum, and polysilicon. Metal-film sheet resistances range from approximately 1 Ω/□ to 500 Ω/□, depending on the film thickness, which is typically in the range of 250 Å to 1 μm thick. The resistivity of undoped or lightly doped polysilicon can be several thousand Ω · cm, and consequently, deposited polysilicon

Figure 4.27 Integrated circuit resistor layout. A. Layout for a large resistor. B. Representative resistor terminations. The termination contributes a fraction of a square, as noted, to the total number of squares.

SOURCE: B from R. A. Colclaser, *Microelectronics Processing and Device Design,* Chap. 9. Copyright © 1980 by John Wiley & Sons, New York. Reprinted by permission of John Wiley & Sons, Inc.

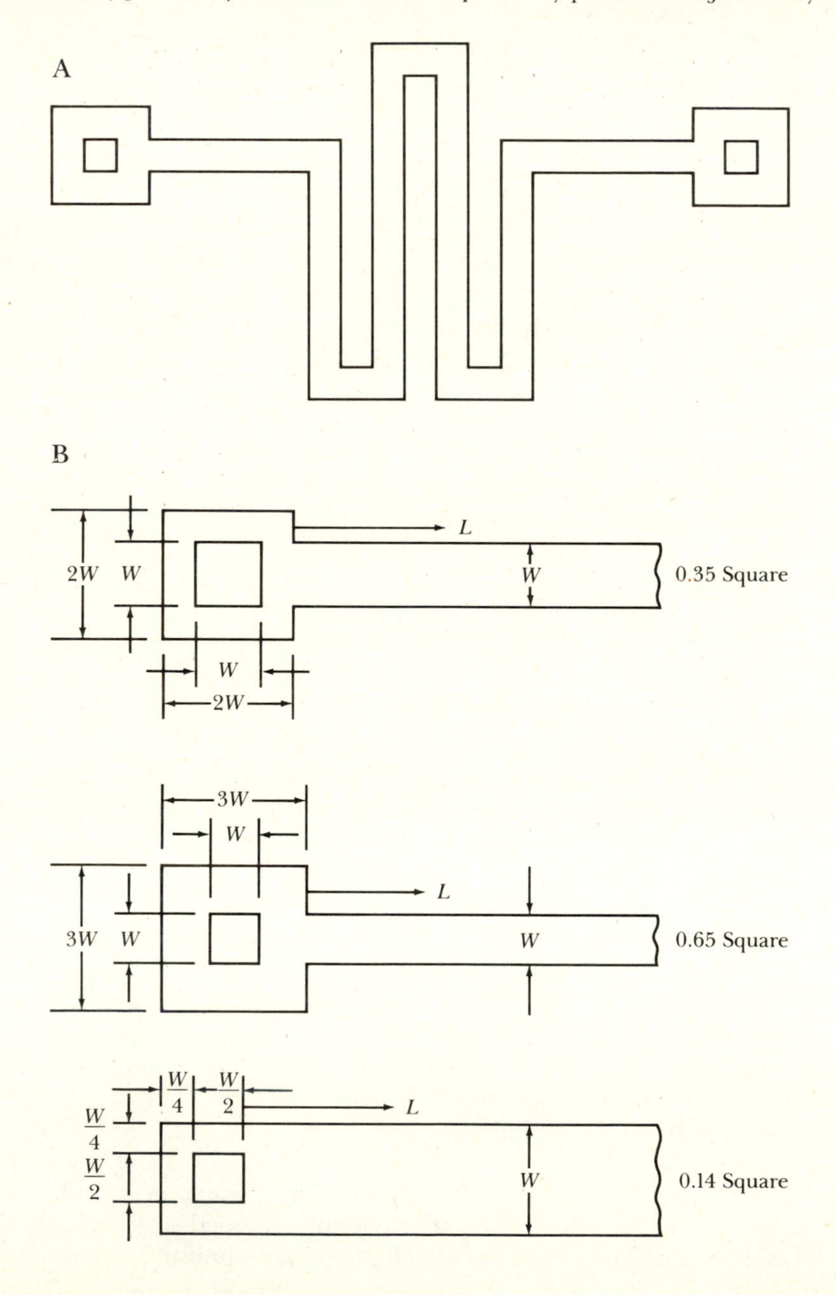

layers can be used for relatively compact resistors with large values. Patterned ion-implant regions are also used for resistors.

4.6.3 Capacitors

One way to realize an integrated circuit capacitor is to utilize the space-charge layer capacitance of a *p-n* junction. A disadvantage of this approach is that the value of the resulting capacitor is voltage dependent and the dc voltage must always be such that the junction is reverse biased. On the other hand, an advantage of this fabrication method for capacitors is that a standard BJT processing sequence can be used without additional steps. Either the emitter-base junction or the collector-base junction can be used, with the collector-base junction having a larger reverse voltage rating and the emitter-base junction having a larger capacitance per unit area.

A better-quality capacitor, in terms of leakage current and voltage dependence, results if an MOS structure is fabricated using heavily doped silicon as the bottom capacitor plate. As discussed in Chapter 3, the MOS capacitor can be considered as two capacitors in series, one arising from the oxide dielectric layer and the other from the semiconductor depletion or inversion layer. For a heavily doped semiconductor layer, the semiconductor capacitance is sufficiently large that the series combination is essentially equal to the capacitance arising from the oxide dielectric. Consequently, for an oxide layer of thickness d_{OX}, the capacitance per unit area is given by Eq. (3.44), repeated here for convenience,

$$C_{OX} \simeq \epsilon_{OX}/d_{OX} \tag{4.69}$$

This value is relatively independent of voltage. Cross-sectional views of junction-isolated MOS capacitors and junction capacitors are shown in Figure 4.28.

Figure 4.28 Integrated circuit capacitors. A. Junction isolated collector-base capacitor. B. Junction isolated MOS capacitor.

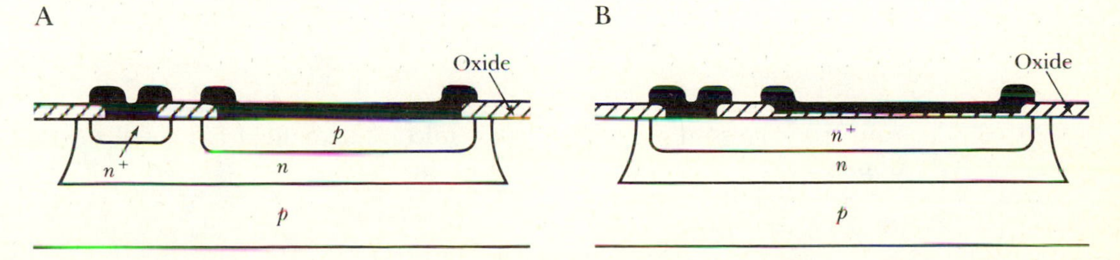

EXAMPLE 4.9

Consider an MOS capacitor for which the bottom electrode is doped *n*-type with a concentration of 10^{19} cm^{-3}. The oxide thickness is 700 Å and the device area is 500 μm × 500 μm. Calculate the capacitance at +5 V and −5 V.

Solution At +5 V the device is in inversion and the capacitance is given by Eq. (4.69) multipled by the area. Therefore, at +5 V,

$$C = \frac{(3.9)(8.85 \times 10^{-14})}{700 \times 10^{-8}}(500 \times 10^{-4})^2$$

$$C = 123 \text{ pF}$$

At −5 V, the series effect of the depletion layer capacitance should be considered and the total capacitance is given by Eq. (3.50),

$$C = \frac{1}{d_{OX}/\epsilon_{OX} + x_D/\epsilon_S}$$

The depletion-layer width, x_D, is obtained by adapting Eq. (3.15) to the case for *n*-type material, resulting in

$$x_D = \left(\frac{-2\epsilon_S\psi_S}{qN_D}\right)^{1/2}$$

where ψ_S is, from Eq. (3.8),

$$\psi_S = -5 \text{ V} - V_{OX}$$

Expressions for V_{OX} and ψ_S can be found by applying Gauss's law and Poisson's equation as follows.

From Gauss's law, the following relationship is found between the electric fields in the oxide and the semiconductor at the boundary $x = 0$:

$$\epsilon_{OX}\mathscr{E}_{OX} = \epsilon_S\mathscr{E}_S$$

where

$$\mathscr{E}_{OX} = V_{OX}/d_{OX}$$

And from Poisson's equation, the semiconductor field is related to the doping as

$$\frac{d\mathscr{E}_S}{dx} = \frac{\rho}{\epsilon_S} = \frac{qN_D}{\epsilon_S}$$

Solving for $\mathscr{E}_S$ yields,

$$\mathscr{E}_S = \frac{qN_D}{\epsilon_S}(x - x_D)$$

At the boundary, the semiconductor field is given by

$$\mathscr{E}_S(0) = \frac{-qN_D x_D}{\epsilon_S}$$

and the oxide field is therefore equal to

$$\mathscr{E}_{OX} = \frac{-qN_D x_D}{\epsilon_{OX}}$$

From the oxide field, the oxide voltage drop is found to be

$$V_{OX} = \frac{-qN_D x_D d_{OX}}{\epsilon_{OX}}$$

which yields the value of ψ_S:

$$\psi_S = -5 + \frac{qN_D x_D d_{OX}}{\epsilon_{OX}}$$

Therefore, the expression for the depletion width may be written as,

$$x_D = \left[\frac{2\epsilon_S[5 - qN_D x_D d_{OX}/\epsilon_{OX}]}{qN_D}\right]^{1/2}$$

which, after manipulation, yields the following quadratic expression for x_D:

$$x_D = -\frac{\epsilon_S d_{OX}}{\epsilon_{OX}} + \left[\left(\frac{\epsilon_S d_{OX}}{\epsilon_{OX}}\right)^2 + \frac{2(5)\epsilon_S}{qN_D}\right]^{1/2}$$

It is convenient to rewrite the above expression as

$$\frac{C_{OX} x_D}{\epsilon_S} = -1 + \left[1 + \frac{2(5)(\epsilon_{OX})^2}{\epsilon_S q N_D (d_{OX})^2}\right]^{1/2}$$

Therefore, since the total MOS capacitance can be written as

$$C = \frac{1}{1/C_{OX} + x_D/\epsilon_S} = \frac{C_{OX}}{1 + C_{OX}x_D/\epsilon_S}$$

the total capacitance at −5 V may be written as

$$C = C_{OX}\left[1 + \frac{2(5)(\epsilon_{OX})^2}{qN_D\epsilon_S(d_{OX})^2}\right]^{-1/2}$$

It may be recognized that this expression is the n-type version of Eq. (3.51).

Carrying out the numerical calculation results in

$$C = C_{OX}\left[1 + \frac{(10)(3.9)^2(8.85 \times 10^{-14})}{(700 \times 10^{-8})^2(1.6 \times 10^{-19})(10^{19})(11.9)}\right]^{-1/2}$$

or $C = 0.993C_{OX} = 122$ pF. Over the range from +5 V to −5 V, the capacitance changes by approximately 1%.

The capacitor in the above example is quite a large device as far as integrated circuit dimensions are concerned. In practice, integrated circuit capacitors much larger than 100 pF are not used because of the large area requirements.

4.6.4 Inductance

Inductive patterns, such as the spiral shown in Figure 4.29, find occasional use in high-frequency circuits. Such inductors are not used in ordinary integrated circuits because the resulting inductor values, which are several nanohenries, are too low to be of practical value in circuit design. However, in microwave circuits, where frequencies are in the GHz range, integrated circuit inductors may be used effectively. For example, a six-turn spiral inductor has been used on a single-chip, 12-GHz, GaAs integrated circuit developed for reception of direct-broadcast satellite television signals [21]. The transistors on this chip were MESFETs, rather than BJTs or MOSFETs, since the MESFET is the standard device for GaAs integrated circuits.

Another approach for achieving passive devices, including inductance, in microwave integrated circuits is to use transmission lines as resonant elements. In this case, the distributed resistance, capacitance, and inductance of appropriately designed transmission lines are used to achieve desired circuit properties. The transmission lines often take the form of microstrip, which is a transmission line consisting of a conducting film over a dielectric with a counter-electrode on the other side of the dielectric. An example of a microstrip line on GaAs is shown in Figure 4.30 [22]. The dielectric in this case

Figure 4.29 A spiral inductor for microwave integrated circuits.

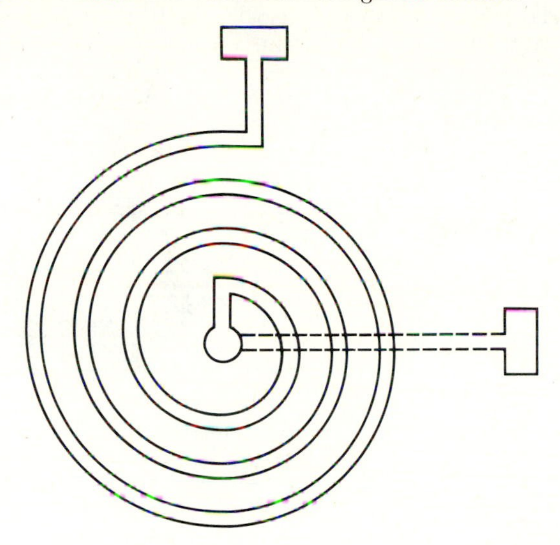

is simply a semi-insulating layer of GaAs, and the bottom conductor consists of heavily doped GaAs. A higher-quality, lower-loss microstrip would use metal conductors for both the top and bottom strips. By appropriately designing the geometric shape of coupled microstrip lines and providing appropriate terminations, high-quality resonant circuits are achieved without the use of spiral inductors or MOS capacitors. For example, Figure 4.31 shows a microstrip band-pass filter consisting of six coupled microstrip segments, each $\lambda/2$ in length where λ is the wavelength at the center frequency [22].

Figure 4.30 Microstrip transmission line on GaAs.
SOURCE: Reprinted, by permission, from J. Frey, "Hybrid and Monolithic Microwave Integrated Circuits," in *Microwave Integrated Circuits*, ed. J. Frey (Dedham, Mass.: Artech House, 1975), pp. xvii–xxi. © 1975 by Artech House, Inc.

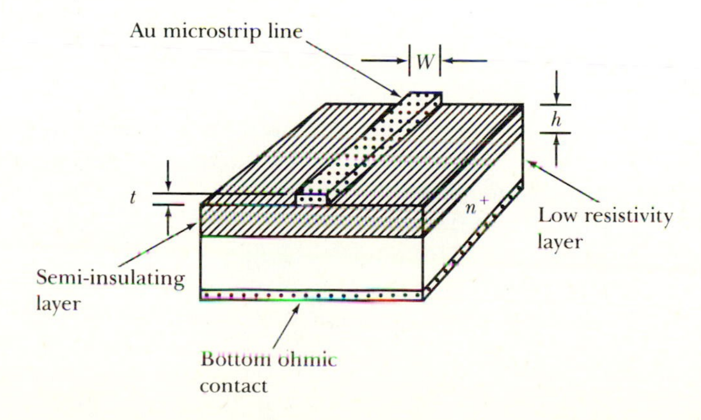

Figure 4.31 Top view of a microstrip, parallel-coupled, band-pass filter designed for 50 GHz. S = separation of strips $\approx W/10$; W = strip width = 8.3×10^{-4} cm; m = filter width $\cong$ .1 cm; $L \cong$ .35 cm. Total surface area = .035 cm^2.

SOURCE: From J. Frey, "Hybrid and Monolithic Microwave Integrated Circuits," in *Microwave Integrated Circuits,* ed. J. Frey (Dedham, Mass.: Artech House, 1975), pp. xvii–xxi.

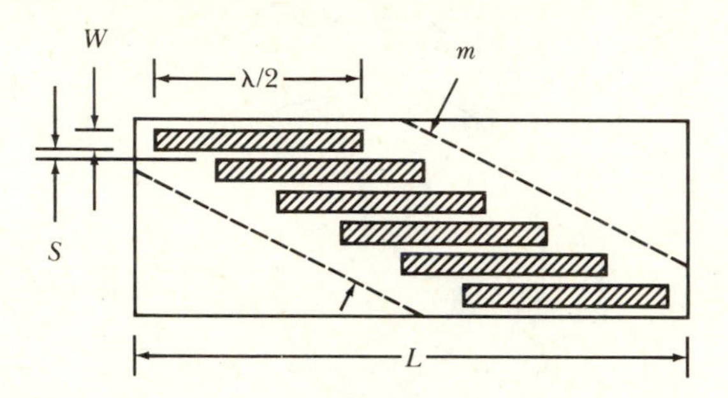

Circuits with transmission-line components, which of necessity use structures with dimensions ranging from about that of a wavelength to much larger than a wavelength, are often referred to as distributed circuits. In contrast, spiral inductors and thin-film capacitors, which are much smaller than a wavelength, are referred to as lumped elements. Lumped circuits use less area than distributed circuits; in the S band (2.0–4.0 GHz) the use of lumped rather than distributed elements may reduce size by a factor of 10 [23].

The trade-off in going to lumped elements involves increased sophistication in fabrication technology and, in some cases, questions of device quality. Consider, for example, the spiral inductor of Figure 4.29. The current, to a first approximation, flows within about one skin depth of the surface. One measure of inductor quality is the Q factor, which is the device reactance divided by the resistance, or

$$Q = \omega L/R \tag{4.70}$$

Because of the skin-depth limitation on the current flow, a large surface area is required for low R and high Q. For a Q greater than 50 at 2 GHz, which is a modest value for Q, this requirement leads to spiral-inductor aluminum-film heights of about 8 μm, which is considerably thicker than typical integrated circuit metalizations [24].

An entirely different method of achieving on-chip inductance, which is applicable to lower frequencies, is by use of a gyrator. A gyrator is a two-port circuit with parameters such that if an admittance $Y(s)$ is connected to port 2, then the input impedance at port 1 is

$$Z_{\text{in}}(s) = Y(s)R^2 \tag{4.71}$$

where R is positive and real and is called the gyrator resistance [25]. Therefore, if a capacitor is connected to port 2 such that $Y(s) = sC$, then the input impedance is sCR^2, which corresponds to an inductor of value CR^2.

Several circuit techniques have been used to realize the gyrator; one common method uses two operational amplifiers [25]. High-quality inductors result from this method, with Q factors greater than 1000 and good frequency response up to the limits of the operational amplifiers. An obvious disadvantage is that several dozen interconnected devices are required to mimic one inductor. Consequently, there is considerable motivation to use integrated circuit designs that do not require inductors.

4.6.5 *p-n* Junction Diodes

Several options exist for incorporating *p-n* junction diodes in the standard fabrication sequence. For example, the base-collector junction may be used with the emitter open, shorted to the base, or absent. Likewise, the base-emitter junction may be used as a diode with an open collector, with the collector shorted to the base, or with the collector shorted to the emitter.

The diode turn-on voltages, breakdown voltages, and switching speeds vary considerably among the several options [13]. The fastest *p-n* diode among the configurations just listed is the base-emitter junction with the collector shorted to the base, since a small junction area is combined with a nonsaturating base-collector junction. However, the highest breakdown voltages are associated with the collector-base diodes because of the lower doping levels in the collector.

4.6.6 *p-n-p* Bipolar Transistors

So far, this chapter has discussed the fabrication of *n-p-n* bipolar transistors but not *p-n-p*. The *n-p-n* device does play a dominant role in bipolar integrated circuitry principally because of that device's higher electron mobility. Also, ohmic contacts to n-type material require an n^+ region under the contact. For *n-p-n* BJTs, an n^+ ohmic contact to the collector is achieved simultaneously with the emitter diffusion. However, an n^+ contact to the base of a *p-n-p* transistor would require an extra step.

Nevertheless, *p-n-p* transistors are sometimes needed. For example, the output stage of an operational amplifier requires a low output resistance and must provide zero dc volts out for zero input voltage. One configuration that serves this purpose is the complementary emitter follower, also called a class B push-pull amplifier, which uses both an *n-p-n* and a *p-n-p* transistor, as shown in Figure 4.32.

To keep the number of processing steps to a minimum, it is desirable to fit *p-n-p* transistors into a standard *n-p-n* process. One commonly used approach is the lateral *p-n-p* BJT shown in Figure 4.33, in which the base width

Figure 4.32 A simple amplifier output stage incorporating a *p-n-p* BJT (complementary emitter follower, or class B push-pull amplifier).

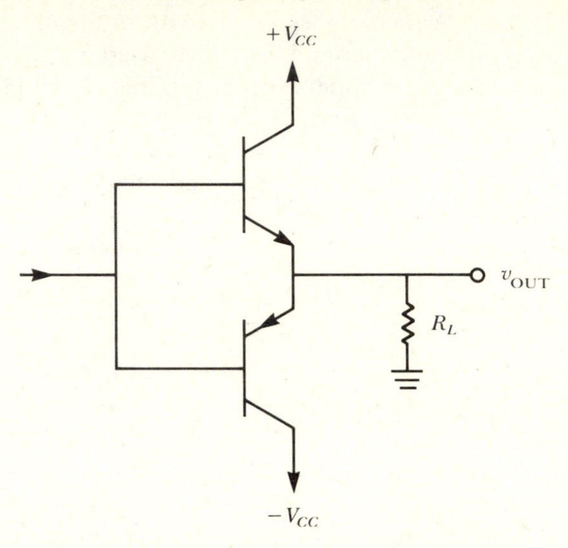

is determined by the lateral distance between the diffused or implanted *p*-type collector and emitter regions. Holes are injected laterally from the emitter to the collector. The wrap-around collector shown in this figure provides improved carrier collection and current gain since it is primarily the emitter sidewalls that are involved in the transistor action. Note that there are two parasitic *p-n-p* transistors, one between the collector and substrate and the other between the emitter and substrate. The buried n^+ layer reduces the current gain of these parasitic devices and also reduces the resistance between the base contact and active base region.

Figure 4.33 A lateral *p-n-p* BJT layout.

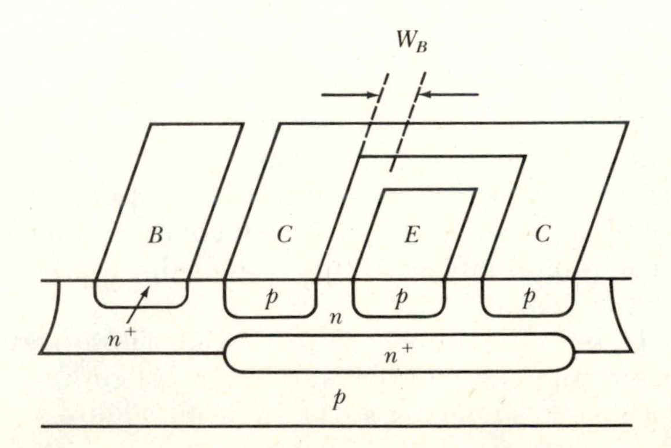

Because the process design is optimized for *n-p-n* devices rather than *p-n-p* ones, and because the layout leads to larger base widths and larger parasitics, the current gains of lateral *p-n-p* transistors are somewhat smaller than those of the *n-p-n* device. Furthermore, the frequency response is also diminished so that the gain-bandwidth product is often an order of magnitude or more smaller for *p-n-p*s than for *n-p-n*s.

As a result of the lateral nature of the device, the lateral *p-n-p* BJT does not provide a large cross-sectional area, unless a very large amount of chip area is used. So it is not well suited for high current applications. An alternative *p-n-p* layout that is more suitable for high currents is the substrate *p-n-p* BJT. In this device the *p*-type substrate serves as the collector, and a *p*-type diffusion establishes the emitter during the base diffusion step of the *n-p-n* processing sequence. As is the case for the lateral *p-n-p*, the *n*-type epitaxial layer between the emitter and collector is the base region, but in the substrate *p-n-p* the current flow is vertical. Since the substrate is usually connected to the most negative voltage in the circuit, the substrate *p-n-p* is not as versatile as the lateral *p-n-p* in terms of circuit applications. A good example of a circuit for which the substrate *p-n-p* would be the best choice is the push-pull stage shown in Figure 4.32, since a large current handling ability is required and the collector is connected to $-V_{CC}$.

Finally, it should be noted that for simplicity, the resistor, capacitor, and *p-n-p* layouts shown in this section have used junction isolated layouts and standard processing. However, the more advanced techniques described in Sections 4.4 and 4.5 may also be used to advantage for these devices. For example, Figure 4.34 illustrates device cross sections for deep-groove, polysilicon self-aligned *p-n-p* transistors, *n-p-n* transistors, and *p*-type resistors [26].

4.7. BJT Gate Configurations

4.7.1 Transistor-Transistor-Logic

Transistor-transistor-logic (TTL) gates have long played an important commercial role in digital integrated circuits. The TTL logic family originated in the 1960s and still finds considerable use over 20 years later. During that time, several versions of TTL have been developed, with the general trend being toward increased speed and circuit complexity. The names attached to successive generations of TTL—namely Schottky-clamped TTL, low-power Schottky-clamped TTL, advanced Schottky-clamped TTL, and advanced low-power Schottky TTL—are indicative of the circuit development that followed the first standard TTL gates. Advanced TTL versions contain approximately 20 devices per simple gate, including transistors, resistors, and diodes. Although the resulting gates are fast, with gate delays of approximately a few

Figure 4.34 Schematic of deep-groove-isolated self-aligned bipolar structures. A. *n-p-n* transistor. B. *p-n-p* transistor. C. Resistor.

SOURCE: From D. Tang, P. Solomon, T. Ning, R. Isaac, and R. Burger, "1.25 μm Deep-Groove-Isolated Self-Aligned Bipolar Circuits," *IEEE Journal of Solid State Circuits,* SC-17 (1982), 925–931. © 1982 IEEE.

A

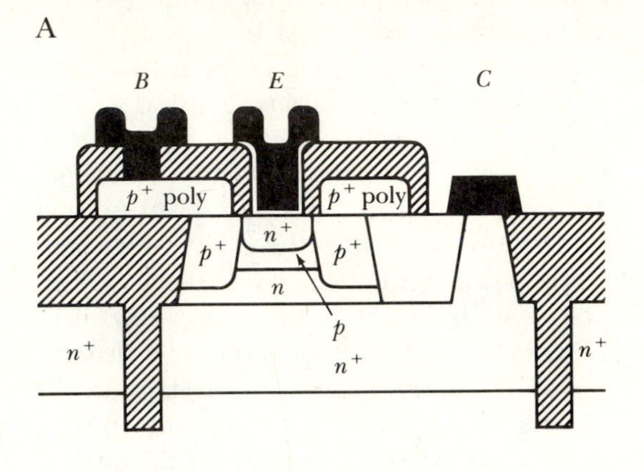

B

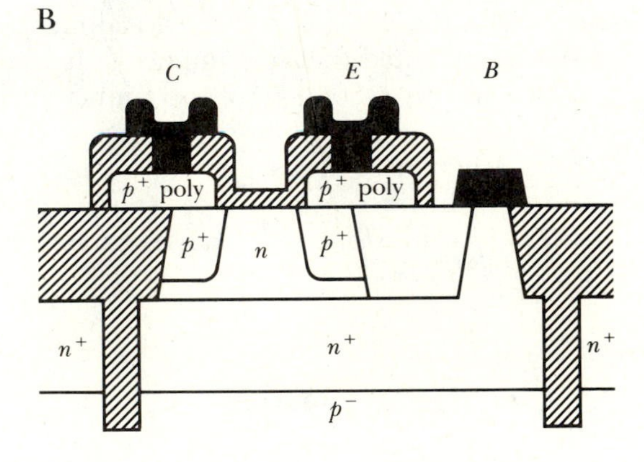

C

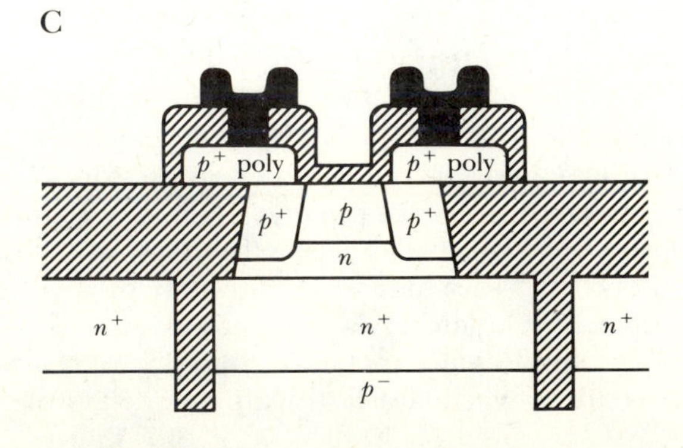

Figure 4.35 The TTL gate concept. A. Simplified NAND gate. B. Multiple emitter input transistor. (E_A is the emitter for input A and E_B is the emitter for input B.)

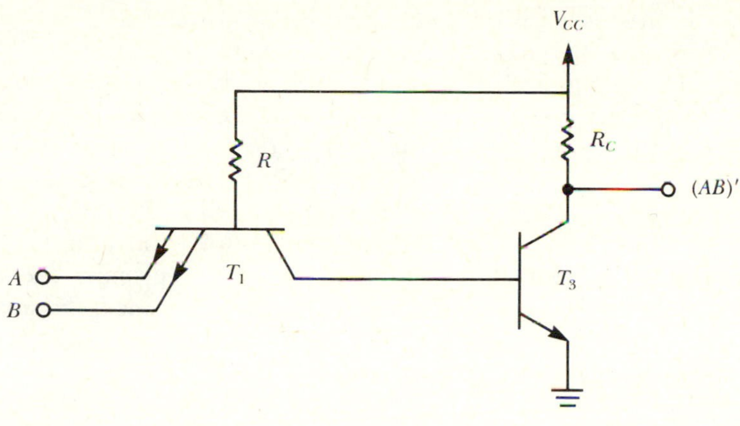

B

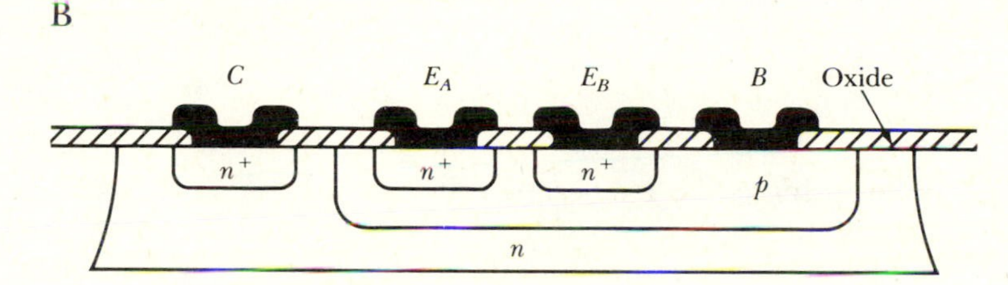

nanoseconds, they are not useful for chips with high gate densities. More recently, simpler but slower compact TTL gates have been developed for gate array applications.

The basic TTL gate concept is shown in simplified form in Figure 4.35 for the case of a 2-input NAND gate. Gate inputs are provided by the multiple emitters of transistor T_1. If either of the inputs, A or B, is at low voltage, near ground, then T_1 is on and T_3 is cut off. Consequently the gate output is high. However, if both inputs are at a high voltage, then the base-emitter junction of T_1 is reverse biased while the base collector junction is forward biased. Under these conditions, T_1 operates in the inverse region and provides base current for T_3, driving T_3 into saturation and making the gate output low.

The circuit shown in Figure 4.35 suffers from two speed limitations. The pull-up time is limited by an appreciable RC time constant owing to the load resistance, R_C, and the capacitive load on the gate. Also, saturated BJTs lead to storage delay times as excess electrons are swept out of the base. The first problem is addressed by providing an active pull-up device to provide a larger charging current for capacitive loads, and the second is addressed by using Schottky clamps on base-collector junctions.

An active pull-up configuration is shown in the standard TTL NAND gate shown in Figure 4.36. Here the combination of an active pull-up device, T_2, and the associated small load-resistor, R_{C2}, allows larger load-charging currents and shorter pull-up times. The operation of this circuit is best appreciated by first considering the transfer characteristics between the gate output and the base voltage on transistor T_4, which directly drives the output devices. When V_{B4} is at ground potential, both T_3 and T_4 are off and transistor T_2 is biased in the forward active region. Neglecting the voltage drop across R_{C4}, which is appropriate since the base current I_{B2} is small, results in

$$V_{\mathrm{OUT}} \simeq \mathrm{V}_{CC} - \mathrm{V}_{BE2} - V_D \tag{4.72}$$

where V_D is the diode voltage drop. For a 5-V power supply and typical forward-bias junction voltages for silicon,

$$V_{\mathrm{OUT}} \simeq 5 - 0.7 - 0.7 = 3.6 \text{ V} \tag{4.73}$$

As V_{B4} is increased from zero volts, there is little change in output voltage until the junction cut-in voltage at around 0.65 V. At this point, T_4 enters the forward active region and the output voltage begins to drop, following this expression:

$$V_{\mathrm{OUT}} = V_{CC} - R_{C4}(I_{\mathrm{B}2} + I_{C4}) - V_{BE2} - V_D \tag{4.74}$$

However, transistor T_3 remains off until V_{B4} is large enough to turn on both T_3 and T_4. As V_{BE4} continues to increase, T_4 and T_3 are driven into saturation. Also, T_2 is driven into cut-off because the voltage difference between the collectors of T_4 and T_3 is not enough to turn on both the diode and the base-emitter junction of T_2. Consequently, the output is isolated from the power supply, V_{CC}, and V_{OUT} is approximately 0.1 to 0.2 V, as determined by the voltage across the saturated transistor. Note that if diode D were not present, T_2 would be on and the output state would be indeterminate.

The overall transfer characteristic of the gate is obtained by considering the relationship between V_{B4} and the input voltage, V_{IN}. When V_{IN} is low, transistor T_1 is saturated and the voltage from collector to emitter is approximately 0.1 V. For small V_{IN}, then, V_{B4} is given by

$$V_{B4} \simeq 0.1 + V_{\mathrm{IN}} \tag{4.75}$$

Figure 4.36 Standard TTL gate and transfer characteristic. A. Circuit schematic. B. Transfer characteristic.

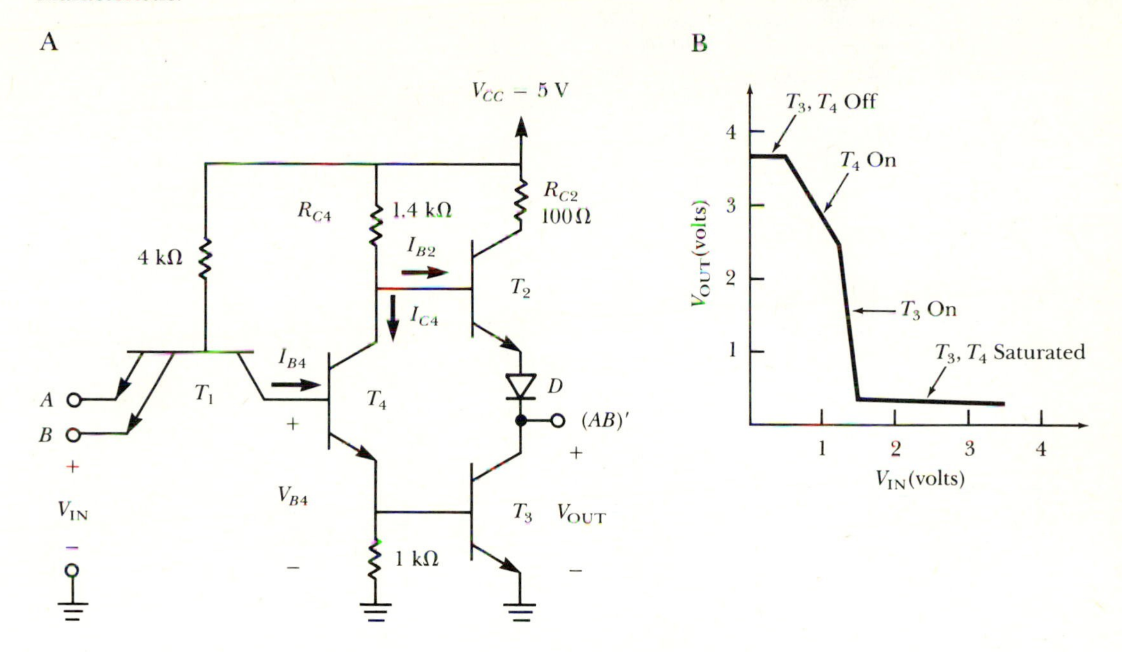

However, as V_{IN} is increased, eventually both T_3 and T_4 are saturated and V_{B4} is clamped at about 1.5 V by the sum of base-emitter voltages for these two transistors, allowing 0.75 V for each saturated transistor. The overall transfer characteristic for the gate is shown in Figure 4.36 [27].

An example of an advanced Schottky-clamped TTL NAND gate is shown in Figure 4.37 [28], along with a much simpler TTL gate that is suitable for gate arrays and designed to operate on a low supply voltage. The latter has been used for chips with several hundred gates [29].

4.7.2 Emitter-Coupled Logic (ECL)

ECL represented the fastest commercial digital switching logic for several years and is the integrated circuit technology on which the first generation of supercomputers was based. Although higher speeds may be obtained by going to GaAs-based circuits or cryogenic temperatures, ECL offers the advantage of a room temperature silicon chip based on fairly standard integrated circuit processing technology.

A simplified version of an ECL OR/NOR gate is shown in Figure 4.38. When either of the inputs is sufficiently high, the input transistor will be on and the transistor connected to V_R is off, such that the output at R_{C2} is high. And if both inputs are low, the input transistors are off and the output

Figure 4.37 Two examples of TTL variations. A. Advanced Schottky clamped TTL. B. Low-power TTL for gate arrays.
SOURCE: A reproduced with permission from D. A. Hodges and H. G. Jackson, *Analysis and Design of Digital Integrated Circuits* (New York: McGraw-Hill, 1983); B from S. K. Wiedmann, "Advancements in Bipolar VLSI Circuits and Technologies," *IEEE Journal of Solid-State Circuits,* SC-19 (1984), 282–291. © 1984 IEEE.

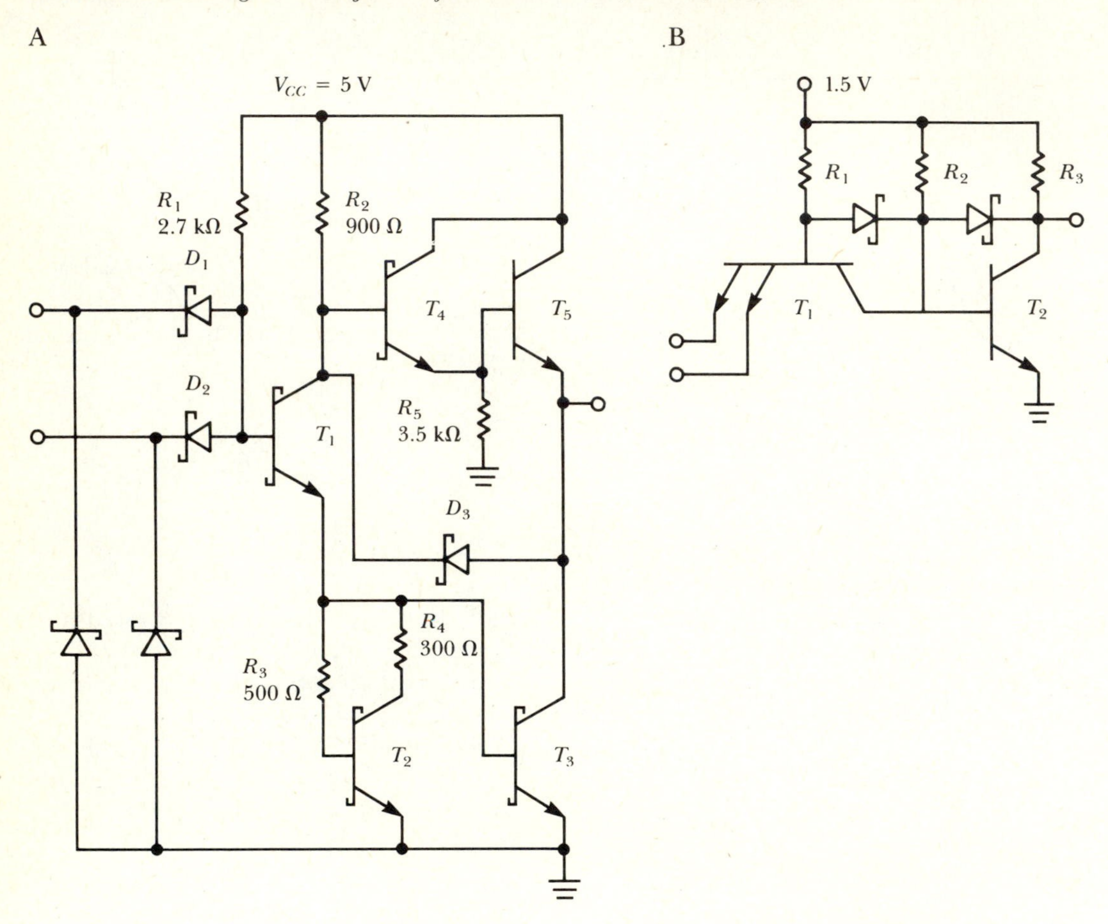

transistor is on, such that the output at R_{C2} is low. Consequently, the OR function is obtained from the output at R_{C2}. If the output is taken at R_{C1}, a NOR gate results. The reference voltage, V_R, the power supply voltage, and the resistor values are chosen such that the transistors are never saturated, thereby avoiding storage delay times caused by excess minority carrier charge in the base.

ECL gates from the commercial 100 K series exhibit typical gate delays of 0.75 ns with a power dissipation of 40 mW per gate. In comparison, advanced low-power Schottky TTL gate delays in the 74ALS series are about 4

Figure 4.38 Simplified version of an ECL NOR gate

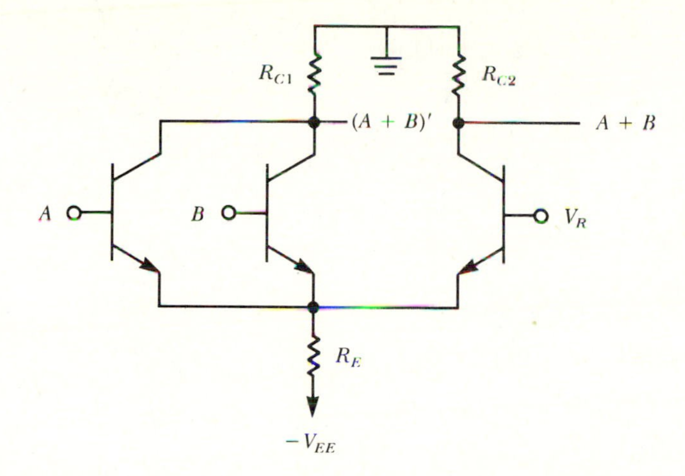

ns and gate power dissipations are 1 mW. Lower-power ECL designs have also been developed, using advanced fabrication techniques, that allow high-density packing such as gate array chips with several thousand gates.

4.7.3 Integrated Injection Logic (I^2L)

Integrated injection logic was the first bipolar technology successfully developed for LSI and VLSI levels of gate densities. The key to the high packing density of I^2L is partly the unique device layout in which the base regions of *n-p-n* transistors are shared with the collector of an adjacent *p-n-p* transistor, and partly that resistors are not required within the gate circuitry. Furthermore, individual device isolation is not required. The merged-transistor layout leads to an alternative name for the technology, merged-transistor-logic, or MTL.

A cross section of an I^2L gate and the associated circuit schematic is shown in Figure 4.39. In this layout the lateral *p-n-p* transistor has two collectors and acts as a current source, or current injector, for the two *n-p-n* transistors. The *n-p-n* transistors differ from the standard layout in that their emitters are shared, residing in the *n*-epitaxial layer, and the collectors are at the surface. Consequently, multiple collector *n-p-n* transistors are possible, and since the emitters are at a common potential, transistor isolation is not required. Although there are different I^2L layouts, common features may be observed as follows:

1. A *p-n-p* current source supplies base current to one or more *n-p-n* transistors.

Figure 4.39 Integrated injection logic gate. A. Layout. B. Associated circuit schematic.

A

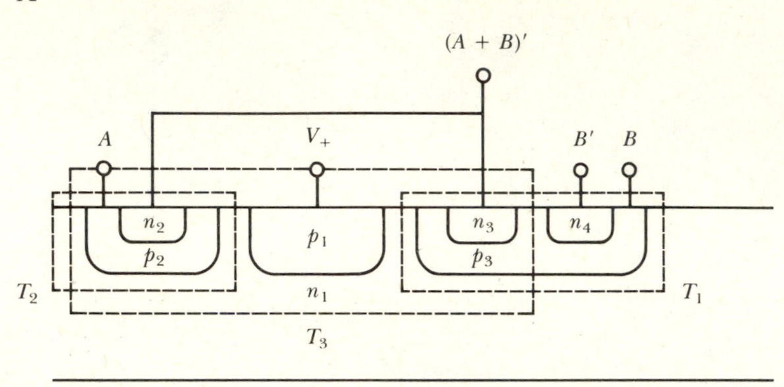

B

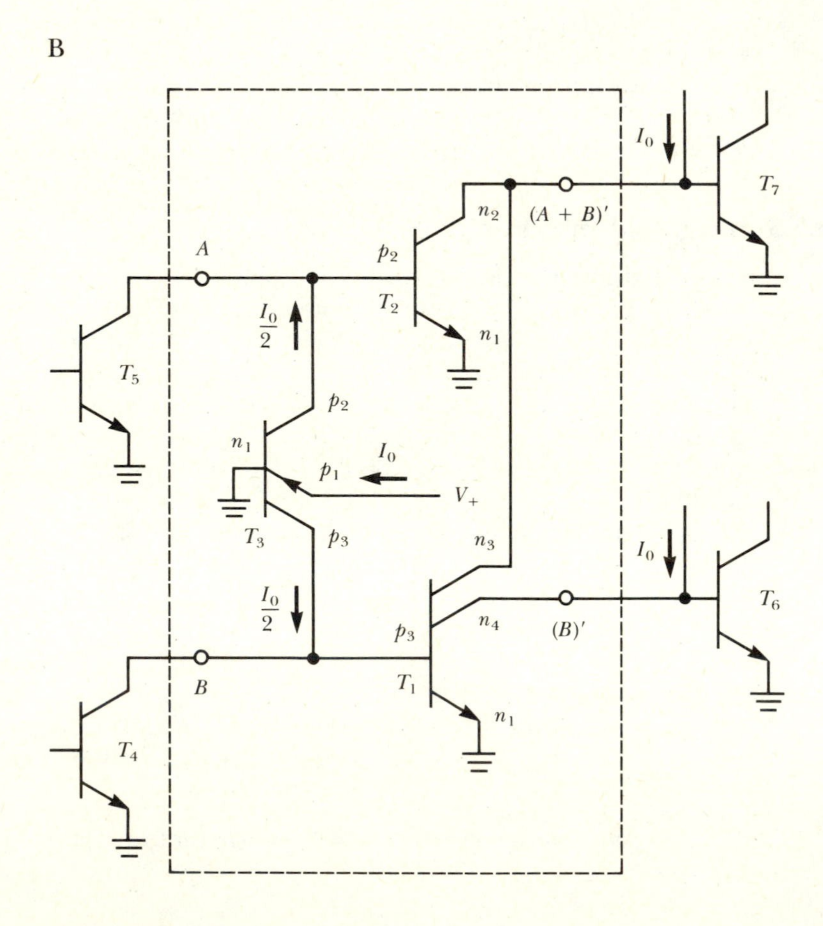

2. The *n-p-n* transistors are run backwards with the collectors at the surface.
3. Transistor layouts are merged.

Figure 4.39 shows a standard first-generation I^2L layout, but modifications have been developed for improved performance. For example, an oxide guard ring may be used to reduce parasitics and increase speed [29]. As an example of the capabilities offered by advanced I^2L structures, a 16-bit bipolar microprocessor based on I^2L has been developed by Fairchild [30].

The I^2L output voltage swing is quite small compared to standard TTL levels, as can be seen by considering the gate operation for the circuit shown in Figure 4.39. If the transistor on the preceding gate, T_4, is cut off, the current $I_0/2$ is supplied to the base of T_1. Consequently, T_1 is saturated and the output voltage at the collector of T_1 is that of a saturated transistor, $(V_{CE})_{sat}$. However, if T_4 is saturated, then the current $I_0/2$ is sunk through T_4 and transistor T_1 is cut off. In this case, the output voltage is determined by the base-emitter voltage of transistor T_6 in the next gate, which is $(V_{BE})_{sat}$. The output logic swing is therefore

$$V_{swing} = (V_{BE})_{sat} - (V_{CE})_{sat} \simeq 0.7 \text{ V} \tag{4.76}$$

The dc supply voltage must be sufficient to drive the current-injecting *p-n-p* transistors, which are connected to the supply via a series resistor, and can therefore be as low as approximately 1 V.

EXAMPLE 4.10

Consider the I^2L gate of Figure 4.39 with $I_0 = 100\ \mu A$. Assume the *n-p-n* transistors are characterized by these Ebers-Moll parameters:

$$I_{ES} = 10^{-16} \text{ A}$$

$$I_{CS} = 0.867 \times 10^{-16} \text{ A}$$

$$\alpha_F = 0.85$$

$$\alpha_R = 0.98$$

Calculate the low-output voltage at transistor T_1 if only one of the collectors is sinking current and again if both collectors are sinking current. What would happen if there was a third collector on T_1 also sinking current I_0? Assume that the parasitic resistors, R_E, R_B, and R_C are negligible.

Solution When one collector is sinking current I_0, the terminal currents are:

$$I_C = 100\ \mu A$$

$$I_B = 50\ \mu A$$

$$I_E = 150\ \mu A$$

Since the transistor is saturated, the Ebers-Moll equations become, from Eqs. (4.24) and (4.25),

$$150 \times 10^{-6} \simeq 10^{-16}\left[\exp\left(\frac{qV_{BE}}{kT}\right) - (0.98)(0.867)\exp\left(\frac{qV_{BC}}{kT}\right)\right]$$

and

$$100 \times 10^{-6} \simeq 10^{-16}\left[(0.85)\exp\left(\frac{qV_{BE}}{kT}\right) - (0.867)\exp\left(\frac{qV_{BC}}{kT}\right)\right]$$

Solving the two equations for the two unknown voltages yields $V_{BC} = 0.732$ V and $V_{BE} = 0.745$ V. Therefore, when one collector is sinking current, the output is

$$V_{\text{low}} \simeq 0.745 - 0.732 = 0.013 \text{ V}$$

This voltage is referred to as the intrinsic saturation voltage since it does not include voltage drops across the parasitic resistors shown in Figure 4.6. Because of these resistors, the actual gate output voltage, which corresponds to the terminal saturation voltage, will be somewhat higher.

When both collectors are sinking current, the terminal currents are:

$$I_C = 200\ \mu\text{A}$$
$$I_B = 50\ \mu\text{A}$$
$$I_E = 250\ \mu\text{A}$$

Repeating the Ebers-Moll equations yields

$$250 \times 10^{-6} \simeq 10^{-16}\left[\exp\left(\frac{qV_{BE}}{kT}\right) - (0.98)(0.867)\exp\left(\frac{qV_{BC}}{kT}\right)\right]$$

and

$$200 \times 10^{-6} \simeq 10^{-16}\left[(0.85)\exp\left(\frac{qV_{BE}}{kT}\right) - (0.867)\exp\left(\frac{qV_{BC}}{kT}\right)\right]$$

Solving for the two voltages yields $V_{BE} \simeq 0.746$ V and $V_{BC} \simeq 0.712$ V. Therefore, when both collectors are sinking current, the intrinsic saturation voltage increases to

$$V_{\text{low}} \simeq 0.746 - 0.712 = 0.034 \text{ V}$$

If a third collector were added to transistor T_1, then the total collector current could be 300 μA. Since the base current is 50 μA, a transistor current gain of at least 6 is required. However, for the device specified, the common emitter current gain is found from Eq. (4.36) to be

$$\beta = \frac{0.85}{1 - 0.85} \simeq 5.7$$

which is not sufficient to support the required current. So the fan-out of this particular gate is only 2. Generally, I^2L circuits have fan-outs of 2 to 5.

4.7.4 Schottky/I^2L Combinations

In spite of the small logic voltage swing, standard I^2L is not ultrafast because it is saturating logic, with typical gate delays of several nanoseconds. One way to increase speed is to further reduce the output logic swing; to do so the gate outputs to the *n-p-n* collectors can be coupled with Schottky barrier diodes, with the diode cathodes connected to the collectors, as shown in Figure 4.40. For this configuration, the low output voltage is

$$V_{\text{low}} \simeq V_{SBD} - (V_{CB})_{\text{sat}} + (V_{BE})_{\text{sat}} \tag{4.77}$$

where V_{SBD} is the forward voltage drop across the diode. The high ouput voltage is

$$V_{\text{high}} \simeq (V_{BE})_{\text{sat}} \tag{4.78}$$

Therefore, the swing in logic voltage output is

$$V_{\text{swing}} \simeq (V_{CB})_{\text{sat}} - V_{SBD} \tag{4.79}$$

Figure 4.40 Schottky I^2L logic.

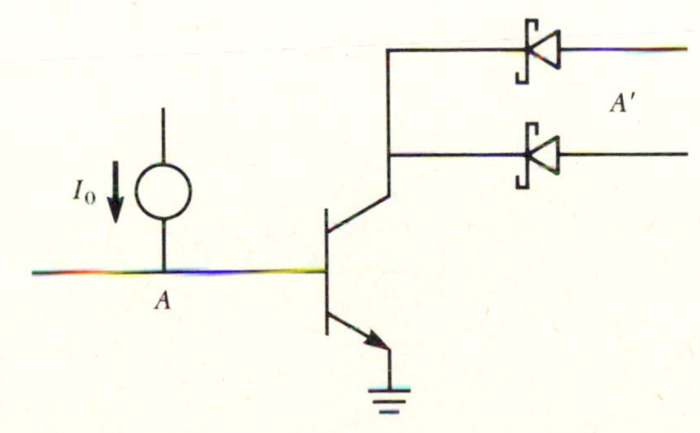

For platinum-silicide diodes, the output swing is about 0.3 V. The diode coupling arrangement at the ouputs also isolates the outputs from each other. Therefore, separate collectors are not required, with some improvement in packing density. This output coupling scheme is often referred to as Schottky I^2L.

When I^2L output-isolation diodes are combined with Schottky-clamp diodes on the base-collector junctions, Schottky transistor logic (STL) results. An example is shown in Figure 4.41. Now, Schottky barrier diodes with different barrier heights are required for clamping and isolation, since the output logic swing is equal to the difference between the forward voltage drops of the two diode types. The low output voltage for STL is

$$V_{\text{low}} = (V_{BE})_{\text{sat}} + V_{SBD1} - V_{SBD2} \tag{4.80}$$

and the high output voltage is $(V_{BE})_{\text{sat}}$. Therefore the output voltage swing is

$$V_{\text{swing}} = V_{SBD2} - V_{SBD1} \tag{4.81}$$

Examples of Schottky diode combinations for STL include titanium for the output diodes and platinum or aluminum for the clamp diodes. Output logic swings are quite small, around 0.2 V. A drawback of STL is that it cannot be fabricated by standard processes in which only one type of Schottky metal is used. The resulting fabrication process is more complicated and consequently yields are lower.

Another way to improve I^2L speed is to use a merged *p-n-p* transistor, rather than a Schottky clamp, to control the level of the *n-p-n* transistor saturation. This method is referred to as integrated Schottky logic (ISL) and offers the advantage of being suited to standard processes. For a self-aligned

Figure 4.41 Schottky transistor logic (STL).

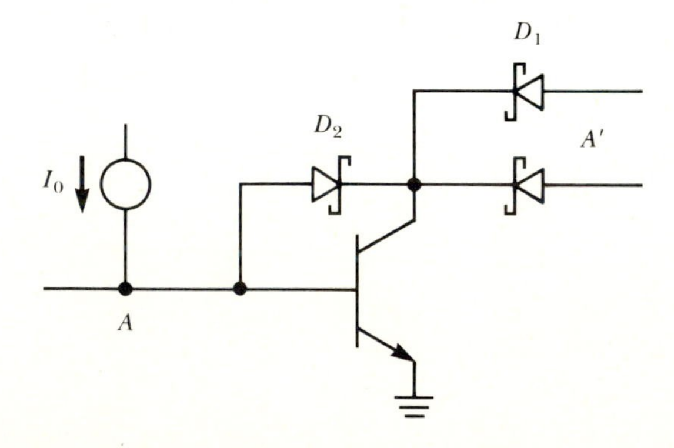

bipolar process with a feature size of 2 μm, gate delays of 1 ns are reported for ISL and 0.9 ns for STL [31].

4.8 Bipolar Scaling Considerations

Some scaling considerations are common to BJTs and MOSFETs [32]. For example, to scale the depletion layer widths, doping is increased as size is decreased in both cases. However, the MOSFET doping level is limited by oxide breakdown (as discussed in Section 3.7), but the BJT doping level is not, so it can be argued that miniaturization can, in principle, be carried further with BJTs.

Bipolar scaling is more complex than MOSFET scaling and a simple set of scaling laws analogous to the MOSFET scaling laws does not exist. An interesting consequence, and complication of bipolar scaling is that device current densities must increase to take advantage of scaling-related improvements in speed. This situation results in part because the junction voltages required to turn on a BJT do not scale down since they are already low. Gate delay times may be expressed in general form as

$$t_D = CV/I = C_o V/J \tag{4.82}$$

where C is the capacitance to be charged and discharged over voltage range V, and where C_o is the capacitance per unit area [3]. Since C_o increases with scaling, owing to decreasing depletion layer widths, the current density, J, must also increase.

Current densities of approximately 10^5 A/cm^2 have been reported in scaled-down bipolar integrated circuit devices [26]. At these levels, high-injection phenomena come into play and devices enter new regimes of operation, with injected minority carrier densities on the order of, or larger than, the equilibrium majority carrier densities. Under these conditions, simple models such as the Ebers-Moll model do not apply.

BIBLIOGRAPHY

Further reading on BJT fundamentals may be found in the following:

Getreu, I. E., *Modeling the Bipolar Transistor* (New York: Elsevier Scientific, 1978).

Neudeck, G. W., *The Bipolar Junction Transistor,* Modular Series on Solid State Devices, vol. 3, R. F. Pierret and G. W. Neudeck, eds. (Reading, Mass.: Addison-Wesley, 1983).

Sze, S. M., *Physics of Semiconductor Devices,* 2nd ed. (New York: Wiley, 1981).

The following contain further reading on applications of bipolar technology to integrated circuits:

Colclaser, R. A., *Microelectronics Processing and Device Design* (New York: Wiley, 1980).

Glaser, A. B., and Subak-Sharpe, G. E., *Integrated Circuit Engineering* (Reading, Mass.: Addison-Wesley, 1977).

Hamilton, D. J., and Howard, W. G., *Basic Integrated Circuit Engineering* (New York: McGraw-Hill, 1975).

Lohstroh J. "Devices and Circuits for Bipolar (V)LSI," *Proceedings of the IEEE,* vol. 69 (1981), 812–826.

Till, W. C., and Luxon, J. T., *Integrated Circuits: Materials, Devices, and Fabrication* (Englewood Cliffs, N.J.: Prentice-Hall, 1982).

Wiedmann, S. K., "Advances in Bipolar VLSI Circuits and Technologies", *IEEE Journal of Solid-State Circuits,* vol. SC-19 (1984), 282–291.

REFERENCES

1. "U.S. Markets Forecast," *Electronics,* Jan. 6, 1977, p. 90.
2. "U.S. Markets Forecast," *Electronics,* Jan. 12, 1984, pp. 138–139.
3. P. M. Solomon, "A Comparison of Semiconductor Devices for High-Speed Logic," *Proceedings of the IEEE,* 70 (1982), 489–509.
4. J. J. Ebers and J. L. Moll, "Large-Signal Behavior of Junction Transistors," *Proceedings IRE,* 42 (1954), 1761–1772.
5. J. M. Early, "Effects of Space Charge Layer Widening in Junction Transistors," *Proceedings IRE,* 40 (1952), 1401.
6. B. G. Streetman, *Solid State Electronic Devices,* 2nd ed. (Englewood Cliffs, N.J.: Prentice-Hall, 1980).
7. L. J. Giacoletto, "Study of *p-n-p* Alloy Junction Transistors from DC through Medium Frequencies," *RCA Review* 15 (1954), 506–562.
8. H. K. Gummel and H. C. Poon, "An Integral Charge Control Model of Bipolar Transistors," *Bell System Technical Journal,* 49 (1970), 827–851.
9. A. Vladimirescu, K. Zhang, A. R. Newton, D. O. Pederson, A. Sangiovanni-Vincentelli, "SPICE Version 2G.6 User's Guide," Electronics Research Laboratory, University of California, Berkeley, 1983.
10. L. W. Nagel, "SPICE2: A Computer Program to Simulate Semiconductor Circuits," Memorandum No. ERL-M520, Electronics Research Laboratory, University of California, Berkeley, 1975.
11. R. M. Burger and R. P. Donovan, *Fundamentals of Silicon Integrated Circuit Technology,* vol. 2, *Bipolar and Unipolar Transistors,* (Englewood Cliffs, N.J.: Prentice-Hall, 1968).
12. I. E. Getreu, *Modeling the Bipolar Transistor* (New York: Elsevier Scientific, 1978).
13. D. J. Hamilton and W. G. Howard, *Basic Integrated Circuit Engineering* (New York: McGraw-Hill, 1975).
14. A. Hayasaka, Y. Tamaki, M. Kawamura, K. Ogiue, and S. Ohwaki, "U-Groove Isolation Technique for High Speed Bipolar VLSI's," *IEEE IEDM Technical Digest* (1982), 62–65.
15. W. C. Till and J. T. Luxon, *Integrated Circuits: Materials, Devices, and Fabrication* (Englewood Cliffs, N.J.: Prentice-Hall, 1982).
16. T. H. Ning, R. D. Isaac, P. M. Solomon, D. Duan-Lee Tang, Hwa-Nien Yu, G. C. Feth, and S. K. Wiedmann, "Self-Aligned Bipolar Transistors for High-Performance and Low-Power-Delay VLSI," *IEEE Transactions on Electron Devices,* ED-28 (1981), 1010–1013.

17. Z. Yu, B. Ricco, and R. W. Dutton, "A Comprehensive Analytical and Numerical Model of Polysilicon Emitter Contacts in Bipolar Transistors," *IEEE Transactions on Electron Devices,* ED-31 (1984), 773–784.
18. M. Suzuki and S. Horiguchi, "A 333 ps/800 MHz 7 K-Gate Bipolar Macrocell Array Employing 4 level Metalization," *IEEE Journal of Solid State Circuits,* SC-19 (1984), 474–479.
19. J. C. Irvin, "Resistivity of bulk silicon and of diffused layers in silicon," *Bell System Technical Journal,* 41 (1962), 387–410.
20. R. A. Colclaser, *Microelectronics Processing and Device Design* (New York: Wiley, 1980), Chap 9.
21. "Single GaAs Chip Holds Circuitry for TV Reception," *Electronics,* March 8, 1984, pp. 81–82.
22. J. Frey, "Hybrid and Monolithic Microwave Integrated Circuits," in *Microwave Integrated Circuits,* ed. J. Frey (Dedham, Mass.: Artech House, 1975), pp. xvii–xxi.
23. C. Aitchison, R. Davies, I. D. Higgins, S. R. Longley, B. H. Newton, J. F. Wells, and J. C. Williams, "Lumped Circuit Elements at Microwave Frequencies," *IEEE Transactions on Microwave Theory and Techniques,* MTT-19 (1971), 928–937.
24. D. A. Daly, S. P. Knight, M. Caulton, and R. Ekholdt, "Lumped Elements in Microwave Integrated Circuits," *IEEE Transactions on Microwave Theory and Techniques,* MTT-15 (1967), 713–721.
25. J. V. Wait, L. P. Huelsman, and G. A. Korn, *Introduction to Operational Amplifier Theory and Applications* (New York: McGraw-Hill, 1975), Chap. 4.
26. D. Tang, P. Solomon, T. Ning, R. Isaac, and R. Burger, "1.25 μm Deep-Groove-Isolated Self-Aligned Bipolar Circuits," *IEEE Journal of Solid State Circuits,* SC-17 (1982), 925–931.
27. H. Taub and D. Schilling, *Digital Integrated Electronics* (New York: McGraw-Hill, 1977).
28. D. A. Hodges and H. G. Jackson, *Analysis and Design of Digital Integrated Circuits* (New York: McGraw-Hill, 1983).
29. S. K. Wiedmann, "Advancements in Bipolar VLSI Circuits and Technologies," *IEEE Journal of Solid-State Circuits,* SC-19 (1984), 282–291.
30. S. Mor, H. Hingarh, M. Vora, D. Wilnai, and T. Longo, "A 16b Microprocessor for Realtime Applications," *IEEE ISSCC Digest of Technical Papers* (1983), 28–29.
31. F. W. Hewlett and D. A. Erickson, "STL Versus ISL: An Experimental Comparison," *IEEE Journal of Solid-State Circuits,* SC-19 (1984), 195–206.
32. R. W. Keyes, "Physical Limits in Digital Electronics," *Proceedings of the IEEE,* 63 (1975), 740–767.

PROBLEMS

1. Consider a junction-isolated bipolar transistor of minimum area with a single base stripe, single emitter stripe, and single collector stripe. The minimum line width, m, is 2 μm, and the epitaxial layer is 2 μm thick. How many isolated devices could be put on a chip whose area is 1 cm^2? If the chip is to dissipate no more than 2 W, how much power per device is allowed? Repeat for the case of $m = 5$ μm and $m = 0.5$ μm, again assuming that the epitaxial layer has thickness equal to m.

2. Derive an expression for the BJT emitter injection efficiency in terms of the base length, the minority carrier mobilities and diffusion lengths, the base doping, and the emitter doping. [See Eq. (4.34).]

3. Show that the BJT common emitter current gain, β, is equal to the carrier lifetime divided by the base transit time.

4. Consider a saturated BJT, as in Figure 4.8, driven by a base current that switches abruptly from I_B to $-I_B$. Use the charge-control model to solve for the resulting storage delay time.

5. Consider two different integrated circuit diodes under forward bias. The first diode consists of the base-emitter junction with the collector shorted to the base. The second diode consists of the base-emitter junction with the collector open. Use the Ebers-Moll model to develop qualitative sketches of the resulting minority carrier distributions in the base under forward bias. Which diode would have the lower turn-on voltage? Which diode would be faster for switching applications? Justify your answers.

6. Design a 10-pF silicon MOS capacitor with an oxide thickness of 500 Å. How many masking steps are required? As the designer, what would you specify as the maximum allowed voltage?

7. Consider the two parasitic *p-n-p* transistors associated with a lateral *p-n-p* device. Under which of the four possible biasing conditions on the lateral *p-n-p*, if any, will the parasitic transistors be activated?

8. A 1-kΩ resistor is to be formed by a base diffusion that has a sheet resistance of 200 Ω/□. Design a layout for a minimum-area resistor if the maximum allowed power dissipation for the device (per chip surface area) is 20 W/cm^2 and the voltage drop across the resistor is to be 5 V. Include resistor contacts.

9. A lateral *p-n-p* BJT has a parasitic capacitance, C_{BS}, between the base and substrate, rather than the collector-substrate capacitance, C_{CS}, associated with the standard *n-p-n* transistor. Calculate the cut-off frequency, f_T, at room temperature, for the common emitter small-signal current-gain for a lateral *p-n-p* with the following characteristics:

$$\beta_0 = 20$$
$$C_{BC} = 2 \text{ pF}$$
$$(C_{BE})_{\text{scl}} = 0.6 \text{ pF}$$
$$C_{BS} = 3.5 \text{ pF}$$

$r_{CC} = 75\ \Omega$
$r_{BB} = 150\ \Omega$
$W_B = 5\ \mu\text{m}$
$I_C = 0.25\ \text{mA}$
$D_B = 10\ \text{cm}^2/\text{s}$

10. Consider the I²L inverter shown in Figure 4P.1. The value for I_0 is 10 μA, and the *n-p-n* transistor is characterized by the Ebers-Moll parameters below:

$I_{ES} = 10^{-16}$ A
$I_{CS} = 0.947 \times 10^{-16}$ A
$\alpha_F = 0.90$
$\alpha_R = 0.95.$

Calculate the output logic swing at collector C_1, if only C_1 is connected to a subsequent gate. Assume the following gate is a similar inverter, which has a fan-out of 1. Repeat the problem if all three collectors are connected to such inverters.

Figure 4P.1 The I²L inverter for Problem 10.

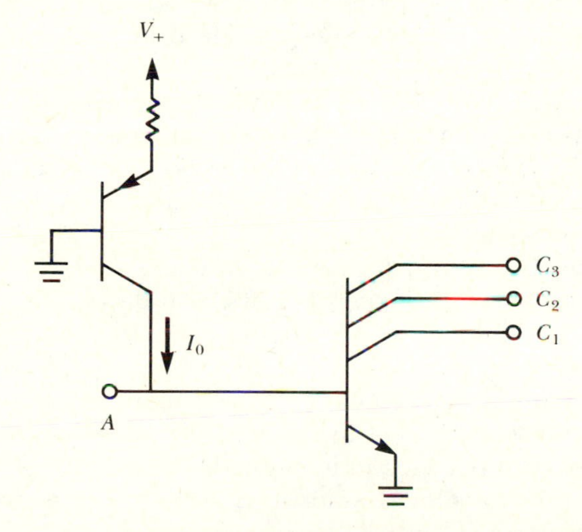

5 Basic Principles of Integrated Circuit Layout

5.1 Introduction

The layout of an integrated circuit, or the actual placement and arrangement of an electronic circuit as an integrated structure, is a design process, and consequently unique layout solutions do not exist. Several layout designs may meet a given set of specifications but may differ in terms of chip area, chip heating, and parasitic effects. The final choice of a layout often involves design trade-off decisions.

However, several general considerations for integrated circuit design should be mentioned at the outset.

1. Resistors: In discrete circuits a 10-Ω, $\frac{1}{4}$-W resistor with 5% precision is the same size and approximate price as a 1-MΩ, $\frac{1}{4}$-W resistor with 5% precision. But in standard-process integrated circuits, large resistors are more costly because of the large amounts of chip real estate they use. Also, small resistors are subject to large variations in value, as discussed in Section 4.6.2. In standard-process bipolar integrated circuits, for example, resistor values tend to be limited to the range between 50 Ω and 100 kΩ. Much larger resistors require additional fabrication methods involving formation of a high-resistivity layer by methods such as shallow ion implantation or polysilicon deposition.
2. Capacitors: Capacitors are also area intensive, and unless they are quite small they should be avoided. When capacitors are used in integrated circuit layouts, their values are generally $\leq$100 pF.
3. Inductors: Except in certain microwave frequency applications, inductors are not used in integrated circuit layouts.
4. Transistors: Since transistors are part of the standard integrated circuit process and usually require relatively small amounts of area, they are heavily used in integrated circuit design.
5. Component trimming: Although component trimming is possible for monolithic silicon integrated circuits, as discussed in Section 6.7, the most commonly used fabrication procedures do not incorporate postfabrication adjustment of component values. This limitation is partly offset by

the fact that matched components are fairly easily obtained. In contrast, it is common practice to trim resistors on hybrid thick-film integrated circuits.

6. Isolation regions: The number of isolation regions should be minimized, to the extent that doing so does not result in a cumbersome layout, since isolation regions also take up chip area. For example, if *n-p-n* bipolar junction transistors (BJTs) share a collector, they can be placed in the same isolation area. Also, diffused resistors can be placed in common isolation regions.
7. Geometry: For a given integrated circuit process, well defined geometric design rules must be followed.
8. Power: The layout should ensure that the chip will not overheat or develop hot spots.

The elementary layout examples in this chapter demonstrate the application of several of these general considerations.

5.2 Power-Density Calculations

5.2.1 General Methods

One factor to be considered in integrated circuit layout is device power dissipation and chip temperature. In low-power circuits, such as complementary metal-oxide-semiconductor (CMOS) digital integrated circuits, power dissipation is not a limiting factor and device size is determined by lithographic constraints and current ratings. However, in circuits where power dissipation is appreciable, the device size may be determined by thermal constraints. Consequently, such devices may be much larger than if lithographic capabilities were the only concern.

Consider an integrated circuit chip mounted on a heat sink as shown in Figure 5.1. The substrate refers to the semiconductor wafer on which the integrated circuit resides, and this discussion focuses on a particular integrated circuit device that is dissipating power P. Since the wafer substrate is several hundred micrometers thick and the device, as formed by diffusion or ion implantation, is in the top few micrometers, it is not a great exaggeration to show the device as residing on the surface of the substrate. Heat flow from the device is three-dimensional, flowing both laterally and vertically in the substrate. To a first approximation, however, the heat flow may be considered one-dimensional, flowing entirely toward the heat sink.

Assuming an infinite heat sink, which maintains a temperature equal to the ambient, and a substrate thickness ΔX, the rate of heat removal from the device, dQ/dt, can be expressed as

$$dQ/dt = \mathcal{K} A(\Delta T/\Delta X) \tag{5.1}$$

Figure 5.1 Power-density considerations for an integrated circuit device.

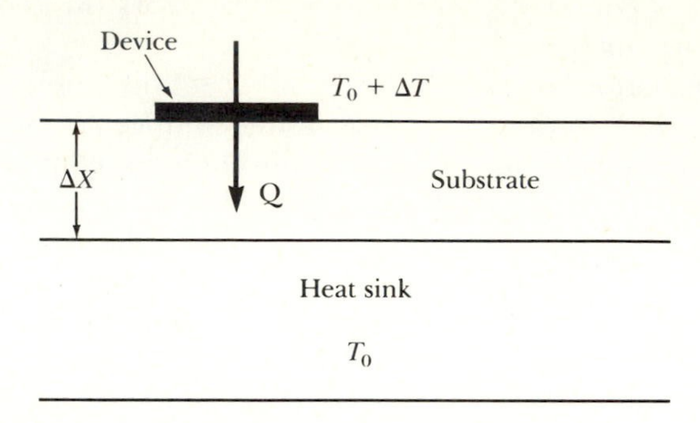

where $\mathcal{K}$ is the thermal conductivity of the substrate, the device area is A, and the device temperature is ΔT above the ambient. Therefore the power dissipation per area from the device relates to the temperature gradient across the substrate according to the expression,

$$P/A = \mathcal{K}(\Delta T/\Delta X) \tag{5.2}$$

For a given substrate thickness and thermal conductivity, then, there is a well-defined relationship between the maximum tolerated increase in temperature, ΔT, and the maximum allowed power dissipation per square centimeter.

EXAMPLE 5.1

Consider a silicon chip 0.25 mm thick and mounted on a copper substrate. The copper substrate is considered to act as an infinite heat sink, and the thermal conductivity of silicon is 1.5 W/cm · K. What is the maximum allowed power dissipation per square centimeter for an integrated circuit device if the maximum allowed temperature on the chip is to be 10°C above ambient? Repeat for the case of 100°C.

Solution For a given maximum allowed temperature gradient, the maximum allowed power dissipation is

$$(P/A)_{max} = \mathcal{K}(\Delta T/\Delta X)_{max}$$

For the case of 10°C temperature rise,

$$(P/A)_{max} = 1.5(10/0.025) = 600 \text{ W/cm}^2$$

So if the device is 100 μm by 100 μm, for example, the maximum allowed power dissipation is 60 mW.

If the allowed temperature rise were increased to 100°C, the maximum power density would be 6000 W/cm^2.

The previous example shows that a silicon chip running hot with a good heat sink can dissipate a considerable amount of power per unit area. Such conditions apply to single-chip power and audio amplifiers, which may dissipate powers in the range of 20 to 40 W. Figure 5.2 shows temperature profiles across a bipolar chip that contains an output-stage power transistor in the upper left-hand corner of the die [1]. As was the case for Ex. 5.1, the chip is 0.25 mm thick and is mounted on a copper heat sink. The simulated temperature profiles across the chip are based on three-dimensional calculations for the case when the transistor is dissipating 40 W. Since the power transistor, whose position along the x-axis is shown by a heavy black line, has

Figure 5.2 Temperature profile for a bipolar integrated circuit die with a power transistor. The simulation is performed along line A, shown in the inset, and the profiles show the temperature on the chip surface relative to the heat sink.
SOURCE: From P. Antognetti, G. Bisio, F. Curatelli, and S. Palara, "Three-Dimensional Transient Thermal Simulation: Application to Delayed Short Circuit Protection in Power IC's," *IEEE Journal of Solid-State Circuits,* SC-15 (1980), 277–281. © 1980 IEEE.

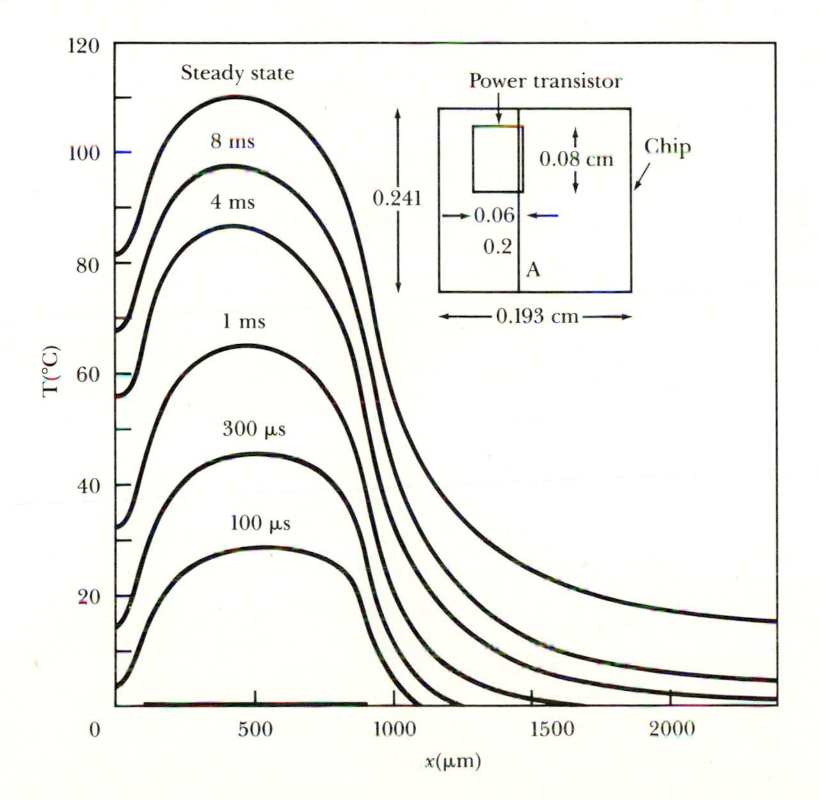

dimensions of about 800 μm by 600 μm, the average device power density is approximately 8.3×10^3 W/cm^2. In steady state, the resulting peak temperature on the chip, near the center of the transistor, is reported to be 110°C. By comparison, using Eq. (5.2) and the methods of Ex. 5.1, the predicted power density for this temperature would be somewhat smaller—6.6×10^3 W/cm^2. A somewhat smaller answer is expected here because the calculation techniques of Ex. 5.1 do not account for lateral heat dissipation.

The previous discussion has assumed a high-quality heat sink to be in intimate contact with the wafer. More typically, however, integrated circuits are mounted in packages that are not even close to such an ideal situation. For example, a 40-pin, dual-in-line package (DIP) can tolerate about 2 W of dissipated power when board mounted and air cooled with a fan. If the chip is 5 mm by 5 mm, then the corresponding power density averaged over the chip is 8 W/cm^2. Since not all the area on the silicon chip is actively involved in generating power, another factor of 2 or 3 is in order for calculating allowed power density for a given device. Clearly, considerable variations in $(P/A)_{max}$ result, depending on the chip mounting, the chip size, and the circuitry's tolerance for high temperatures. Our calculations in the next section use a representative value for $(P/A)_{max}$ of 20 W/cm^2 for designing device layout. Since not all the chip area is allocated to active devices, the average power density over the chip will be considerably less than 20 W/cm^2.

5.2.2 A Bipolar-Inverter Example [2]

A first simple layout example that illustrates several of the concepts in the previous section is a bipolar inverter with a resistor load and a speed-up capacitor. The use of a speed-up capacitor is, in fact, a circuit design technique occasionally applied in discrete-device pulse-circuits to improve rise- and fall-times, but generally this technique is not used in monolithic integrated circuits because of area considerations. However, this example includes the capacitor to show several circuit element types in one simple layout and to demonstrate the area requirements of different integrated circuit devices.

Figure 5.3 shows the inverter to be designed as a silicon integrated circuit. Since the circuit elements must be electrically isolated from one another, one design approach would be to use a separate isolation region for each of the four devices. However, keeping in mind the general design principle of reducing the number of isolation regions, it is preferable to place the resistors in a common isolation region, since resistors may be self-isolated. Consequently, three isolation regions result, as shown in Figure 5.3.

Next, we turn our attention to the design of each device, leaving the relative placement and interconnection of the devices as the last step. For simplicity, we use the standard BJT process with *p*-type isolation walls, as discussed in Section 4.4.1, and we assume an air-cooled DIP chip package.

Figure 5.3 Bipolar-inverter example with speed-up capacitor. A. Circuit schematic. B. Isolation regions.

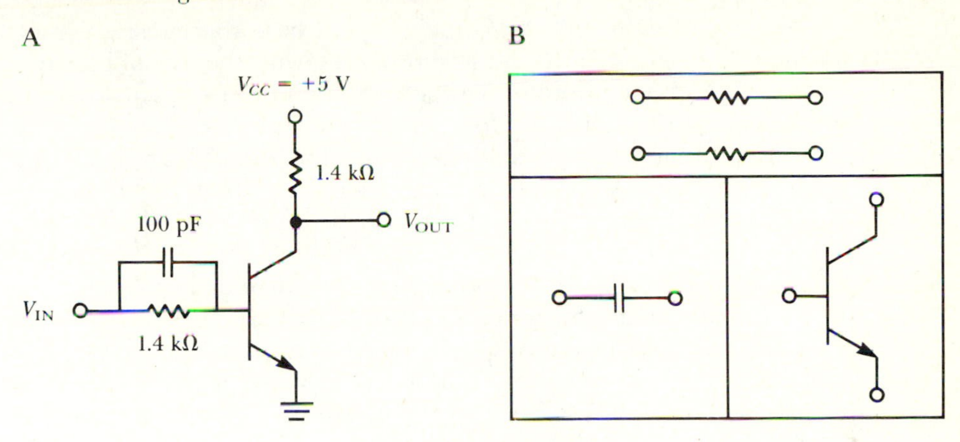

In general, either a *p-n* junction capacitor or an MOS capacitor can be used. But in this case, since the capacitance value is to be fixed at 100 pF and the voltage across the capacitor is variable, the MOS capacitor is clearly the appropriate choice. Recall from Ex. 4.9 that an MOS capacitor with an n^+ silicon layer for the bottom electrode results in a capacitor value that is largely independent of capacitor voltage. If we choose a capacitor oxide thickness of 1000 Å, then the capacitor area is calculated using Eq. (4.69) as follows:

$$A = 100 \text{ pF} \frac{1000 \times 10^{-8} \text{ cm}}{(3.9)(8.85 \times 10^{-14} \text{ F/cm}^2)}$$

or $A = 2.90 \times 10^{-3}$ cm^{-2}. For a square device, the length and width of the capacitor oxide layer is therefore $L = W = 538$ μm.

The device cross section will be of the general form shown in Figure 4.28B, and the following photolithographic steps, or mask levels, can be identified:

1. Isolation diffusion mask
2. n^+ bottom-layer mask
3. Oxide definition mask
4. Contact window mask
5. Metalization mask

The n^+ layer can be established during the emitter diffusion and therefore does not represent a step additional to the standard bipolar fabrication process. However, the definiton of the capacitor oxide layer, in which the oxide is etched away to allow oxide regrowth to a specified thickness, does represent an additional step.

For the resistors, the base diffusion can be used to establish p-type resistors that are self-isolated in an n-type region, as discussed in Section 4.6.2. For this example, let us assume a typical base sheet-resistance of 200 Ω/□. Then each resistor must be seven squares, with the size of the squares determined by either photolithographic or power-density considerations. If the minimum diffusion width were 5 μm, for example, then according to photolithographic limits alone, the resistor's minimum size would be 5 μm by 35 μm. However, a quick power calculation shows that the resulting power dissipation per square centimeter would be quite large for an air-cooled DIP mounted chip.

When the transistor is saturated, the current through the load resistor reaches its maximum value of approximately 5 V/1.4 kΩ = 3.6 mA. The resulting power dissipation is 18 mW. If the resistor were 5 μm × 35 μm, then the resulting power density would be 10 kW per square centimeter, which is even higher than for the power transistor in Figure 5.2, which is mounted on a copper heat sink. Therefore the resistor size in this example will be determined by power-density calculations.

Consider the load resistor design that results from taking $(P/A)_{max} = 20$ W/cm^2, as discussed in the previous section. Using this guideline, the resistor area is

$$A = \frac{18 \text{ mW}}{20 \text{ W/cm}^2} = 9 \times 10^{-4} \text{ cm}^2$$

Since $L = 7W$, the length and width are calculated as, $7W^2 = 9 \times 10^{-4}$ cm^2, such that $W = 113$ μm and $L = 794$ μm. Assuming that the input voltage ranges up to 5 V, the maximum power dissipation of the input resistor will be slightly less than that of the load resistor because of the base-emitter voltage drop. However, the maximum power dissipations for the two resistors will be quite close, so for simplicity, in this example we take the two resistors to be of equal physical size.

Referring to the resistor cross sections shown in Figures 4.25 and 4.26, we can identify the following mask levels for the resistor fabrication:

1. n^+ buried-layer mask
2. Isolation diffusion mask
3. Base diffusion mask
4. n^+ ohmic contact mask
5. Contact window mask
6. Metalization mask

This mask sequence does not add any new steps to the standard process, since the n^+ ohmic contact to the "sea" of n-type material around the p-type resistors can be established during the emitter diffusion.

Power-dissipation considerations may also be used to determine the minimum transistor size in this example. Since the power dissipated in the transistor is approximately equal to the product $I_C V_{CE}$, and $V_{CE} = 5 - I_C(1.4\ \text{k}\Omega)$, the transistor power dissipation can be written as $P = 5I_C - (I_C)^2(1.4\ \text{k}\Omega)$. A first derivative test yields a maximum in power at $I_C = 1.79$ mA and $V_{CE} = 2.5$ V. Therefore the peak power dissipation in the transistor is 4.5 mW, and the corresponding minimum area is

$$A = \frac{4.5\ \text{mW}}{20\ \text{W/cm}^2} = 2.25 \times 10^{-4}\ \text{cm}^2$$

Since most of this power is dissipated in the active portion of the transistor—that is, in the portion under the emitter—a conservative design decision would make the emitter area equal to 2.5×10^{-4} cm^2. Also, since an elongated emitter reduces emitter crowding effects, a reasonable approach would use an emitter elongation factor of 2. Consequently, the emitter dimensions will be set at approximately 106 μm by 212 μm. And once the emitter dimensions are fixed, the junction isolated BJT layout shown in Figure 4.14 allows the rest of the transistor design to follow, given the minimum dimension m, which separates adjacent features.

For the BJT, referring to the process illustrated in Figure 4.15, we can identify the following mask levels:

1. n^+ buried-layer mask
2. Isolation diffusion mask
3. Base diffusion mask
4. n^+ emitter diffusion mask
5. Contact window mask
6. Metalization mask

Adding all the mask steps in their proper order produces the following combined mask list for this integrated circuit:

1. Buried n^+ layer definition (n^+ diffusion)
2. Isolation region definition (p^+ diffusion)
3. Base and resistor definition (p diffusion)
4. Emitter, capacitor bottom plate, and n^+ contact definition (n^+ diffusion)
5. Capacitor oxide definition (oxide etch, followed by oxidation)
6. Contact window definition (oxide etch)
7. Metalization (metal etch)

Some additional layout considerations are as follows: The p-type base diffusion should also include the isolation regions. Recall that boron tends to be depleted at the Si-SiO_2 interface, so the added acceptors later in the fabrication sequence are helpful. Also, n^+ regions must be used where ohmic contacts to n-type silicon are required, and contact cuts are included on the capacitor oxide definition mask to reduce the contact-cut etch time. And finally, bonding pads are required on the chip to provide electrical access to the circuit. Typical pad dimensions for wire-bonded leads are approximately 100 μm by 100 μm.

Figure 5.4 shows one possible inverter layout that illustrates the relative size and position of the four devices, and Figure 5.5 shows the corresponding mask set. The patterns shown in these figures were generated on an interactive graphics terminal using an integrated circuit design software

Figure 5.4 A layout for the bipolar-inverter example.

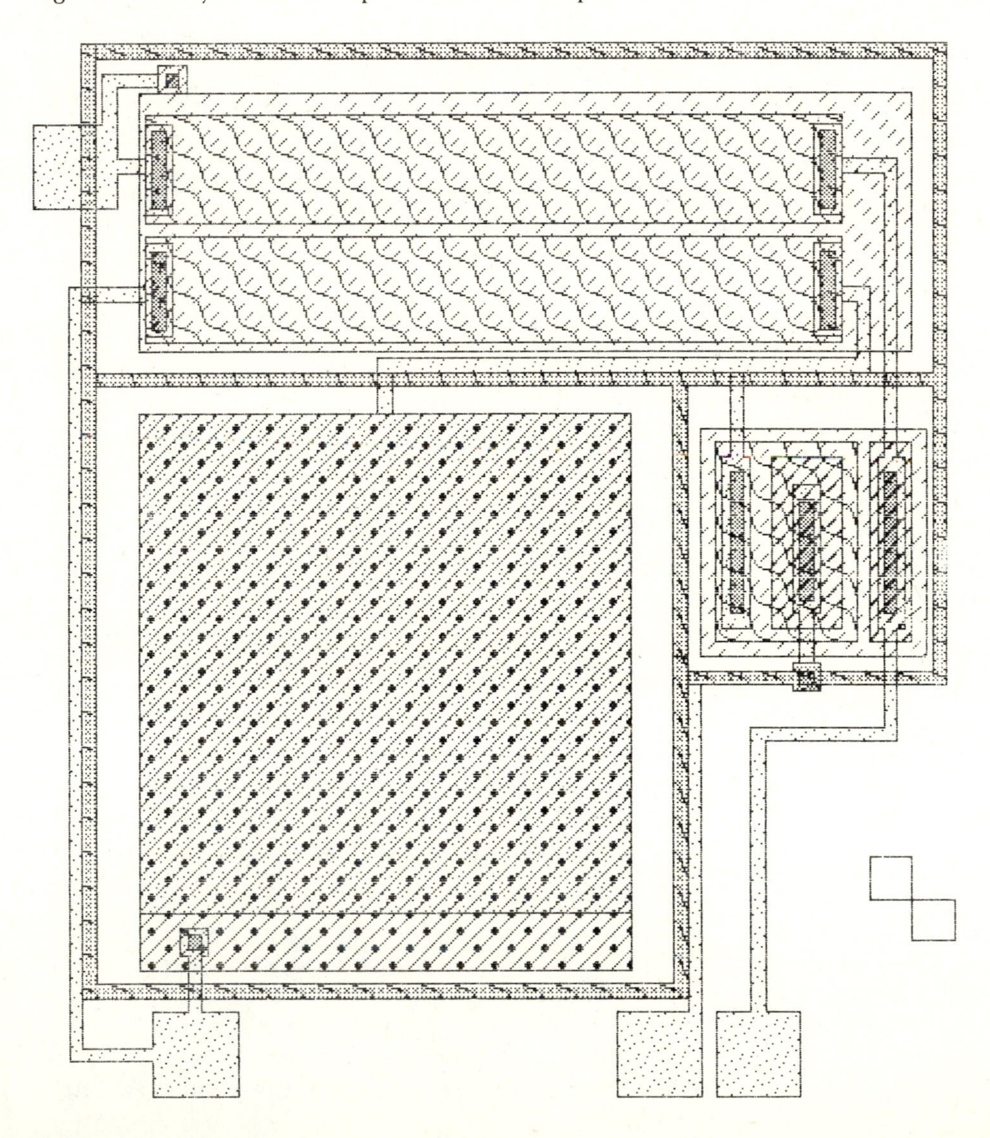

package, as described in Section 6.9. Taking each mask as a layer in a composite drawing of the integrated circuit, a computer-aided layout design allows the layer under construction to be viewed alone or in combination with some or all of the other layers. Using an interactive graphics terminal helps greatly in investigating alternative layout designs and is therefore useful even for simple circuits such as the inverter in question. For complex circuits with irregular structures, computer-aided design is virtually essential.

Assuming negative photoresist, the enclosed areas on the masks are

Figure 5.5 (A–G) Bipolar inverter mask set. A. Buried layer.

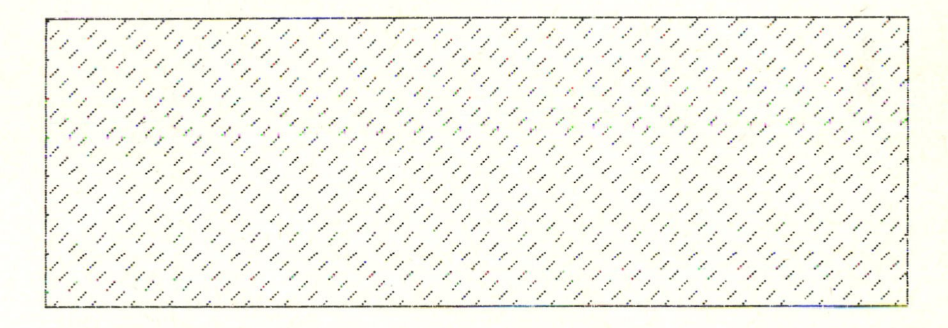

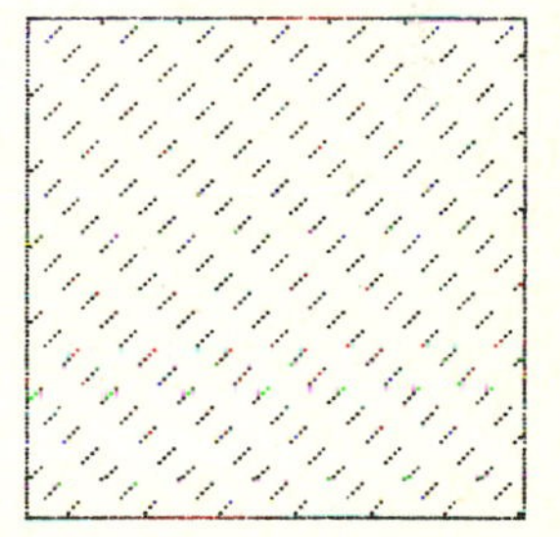

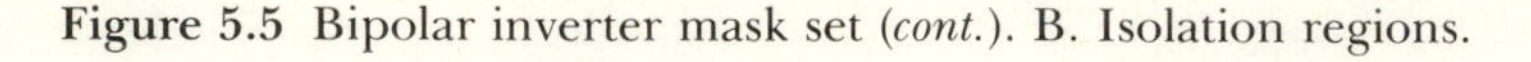

Figure 5.5 Bipolar inverter mask set (*cont.*). B. Isolation regions.

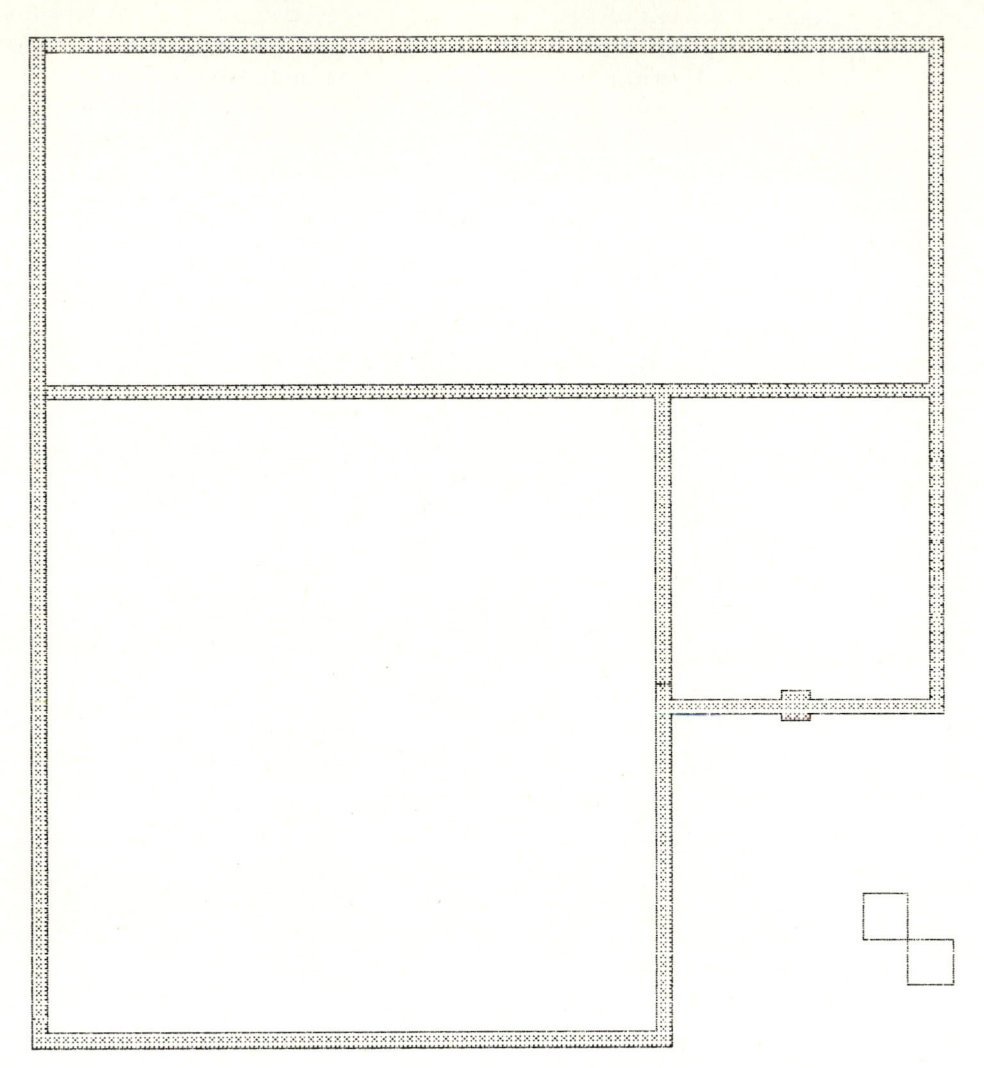

opaque, except for those on the metal mask, which are transparent. The capacitor's large size relative to the transistor emphasizes that large capacitors are not practical for monolithic integrated circuits. Also, a striking feature of this example is that the transistor and resistor dimensions, and therefore to a large extent the circuit dimensions, are determined by power-dissipation considerations rather than photolithographic limitations. This feature demonstrates the importance of reducing power dissipation as much as possible

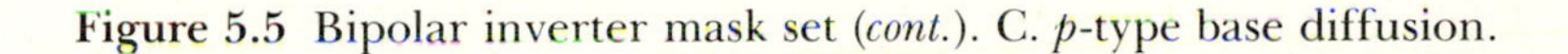

Figure 5.5 Bipolar inverter mask set (*cont.*). C. *p*-type base diffusion.

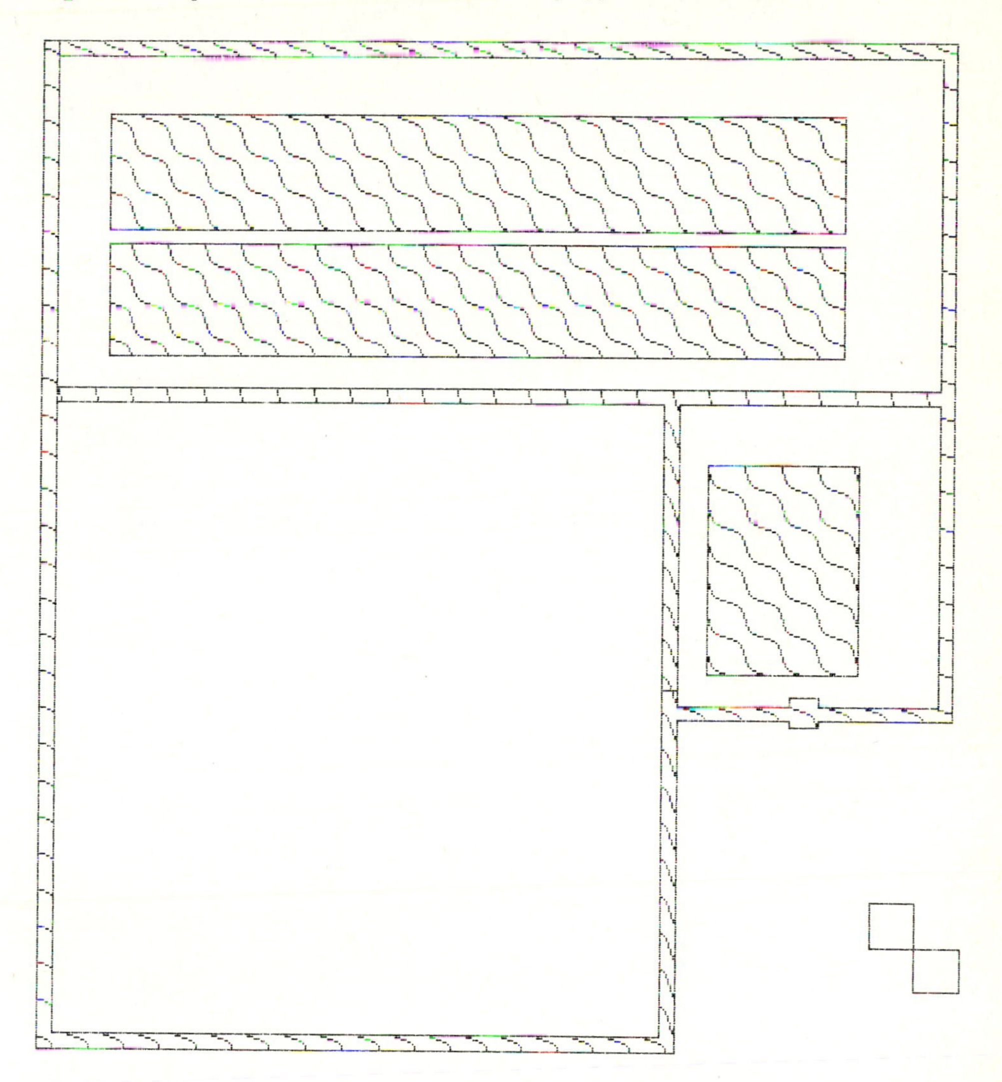

in the original circuit design. The circuit area resulting in this example is approximately 1 mm by 1 mm, and the device count is only 4! Clearly high device densities go hand in hand with low power dissipation per device.

Now that a layout has been obtained, it is useful to reexamine the chip power-dissipation associated with this circuit design. The value of 20 W/cm^2 for $(P/A)_{max}$, used in the device-layout designs for this example, was obtained in Section 5.2.1 based roughly on a chip size of 5 mm by 5 mm. If an array

Figure 5.5 Bipolar inverter mask set (*cont.*). D. n^+ emitter diffusion.

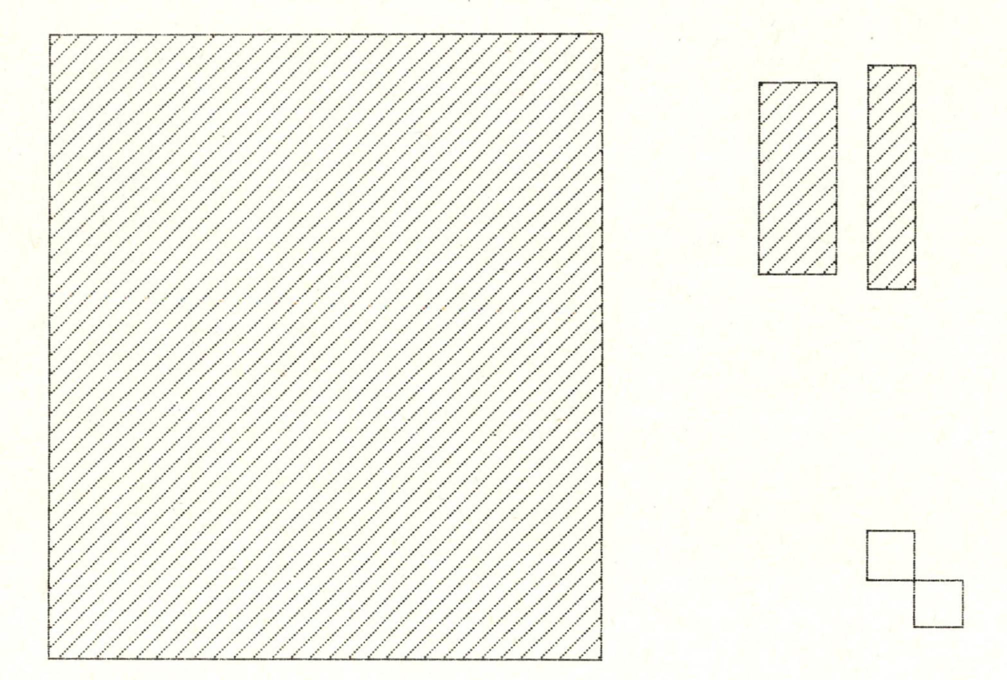

of inverters were placed on a chip of this size, there would be about 25 inverters on the chip. And since the maximum power dissipation per inverter is approximately 30 mW when the transistor is driven into saturation, the total chip power would be approximately 0.75 W.

Another check on the allowed chip power may be obtained from calculations based on the thermal resistance associated with heat flow from the device to the ambient. The difference between device temperature and ambient temperature is equal to the product of the thermal resistance and the

Figure 5.5 Bipolar inverter mask set (*cont.*). E. Patterning for oxide regrowth.

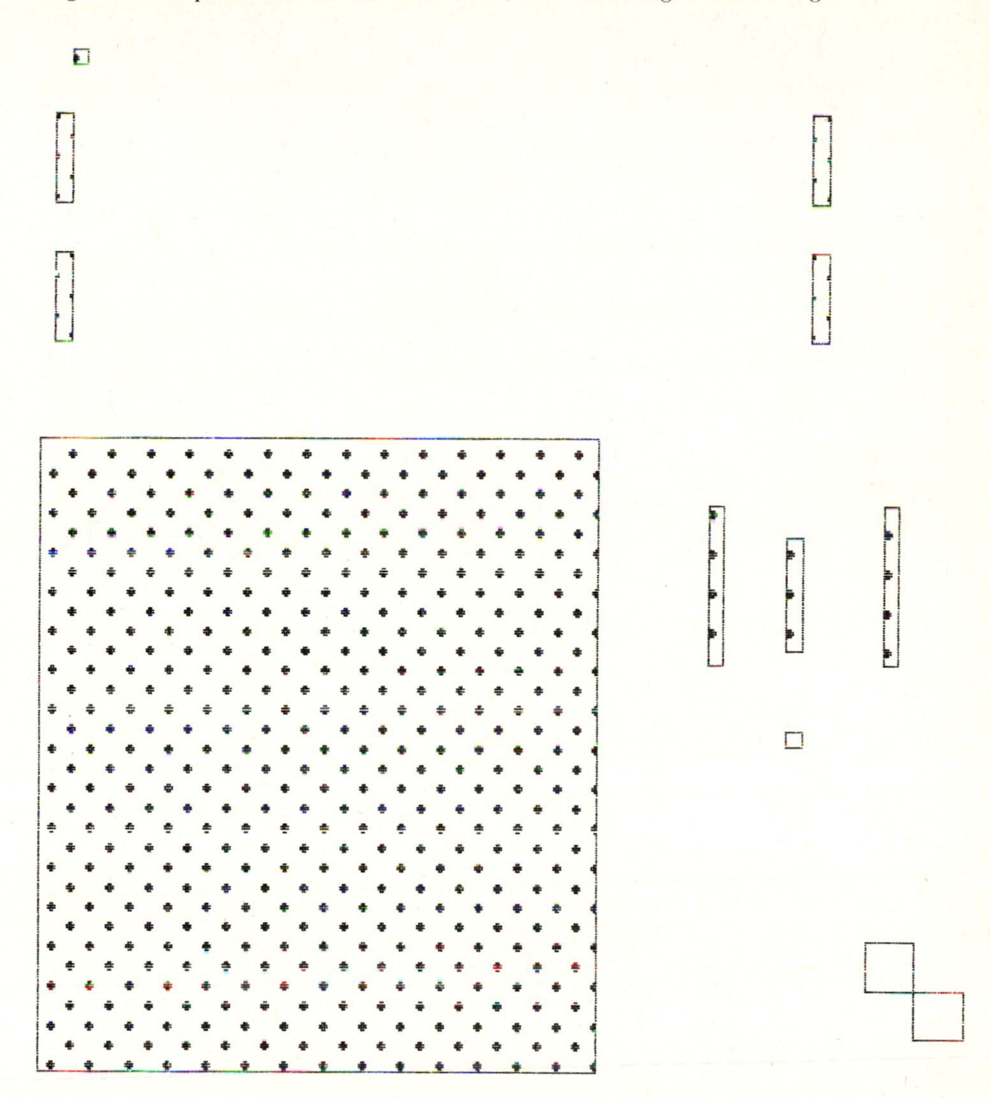

chip power-dissipation. Values for junction-to-ambient thermal resistance range from about 40° C/W to 250° C/W for common air-cooled integrated circuit packages. For a maximum device, or junction, temperature of 150°C (use of 150°C as the maximum junction temperature for power rating is widespread in the industry), the corresponding allowed chip power-dissipation ranges from about 0.5 W to 3 W for a room temperature ambient. However, the thermal resistance method of calculating chip temperature is somewhat oversimplified because it does not account for localized hot spots on the chip, as shown

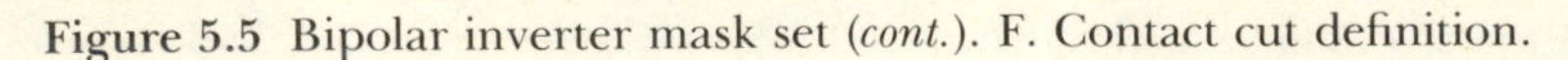
Figure 5.5 Bipolar inverter mask set (*cont.*). F. Contact cut definition.

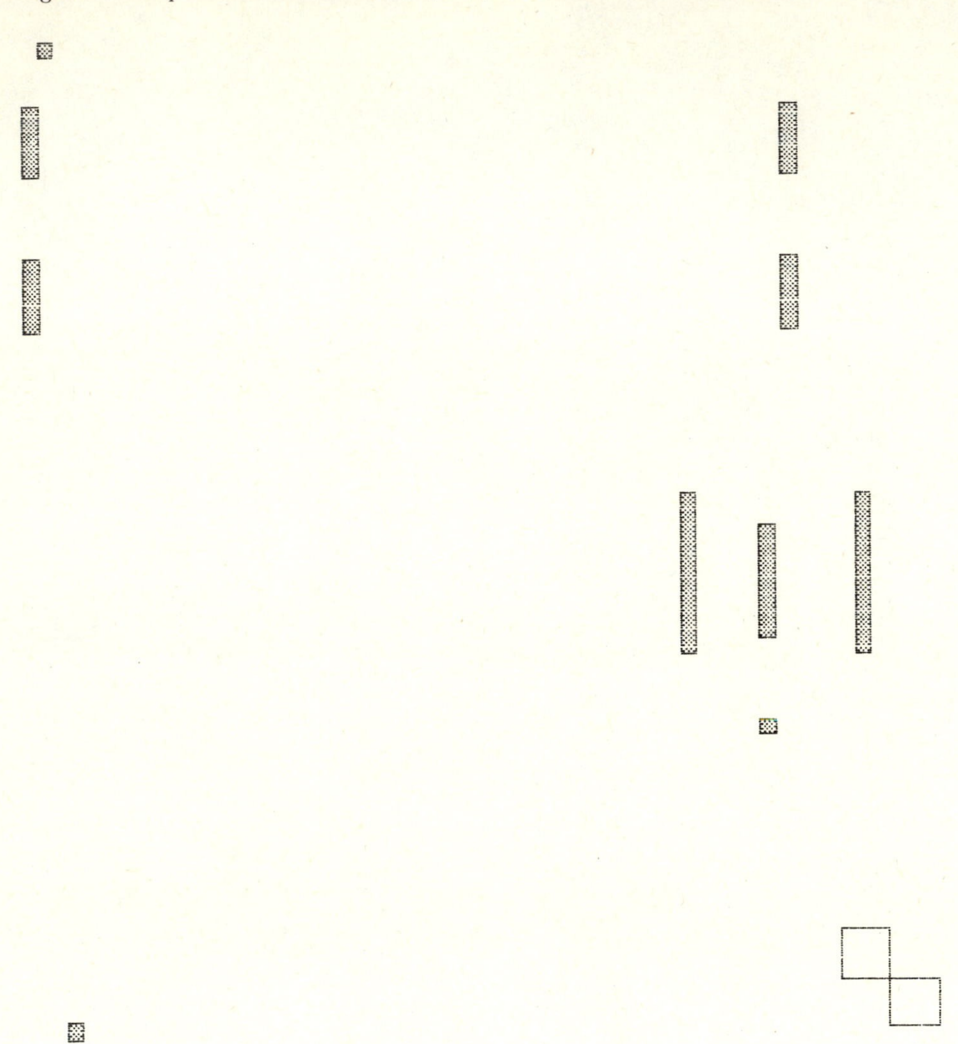

in Figure 5.2 and discussed further in Section 6.6. Also it should be noted that the ambient temperature is often above room temperature, which requires a derating of the maximum allowed power.

For smaller die, larger values of $(P/A)_{max}$ can be used without exceeding the heat dissipating ability of a given integrated circuit package. As a specific example, the 741 operational amplifier has a die area between 0.01 and 0.02 cm^2, depending on the manufacturer and on the particular layout used. When mounted in a DIP, a representative maximum power rating for a 0.012 cm^2 741 is 310 mW. This corresponds to a maximum power density of about 25 W/cm^2 when averaged over the entire chip. And for certain devices on this

Figure 5.5 Bipolar inverter mask set (*cont.*). G. Metalization patterning.

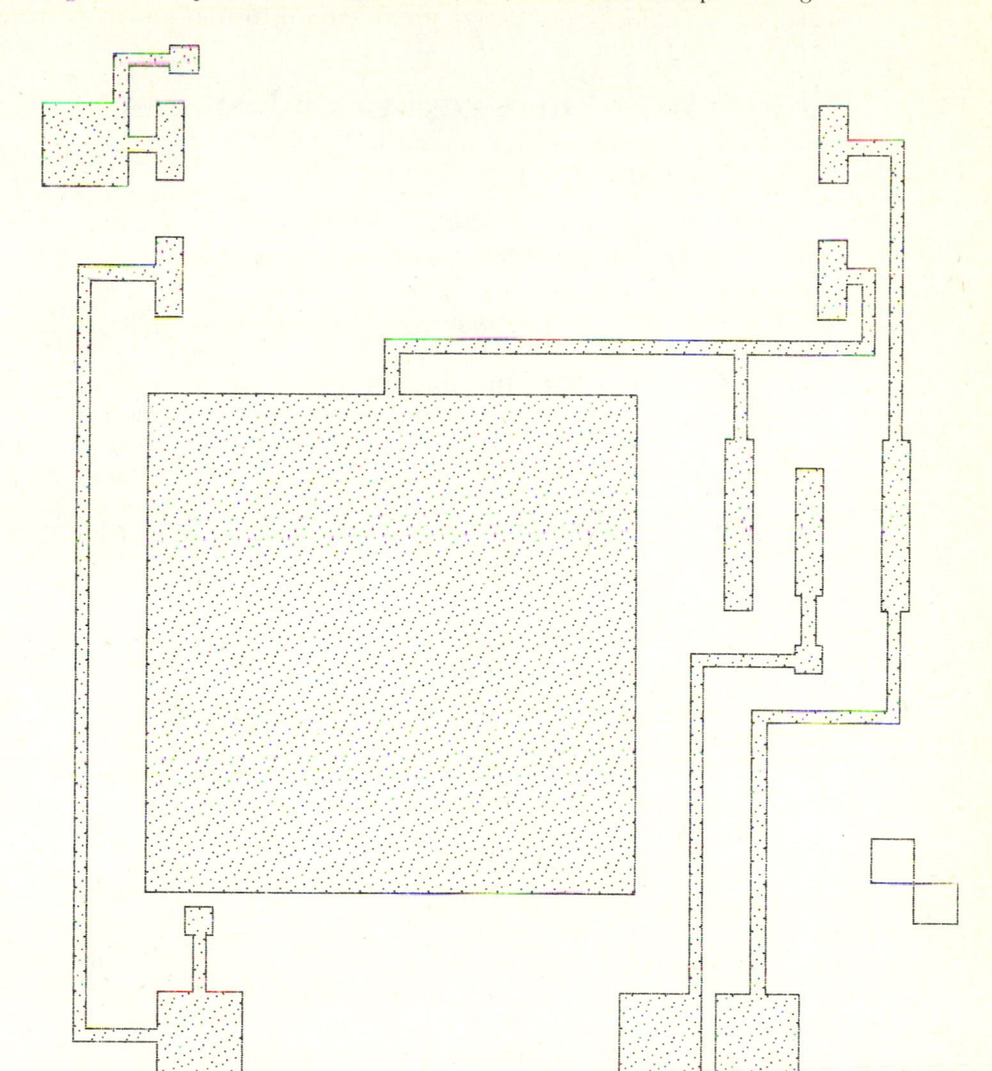

chip, such as the transistors in the output stage, the power density is much larger than the chip average. In such a case, the correlation between the spatial temperature distribution near the surface of the die and the spatial power distribution is complex because of the thermal interaction between the die and the package. For example, the material underneath the die, and in intimate contact with it, provides a path for lateral heat flow that affects the temperature on the top surface of the die. The whole chip tends to rise to the same temperature as the output stage, but gradients of temperature do exist, with the output stage being hotter than the rest. The implications of such gradients for layout design are discussed further in Section 6.6.

5.3 Solutions to the Cross-Over Problem

5.3.1 General Methods

More complicated layouts raise the question of how to deal with crossing wires, that is, conducting paths on the chip that overlap. In this section we consider several ways to solve the problem.

An obvious method for dealing with cross-overs is to use double or triple metalization layers that are separated by an insulator deposited by chemical vapor deposition (CVD). In this approach, which is shown in Figure 5.6, metal lines may be used for both of the crossing signal paths. An advantage is that the metalization layer provides the best-quality signal path because of the metal's high conductance. The disadvantage is that extra processing steps are added. Highly complex layouts often use multiple metalizations. For example, a VLSI bipolar chip has been reported with three layers of metalization that allowed 5.8 m of signal path on the chip [3].

Alternatively, a nonmetallic path may be used for one of the crossing signals. A diffusion path in the silicon, for example, offers a buried signal path that may be effective if the distance is not too long and the signal current not too high. Such a buried diffusion layer may be intentionally added to allow the cross-over, as shown in Figure 5.7A. However, in many cases, a diffusion path associated with an already-existing device may be used. As a specific example, Figure 5.7B shows that if one of the signal paths contains a resistor, then the layout may be planned such that the cross-over occurs over the resistor. Also, buried n^+ layers associated with devices such as the BJT can be used to provide a buried path for one of the crossing signals, as Figure 5.7C shows.

Polysilicon offers a third possible signal path, which is useful in solving the cross-over problem in integrated circuits that use polysilicon as part of the device fabrication process. By depositing polysilicon on an oxide-covered wafer, the underlying substrate is isolated from the polysilicon path, and a subsequent CVD oxide insulates the polysilicon from overlying metal paths.

Figure 5.6 Multiple metalization cross-over.

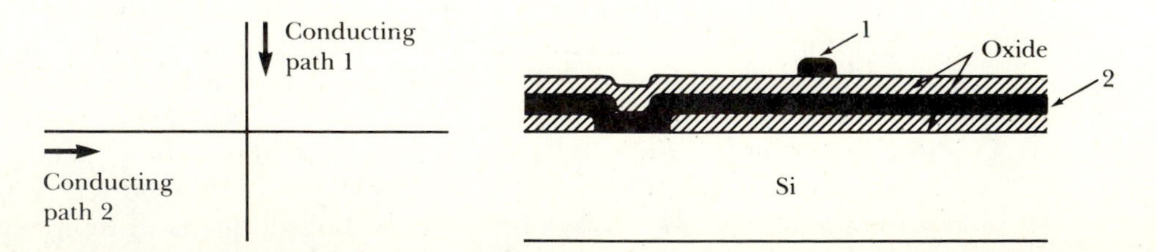

Figure 5.7 Buried paths for the cross-over. A. An added n^+ region. B. An existing resistor. C. An existing buried n^+ layer.

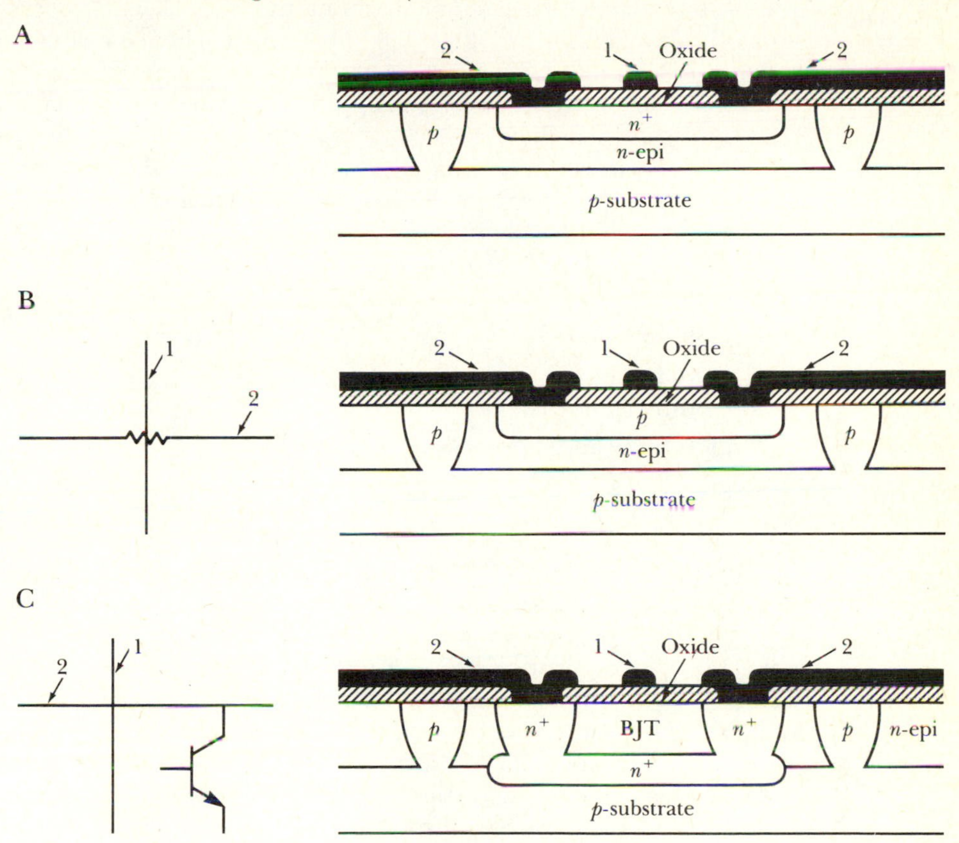

The resistivity of diffusion and polysilicon paths is usually more than 100 times greater than the resistivity of the metal paths. Therefore they should not be used in cases where path resistance is a critical issue. For example, power supply lines and ground lines are metal, since they carry high currents and run over appreciable distances. A high-resistance path may also limit speed because of resistor-capacitor (RC) time-constants associated with charging and discharging the load capacitance and interconnect-line parasitic-capacitance. Furthermore, because of the junction space-charge-layer capacitance, the diffusion path has a larger capacitance per unit area than do either metal or polysilicon lines.

In combination then, diffusion paths, polysilicon paths, and metalization can provide three alternative conducting paths. However, the three paths are not of equal quality. Finally, careful layout should keep the number of cross-overs to a minimum. Often, a line can be rerouted or a topology changed to eliminate a cross-over by layout alone.

5.3.2 A Bipolar Flip-Flop Example [4]

As a layout example of a circuit containing unavoidable cross-overs, consider the RS flip-flop shown in Figure 5.8. The circuit schematic shows 14 circuit elements, consisting of 8 transistors and 6 resistors. One of the first steps in the layout design is to decide on the appropriate number of isolation regions. Since transistors that have collectors in common can share an isolation region, and since resistors can share an isolation region, the minimum number of isolation regions for this circuit is seven, with device groupings as follows:

Region 1: R_1, R_2, R_3, R_4, R_5, and R_6
Region 2: T_1
Region 3: T_2 and T_3
Region 4: T_4
Region 5: T_5
Region 6: T_6 and T_7
Region 7: T_8

Unlike the simple inverter example, this one contains several unavoidable cross-overs. In Figure 5.8, five cross-overs are noted and labeled numerically. In addition, the ground is trapped inside the circuit, which also represents a cross-over problem since, for standard DIP mounting, all circuit connection points must be brought to bonding pads at the periphery of the circuit. A brute-force solution to the cross-over problem would be to use a double metalization layer. However, double metalization is unnecessary and inappropriate for a circuit of this size, since all the cross-overs can be accommodated by using existing buried conducting paths.

Three of the cross-over problems can be eliminated by using resistor cross-over points, as described in Section 5.3.1. As seen in Figure 5.9, this method allows the flip-flop outputs to be brought to the periphery over resistors R_1 and R_6. The feedback cross-overs can be eliminated by using the buried n^+ layer associated with the transistor pair T_6 and T_7, as shown in Figure 5.10, where the notation C_{BUR} refers to the buried n^+ collector conduction path. Since the buried layer is highly conductive, it represents a nearly equipotential region and can be contacted at any convenient point, thereby providing a buried conduction path insulated from the overlying metalization layer. Likewise, the trapped ground line can be freed by combining a resistor cross-over and a transistor buried-layer cross-over, as Figure 5.11 shows.

A mask set corresponding to a possible layout for this flip-flop is shown in Figure 5.12, which consists of the buried-layer mask, the isolation region mask, the base diffusion mask, the emitter diffusion mask, the contact cut mask, and the metalization mask. Assuming negative photoresist, the metalization mask is shown as a reverse image.

Figure 5.8 Circuit schematic and truth table for an *RS* flip-flop. A. Schematic. B. Truth table.

A

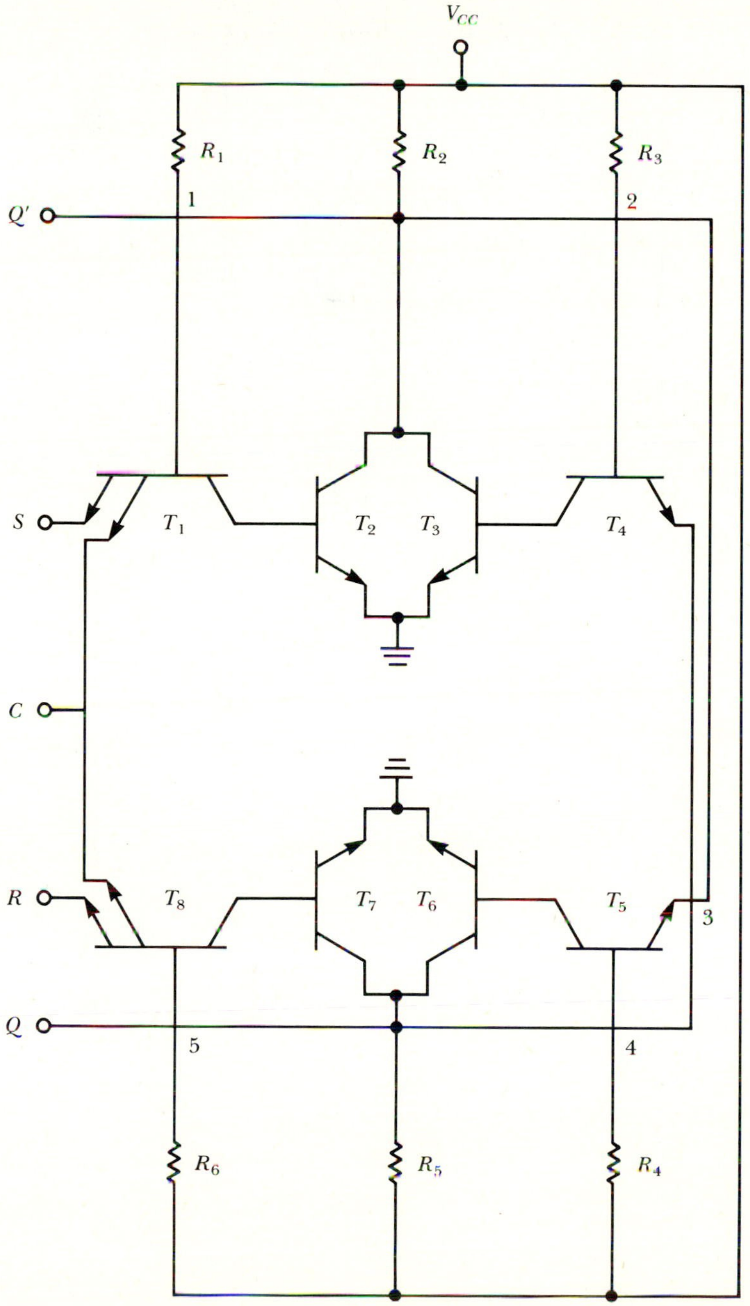

B

Present state		Next state
$R(t)$	$S(t)$	$Q(t+1)$
0	0	$Q(t)$
0	1	1
1	0	0
1	1	Not allowed

Figure 5.9 Use of resistors as cross-over paths in the *RS* flip-flop.

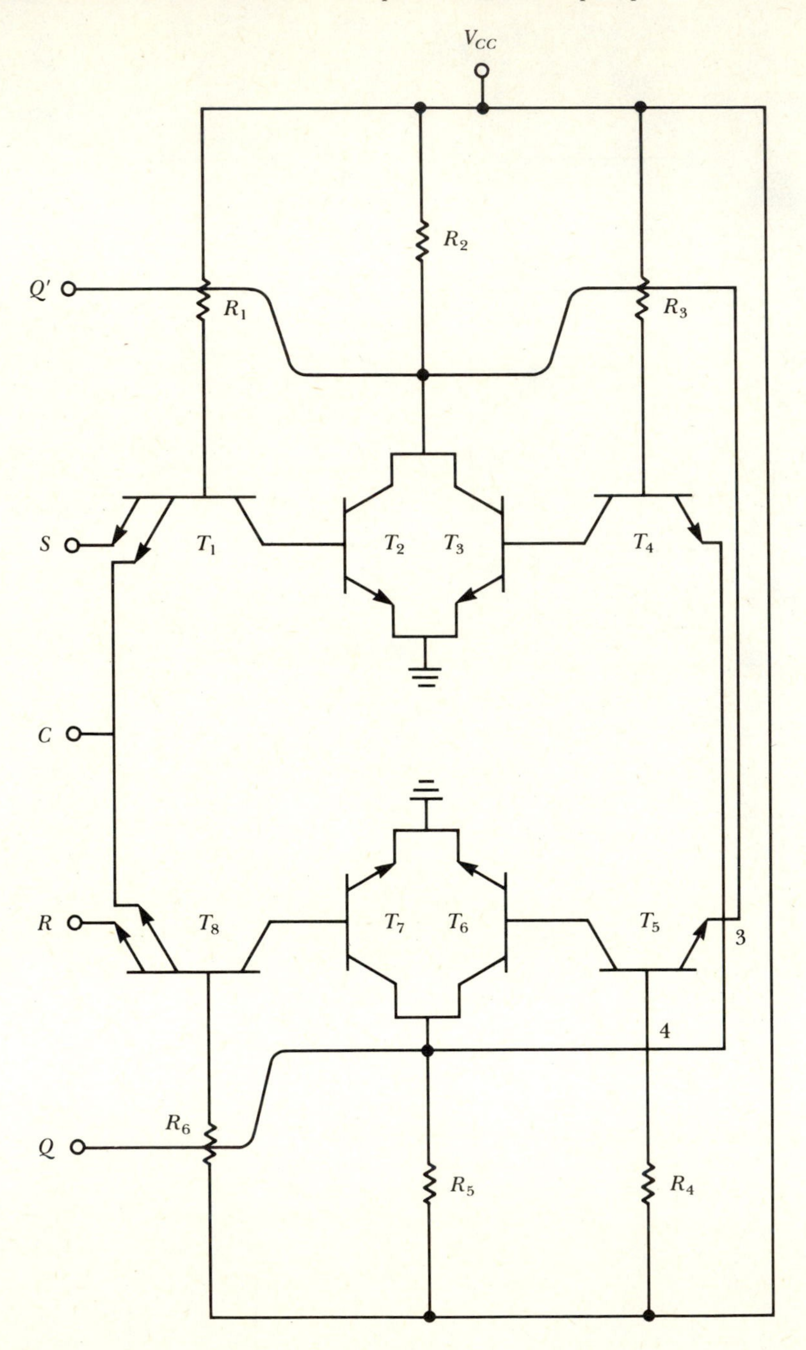

Figure 5.10 The BJT n^+ buried layer as a cross-over aid in the *RS* flip-flop.

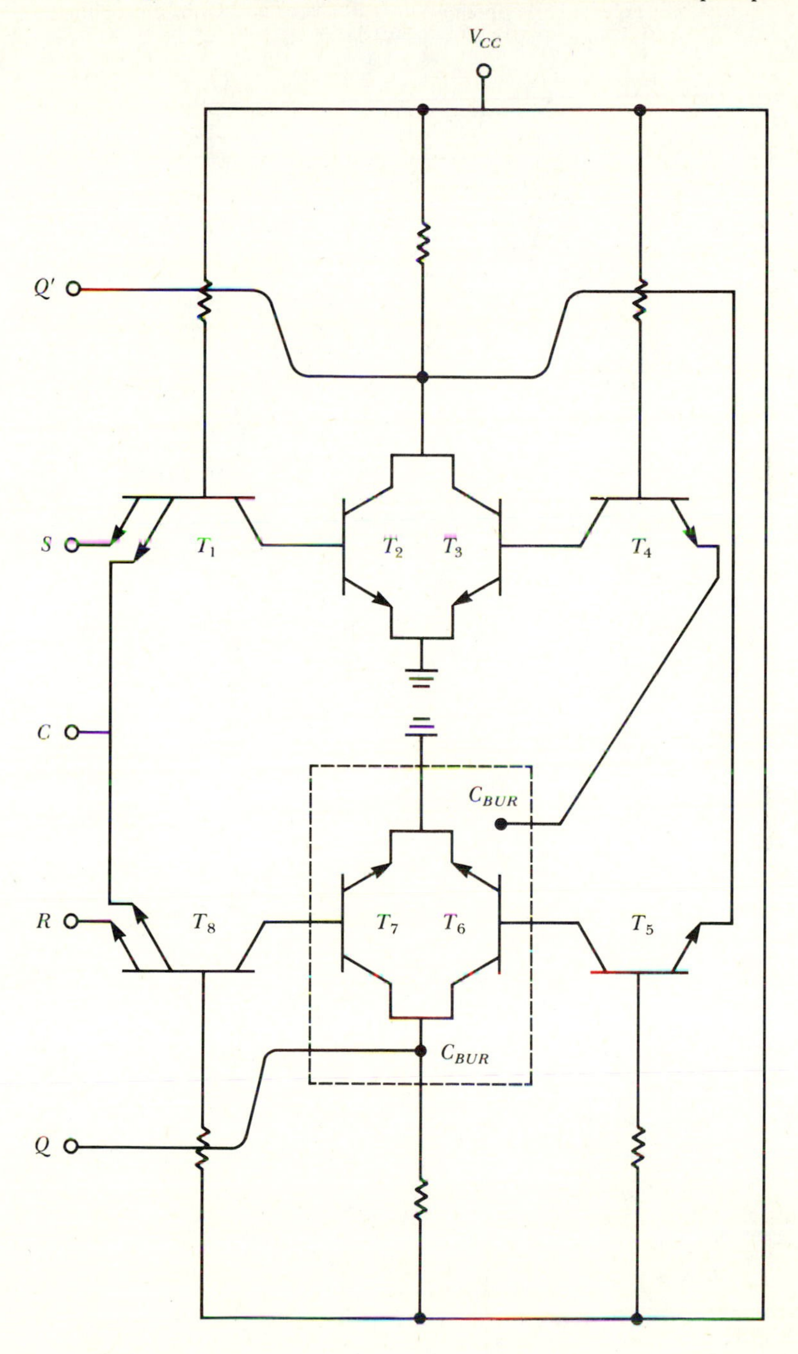

Figure 5.11 Routing of the ground lead on the *RS* flip-flop.
SOURCE: Following the method of A. B. Glaser and G. E. Subak-Sharpe, *Integrated Circuit Engineering* (Reading, Mass.: Addison-Wesley, © 1977).

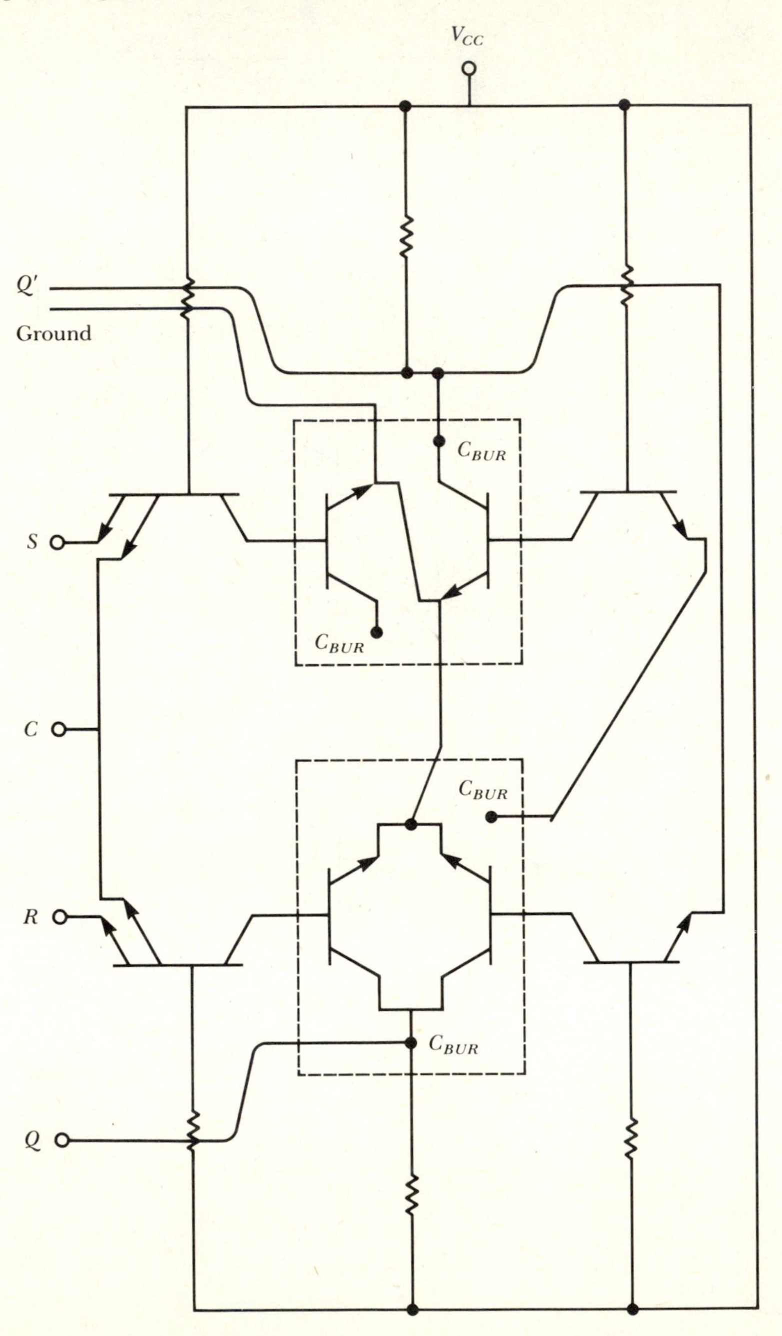

Figure 5.12 (A–F) Mask set for the *RS* flip-flop. A. Buried-layer mask. B. Isolation region mask.

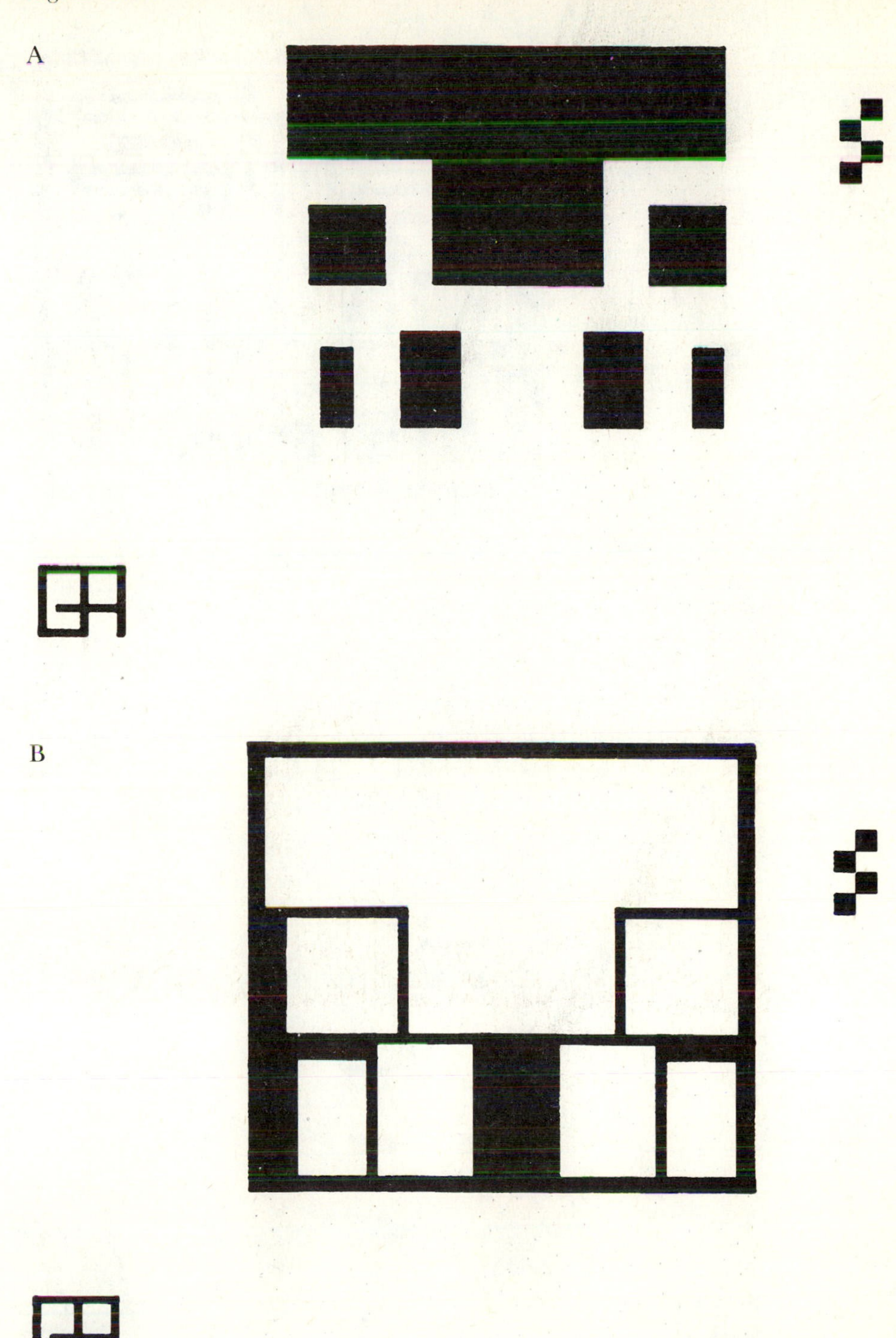

Figure 5.12 Mask set for the *RS* flip-flop (*cont.*). C. Base diffusion mask. D. Emitter diffusion mask.

C

D

Figure 5.12 Mask set for the *RS* flip-flop (*cont.*). E. Contact cut mask. F. Metalization mask. (Reverse image.)

E

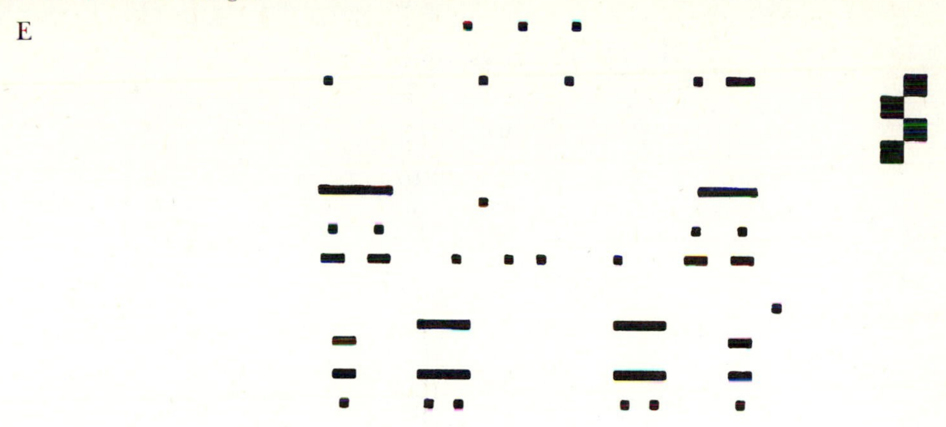

F

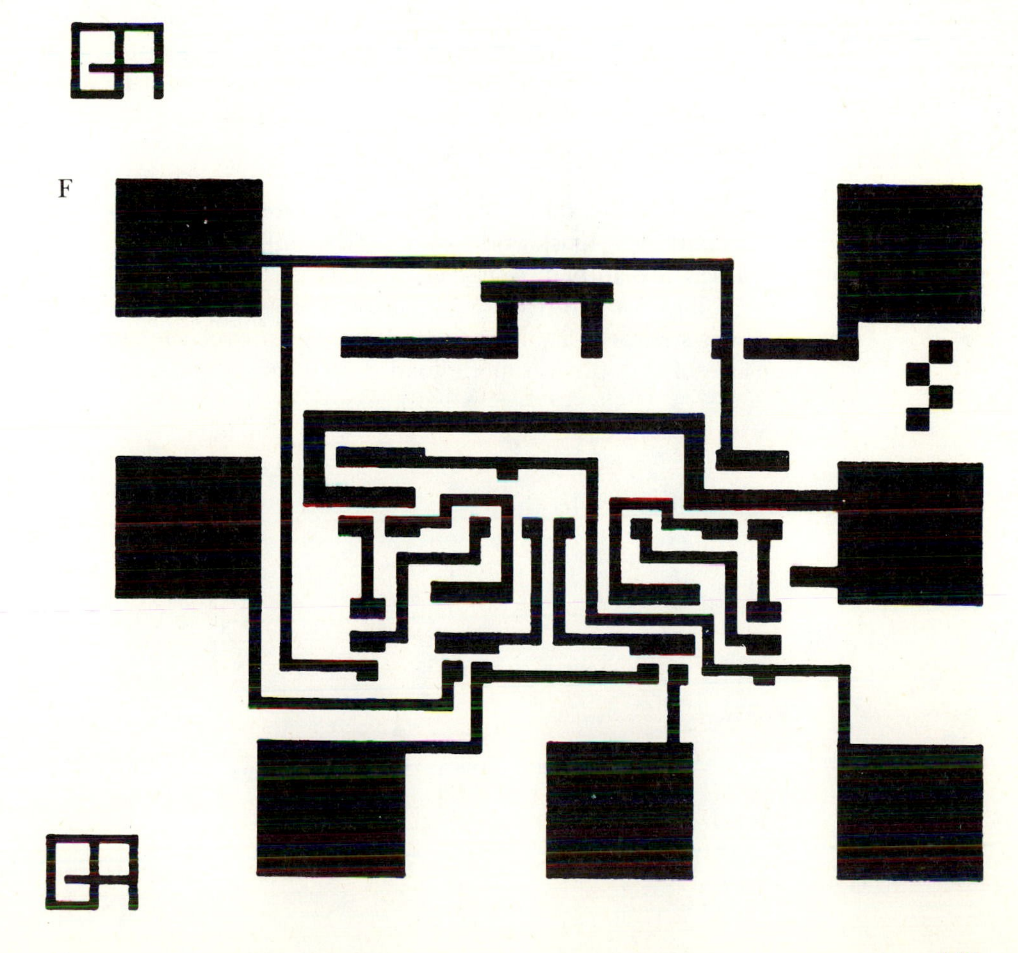

5.4 Multiple Data Paths

5.4.1 NMOS Inverter Example

Section 5.3.1 noted that n^+ diffusion layers, polysilicon layers, and metal layers can be used in combination to allow three distinct conducting paths. This approach, demonstrated by example in this section, is particularly useful in MOS circuit design.

Section 3.9.1 discusses several n-channel MOS (NMOS) circuit options for the basic logic inverter. NAND and NOR gates are straightforward extensions of the inverter, since the NOR gate results from placing driving transistors in parallel and the NAND gate from placing driving transistors in series. For a depletion-mode load, the circuit schematic for an NMOS inverter is shown in Figure 5.13.

Figures 5.14 and 5.15 show an inverter layout and the sequential processing and photolithographic steps for a self-aligned polysilicon-gate NMOS process, as described in Section 3.6.2. The initial step is to define the field oxide, as Figure 3.18 shows in more detail, and in the process to expose those areas of silicon that will receive implantations or diffusions. (For simplicity, the bird's beak and channel-stop implant are not shown in Figure 5.14.) Next, the n-type ion implantation region, which establishes the negative threshold voltage for the depletion-mode load, is patterned and the ion implantation is carried out. The process illustrated here uses photoresist for the ion implantation mask and assumes that only the depletion-mode transistor requires the implant. In fact, it may also be necessary to tailor the threshold voltage of the driving transistor with a second implant. After the photoresist is stripped, a thin gate oxide is grown over the entire surface.

The use of a depletion-mode load device makes it necessary to provide

Figure 5.13 NMOS inverter circuit.

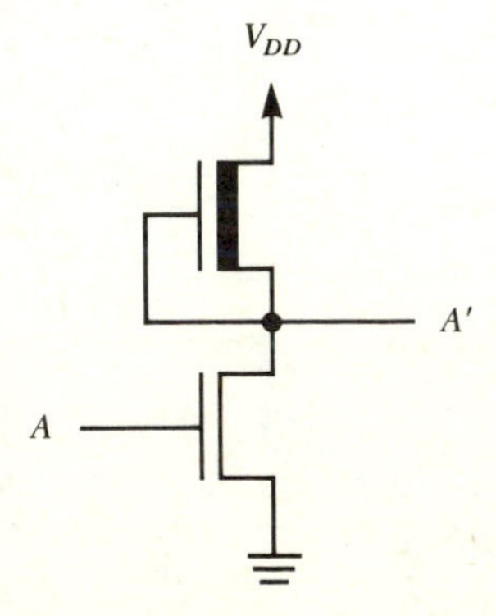

Figure 5.14 NMOS inverter layout with polysilicon input and polysilicon output (see also Figures 5.13 and 5.15).

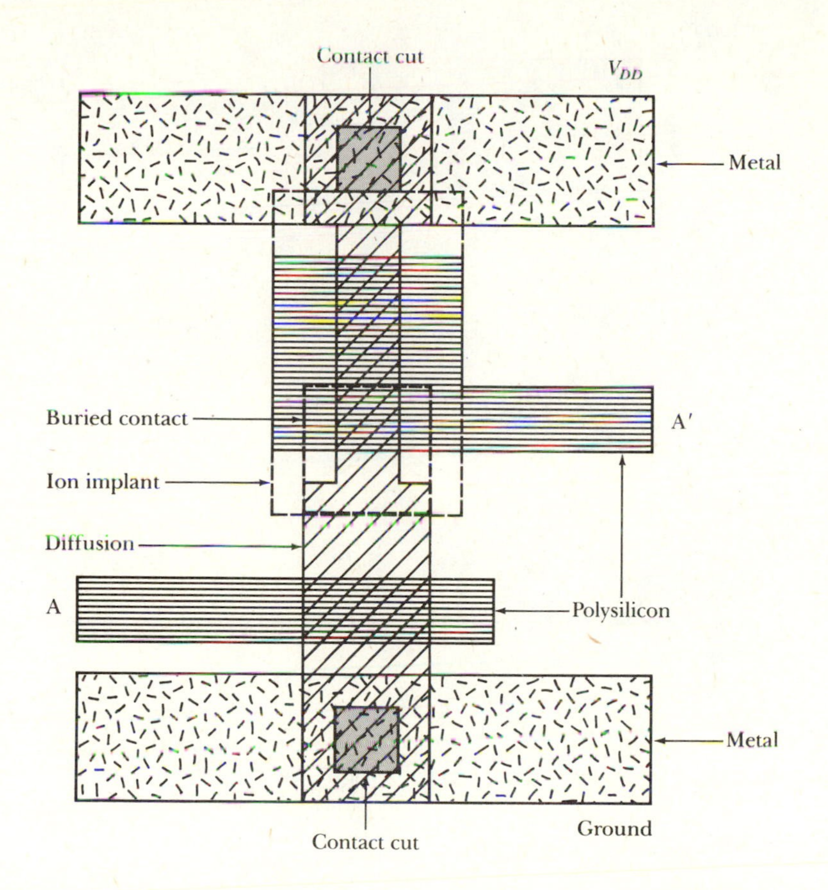

for a connection between the gate and source of the load transistor. The layout illustrated in Figure 5.15 uses a buried contact to accomplish this task compactly. Prior to the chemical vapor deposition of a polysilicon layer, the gate oxide is etched, as shown in Figure 5.15C. Subsequently, the n^+ polysilicon acts as a doping source for the silicon exposed in the opened area of the gate oxide, as shown in Figure 5.15D. As a consequence, an ohmic contact exists between the polysilicon and the underlying n^+ region.

The polysilicon is next etched to define the gates for the two transistors and also, in this particular layout, to provide conduction paths for the input and output signals. After the patterning of the polysilicon, an n^+ diffusion (or implant) is performed. This step requires no mask, since the field oxide and polysilicon define the diffusion regions. The diffusion's purpose is not only to define the source and drain of the transistors, but also to provide a

Figure 5.15 (A–H) NMOS inverter process steps and patterning. A. Field oxide definition. (See Figure 3.18 for more details.) B. Ion implantation patterning to set the threshold voltage of the depletion-mode transistor.

A

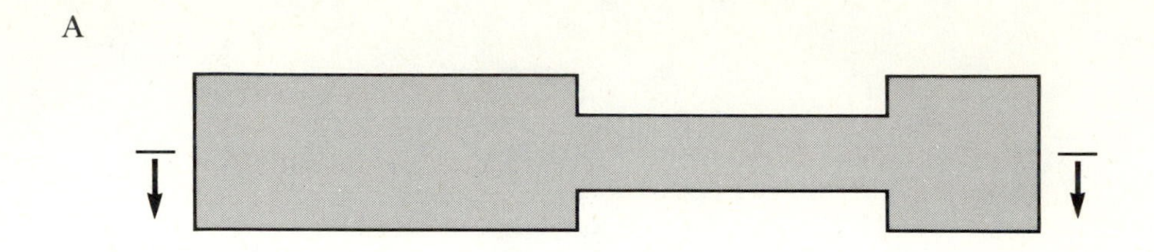

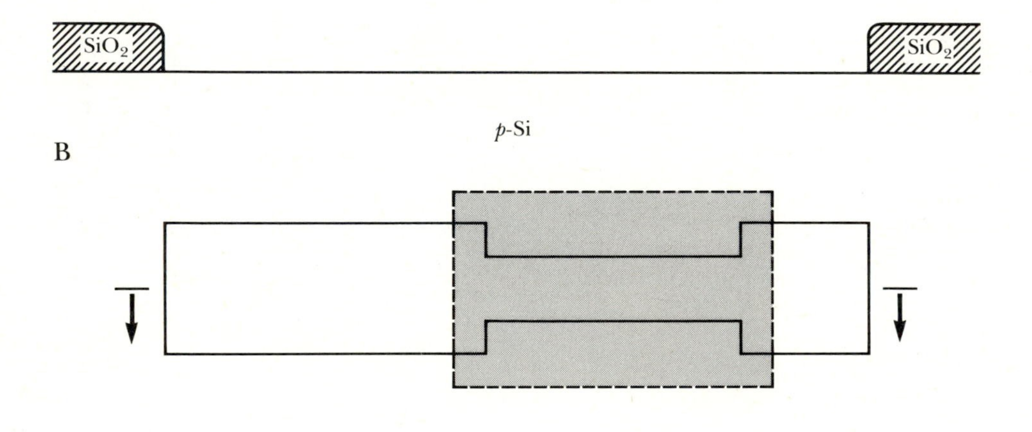

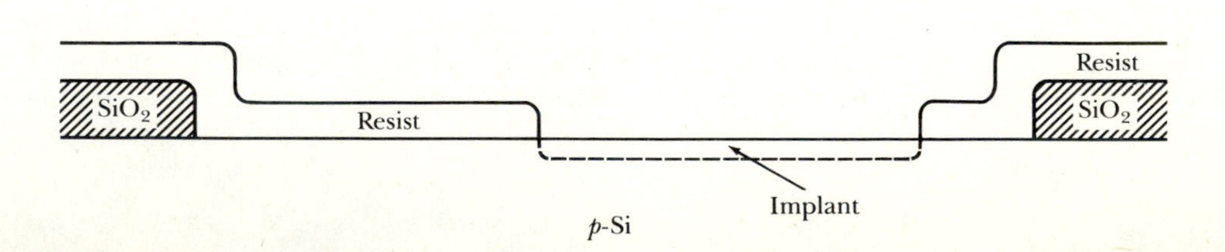

Figure 5.15 NMOS inverter process steps and patterning (*cont.*). C. Buried-contact definition. D. Polysilicon deposition and doping.

C

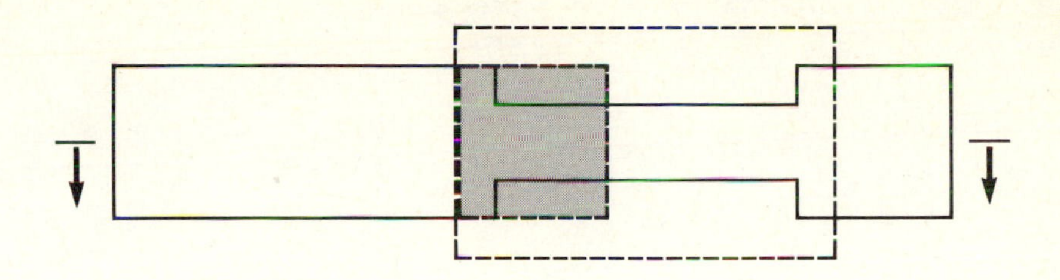

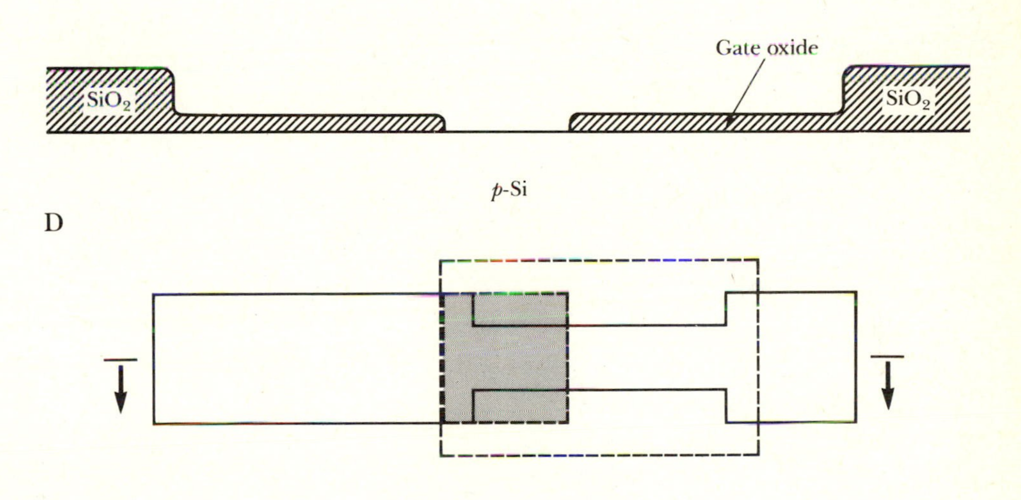

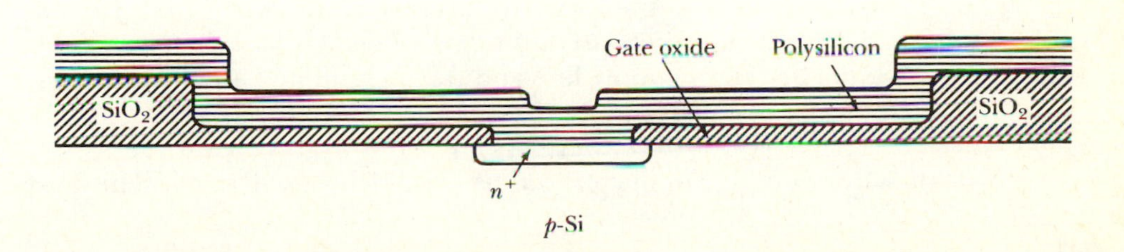

Figure 5.15 NMOS inverter process steps and patterning (*cont.*). E. Polysilicon patterning.

E

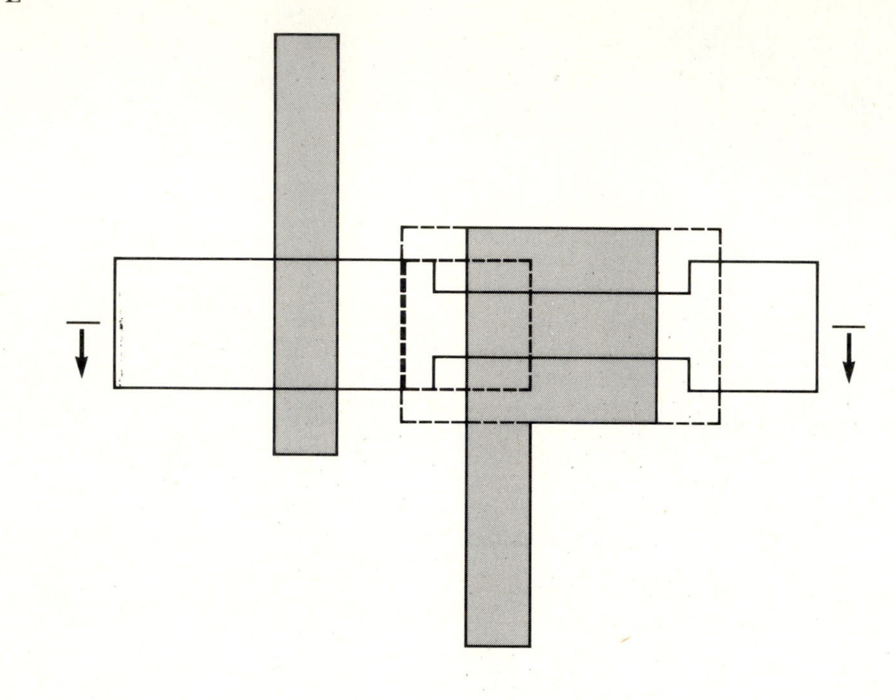

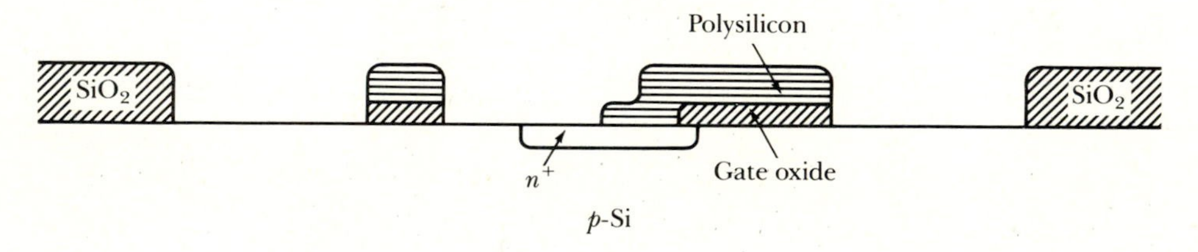

conducting path that interconnects the two transistors. Note that the buried-contact n^+ region provides a buried equipotential layer as a common connecting point for the gate and source of the load transistor, as well as for the drain of the driving transistor. Depending on the details of the fabrication process, the n^+ region under the buried contact may be deeper than the diffused source and drain regions, or as is shown in Figure 5.15, not as deep. Finally, the contact cuts are defined and etched and the entire wafer is metalized. The power supply line and the ground line are defined by selectively etching the metal layer.

That no cross-overs were encountered in this layout raises the question of why multiple conducting paths should be used at all. The answer is in

Figure 5.15 NMOS inverter process steps and patterning (*cont.*). F. n^+ diffusion or implant.

F

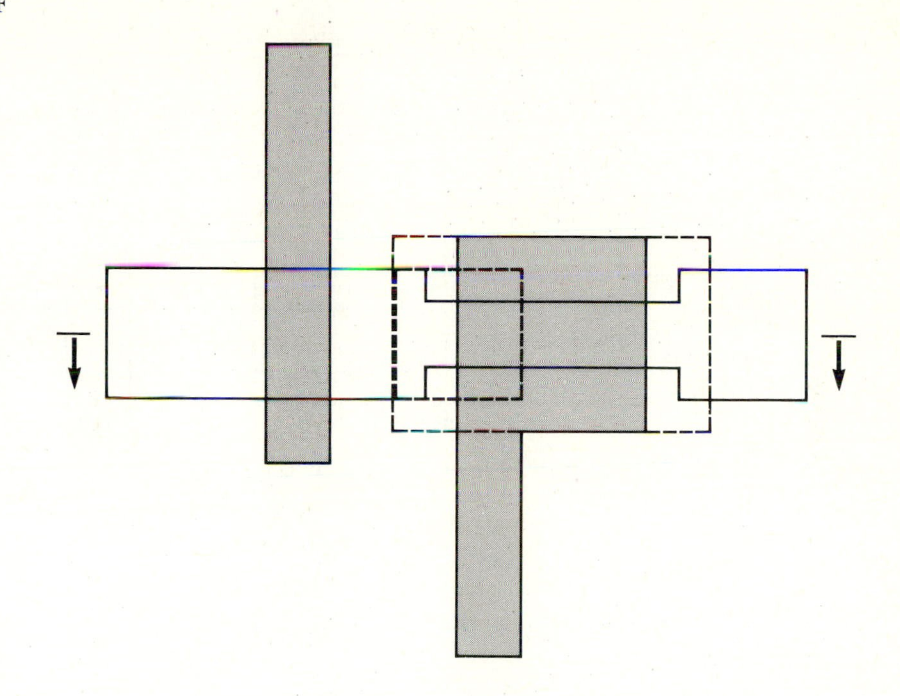

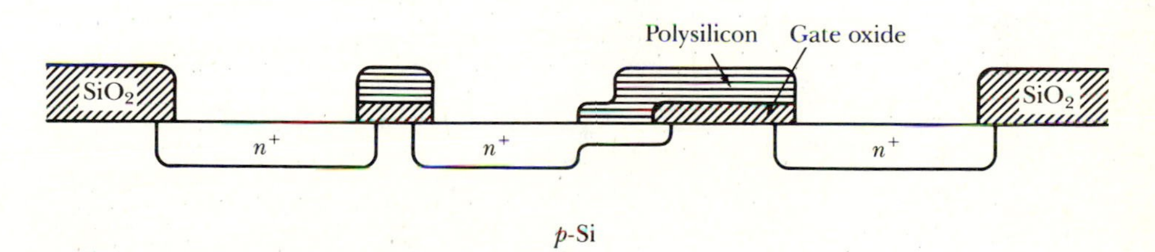

part that basic MOS gates are generally part of a larger and complex circuit where cross-overs are a consideration, and in part that the multiple paths allow a more compact layout for the gate itself. If metalization were the only option for interconnecting devices and providing data paths, the gate would require more area in the form of additional space for the extra metal contacts.

A disadvantage of the buried-contact scheme shown in Figure 5.15 is that the depletion-mode transistor is not a self-aligned gate MOS field-effect transistor (MOSFET). The buried-contact mask defines the source diffusion of the load transistor, and the polysilicon mask defines the drain diffusion. Therefore the channel length is subject to mask misalignment errors.

Figure 5.15 NMOS inverter process steps and patterning (*cont.*). G. Contact cut patterning.

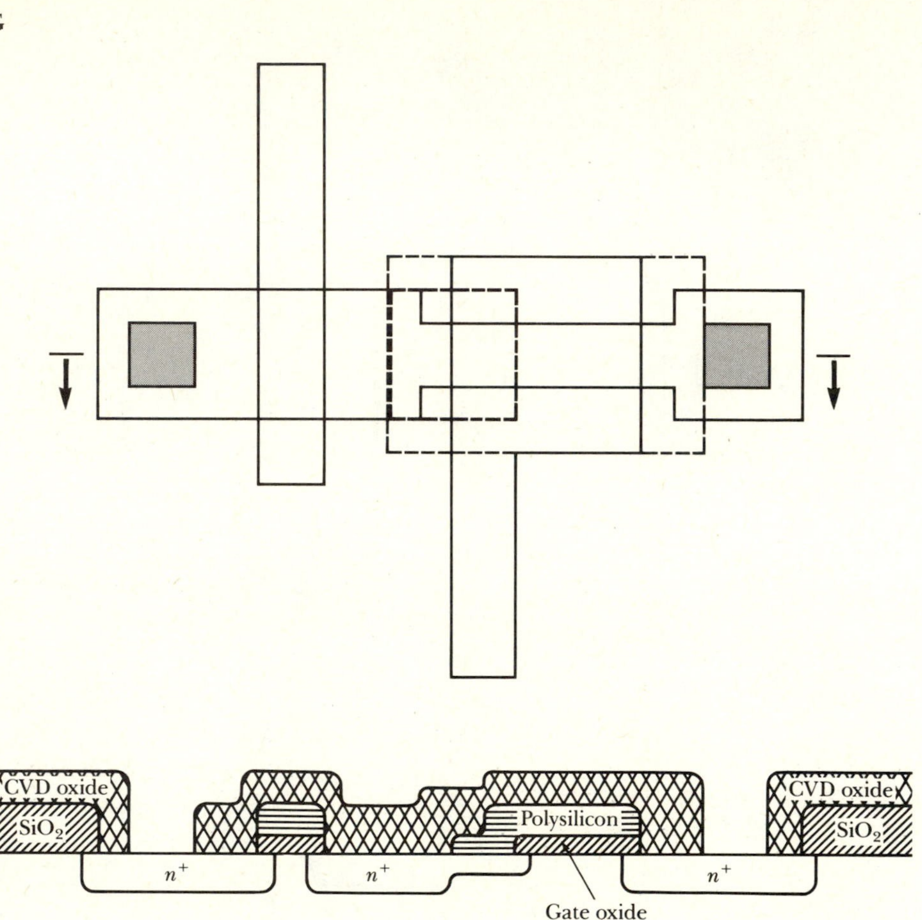

However, a self-aligned load transistor and a buried-contact are mutually compatible if the gate output goes on the diffusion layer as shown in Figure 5.16 [5]. The inverter in Figure 5.16 also differs from the inverter in Figure 5.14 in that the Z_{PU}/Z_{PD} ratio is 6, rather than 4.

5.4.2 Stick Diagrams [6]

The layouts of Figures 5.14 and 5.16 show many details of the overall design, including the geometry of each transistor; the relative placement of the devices; the load transistor's identity as a depletion-mode device; and the types of conducting paths for the input signal, output signal, power supply, and

Figure 5.15 NMOS inverter process steps and patterning (*cont.*). H. Metalization patterning.

H

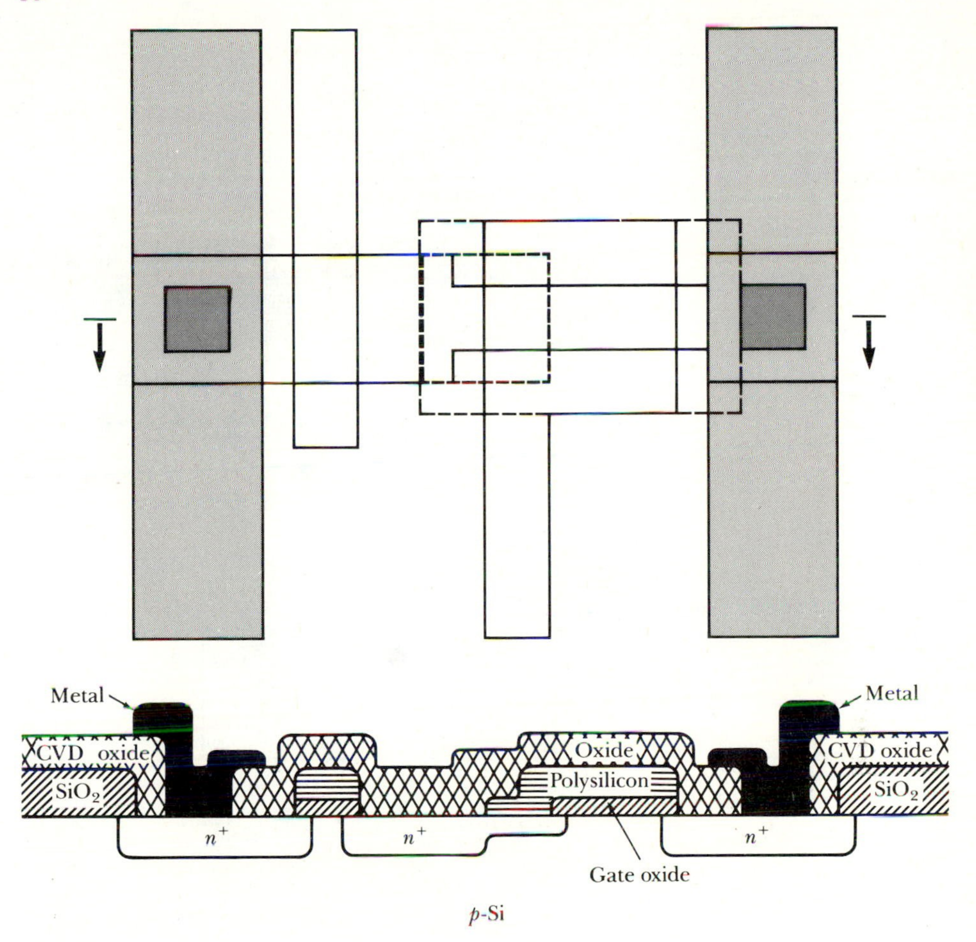

ground. These layouts allow the determination of the circuit schematic and the processing technology as well as the actual layout. However, the layouts are cumbersome to use as general representations of MOS circuits.

Stick diagrams offer a simpler symbolic representation of NMOS circuits. Although much easier to work with than the full layouts, they present much of the same information as to circuit topology. Stick-diagram representations show conducting paths as color- or pattern-coded lines with different codes for metal, polysilicon, and n^+ diffusion paths. In addition, ion implantation regions and contact cuts are shown schematically instead of by actual dimensions. Geometric information can be added by, for example, indicating the Z ratio for transistors.

Figure 5.16 NMOS inverter layout variation, with polysilicon input and diffusion-layer output. The same pattern codes are used as in Figure 5.14.

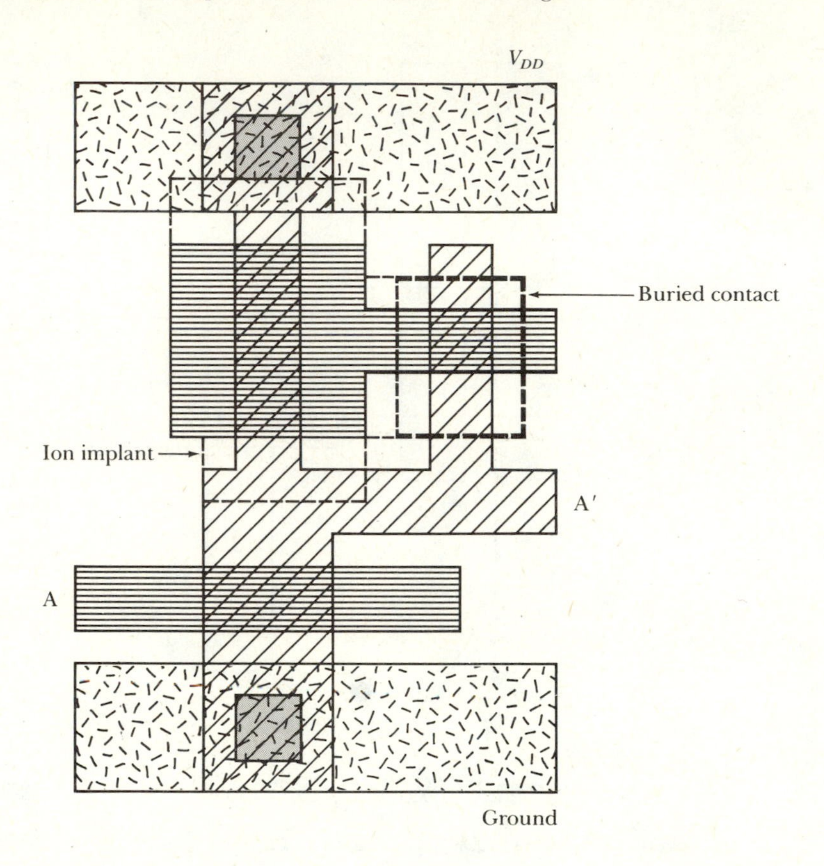

Figure 5.17 shows three levels of representation for an enhancement-mode MOSFET, namely the circuit symbol, the stick diagram, and the layout. Stick diagrams, and variations on them, can allow a designer to control circuit topology while letting a computer worry about the details of satisfying design rules and other geometric considerations such as layout compaction.

Note that when a diffusion line crosses a polysilicon line on a stick diagram, a transistor results. Recall that in the fabrication sequence, the diffusion step follows the polysilicon definition step and the polysilicon, with the underlying gate oxide, acts as a mask to the diffusion. Consequently the diffusion establishes an n^+ source and drain to contact the ends of the MOSFET channel beneath the polysilicon gate. In contrast, polysilicon lines and metal lines are separated by a thick CVD oxide. Therefore, when polysilicon and metal stick-diagram lines cross, no interconnection results between the two paths unless a contact cut is explicitly shown. Likewise, diffusion and

Figure 5.17 Enhancement-mode MOSFET representations. A. Circuit symbol. B. Stick diagram. C. Layout.

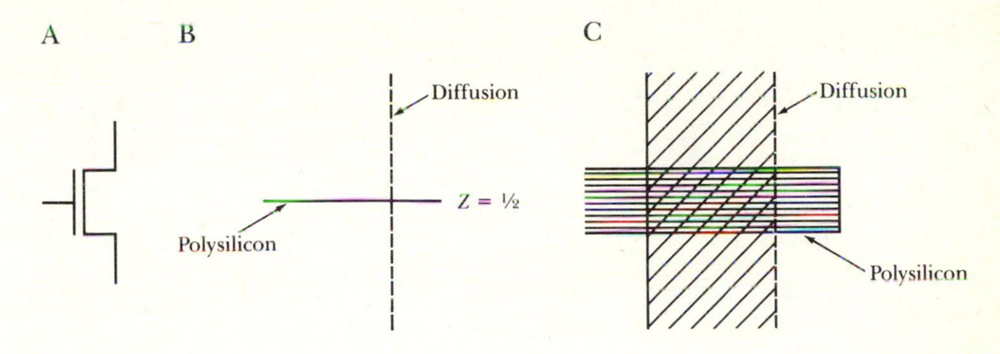

metal stick-diagram lines are separated by a thick oxide layer and are independent of each other, except for parasitic capacitance effects, unless a contact cut is shown.

Figure 5.18 shows a NAND gate in stick-diagram representation. Note that an X distinguishes the buried-contact from the regular contact cut with its solid circle symbol. Stick diagrams, gate, and circuit schematic symbols are also sometimes mixed as shown in Figure 5.19, in which two gates, a NAND gate and an inverter, are separated by a pass transistor that connects the output of the NAND gate to the input of the inverter when a clock signal, φ, is high. In this case, the NAND gate uses the diffusion path as the output, as shown in the layout of Figure 5.16. A buried contact, not shown, is used

Figure 5.18 Stick diagram for the NAND gate. A. Gate schematic. B. Stick diagram.

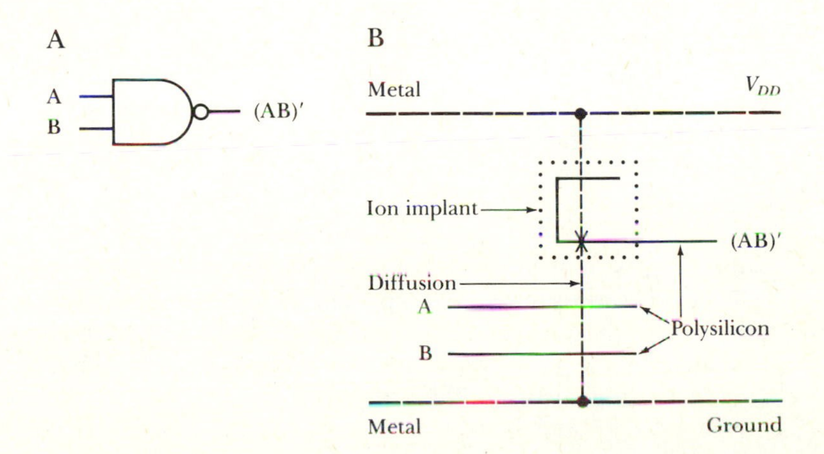

Figure 5.19 Mixed symbol representation example. A. Circuit. B. Symbolic representation.
SOURCE: From C. Mead and L. Conway, *Introduction to VLSI Systems.* © 1980, Addison-Wesley Publishing Company, Inc., Reading, Mass. Reprinted with permission.

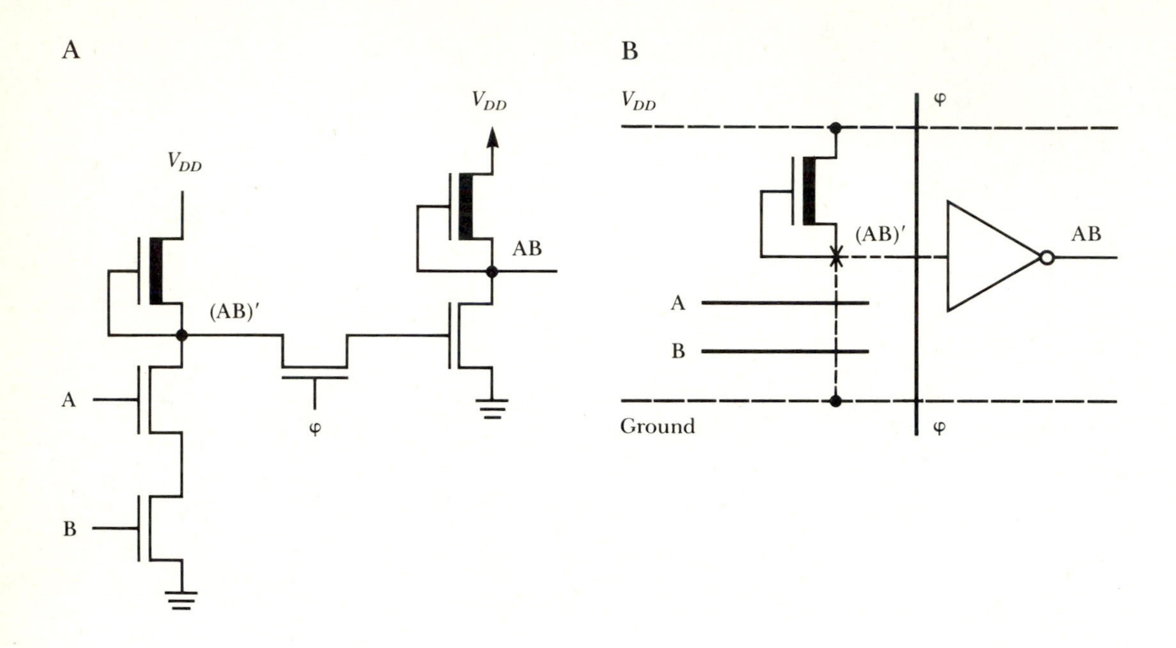

to connect the polysilicon input of the inverter to the diffusion path output of the NAND gate.

5.5 Geometric Design Rules

Several geometric design rules govern designing a layout or going from a specified stick diagram to a layout. For example, a certain minimum distance must separate adjacent metal lines and contact cuts must be of a certain minimum size. Such design rules are clearly process dependent and are generally expressed in dimensional units, such as micrometers. Design rules are often considered proprietary information, and a variety of design rule sets are in use.

However, it is useful, particularly for pedagogical purposes, to use dimensionless design rules that are expressed in terms of a basic length unit λ. This minimum dimension represents a fundamental resolution of the process and is determined by processing variables such as misalignment between masks, degree of overetching, and wafer distortion. These variables cause a geometrical feature on the wafer to deviate from its assigned position in the layout, and λ may be taken to be the maximum deviation of a feature on the wafer from its intended position in the artwork. Lambda-based design rules for NMOS were popularized by Mead and Conway [7], and lambda-based design rules for CMOS are also used for designs implemented by MOSIS. (MOSIS refers to the MOS Implementation System, a service at the Information Sciences Institute of the University of Southern California. MOSIS turns out small-volume runs of integrated circuits for a variety of customers by acting as an interface between designers and the silicon foundry [8].)

The use of simplified, dimensionless design rules allows the layout design to proceed without detailed regard for wafer fabrication details or processing limitations. On the other hand, a set of design rules sufficiently general to hold over a range of values for λ and for different processing lines must be quite conservative. Consequently, the resulting layouts will not necessarily be optimum in terms of area and performance. Also, a set of design rules can only be expected to hold over a reasonable range of λ. For example, a design rule appropriate for $\lambda = 2$ μm may not be appropriate for $\lambda = 0.5$ μm. This section gives examples of both dimensioned and dimensionless design rules, and shows the application of design rules to a circuit.

Geometric design rules take the form of specifications for the minimum width of feature X, W_X, or the minimum separation between features X and Y, S_{XY}, or the minimum extension of feature X beyond feature Y, E_{XY}. Table 5.1 lists two sets of nearly identical lambda-based NMOS design rules from different references. The second set is slightly more conservative in terms of

Table 5.1 Dimensionless NMOS design rules, expressed in terms of the minimum dimension λ

	Reference 7	Reference 5
W_D	2	2
W_C	2	2
W_P	2	2
W_M	3	3
S_{DD}	3	3
S_{CC}	2	2
S_{PP}	2	2
S_{MM}	3	3
S_{IG}	1.5	2
S_{CG}	2	2
S_{PD}	1	1
E_{PD}	2	2
E_{IG}	1.5	2
E_{DC}	1	1
E_{PC}	1	1
E_{MC}	1	1

patterning the ion implantation step [5,7]. In this table, the subscripts C, D, G, I, M, and P, refer to contact cut, diffusion, gate, ion implant, metal, and polysilicon features, respectively.

This discussion will not give a detailed rationale for the design rules in Table 5.1, but a few comments are in order. The minimum width for the metal lines is shown as being larger than for either the polysilicon or diffusion lines. This difference occurs because the metalization is overlayed on an uneven surface, as this processing step occurs late in the fabrication sequence. The cumulative effects of the several etches, oxide growths, and depositions prior to the metalization may produce appreciable steps in the wafer surface, as discussed in Chapter 2. The lack of a planar surface can have a large effect on the quality of photolithography because of sideways reflections from hills and valleys. A wider metal line helps to ensure continuity of the conducting path over these steps. Also, diffusion paths should be more widely separated than polysilicon paths, because the diffusion layers involve associated depletion layers, as well as a certain amount of lateral diffusion.

The design rules also mandate a minimum length and width for the MOSFET channel. Since a transistor results when a diffusion path crosses a polysilicon path, the minimum channel length is equal to the minimum polysilicon dimension of 2λ. Likewise, the minimum channel width is equal to the minimum diffusion width, which is also 2λ. When a transistor is formed, the polysilicon line must extend beyond the diffusion path to ensure that the entire diffusion path is under control of the gate. Without such an extension, mask misalignment between the polysilicon and the oxide patterning could allow the diffusion path to short the transistor around the end of the gate. When two geometric features are on different levels and their relative placement is critical to the design, then 2λ should be allowed to account for a worst-case misplacement of λ on each level. This is the case here, so E_{PD} is equal to 2λ.

Since the metal lines are separated from the diffusion paths and the polysilicon paths by thick oxides, Table 5.1 gives no values for S_{PM} and S_{DM}. However, in practice, metal paths running parallel to diffusion or polysilicon paths, for appreciable distances, should be separated from the latter by a minimum value of λ to reduce parasitic effects caused by the capacitance between the metal path and the underlying polysilicon or diffusion paths. Also, although it is technically possible, it is not advisable to make a contact cut to a polysilicon layer over a polysilicon gate. Aluminum can spike through the polysilicon, along the grain boundaries between the crystallites, and react with the thin gate oxide so that the gate is shorted to the substrate.

Contact cuts through the thick CVD oxide should have a minimum dimension of 2λ by 2λ. Diffusion, polysilicon, and metal patterns should overlap the contact cut by λ. In this way, a worst-case misalignment of 2λ between the two levels would still maintain at least some contact area, although the contact resistance would increase. The increased contact resistance would be undesirable, but not catastrophic to the design. Consequently the overlaps E_{PC}, E_{MC}, and E_{DC} are set at λ rather than 2λ. This is in contrast to the case described earlier for E_{PD} in which a failure of the overlap would be catastrophic to the design. The design rules for buried-contact cuts through the gate oxide must prevent the occurrence of parasitic transistors that could result from misalignment of the buried-contact mask and the polysilicon mask. If the polysilicon strays beyond the buried contact cut into the intended diffusion path, an unwanted enhancement-mode MOSFET is established in the path. Therefore, the buried gate-oxide contact cut should extend a distance 2λ in the direction of the diffusion conducting path [5]. In addition, the buried-contact cut should extend beyond the polysilicon-diffusion overlap by λ in the other three directions to maintain a minimum contact area between the polysilicon and the diffusion paths.

Ion implantations for depletion-mode loads must also extend beyond the actual channel region. Without this precaution, misalignment between the ion implant mask and the oxide patterning mask could result in a portion of the channel not being modified by the implant. Consequently, an unwanted parasitic enhancement-mode MOSFET would be either in series or parallel with the depletion-mode device, depending on the direction of misalignment. Likewise, depletion-mode ion implantation should be separated from the gates of enhancement-mode transistors to make sure that the implant does not stray into enhancement-mode channels.

Several of the dimensionless design rules from the second column of Table 5.1 are illustrated in Figure 5.20 for an NMOS NAND gate with polysilicon inputs and a diffusion path output. This figure uses the same pattern code as in Figures 5.14 and 5.16. The layout extends the *n*-type ion implant to include the buried-contact region, thus facilitating the interconnection between the *n*-type diffusion layer and the polysilicon layer.

Figure 5.20 shows that the ratio of Z_{PU} to Z_{PD} is doubled for the NAND gate as compared to the inverter shown in Figure 5.16. The NAND gate pull-down transistors are in series, so it is necessary to either double Z_{PU} or to halve Z_{PD} for each driving transistor (or change both by a fractional amount) to maintain the same logic voltage swing as in the inverter. The longer load transistor shown in figure 5.20 means a slower pull-up time. This is in contrast to the NOR gate where the driving transistors are in parallel and the same Z_{PU} to Z_{PD} ratios can be used as in the inverter. For this reason, in NMOS the NOR gate is preferred to the NAND gate as a basic building block.

Finally, it is emphasized again that the lambda-based dimensionless design rules do not give as compact a layout as would a set of dimensioned design rules finely tuned for a given process line. Table 5.2 shows an example of such dimensioned design rules for an NMOS process, and clearly they do not lend themselves to a simple λ interpretation. For example, the table gives the minimum diffusion width as 5 μm. Taking this to be 2λ, as given in both columns of Table 5.1, would yield a basic length unit of λ = 2.5 μm.

Figure 5.20 Application of dimensionless geometric design rules to an NMOS NAND gate layout.

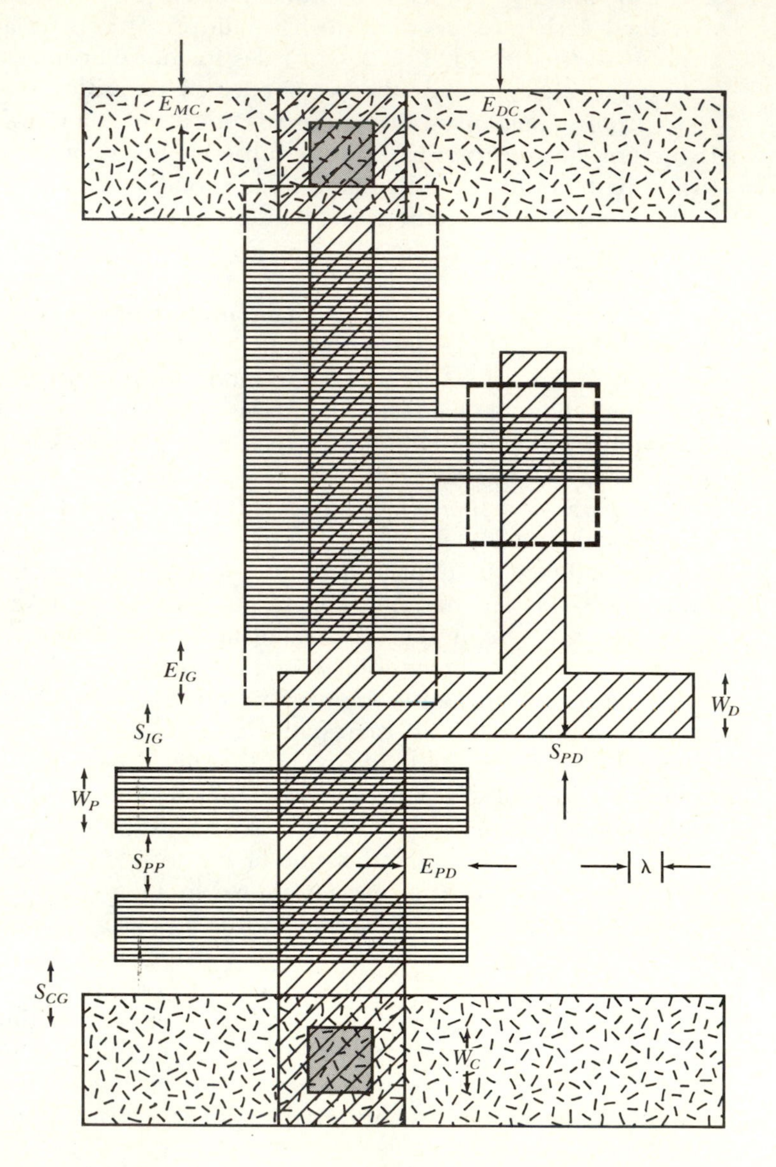

Table 5.2 An example of dimensioned design rules that do not lead to a simple λ interpretation

	Minimum dimension (μm)	Implied λ from Table 5.1 dimensionless design rules (μm)
W_C	4	2.0
W_D	5	2.5
W_M	6	2.0
S_{DD}	4	1.3
S_{MM}	4	1.3
S_{CC}	3	1.5
E_{DC}	1	1.0
E_{MC}	1	1.0

On the other hand, the separation between metal lines is given as 4 μm, which implies a λ of 1.33 μm. And the metal and diffusion extension beyond contact cuts is given as 1 μm, which would imply a λ of only 1 μm!

If dimensionless design rules were applied to this particular NMOS process, a value of 2.5 μm would have to be used to accommodate the worst case of diffusion paths, and as a consequence, other minimum dimensions would be much larger than necessary. The result would be a larger die size, which has economic implications. First of all, the number of chips per wafer decreases. Also the yield decreases as the die size increases, so that the fraction of chips which are good also decreases. Therefore, for competitive products that are to be produced in large lots, the use of finely tuned, dimensioned design rules offer an appreciable advantage.

Bonding pads are also subject to geometric constraints. Table 5.3 gives an example of design rules for peripheral bonding pads. Also, adjacent chips on wafers are separated by scribe lines to provide a space for sawing, scribing, or laser cutting the wafer into individual die. Typically, scribe lines are on the order of 150 μm wide.

Table 5.3 A sample set of design rules for bonding pads

Minimum size	125 μm × 125 μm
Minimum spacing between pads	50 μm
Minimum spacing between pad and another metal line	25 μm
Minimum spacing of passivation cut to pad edge	5 μm

BIBLIOGRAPHY

The following sources are recommended for further reading on MOS layout principles and methods:

Carr, W. N., and J. P. Mize. *MOS/LSI Design and Applications.* New York: McGraw-Hill, 1972.

Glasser, L. A., and D. W. Dobberpuhl. *Design and Analysis of VLSI Circuits.* Reading, Mass.: Addison-Wesley, 1985.

Mavor, J., M. A. Jack, and P. B. Denyer. *Introduction to MOS LSI Design.* Reading, Mass.: Addison-Wesley, 1983.

Mead, C., and L. Conway. *Introduction to VLSI Systems.* Reading, Mass.: Addison-Wesley, 1980.

Mukherjee, A. *Introduction to NMOS and CMOS VLSI Systems Design.* Englewood Cliffs, N.J.: Prentice-Hall, 1986.

For further reading on bipolar layout techniques, the following sources are suggested:

Glaser, A. B., and G. E. Subak-Sharpe. *Integrated Circuit Engineering.* Reading, Mass.: Addison-Wesley, 1977.

Hamilton, D. J., and W. G. Howard. *Basic Integrated Circuit Engineering.* New York: McGraw-Hill, 1975.

Weidmann, S. G. "Advancements in Bipolar VLSI Circuits and Technologies." *IEEE Journal of Solid-State Circuits,* SC-19 (1984), 282–291.

REFERENCES

1. P. Antognetti, G. Bisio, F. Curatelli, and S. Palara, "Three-Dimensional Transient Thermal Simulation: Application to Delayed Short Circuit Protection in Power IC's," *IEEE Journal of Solid-State Circuits,* SC-15 (1980), 277–281.
2. D. J. Hamilton and W. G. Howard, *Basic Integrated Circuit Engineering* (New York: McGraw-Hill, 1975).
3. A. H. Dansky, "Bipolar Circuit Design For VLSI Gate Arrays," *Proceedings of the IEEE International Conference on Circuits and Computers,* Port Chester, N.Y., Oct. 1980, 674–677.
4. A. B. Glaser and G. E. Subak-Sharpe, *Integrated Circuit Engineering* (Reading, Mass.: Addison-Wesley, 1977).
5. J. Mavor, M. A. Jack, and P. B. Denyer, *Introduction to MOS LSI Design* (Reading, Mass.: Addison-Wesley, 1982).
6. J. D. Williams, "Sticks—A New Approach to LSI Design," M.S.E.E. thesis, Department of Electrical Engineering and Computer Science, Massachusetts Institute of Technology, June 1977.
7. C. Mead and L. Conway, *Introduction to VLSI Systems* (Reading, Mass.: Addison-Wesley, 1980).
8. A. Mukherjee, *Introduction to NMOS and CMOS VLSI Systems Design* (Englewood Cliffs, N.J.: Prentice-Hall, 1986).

PROBLEMS

1. This problem reconsiders the minimum-area junction-isolated BJTs of Prob. 4.1. Assume the chips are allowed to run hot and have good heat

sinks, so the maximum device power density is 250 W/cm^2. Again, the epitaxial thickness is assumed to be equal to the minimum line width, m, and the emitter elongation factor is taken to be $k = 1$. Also,

$$N_{BASE} = 1 \times 10^{15}\ \text{cm}^{-3}$$
$$V_{CE} = 1\ \text{volt}$$
$$\text{base hole mobility} = 450\ \text{cm}^2/\text{V} \cdot \text{s}$$
$$\text{device temperature} = 85°\text{C}$$
$$\beta = 100$$

Consider the three cases of $m = 5\ \mu$m, $m = 2\ \mu$m, and $m = 0.5\ \mu$m.

a. As the device designer, designate the maximum device current rating. Take into account both thermal considerations and high injection β degradation. Assume that emitter-crowding and the onset of high-injection occur at about the same current level. Apply the maximum power-density constraint to the active portion of the transistor, that is, the portion under the emitter.

b. At your current rating, what would be the power dissipated in a chip 1 cm by 1 cm, assuming the maximum device densities calculated in Prob. 4.1?

2. A popular form of emitter-coupled logic (ECL) is the 10K series that is produced by several manufacturers. A gate schematic is shown in Figure 5P.1.

Figure 5P.1 A 10K ECL logic gate.

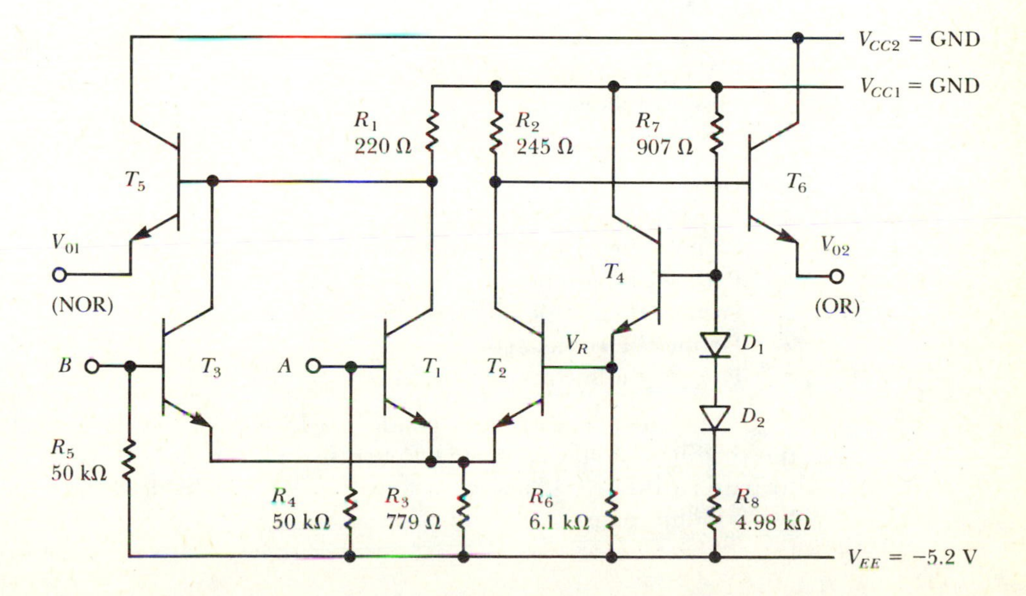

a. Assuming a standard junction-isolated bipolar-process, what is the minimum number of isolation regions required for the layout of this gate? For each isolation region, list the devices contained.
b. For the circuit as drawn in Figure 5P.1, how many cross-over points can you find? Circle them on the schematic and include a copy of the schematic with your solutions.
c. Considering each of the cross-overs, can the layout on silicon be done so as to avoid double metalization or additional n^+ structures?
d. Assume that all resistors are established by the base diffusion, which has a sheet resistance of 200 Ω/□, and that inputs A and B may vary between 0 and −5 V. Find the minimum allowed length and width for the resistors numbered 5, 7, and 8. Take the maximum device power density to be 20 W/cm^2.

3. Consider the simplified transistor-transistor-logic (TTL) NAND gate shown in Figure 4.35. Take the value of resistor R to be 4 kΩ and the value of R_C to be 1.4 kΩ. Assume that the power supply is 5 V and that the resistors are formed by a 200-Ω/□ base diffusion. The input voltage varies between 0 and 5 V. Generate plots of a mask set and composite layout for a junction-isolated process if $(P/A)_{max}$ is 20 W/cm^2. Indicate for each mask which portions of the mask are opaque and which are transparent, assuming negative photoresist. Also include alignment marks on the mask artwork.

4. In this exercise, a mask set is to be designed for the CMOS inverter shown in Figure 3.31. The fabrication process is to be similar to the NMOS gate example in Section 5.4.1, but with two exceptions. Rather than form a depletion-mode load, the ion implant establishes an n-type well in the p-silicon wafer, and the p-channel transistor will reside in this well. Also, both p and n diffusions are required. The key processing steps are as follows:

    1. Define n wells (mask 1).
    2. Define the field oxide pattern (mask 2).
    3. Deposit polysilicon (no mask).
    4. Pattern the polysilicon (mask 3).
    5. Pattern the n-type diffusion (mask 4).
    6. Pattern the p-type diffusion (mask 5).
    7. Define the contact cuts (mask 6).
    8. Pattern the metal layer (mask 7).

    This exercise does not include masks for ion implants that adjust the threshold voltages of the two transistors, nor does it include a mask to pattern the over-glass, or passivation layer, which covers everything but bonding pads.

The specific tasks in this exercise are as follows.

a. Generate the artwork for each mask relative to an alignment mark, and also generate a composite layout. Use the design rules of the second column of Table 5.1, and take $\lambda = 2\ \mu m$. Also, take $W/L = 3$ for each transistor. Do not include bonding pads connected to the gate. Indicate for each mask which portions of the mask are clear and which are opaque, assuming negative photoresist.
b. Take the threshold voltage for the *n*-channel device to be $+1$ V and the threshold voltage for the *p*-channel device to be -1 V. The power supply is 5 V. If a 15-gate ring-oscillator is fabricated with the gates of part A, estimate the expected oscillation frequency. Take the gate oxide thickness to be 400 Å, the hole mobility to be 400 $cm^2/V \cdot s$ and the electron mobility to be 1000 $cm^2/V \cdot s$.
c. Estimate how many of your gates could be put on a 1 cm by 1 cm chip (ignoring the problems associated with interconnecting all those gates).

6 Design Methods

6.1 Introduction

Circuit design has long been considered at least partly an art. In the absence of a single formal theory that leads to unique design solutions, successful circuit design has traditionally relied on the skill and ingenuity of an individual designer or design team. This reliance is evident in that many basic circuits carry the names of the individuals or groups that invented them; examples are Widlar current sources, Schmitt triggers, Wien-bridge oscillators, Darlington pair amplifiers, and Manchester carry adders [1, 2, 3]. Unfortunately, guaranteed formulas for ingenuity cannot be packaged between the covers of a textbook. However, several methodologies have emerged in modern integrated circuit design, and many of them are decidedly different from traditional design approaches that assume the use of small-scale integrated circuits. The intent of this chapter is to introduce, by example, a selected sampling of basic design methods, thereby demonstrating some underlying characteristics of integrated circuit design.

A distinction can be drawn between circuit design and layout design. As discussed in the operational amplifier example of Section 6.6, a given circuit design can often be realized by several quite different layouts. Very large-scale circuit design, however, strongly couples circuit design and layout design in that the circuit design should lend itself well to layouts characterized by order and regularity. Indeed, large-scale circuit design differs fundamentally from small-scale circuit design because it requires systematic, structured layout. If a small-scale circuit is designed and laid out in a highly unstructured way and does not work correctly, it is fairly easy to implement iterations because there are relatively few devices in the circuit. However, if the circuit has several thousand devices, a highly structured design is more likely to work the first time, and it is easier to troubleshoot if it does not.

This chapter presents several basic circuit structures that allow the storage and manipulation of data. Many of these structures take the form of basic cells that are readily combined to form larger systems or subsystems so that regularity is built into the design. A preponderance, but not all, of the examples will be in the context of digital metal-oxide-semiconductor (MOS) circuits. In addition, the chapter discusses input/output circuits, as well as some of the design opportunities and problems in linear bipolar monolithic amplifiers and analog-to-digital converters. Finally, the chapter closes by reviewing the important role of computer aids in integrated circuit design.

6.2 MOS Logic

6.2.1 Two-Phase-Clock Registers

Although it is possible to design digital integrated circuits using the same methods that are used when building circuits from basic transistor-transistor-logic (TTL) building blocks such as NAND gates and flip-flops, doing so may not be advisable in terms of obtaining a desirable layout. Furthermore, a designer faced with a blank piece of silicon has greater freedom than one who must generate a design using standard small-scale TTL-type parts. For example, an important feature of many LSI and VLSI digital circuits is the multiphase clock. In this section, for purposes of comparison, the design of a 4-bit shift register will be considered both for a TTL-type building-block approach and for a two-phase-clock approach.

A conventional design for a 4-bit shift register that uses type-*D* flip-flops is shown in Figure 6.1 [2]. The characteristics of a *D* flip-flop are such that, after the triggering transition of the clock, the flip-flop output *Q* goes to the state present at its input *D* just before the clock transition. Consequently, data flows sequentially through the register, as Figure 6.1 shows. Variations on such registers are useful for the temporary storage of data at various points in digital systems.

Type-*D* flip-flops can be realized with NAND gates, as shown in Figure 6.2 [2]. If the NAND gates are of the enhancement-mode/depletion-mode NMOS design shown in Figure 5.20, then the total device count is arrived at as follows:

5 two-input NAND gates with 3 MOSFET's each	=	15 transistors
1 three-input NAND gate	=	4 transistors
		19 transistors/flip-flop

Therefore, the 4-bit shift register shown in Figure 6.1 requires 76 transistors.

Next, consider an alternative design that utilizes a nonoverlapping two-phase clock, as shown in Figure 6.3 [3]. Note that the two clock periods are equal but have nonoverlapping high signals. That is, φ_1 is always low when φ_2 is high, and φ_2 is always low when φ_1 is high. Both phases may be simultaneously low, since the low times are longer than the high times, but they are never simultaneously high. When φ_1 is high, the input is connected to

Figure 6.1 Shift register based on a small-scale integration design approach. A. Circuit configuration. B. Data flow.

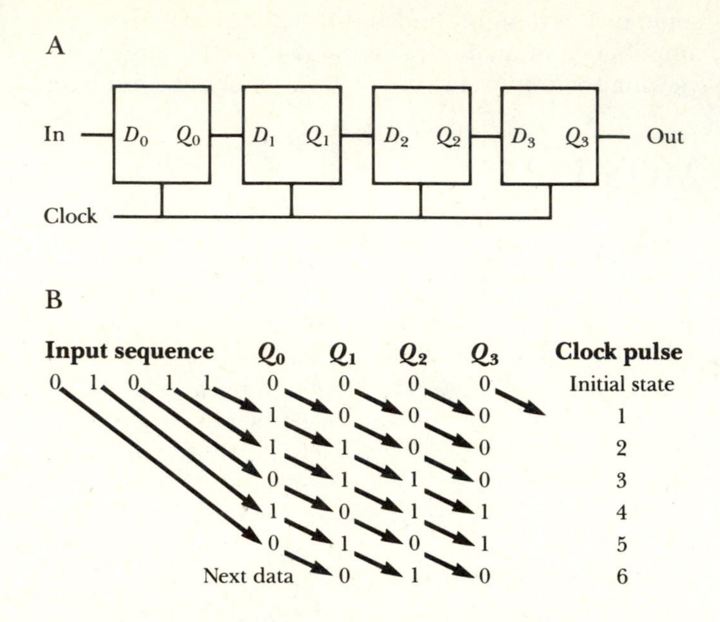

Figure 6.2 NAND gate implementation of a type-D flip-flop.
SOURCE: From H. Taub and D. Schilling, *Digital Integrated Circuits* (New York: McGraw-Hill, © 1977). Reproduced with permission.

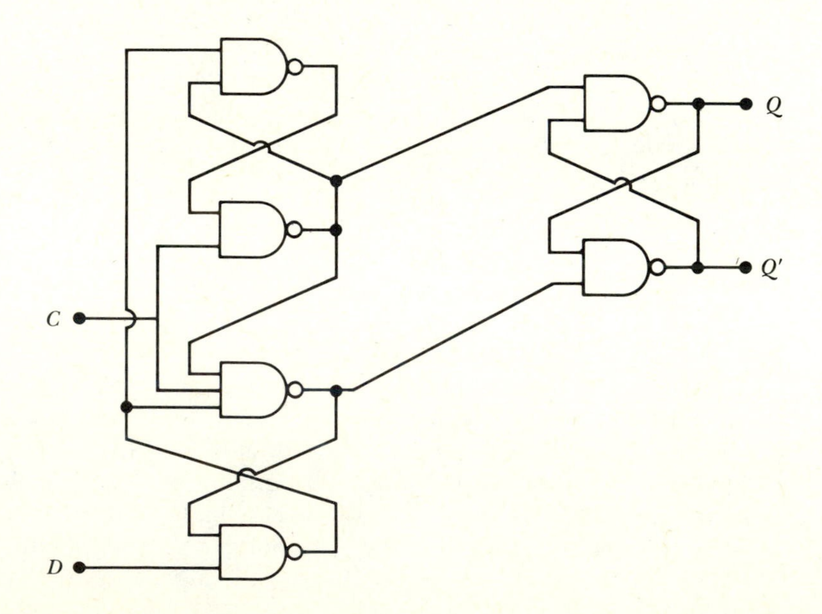

Figure 6.3 A two-phase-clock dynamic shift register.

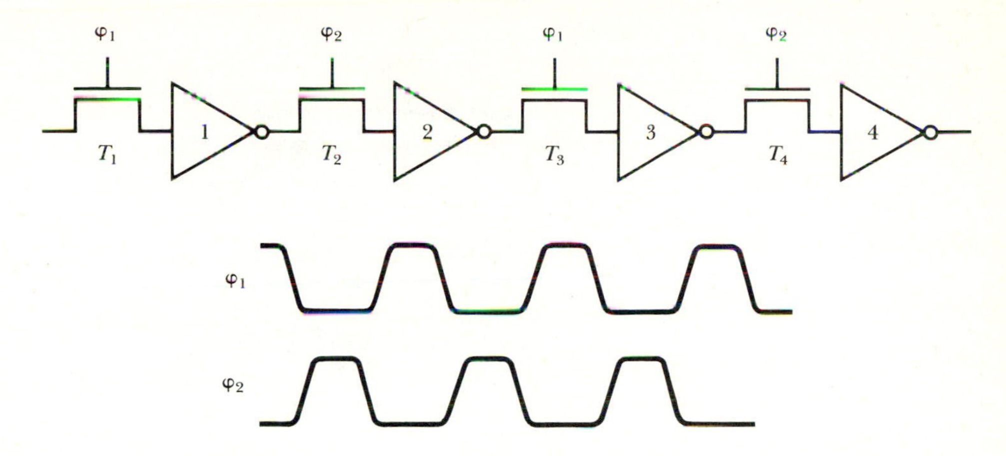

inverter 1 via the n-channel enhancement-mode pass transistor, T_1. If, for example, the input is high, then the output of inverter 1 is driven low. However, inverter 2 does not respond until φ_2 is high, at which point the output of inverter 2 goes high. During this portion of the clock cycle, the first inverter is separated from the input because of the nonoverlapping nature of the two-phase clock. At this point, the high bit at the input has been shifted to the output of inverter 2, which corresponds to one shift of the register.

Clearly, the pass transistor's role in this design is to allow data to flow through the register in a controlled and unambiguous fashion. The inverters, on the other hand, are required to periodically restore the voltage levels of the transmitted data. Signals are degraded as they pass through the register, not only because of capacitive discharge with time but also because of voltage drops across the pass transistors. As Figure 6.4 shows, a minimum gate-to-source voltage equal to the threshold voltage, V_T, is required to establish a conducting channel. Therefore, if the logic level to be passed is high, at 5 V, then the high signal beyond the pass transistor is at a voltage of $(5 - V_T)$ V, and consequently the pass transistor degrades the logic level. However, provided the voltage level after the pass transistor is still within the inverter noise margins, the inverter chain serves to restore the signal as it progresses through the register.

Since inverters fed by pass transistors must function reliably with degraded logic levels, they require better transfer characteristics than ordinary inverters. Consequently their layout must allow for a higher Z_{PU}/Z_{PD} ratio, as discussed in Section 3.9.1. A typical ratio for inverters used with pass transistors is in the range of 7 to 8, and for inverters that have "good" logic-level inputs, typical ratios are in the range of 4 to 5.

Considering 1 bit of the two-phase-clock shift register to consist of two

Figure 6.4 Origin of the pass-transistor voltage drop.

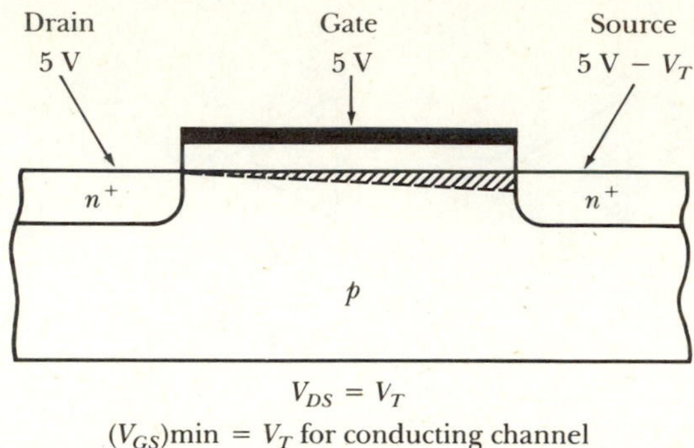

inverters and two pass transistors, the device count for the 4-bit register of this design is as follows:

8 inverters with 2 MOSFETs each	=	16 transistors
8 pass transistors	=	8 transistors
		24 transistors

Consequently, the device count is only about one-third that of the conventional flip-flop–based design.

The price paid for this reduction in devices is twofold. The clocking is more complex since a two-phase clock is required rather than a single-phase clock. Secondly the flip-flop is dynamic; to avoid loss of data, the clock must run faster than a minimum frequency. Clearly, if the clocking sequence were entirely halted at any point in the data transfer, leakage would cause the gates of driving MOSFETs at high logic levels to eventually discharge. Consequently the clock frequency must be sufficiently high to ensure that data is passed along before levels drop enough to cause reliability problems. Therefore, dynamic MOS logic circuitry has both an upper and lower limit on clock frequency. For example, the Motorola 6800 8-bit microprocessor requires a clock frequency in the range of 100 kHz to 1 MHz. Upper clock-frequency limits are determined by the times required to charge and discharge capacitive loads, as discussed in Section 3.10. Lower clock frequencies are determined by the discharge times associated with leakage through transistors in the off state.

The shift register concept illustrated in Figure 6.3 is readily expanded to handle words, as indicated in Figure 6.5, which shows the same shift register design implemented for a 16-bit word. A modified stick diagram for this two-phase-clock shift register emphasizes the regularity of the design. As illustrated, the clock lines that control the flow of data through the register

Figure 6.5 Modified stick diagram for a 16-bit-word shift register.

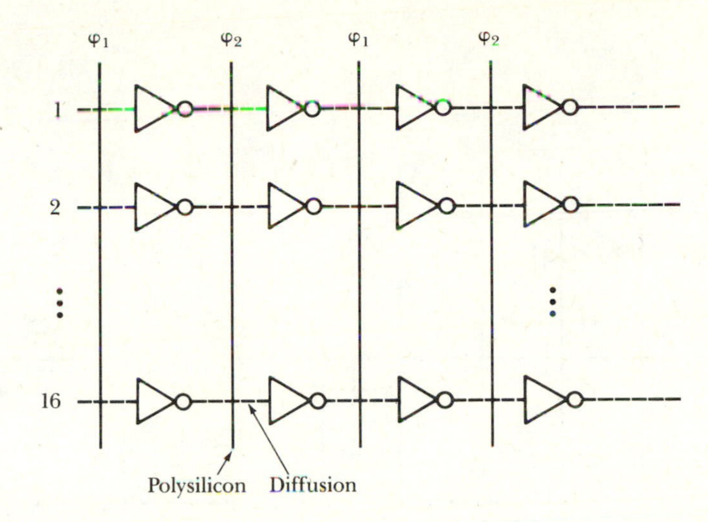

are implemented in polysilicon and the interconnection between inverters is along a diffusion path. This arrangement allows an efficient layout, since the pass transistors are formed automatically when the diffusion lines cross the clock lines.

Variations on the two-phase-clock-driven register, including dynamic registers with automatic refresh and stack registers, are addressed in the problems at the end of the chapter.

6.2.2 Programmable Logic Arrays

The preceding section shows an alternative shift register design that uses appreciably fewer devices than a design based on standard logic gates. Alternatives to standard gates for the design of combinational logic also exist and offer advantages in terms of attractive layout. One design approach that provides a highly structured layout is the programmable logic array (PLA).

Consider, for example, a logic circuit with three inputs, A, B, and C, and three outputs, Z_1, Z_2, and Z_3. In general the outputs are related to the inputs by Boolean functions, as follows:

$$Z_1 = f_1(A,B,C) \tag{6.1}$$

$$Z_2 = f_2(A,B,C) \tag{6.2}$$

$$Z_3 = f_3(A,B,C) \tag{6.3}$$

In standard random logic design using NOR gates, NAND gates, and the like, the design and layout of such a circuit depends on the nature of f_1, f_2,

Figure 6.6 NMOS PLA circuit.

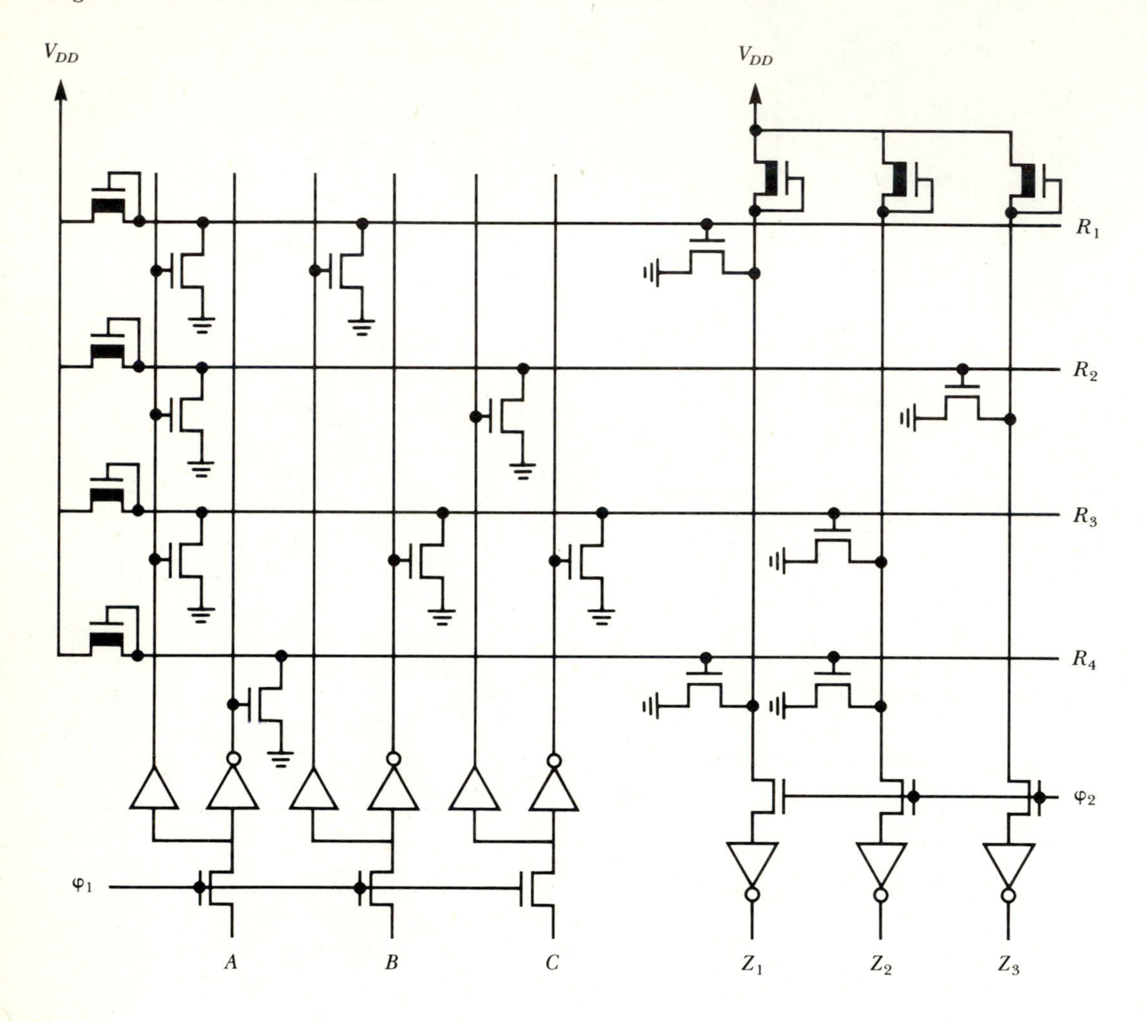

and f_3 and may change considerably if the functional relationships between the outputs and inputs change. However, with a PLA the broad features of the layout are independent of the f_i.

To illustrate, consider the case where the relationships between outputs and inputs are expressed as sums of products:

$$Z_1 = A + A'B' \tag{6.4}$$

$$Z_2 = A + A'BC \tag{6.5}$$

$$Z_3 = A'C' \tag{6.6}$$

where prime indicates the complement of the variable. Figure 6.6 shows a PLA implementation of these functions. (Note the presence of a two-phase clock again to control the flow of data into and out of the logic circuit.)

Figure 6.6 shows that the logic circuit consists of two banks of NOR gates with depletion-mode loads and multiple inputs. The actual logic function of the logic array is determined, or programmed, by the placement of the enhancement-mode driving transistors. As noted on the circuit, the output of the first bank of NOR gates provides inputs R_1, R_2, R_3, and R_4 to the second bank. However, it is also useful to think of this PLA as consisting of two planes, the AND plane and the OR plane, as shown in Figure 6.7. More specifically, the outputs of the first plane of the PLA shown in Figure 6.6 may be written in the AND product format as follows:

$$R_1 = (A + B)' = A'B' \tag{6.7}$$

$$R_2 = (A + C)' = A'C' \tag{6.8}$$

$$R_3 = (A + B' + C')' = A'BC \tag{6.9}$$

$$R_4 = A \tag{6.10}$$

Also, the Z output expressions in Eqs. (6.4), (6.5), and (6.6) are sums of the above products. Specifically,

$$Z_1 = R_1 + R_4 \tag{6.11}$$

$$Z_2 = R_3 + R_4 \tag{6.12}$$

$$Z_3 = R_2 \tag{6.13}$$

Figure 6.7 PLA architecture.

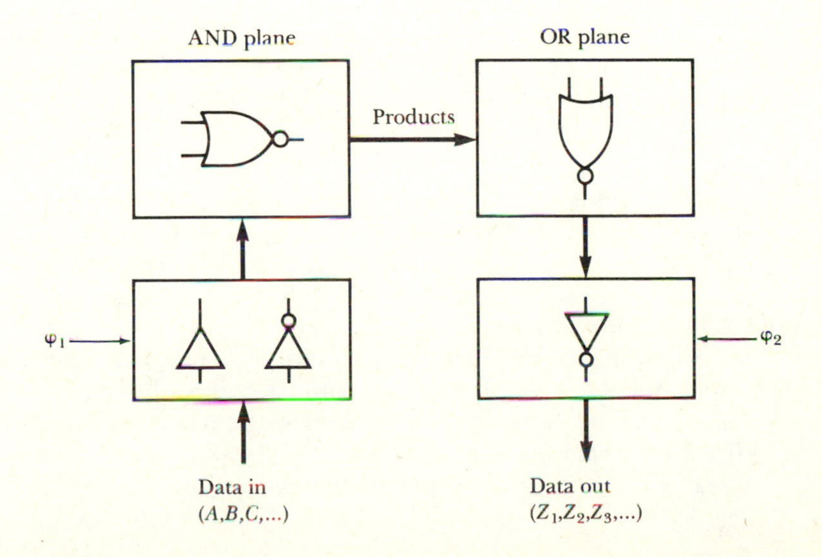

Table 6.1 Truth table for the PLA circuit of Figure 6.6

Inputs				Outputs		
A	B	C	Product term	Z_1	Z_2	Z_3
0	0	x	R_1	1	0	0
0	x	0	R_2	0	0	1
0	1	1	R_3	0	1	0
1	x	x	R_4	1	1	0

Referring to the circuit in Figure 6.6 shows that the OR plane indeed realizes these sums of products. In summary, therefore, the placement of enhancement-mode driving transistors in the AND plane establishes the needed product terms, and the placement of driving transistors in the OR plane provides the desired sum of products. A systematic set of rules for the placement of transistors results from an inspection of the truth table for this example, shown in Table 6.1. To generate the product terms in the AND plane, driving transistors should be associated with the input data lines as follows. For a given product term, a transistor should link that product-term line and each noninverting data-input line for which a 0 appears in the input columns of the table. Conversely, a transistor should link the product-term line to each inverting data-input line for which a logic 1 appears in the table. Next, to generate the sum of products in the OR plane, a transistor should link a given output line with each product term line for which a logic 1 appears in that output column of the table.

Another useful concept that can be used in conjunction with a PLA is the *finite-state machine*. Suppose that there is feedback between the output and input of the PLA, as Figure 6.8 shows. Now the response of the logic array to the inputs A_i, which are clocked into the array on phase φ_1, depends on the value of Y, which was obtained from the output on the previous φ_2 clock phase. If only one feedback line is provided, then there are two possible states corresponding to a logic 0 or 1 on the feedback path. With N feedback lines, there are 2^N possible states. The placement of transistors in the finite-state machine follows from the state-transition table using the previously stated rules; the present-state column is grouped with the input columns, and the next-state column is grouped with the output columns.

EXAMPLE 6.1

Consider a PLA configured as a finite-state machine that has only feedback inputs for the AND plane. As Figure 6.9 shows, there are two feedback paths, Y_0 and Y_1, so there are four states for the machine. The array output is in the form of a 4-bit-word instruction, $Z_3Z_2Z_1Z_0$, which is used for control purposes elsewhere in the system. In this case, the finite-state machine cycles

Figure 6.8 A finite-state machine.

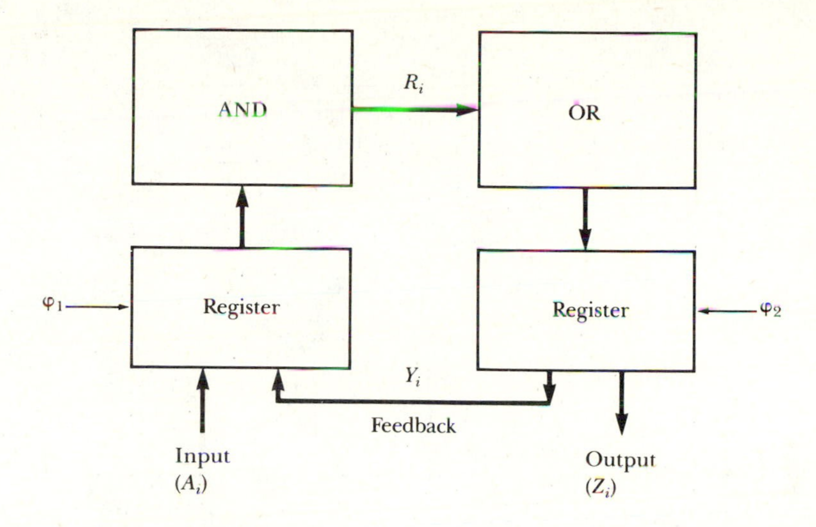

through a fixed series of states, as the state-transition table (Table 6.2) shows, such that the control algorithm sequence is fixed.

In this exercise, the goal is to design a PLA, with feedback, that accomplishes this control sequence and to show a corresponding layout in the form of a modified stick diagram.

Solution Figure 6.10 shows a circuit schematic of the logic array, including input and output registers, that implements this transition table. Note that the

Figure 6.9 Block diagram of the sequential controller of Ex. 6.1.

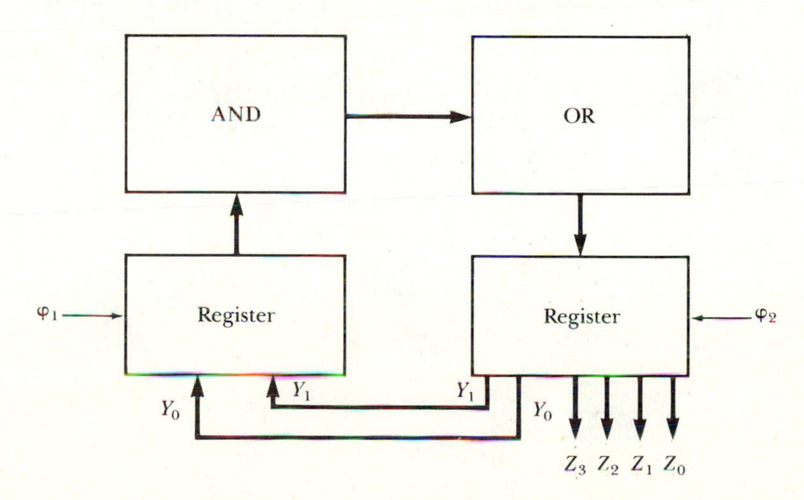

Table 6.2 State-transition table for Ex. 6.1

Present state		Next state		Output			
Y_0	Y_1	Y_0	Y_1	Z_0	Z_1	Z_2	Z_3
0	0	0	1	0	1	1	0
0	1	1	0	0	1	1	1
1	0	1	1	0	0	0	1
1	1	0	0	1	1	1	0

Figure 6.10 A circuit schematic for the sequential controller of Ex. 6.1.

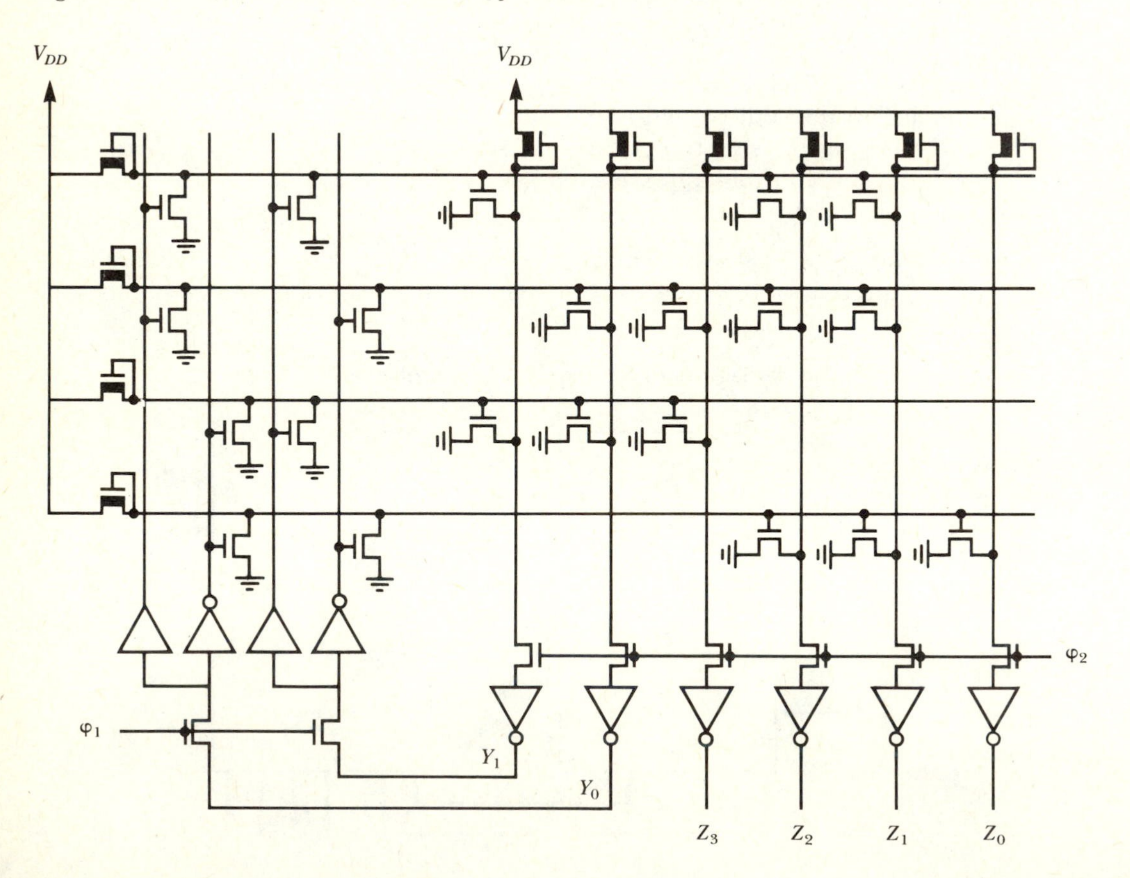

Figure 6.11 Stick-diagram layout for the sequential controller of Ex. 6.1.

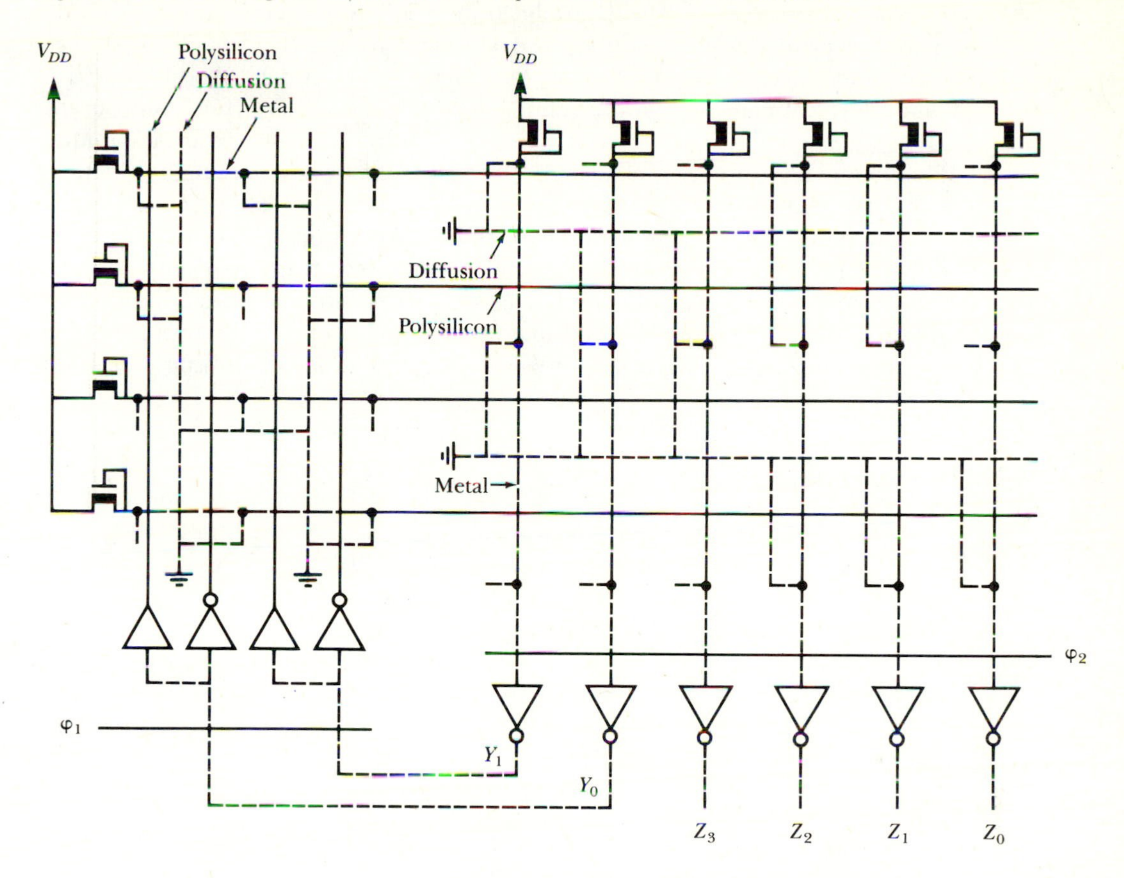

outputs are clocked out of the OR plane on clock phase φ_2, and the new state is clocked into the AND plane on clock phase φ_1. Figure 6.11 shows the silicon stick diagram layout. The input lines to both planes are polysilicon, and the required enhancement-mode transistors are placed by forming a crossing diffusion path that goes to ground. This solution uses to full effect the triple data path described in Section 5.4. The output lines from each plane are metal, the input lines are polysilicon, and the paths running from the source of the driving transistors to ground are diffused.

The basic architecture of the PLA is highly structured and fairly independent of the combinational or sequential logic being implemented. This description is true for the MOS design considered here and also for bipolar versions. The approach may appear to be somewhat inefficient in terms of area, since many open spaces in the array are reserved for driving transistors but are not used for a specific application. Note, however, that the examples in this section are quite simple, instructional ones. A modern PLA application could have 32 inputs, for instance, and well over 100 product lines. At such a level of complexity, the regular layout of the PLA offers considerable advantages over random logic. Moreover, topological compaction techniques have been developed to make more efficient use of chip area when dealing with sparse arrays. For example, *array folding* permits a set of column pairs (or row pairs) to be implemented in the same column (row) of the physical array. This can be accomplished, for instance, by bringing inputs and outputs into both the top and bottom of the array and splitting columns into upper and lower segments. The major drawback to folded arrays is that inputs and outputs to them may involve complex, area-consuming routing of signals.

6.2.3 A Selectable Logic Unit

The PLA provides fixed logic functions which, for the MOS version discussed, are programmed by the placement of diffusion lines in the silicon layout. However, it is often desirable to have a more versatile logic unit that can perform several different logic operations, depending on which software instruction it has received. For example, if the machine code, or op-code, for an exclusive OR operation on a microprocessor is 33C3H (hexadecimal), a portion of that 16-bit word is used to tell the microprocessor's arithmetic-logic unit (ALU) which logic or arithmetic operation it is to perform on the incoming data. This section considers the design and layout of a simple transistor switch array that can be instructed to perform all possible logic operations on two binary inputs, A and B. Since there are 16 possible truth tables associated with two binary variables, a 4-bit control word is used to select the desired logic operation.

Consider the transistor switch array shown in Figure 6.12. The unit can be viewed as a multiplexer, with the output related to the inputs according to the expression

$$Z = S_0A'B' + S_1A'B + S_2AB' + S_3AB \tag{6.14}$$

Figure 6.12 also shows the corresponding truth table, which makes it clear that the output as a function of A and B is determined by the control word $S_3S_2S_1S_0$. For example, if the control word is 0110, or 6H, then the resulting logic function is the exclusive OR. Likewise, if the control word is 1000, or 8H, then the logic function is the AND operation. Therefore, if the sequential controller of Ex. 6.1 is used to drive this selectable logic unit, the unit is instructed to sequentially perform XOR, OR, AND, and NAND operations on the input data.

Figure 6.12 Transistor-switch-array selectable logic unit.

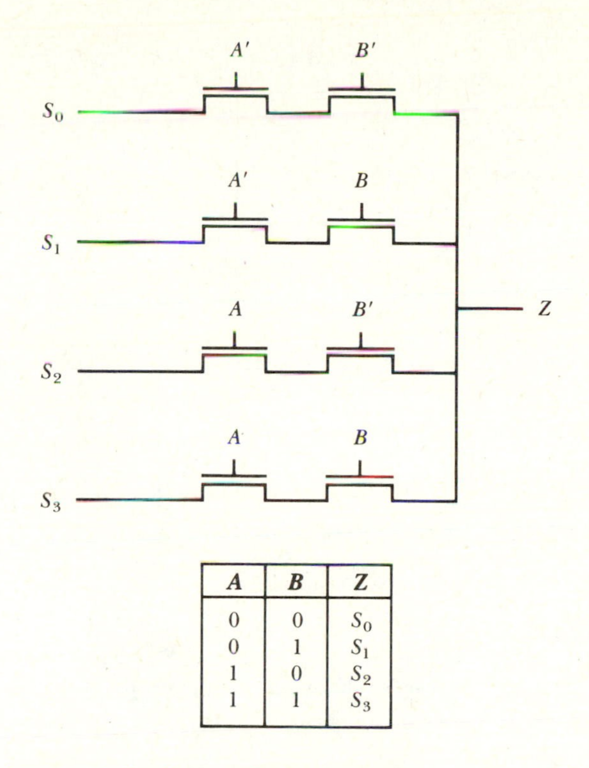

A	B	Z
0	0	S_0
0	1	S_1
1	0	S_2
1	1	S_3

This circuit essentially forms the logic portion of an ALU. To perform arithmetic operations requires the addition of circuitry to handle carry-in and carry-out bits. The carry functions can also be realized with transistor switch arrays, and a general-purpose ALU consisting of three arrays similar to the one in Figure 6.12 has been described which provides the capability for receiving and generating carry bits [3].

Figure 6.13 shows an interesting layout for the selectable logic transistor array. Essentially, the layout simply consists of four parallel polysilicon lines crossed by four parallel diffusion lines. The resulting 16 transistors include 8 that are not needed. However, using an *n*-type ion implant to make these extra 8 devices depletion-mode MOSFETs means that they are always on and are simply part of the signal path. Although this layout does not result in a minimum number of transistors, it is probably near the minimum area. Furthermore, the topology is regular and easily expanded to multiple parallel bit paths. An alternate layout that does not use depletion-mode transistors is considered in the problems at the end of the chapter.

Note that a logic signal must pass through two sequential pass-transistors in the transistor switch array. However, the total degradation of a logic-1 voltage is still only equal to V_T, since each pass transistor has a direct gate

Figure 6.13 Stick-diagram layout for the switch-array selectable logic unit.

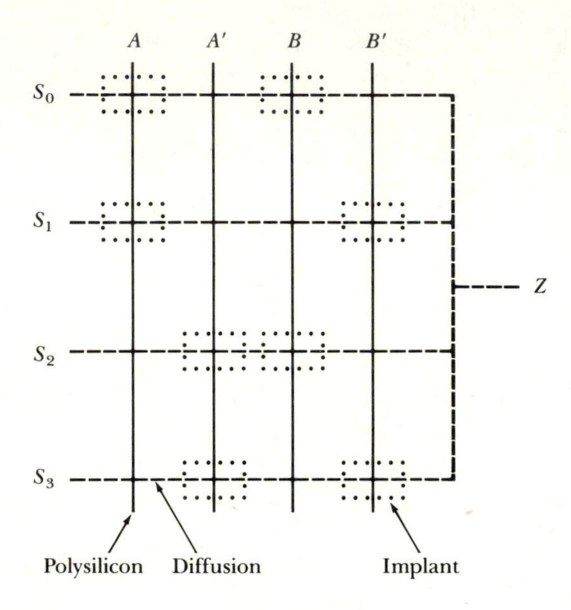

Figure 6.14 Unacceptable use of the pass transistor.

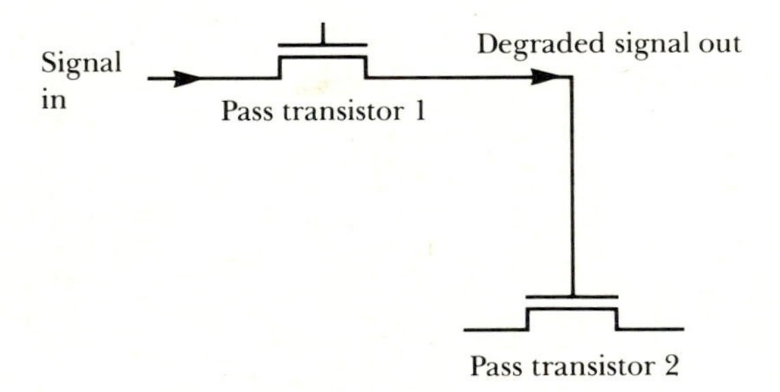

drive. Therefore the design is acceptable. On the other hand, if the output of a pass transistor were used to control the gate of another pass transistor, the output of the second pass transistor would be degraded by $2V_T$ and would possibly not reach a full logic-1 voltage. Figure 6.14 shows an example of such an unacceptable design.

6.2.4 Low-Power, Ratioless Logic

Sections 3.9 and 3.10 discuss the distinctions between ratioed and ratioless logic. Except for the selectable logic unit, the MOS examples considered so far have used ratioed logic and have exhibited nonnegligible quiescent-state power dissipation. Ratioless logic, such as complementary MOS (CMOS), offers the advantage of improved gate transfer characteristics and therefore high noise immunity, as well as rise and fall times that are about equal to each other. In additon, CMOS offers extremely low quiescent-state power dissipation. This feature is an obvious advantage for applications such as static microprocessors, which must maintain internal register and data values when the clock is stopped under battery-operated standby conditions and must be able to resume operation as soon as the clock is restarted. However, power dissipation is also a concern for very high-density integrated circuits where heat dissipation is appreciable. Although CMOS is an obvious candidate for low-power, ratioless logic, note that dynamic NMOS logic circuits can also be used with appropriate clocking of enhancement-mode loads. [4].

Static CMOS logic gates can be realized by expanding the basic inverter discussed in Section 3.9.2. However, to maintain negligible static power-dissipation, each n-type driving transistor must be accompanied by a p-type load transistor, resulting in "totem pole" circuit structures. For example the three-input CMOS NAND gate and the three-input CMOS NOR gate in Figure 6.15 require three p-type transistor loads and three n-type transistor

Figure 6.15 Totem pole CMOS static logic gates. A. Three-input NOR gate. B. Three-input NAND gate.

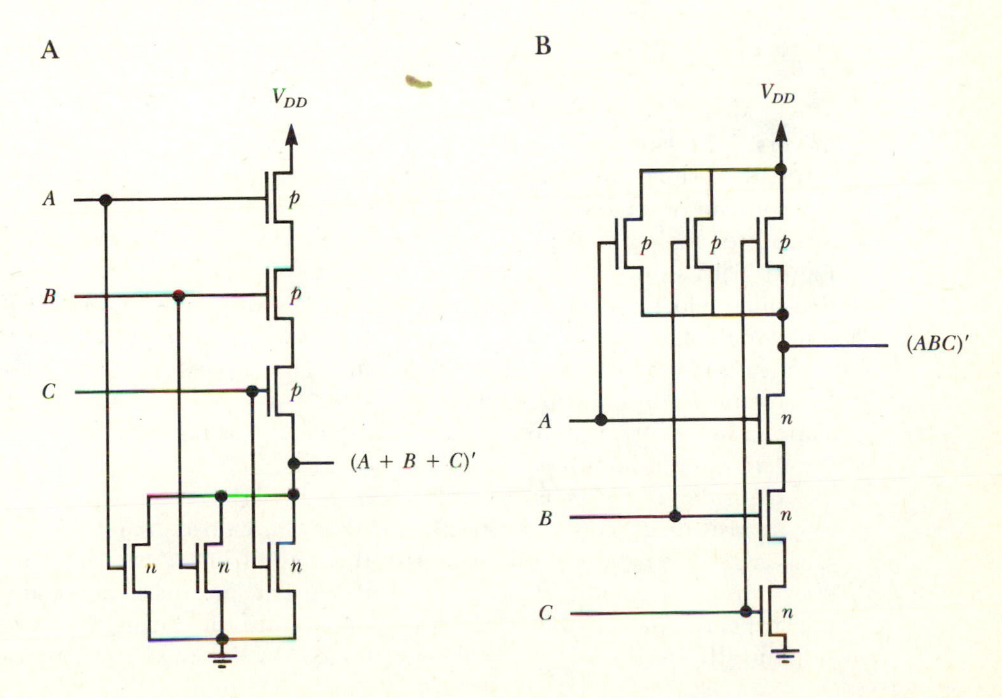

Figure 6.16 A dynamic CMOS gate.

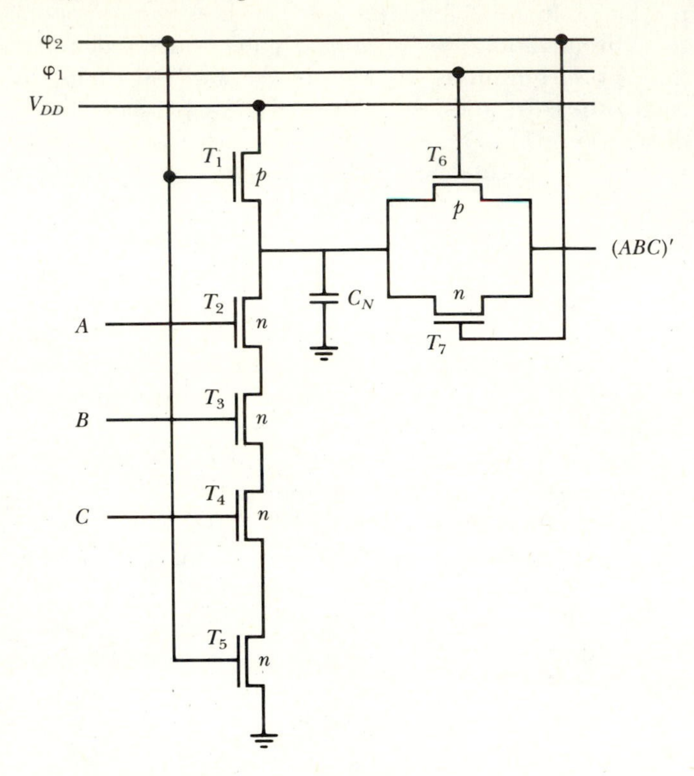

drivers. The same gate functions in ratioed NMOS would require only four transistors, since a common load could be used for the three drivers.

On the other hand, dynamic gates make it possible to achieve ratioless, low-power logic without a separate *p*-type load for each driving transistor. Figure 6.16 shows a three-input dynamic CNOS NAND gate that utilizes a two-phase clock and only one *p*-type load. The output is separated from the following stages by the combination of T_6 and T_7, which play the role of a pass transistor in dynamic CMOS circuits. The capacitor, C_N, represents the capacitance at the output node of the gate. This circuit retains the negligible static power consumption associated with static CMOS, as a consideration of the gate operation shows.

Consider the precharge case when φ_1 is high. Then, since φ_2 must be low, transistor T_5 is off and transistor T_1 is on. Consequently the voltage at C_N is high, regardless of the values of the gate inputs. Also, the static power consumption is negligible regardless of the gate inputs, since transistor T_5 interrupts the path to ground. Both T_6 and T_7 are off during this clock phase, isolating the gate from the following stage. On the next clock phase, when

φ_2 goes high and φ_1 is low, the gate responds to the logic inputs. There are now two possibilities. When all inputs are high, the four series transistors—T_2, T_3, T_4, and T_5—are all on and the capacitance will discharge to ground. Note that the *p*-type load, T_1, is off during this clock phase, which prevents static power consumption. The low logic level at C_N is passed on to the next stage, since T_7 is on. On the other hand, if any one of the gate inputs is low, then the capacitor does not discharge and a logic 1 is passed to the next stage, without voltage degradation, through the *p*-type transistor, T_6, which is strongly in the on state with a $|V_{GS}|$ of 5 V. Since no current path runs to ground, there is essentially zero power dissipation in this state also. Consequently, the steady-state power consumption is negligible in all states. Also, note that the CMOS pass-transistor combination does not cause a degradation of logic level, in contrast to the NMOS case.

The dynamic CMOS gate concept illustrated in Figure 6.16 can be expanded to more complex logic, retaining the general principle of a single clocked *p*-type load and a clocked *n*-type transistor in the ground path to maintain negligible static power consumption. For complex logic structures, the absence of the rather cumbersome totem pole circuit structures found in static CMOS logic is an appreciable advantage.

6.3 Semiconductor Memories

The steady progression of integrated circuit technology is perhaps most evident in the area of semiconductor memory circuits. Beginning in 1970 with Intel's 1K dynamic random-access memory (DRAM) chip, the first commercial DRAM, single-chip memory sizes increased by approximately three orders of magnitude in less than two decades. Semiconductor memories lend themselves well to the regular, structured designs favored by VLSI, since they consist of an array of memory cells arranged in rows and columns, as Figure 6.17 indicates. Peripheral circuitry, which is usually on the same chip, is required to address cells and read and write into the array.

The circuit design associated with the individual memory cells varies according to memory type and within memory types. That is, static random-access memories (SRAMs), read-only memories (ROMs), programmable read-only memories (PROMs), erasable programmable read-only memories (EPROMs), electronically erasable programmable read-only memories (EE-PROMs), and DRAMs all utilize different circuits for the memory cells. Furthermore, several different circuit options are available within a given memory type such as SRAMs.

As an example of a semiconductor memory, consider the DRAM. For the highest-density circuits, a DRAM memory cell consists of a single transistor and a capacitor, as Figure 6.18 shows and as Dennard envisioned in his 1968 patent [5]. The memory mechanism for this cell is electrostatic-charge

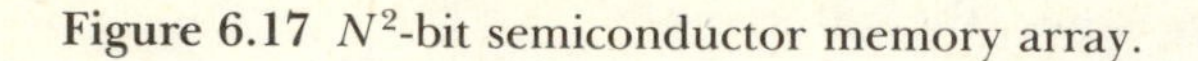

Figure 6.17 N^2-bit semiconductor memory array.

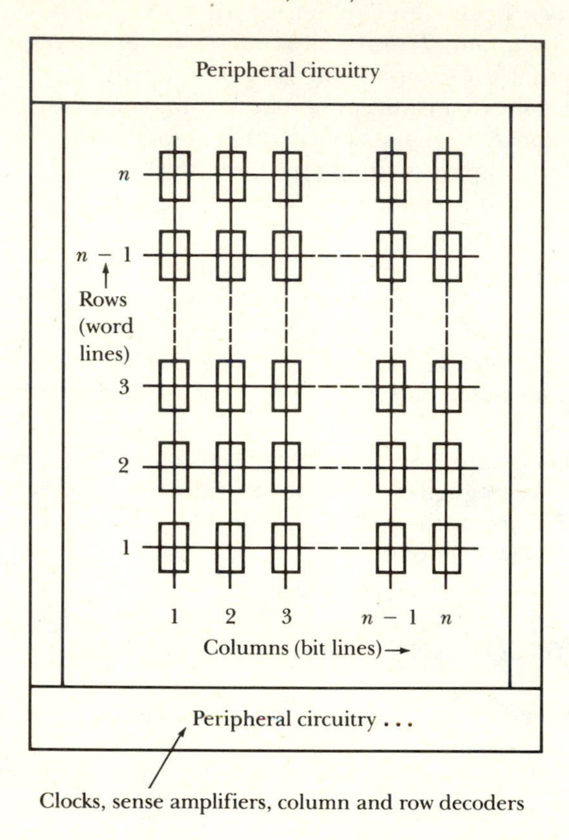

storage and the actual memory device is the capacitor, which registers a logic 1 if it is charged and a logic 0 if it is discharged. The series transistor provides access to the capacitor both for detecting the charge state, or *reading*, and for establishing the charge state, or *writing*. Since the voltage on the capacitor is subject to discharge, the cell must be periodically refreshed. However, the storage time is long enough to allow many memory operations between refresh cycles.

A read operation on the storage capacitor is destructive, since the stored voltage is divided between the storage capacitor and the combined input capacitance of the sensing device and bit line. Therefore a refresh operation must automatically follow and is accomplished by sensing the capacitor voltage with a latch circuit, which functions as a high-gain differential-amplifier. One input of the latch circuit, or sense amplifier, is connected to the bit line addressing the cell of interest, and the other input is connected to a dummy cell that stores a voltage level midway between the two possible voltage levels

Figure 6.18 The single-transistor DRAM cell.

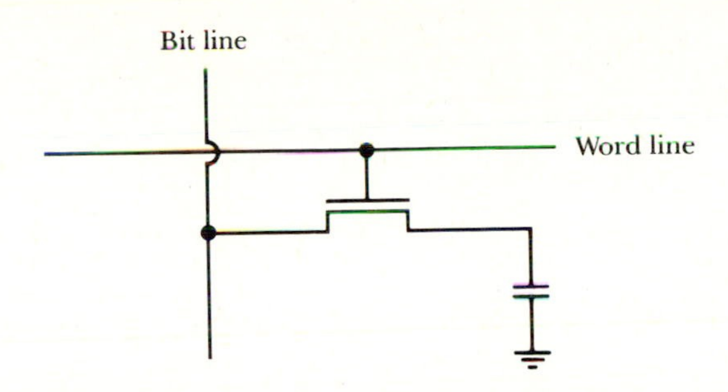

stored on the memory cell's capacitor. The voltage difference is amplified up to full digital levels and fed back along the bit lines to the storage cell, where it is automatically written back onto the capacitor to restore the original signal.

The sensing amplifier may be NMOS, CMOS, or bipolar, but in any case clocking is used. One possible configuration based on NMOS is shown in Figure 6.19 [6]. This circuit is a gated flip-flop with the switching transistors, T_1 and T_2, in a cross-coupled flip-flop arrangement with T_3 and T_4 as loads. Consider first the resetting of the sensing circuit prior to a read operation, beginning with the phases φ_2 and φ_3 high. Bit line A and bit line B are both connected to dummy cells that have voltages of approximately $V_{DD}/2$, and transistor T_5 is turned on to connect nodes A and B. At this point the circuit is balanced and the switching point of the latch is set. Then φ_2 goes low, and the supply voltage and ground of the amplifier are reversed. Consequently, transistors T_1, T_2, T_3, and T_4 are turned off. Shortly after, φ_3 goes low and the nodes A and B are isolated from each other. The circuit is now precharged and ready for a read operation. Next, as φ_1 goes high, a storage cell is connected to one of the bit line inputs and a dummy cell to the other. An imbalance in one direction or the other results. The clock sequence in Figure 6.19 shows voltage levels for reading a 1 and then a 0. For the first case, a storage cell with a logic 1 is connected to bit line A and a dummy cell is connected to bit line B such that node A is at a slightly higher voltage than node B. When φ_2 goes high again, the supply-voltage and ground of the circuit are again reversed, and the flip-flop is activated. Since positive feedback makes the flip-flop inherently unstable, regenerative switching causes the difference in bit line outputs to increase with time, and the refresh is accomplished.

The NMOS sense/refresh amplifier shown in Figure 6.19 is based on an early design and is among the simpler of possible options. Decreasing storage capacitor sizes and power supply voltages call upon the sensing circuit to detect smaller signals. VLSI-level memory designs have used a variety of approaches to preamplify the initial charge imbalance. Some have used over

Figure 6.19 NMOS sense/refresh circuit for a DRAM. A. Circuit schematic. B. Timing diagrams showing the signals for a read/refresh operation of a logic 1 and a logic 0.

SOURCE: Adapted from K. U. Stein, A. Sihling, and E. Doering, "Storage Array and Sense Refresh Circuit for Single-Transistor Memory Cells," *IEEE Journal of Solid-State Circuits*, SC-7 (1972), 336–340. © 1972 IEEE.

A

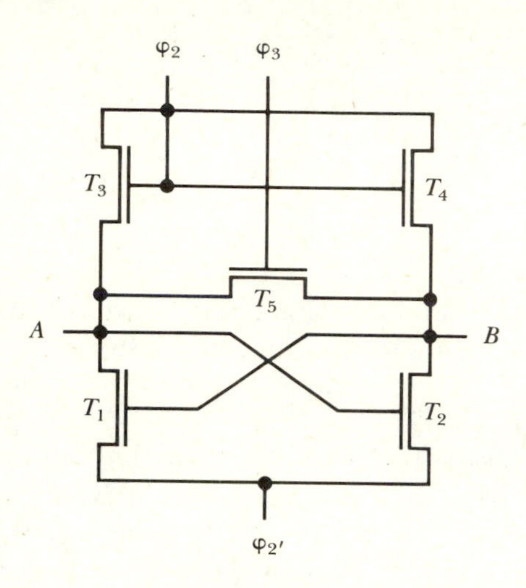

B

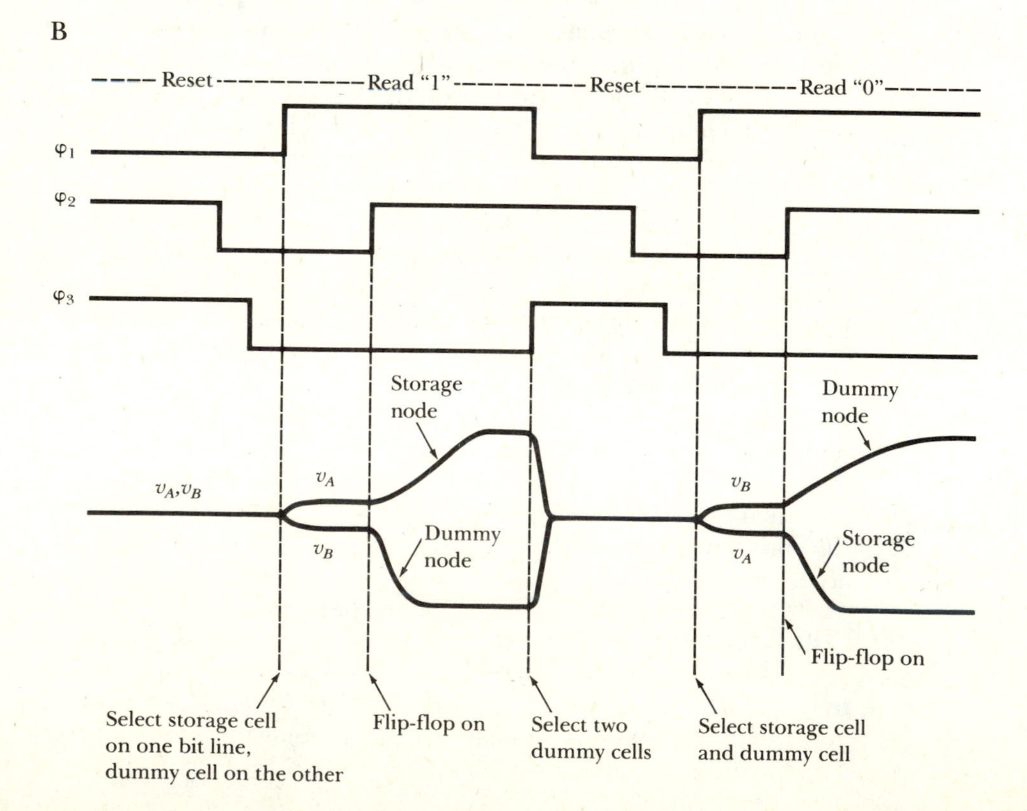

20 additional transistors in the sensing circuit, in addition to a balanced latch similar to the one shown in Figure 6.19 [7].

DRAMs based on CMOS technology offer improved noise immunity and lower power consumption, as would be expected from CMOS in general. Furthermore, CMOS DRAMs offer improved resistance to alpha-particle bombardment. Incident alpha particles cause ionizations in silicon of approximately 10^5 electrons per micrometer of track [8]. Since the charge stored on a storage capacitor in a memory based on 1-μm technology is only about 100 fC, alpha-particle hits can cause transient memory errors. Actually, transient memory errors due to alpha-particle generation of carriers were first reported in 16K memories and became an appreciable concern in the design and packaging of 64K memories. In CMOS designs with shallow wells, most of the hole-electron pairs are generated more or less harmlessly in the substrate, below the active circuitry in the wells.

Figure 6.20 shows a cross-sectional view of a 64K CMOS DRAM memory cell in which the bit line is aluminum and the word line, as well as the top capacitor plate, is polysilicon [9]. (Since the storage cell is still just one transistor and one capacitor, it is the peripheral circuitry that is actually complementary MOS with *p*- and *n*-channel transistors.) Note the efficient layout of the storage cell, with overlapping capacitor and transistor. For this particular design, the MOSFET gate oxide thickness is 250 Å, the storage gate oxide thickness is 150 Å, the channel lengths are 1.2 μm and 1.1 μm for the *n*- and *p*-channel transistors, respectively, and the *p*-type source and drains are placed by ion implantation with a junction depth of 0.4 μm. Figure 6.21 presents a floor plan showing the basic architecture of the chip. Four redundant rows and four redundant columns are built into the design for yield enhancement and are selected as needed by polysilicon fuses.

The chip area for the 64K CMOS DRAM in Figure 6.21 is 0.20 cm^2. Larger memories are achieved by shrinking device and memory cell dimensions and increasing the size of the die. For example, a 1-Mbit CMOS DRAM has been reported that utilizes submicrometer transistors with channel lengths of 0.9 μm and 0.7 μm for the *n*- and *p*-channel transistors, respectively, and that has a chip area of 0.68 cm^2 [10]. Figure 6.22 shows a die photograph of this 1-Mbit memory chip. Another approach for achieving large memories is to etch trenches into the silicon and then form the storage capacitors, or the transistor and storage capacitor, on the sidewalls of the trench. For a given die area, the use of trenches increases the surface area available for devices.

Some examples of other approaches to memory cell design are shown in Figure 6.23. Part A shows a static RAM cell that is an MOS flip-flop with high-resistance polysilicon resistor loads and two additional pass transistors [11]. This cell is used in a high-speed CMOS memory with 18-ns access times. As Figure 6.23B shows, for even higher speeds the bipolar emitter-coupled logic (ECL) approach may be used for a memory cell, yielding a 4.5-ns access time [12]. Figure 6.23C shows a programmable read-only memory cell with a polysilicon fuse, with one level of polysilicon for the word line and another

Figure 6.20 Layout for the storage cell in a CMOS dynamic RAM. A. Circuit schematic. B. Cross-sectional layout.
SOURCE: From R. J. C. Chwang, M. Choi, D. Creek, S. Stern, P. H. Pelley, III, J. D. Schutz, P. A. Warkentin, M. T. Bohr, and K. Yu, "A 70 ns High Density 64K CMOS Dynamic Ram," *IEEE Journal of Solid-State Circuits*, SC-18 (1983), 457–463. © 1983 IEEE.

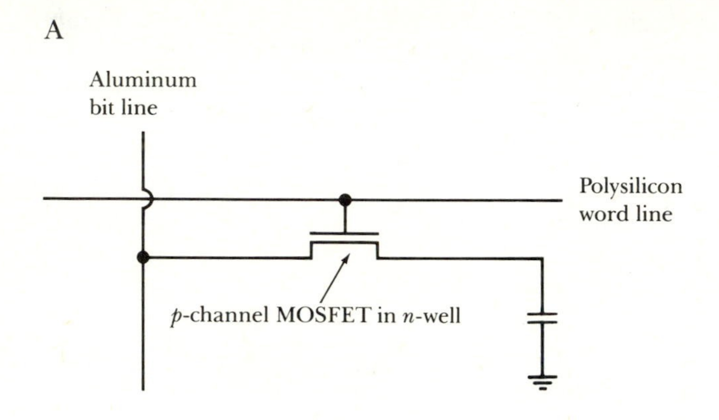

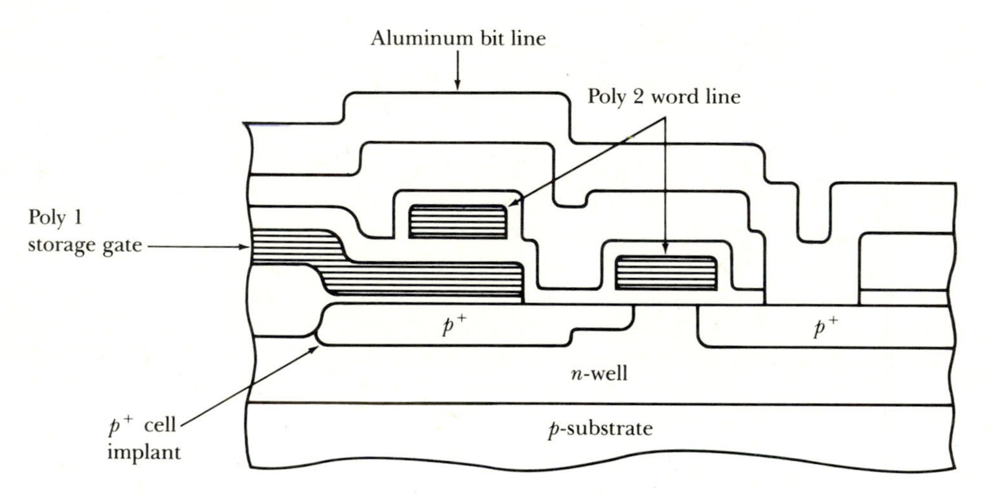

for the fuse [13]. In this case the bipolar memory cell is combined with CMOS logic circuitry on the same chip and the *n-p-n* bipolar junction transistor (BJT) used is inherent to *p*-well CMOS technology. Conversely, a CMOS DRAM has been reported with MOS memory elements and bipolar sensing circuits, again on the same chip [14].

Finally, Figure 6.23D shows a cross section of one version of a floating-gate avalanche-injection (FAMOS) device used in EPROMs. Gate 1 is a floating gate with no electrical contacts, and gate 2 is used for cell selection. To write a 1 in the cell, gate 2 and the drain are raised to sufficiently high

Figure 6.21 64K CMOS DRAM architecture.

SOURCE: From R. J. C. Chwang, M. Choi, D. Creek, S. Stern, P. H. Pelley, III, J. D. Schutz, P. A. Warkentin, M. T. Bohr, and K. Yu, "A 70 ns High Density 64K CMOS Dynamic Ram," *IEEE Journal of Solid-State Circuits*, SC-18 (1983), 457–463. © 1983 IEEE.

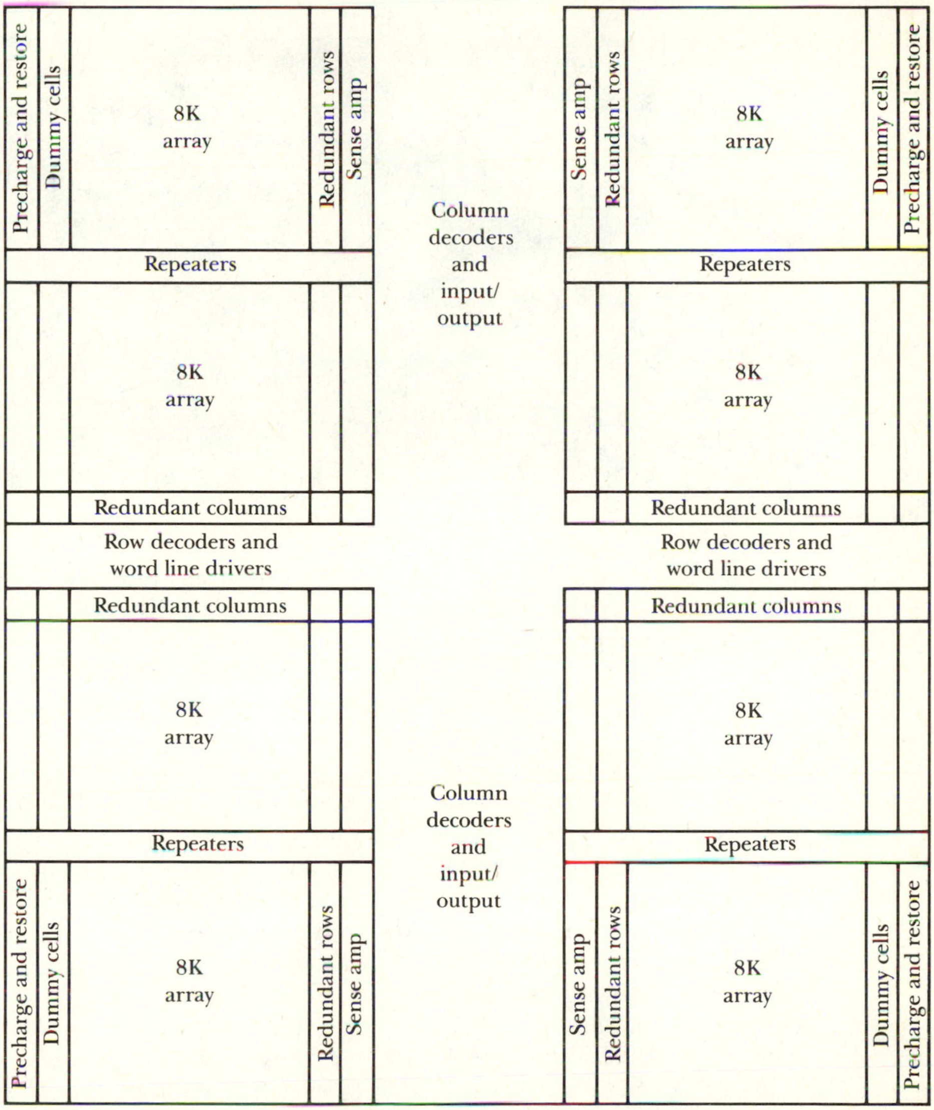

Precharge and restore
Dummy cells
8K array
Redundant rows
Sense amp
Sense amp
Redundant rows
8K array
Dummy cells
Precharge and restore
Column decoders and input/ output
Repeaters
Repeaters
8K array
8K array
Redundant columns
Redundant columns
Row decoders and word line drivers
Row decoders and word line drivers
Redundant columns
Redundant columns
8K array
8K array
Column decoders and input/ output
Repeaters
Repeaters
Precharge and restore
Dummy cells
8K array
Redundant rows
Sense amp
Sense amp
Redundant rows
8K array
Dummy cells
Precharge and restore

Figure 6.22 Die photograph of a 1-Mbit CMOS DRAM. Over 2.2 million devices are integrated on the chip, which has dimensions of 5.97 mm by 11.4 mm.
SOURCE: From R. T. Taylor and M. G. Johnson, "A 1-Mbit CMOS Dynamic RAM with a Divided Bitline Matrix Architecture," *IEEE Journal of Solid-State Circuits,* SC-20 (1985), 894–902. Photo courtesy of Thomson Components-Mostek Corporation.

voltages to cause avalanche breakdown at the drain junction. Hot electrons are injected over the silicon-oxide potential barrier, where they are trapped on gate 1. Subsequently, when the cell is read by applying a 5-V read voltage to gate 2, the device will be off owing to the trapped negative charge on gate 1. However, if the device does not receive a write operation, negative charge is not trapped on gate 1 and the 5-V read voltage will turn the device on. A written device will hold its charge on gate 1 for many years, since the approximately 3-eV potential barrier between gate 1 and the substrate (see Figure 3.2) is sufficiently large that spontaneous emission over the barrier is negligible. The written cell is erased by an ultraviolet source with sufficient energy to create hole-electron pairs in the oxide, which provides a discharge path for the gate. A quartz window on the top of the integrated circuit package provides ultraviolet access to the EPROM.

For the writing mechanism in an EEPROM, cells rely on tunneling between the drain and gate 1. This arrangement requires that a portion of the gate insulator between gate 1 and, in this example, the drain be sufficiently thin, approximately 100 Å, so that a writing voltage of about 15 V causes appreciable charge transfer. Since tunneling is bidirectional, the device is erased by reversing the polarity of the gate field.

Erasable programmable memories are useful during the prototype stages of system development because designers can change programs easily while developing and troubleshooting a product. Furthermore, these memory types are useful in production systems in cases where several versions exist or where

Figure 6.23 A selection of memory cell options. A. 18-ns MOS static RAM with polysilicon resistors. B. 4.5-ns ECL RAM with 15 kΩ/□ ion-implanted resistors. C. PROM cell with polysilicon fuse. D. The EPROM FAMOS device.

SOURCE: Part A is adapted from L. F. Childs and R. T. Hirose, "An 18 ns 4K x 4 CMOS SRAM," *IEEE Journal of Solid-State Circuits*, SC-19 (1984), 545–556, © 1984 IEEE; part B is adapted from J. Nokubo, T. Tamura, M. Nakamae, H. Shiraki, T. Ikushima, T. Akashi, H. Mayumi, T. Kubota, and T. Nakamura, "A 4.5 ns Access Time 1K x 4 Bit ECL RAM," *IEEE Journal of Solid-State Circuits*, SC-18 (1983), 515–520, © 1983 IEEE; part C is adapted from L. R. Metzger, "A 16K CMOS PROM with Polysilicon Fusible Links," *IEEE Journal of Solid-State Circuits*, SC-18 (1983), 562–567, © 1983 IEEE.

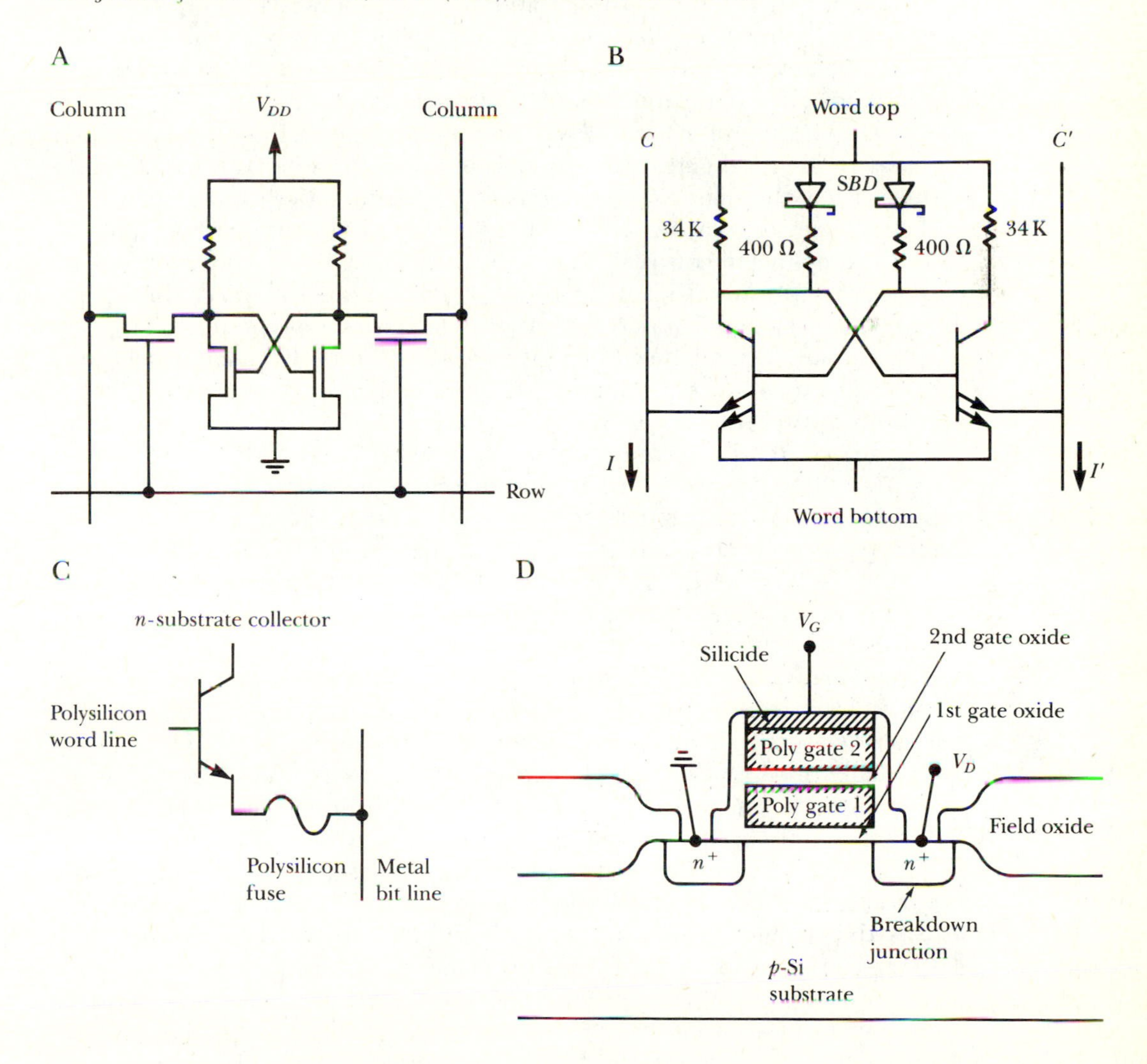

updates are required at a rate that makes the time and expense of mask-programmed ROMs prohibitive.

6.4 Standard Cells and Gate Arrays

Semicustom integrated circuits are ones in which a portion of the design is preordained. In the standard cell approach, the designer draws from a library of predefined and predesigned logic elements to construct the larger circuit. The design process consists of going from the circuit specifications to the relative placement and wiring of the standard cells. In the gate array approach, on the other hand, a partially prefabricated chip already contains an array of identical gates, or logic modules, which are to be linked by interconnect layers to achieve the desired chip function. Both the standard cell and gate array approaches simplify the design process but sacrifice efficiency in the use of the silicon real estate. In some cases, as much as 80% of the chip area for a standard cell layout is taken up by interconnect wiring [15]. And in gate array integrated circuits, only a fraction of the devices present on the chip are actually used in a given design. Figure 6.24 shows schematic layout plans for standard cell and gate array chips.

Both approaches are well suited to automated layout. In the standard cell approach, the problem is similar to printed circuit board layout, with the designer telling the system what kind of cells are needed, such as multiple-input NAND gates, registers, and counters. Then these cells must be assigned positions on the chip in a way that enables all the necessary interconnections, preferably with a minimum of long conducting paths to maintain speed. A variety of algorithms have been developed for autoplacement and autorouting of standard cell layouts.

Gate arrays, also known as uncommitted logic arrays and as masterslice layouts, differ from standard cell layout in that they employ an identical gate pattern regardless of the chip function. Consequently, gate array chips can be partially processed in advance, since only the interconnect masks must be designed to customize a chip for a particular function. As a result, the time lag from circuit specification to circuit delivery may be only a few weeks.

There are many approaches to gate array designs, depending on what the basic gate is. For example, the basic gate may be a simple logic gate such as a NAND gate. Or, more often, the gate may actually be a set of perhaps 4 to 20 components such as transistors and resistors that are replicated in array form across the chip. In this case, the design process also involves the interconnect of components within the cell to achieve a desired function. Alternatively, the gate may be a universal logic module, or ULM, which can realize all possible 2^{2^n} functions of n independent binary input variables, x_1, . . . x_n, depending on the pattern of input connections [16].

As a specific example of a gate array cell, consider the 4-transistor CMOS cell shown in Figure 6.25 [17]. An array formed from this cell requires that

Figure 6.24 Standard cell and gate array floor plans. A. Standard cell. B. Gate array.

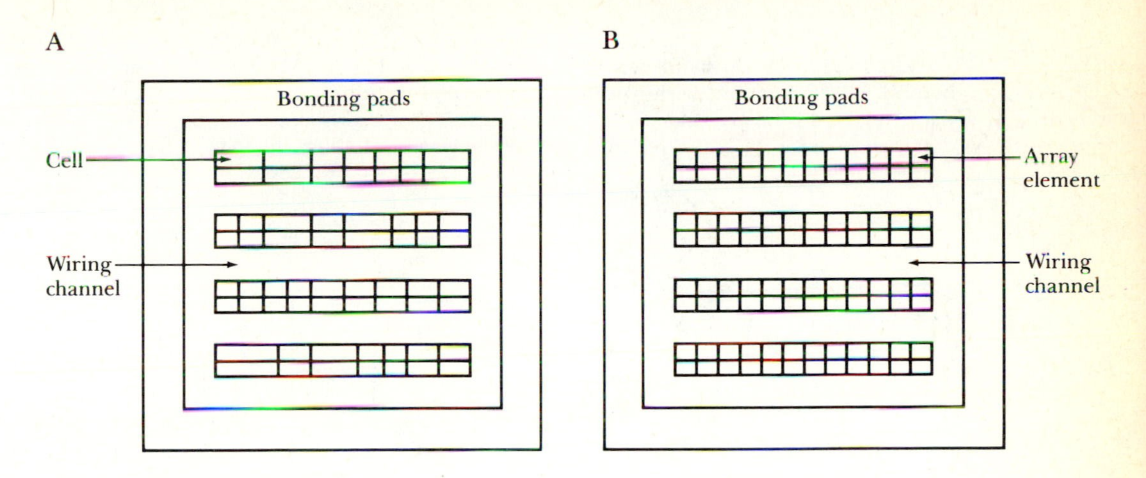

Figure 6.25 A 4-transistor cell for a CMOS gate array. A. Cell schematic. B. Cell layout.

SOURCE: Adapted from B. Alexander, "MOS and CMOS Arrays," Chapter 3 in *Gate Arrays: Design Techniques and Application*, ed. J. W. Read, reprinted with permission by McGraw-Hill, New York, copyright © 1985.

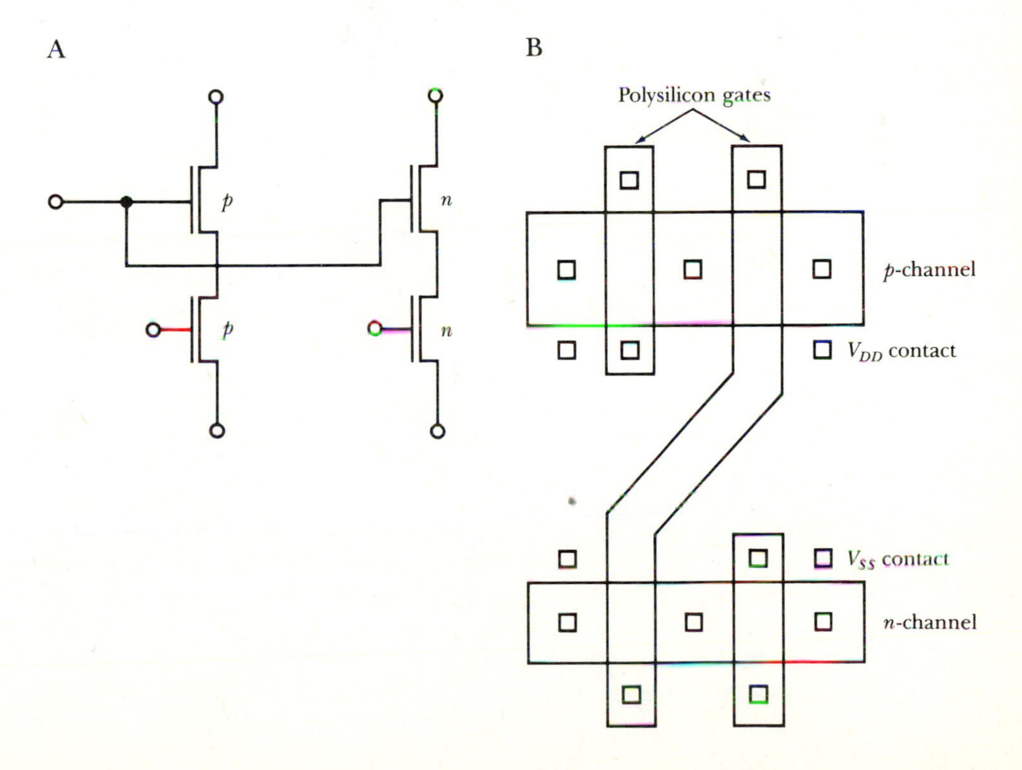

Figure 6.26 The 4-transistor cell of Figure 6.25 wired as a CMOS transmission gate. A. Circuit schematic. B. Cell layout with internal cell interconnections shown schematically as wires.
SOURCE: Adapted from B. Alexander, "MOS and CMOS Arrays," Chapter 3 in *Gate Arrays: Design Techniques and Application*, ed. J. W. Read, reprinted with permission by McGraw-Hill, New York, copyright © 1985.

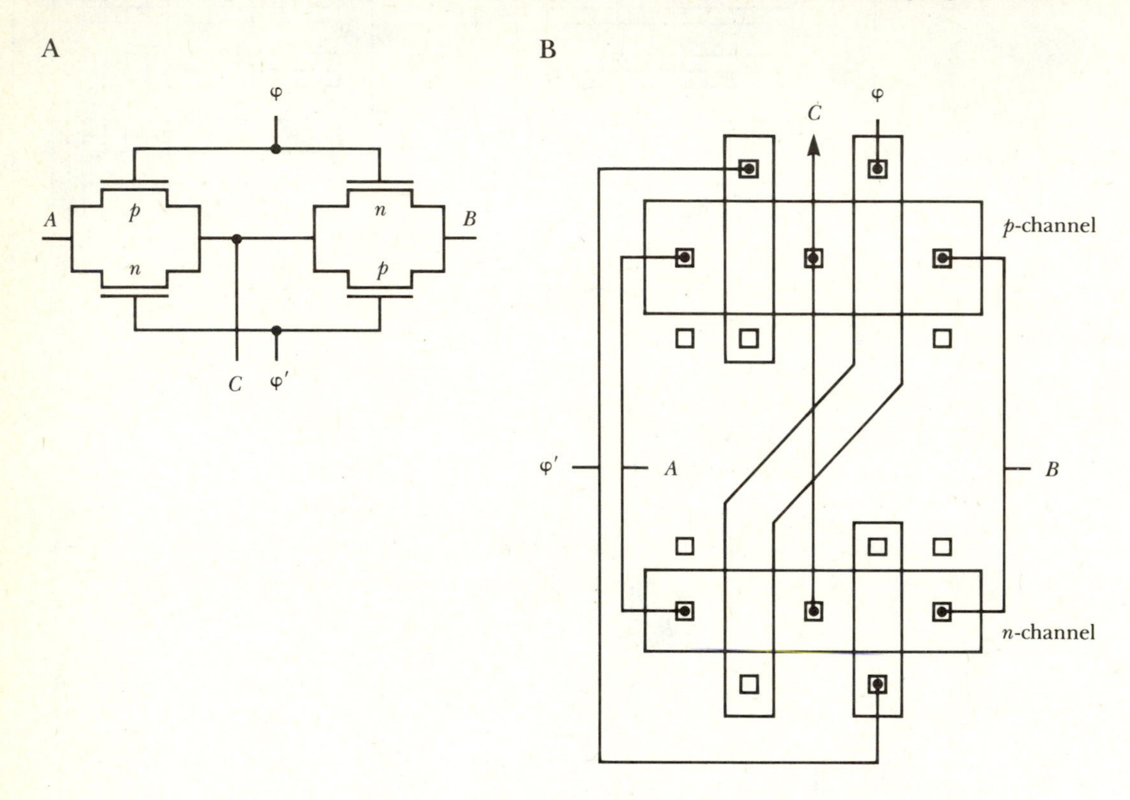

the cell be internally connected, or wired, to form a desired function and then be interconnected to neighboring cells to form the overall circuit. Figure 6.26 shows the internal wiring required to form a CMOS transmission gate, as described in Section 6.2.4, and Figure 6.27 shows the internal cell wiring for a 2-input NAND gate. Once the basic cell functions are defined, transmission gate cells and logic gate cells can then be interconnected in the wiring channels to form custom dynamic logic circuitry. Wiring cross-overs are solved by using multiple metalization or polysilicon underpasses (or cross-overs), depending on the particular technology employed.

A variety of integrated circuit technologies are applied to gate array design with advantages and disadvantages inherent to the chosen technology. ECL gates, as described in Section 4.7.2, are useful for very high-speed arrays. Figure 6.28 shows a 20-component cell used in a 9000-gate ECL masterslice chip that allows internal gate delays as small as 150 ps [18]. This

Figure 6.27 The 4-transistor cell of Figure 6.25 wired as a 2-input CMOS NAND gate. A. Cell schematic. B. Cell layout with internal cell interconnections shown schematically as wires.
SOURCE: Adapted from B. Alexander, "MOS and CMOS Arrays," Chapter 3 in *Gate Arrays: Design Techniques and Application*, ed. J. W. Read, reprinted with permission by McGraw-Hill, New York, copyright © 1985.

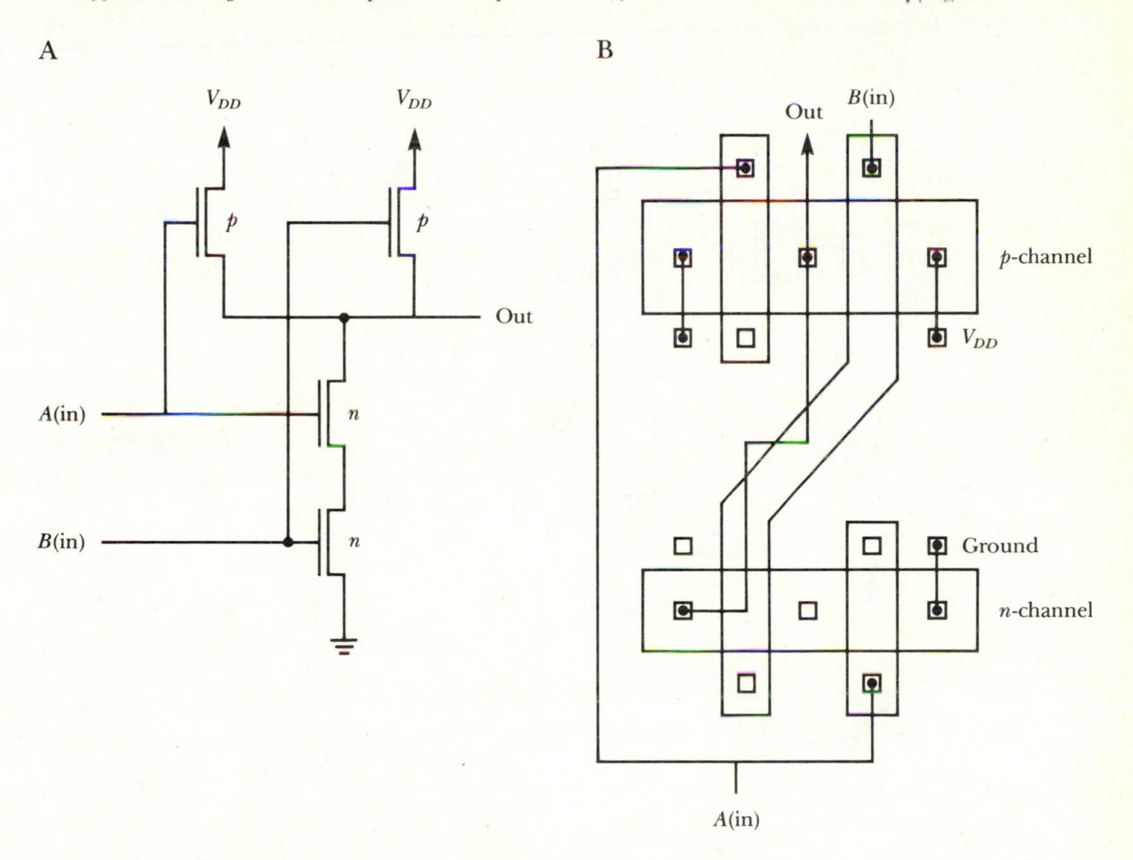

speed is achieved at the cost of high power dissipation, and the chip mounting for this array makes use of a water-cooled heat sink.

An alternate approach to high-speed gate arrays is to use GaAs. Figure 6.29 shows an example of a basic cell used in a MESFET based 2K-gate GaAs array [19]. In this case, the 4-transistor cell consists of one depletion-mode MESFET and 3 enhancement-mode MESFETs. Properties of these devices are described in Section 3.8. The basic cell shown in Figure 6.29 allows implementation of an inverter or a 2- or 3-input NOR gate. Ring oscillator test circuits built from cells on this array show a 42-ps gate delay with a fan-out and fan-in of 1.

As an example of how the individual gate array cells shown in Figure 6.29 can be combined to perform a useful circuit function, consider their

Figure 6.28 Components of an ECL gate array cell. The cell size is 24,300 μm^2. A. Layout. B. Cell schematic.
SOURCE: From W. Bräckelmann, H. Fritzsche, H. Ullrich, and A. Wieder, "A 150 ps 9000 gate ECL Masterslice," *IEEE Journal of Solid-State Circuits,* SC-20 (1985), 1032–1035. © 1985 IEEE.

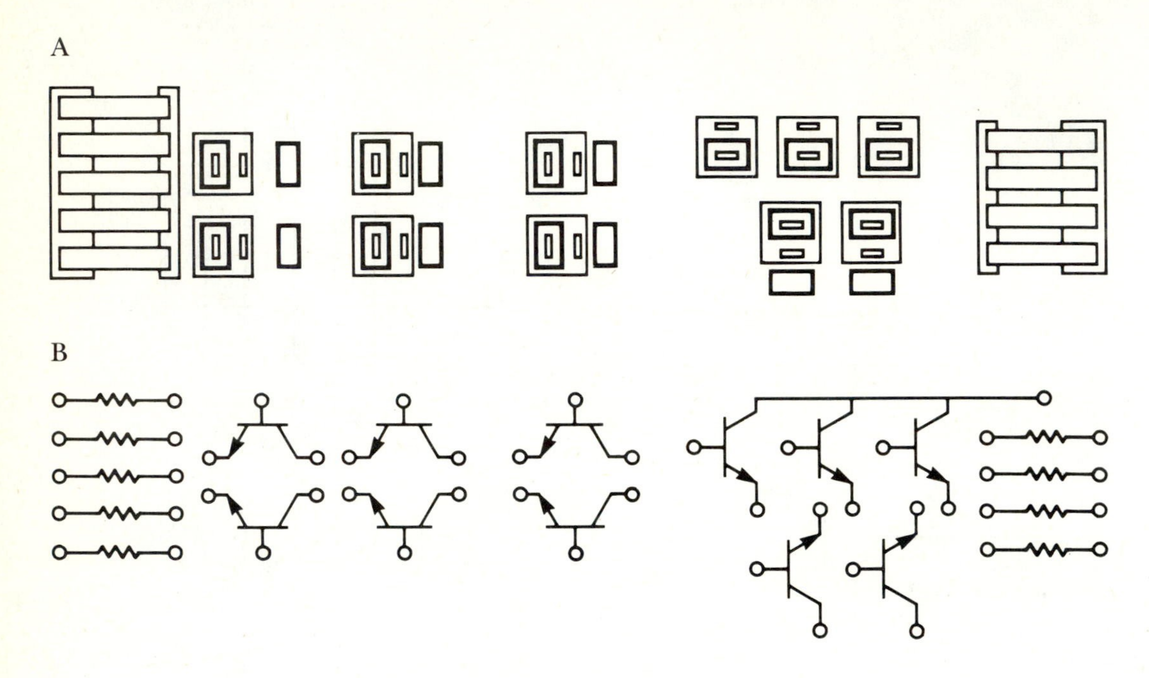

Figure 6.29 Cell schematic for a GaAs gate array with MESFETs. The cell size is 2,016 μm^2. Enhancement (E)- and depletion (D)-mode devices are as noted.
SOURCE: Adapted from N. Toyoda, N. Uchitomi, Y. Kitaura, M. Mochizuki, K. Kanazawa, T. Terada, T. Ikawa, and A. Hojo, "A 2K-Gate GaAs Gate Array with a WN Gate Self-Alignment FET Process," *IEEE Journal of Solid-State Circuits,* SC-20 (1985), 1043–1049. © 1985 IEEE.

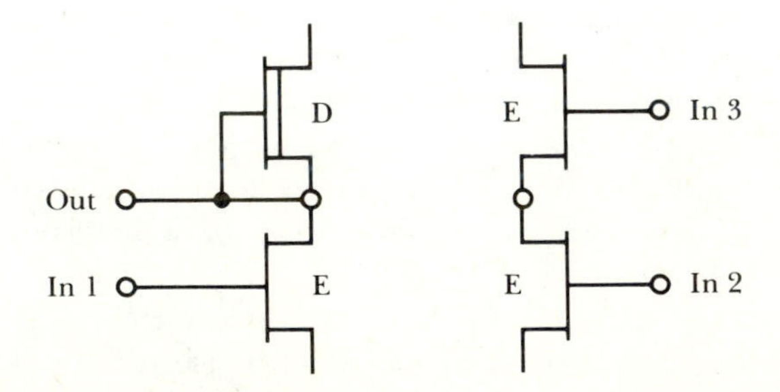

application to an 8 bit by 8 bit parallel multiplier circuit. The resulting circuit consists of 8 half adders, 48 full adders, 64 NOR gates, and 32 input/output buffers, as shown in Figure 6.30B. In turn, the adders are formed from 2- and 3-input NOR gates and inverters using standard digital circuit design, as Figure 6.30A shows. As a result, 728 of the basic cells shown in Figure 6.29 are used to construct the multiplier, utilizing approximately 50% of the gate array area.

6.5 Output Circuitry

6.5.1 Driving Circuitry

Much of the discussion so far has assumed that gate outputs are connected to the inputs of similar gates. Eventually, however, connections must be made to input/output pins, which requires consideration of two factors. First, the capacitance connected to an output pin is usually orders of magnitude larger than that of a typical gate on the chip, and special circuitry is required to drive the load. Second, circuits connected to input/output pins are subject to a variety of insults from the outside world, such as voltage surges and static-charge buildup, and must therefore be protected. This section considers the problem of providing high-current-driving output circuitry.

EXAMPLE 6.2

In Chapter 3, Ex. 3.10, the pull-up and pull-down times for a CMOS inverter are found to be 1 ns and 0.4 ns, respectively, when the inverter is driving an identical gate. Now suppose the CMOS inverter is driving an output pin connected to a 10× oscilloscope probe that has a 15-pF capacitance. What will the pull-up and pull-down times be?

Solution The switching times are directly proportional to the load capacitance. When the output was connected to a similar gate, the load capacitance was estimated to be 0.12 pF. Now that the load is 15 pF, the switching times will scale by the factor 15/0.12 = 125. Consequently the new values of pull-up and pull-down times are 125 and 50 ns, respectively.

Example 6.2 shows that speed advantages achieved by scaling are lost at the output pin unless buffering circuitry is provided between the on-chip logic and the output pins. Such circuitry can be conceptualized as a series of gates, each with a larger current-driving capability than the previous one. For example, consider the chain of gates shown in Figure 6.31. There the MOSFET channels in each gate are wider than in the previous gate by a factor of f. Thus the current-driving ability of each successive gate is larger than that of its predecessor by a factor of f. Assume that the original logic gate, from which the signal to be passed to the output pin originates, has

Figure 6.30 Application of the GaAs gate array cell shown in Figure 6.29 to an 8 bit by 8 bit multiplier. A. Logic diagram of a full adder; each gate corresponds to an appropriately wired cell or cells as shown in Figure 6.29. B. Logic diagram of multiplier. NOR: NOR gate, FA: full adder with NOR gate, FA*: full adder, HA: half adder with NOR gate, HA*: half adder. "S" is the SUM signal and "C" is the carry signal.

SOURCE: Adapted from N. Toyoda, N. Uchitomi, Y. Kitaura, M. Mochizuki, K. Kanazawa, T. Terada, T. Ikawa, and A. Hojo, "A 2K-Gate GaAs Gate Array with a WN Gate Self-Alignment FET Process," *IEEE Journal of Solid-State Circuits,* SC-20 (1985), 1043–1049. © 1985 IEEE.

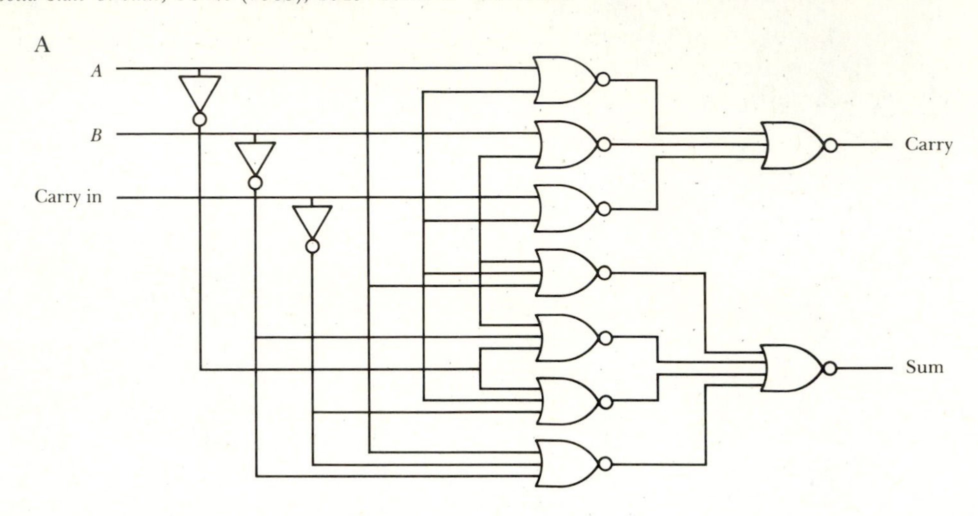

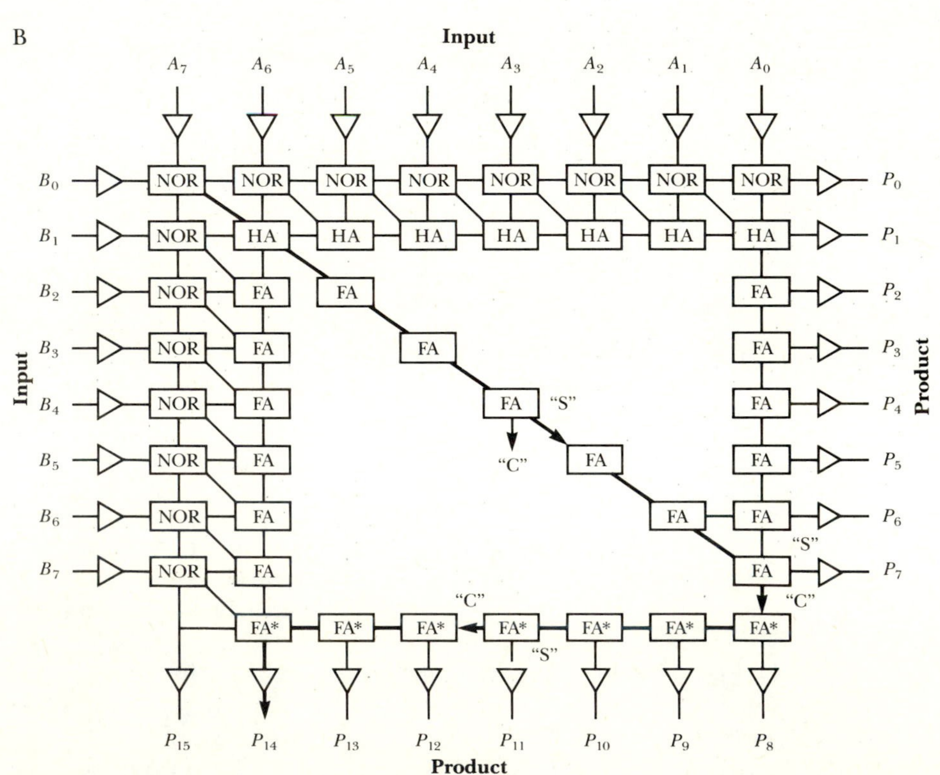

Figure 6.31 A buffer chain for output-pin driving.

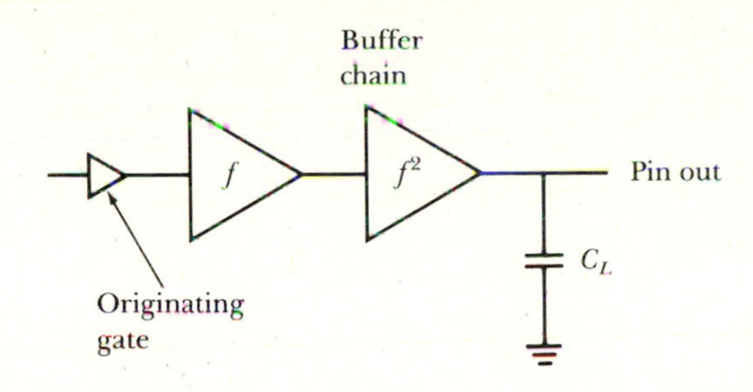

transistor geometry W/L, capacitance C_G, and switching delay τ. Since the input capacitance of a logic gate scales with the area of the MOSFET gate, the input capacitance of the buffer gate immediately following the originating logic gate is fC_G. Therefore the switching time associated with driving this gate is $f\tau$. Likewise, the second buffer gate in the buffer chain has an input capacitance equal to f^2C_G. However, it is being driven by a gate whose current-driving ability is f times larger than that of the original logic gate, so that it also contributes a switching delay of $f\tau$. Therefore, the accumulated switching delay at that point in the chain is $2f\tau$, and so on down the chain. If the load capacitance on the output pin is YC_G, where $Y = f^3$, then the total delay on transmitting the data from the original gate to the output load is $3f\tau$, using 2 buffering gates. Without the buffering inverters, the delay would have been $f^3\tau$, so the improvement is equal to the ratio $f^3/3f$. For example, if the gate in Ex. 6.2 were buffered by a chain such as the one shown in Figure 6.31, with $f = 5$, then the improvement in driving speed would be by a factor of 8.33.

The buffer-chain concept can be generalized as follows [3]: The total delay is $Nf\tau$ where $f^N = Y$; consequently the total delay may be expressed as

$$\tau_{\text{delay}} = (\ln Y)[(f/\ln f)]\tau \tag{6.15}$$

where τ is the delay associated with the originating gate. Equation (6.15) is minimized when f is equal to e, in which case $N = \ln Y$. It is not a sharp minimum, however, and Mead and Conway show that there is very little change in time delay as f varies between 2 and 4 [3].

EXAMPLE 6.3

The gate capacitance of an inverter on an integrated circuit is 0.01 pF, and the delay time is 1.0 ns when the load is an identical gate. A signal is to be transferred from this inverter to an output pin with a capacitive load of 11 pF. For minimum time delay, how many buffering gates should be used? What is the minimum delay time required to drive this output pin?

Also, if the integrated circuit is scaled in size by a factor of 7.4, according to the scaling laws in Section 3.7.1, then what would be the new minimum delay time required to drive the output?

Solution The value of N is determined from

$$N = \ln Y = \ln(11/0.01) = 7$$

which means that 6 buffers are used. The switching delay is

$$\tau_{\text{delay}} = 7(e)\tau = 19 \text{ ns}$$

If the integrated circuit is scaled by a factor $S = 7.4$, then the new gate delay is $\tau' = (1/7.4)$ ns and the gate capacitance is (0.01/7.4) pF. Therefore, the new value of N is

$$N = \ln \frac{11}{0.01/7.4} = 9$$

Now 8 buffers are used, and the switching delay is

$$\tau_{\text{delay}} = 9(e)\tau' = 3.3 \text{ ns}$$

The scaling ratio of output-pin delay times is 5.8, somewhat less than the value of 7.4 for the increased speed ratio of the basic gate.

Buffer gates require output transistors with large W/L ratios. For example, Figure 6.32 shows the circuit schematic and gate layout relative to the output-bonding pad for a CMOS output-buffer gate designed to drive a TTL load and to have a 50-ns delay when connected to a 50-pF load [20]. The channel length for both the n- and p-channel MOSFET is 4 μm, the channel width for the n-channel device is 550 μm, and the channel width for the p-type device is 1000 μm. This case employs a dual polysilicon technology such that a second polysilicon layer interconnects polysilicon gates. This particular layout is referred to as a *tall buffer,* since it keeps the total cell width relatively small compared to the height so that more input/output buffers and pads can be placed on the chip periphery.

Cases where a chip output is to be connected to a data bus shared by other chips require a tristate output such that the possible output states are logic 1, logic 0, and high impedance. With this provision, the chip can actively communicate with the bus by driving it high or low with a 1 or 0, or the chip can be passively connected to the bus, appearing essentially as an open circuit, when another chip is interacting with the bus. Figure 6.33A shows that a "disable" signal activates the high-impedance state, and Figure 6.33B shows an NMOS realization of a tristate pad. The output transistors,

Figure 6.32 A CMOS input/output tall buffer gate. A. Circuit schematic. B. Gate layout relative to output pad.
SOURCE: Adapted from M. L. Perrine, *Michigan State University Co-operative Education Report,* Sept. 1, 1983.

A

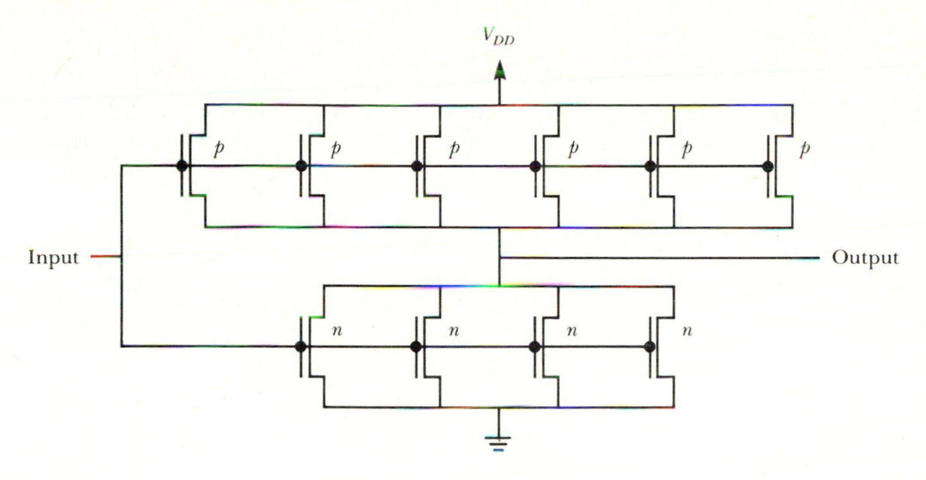

B

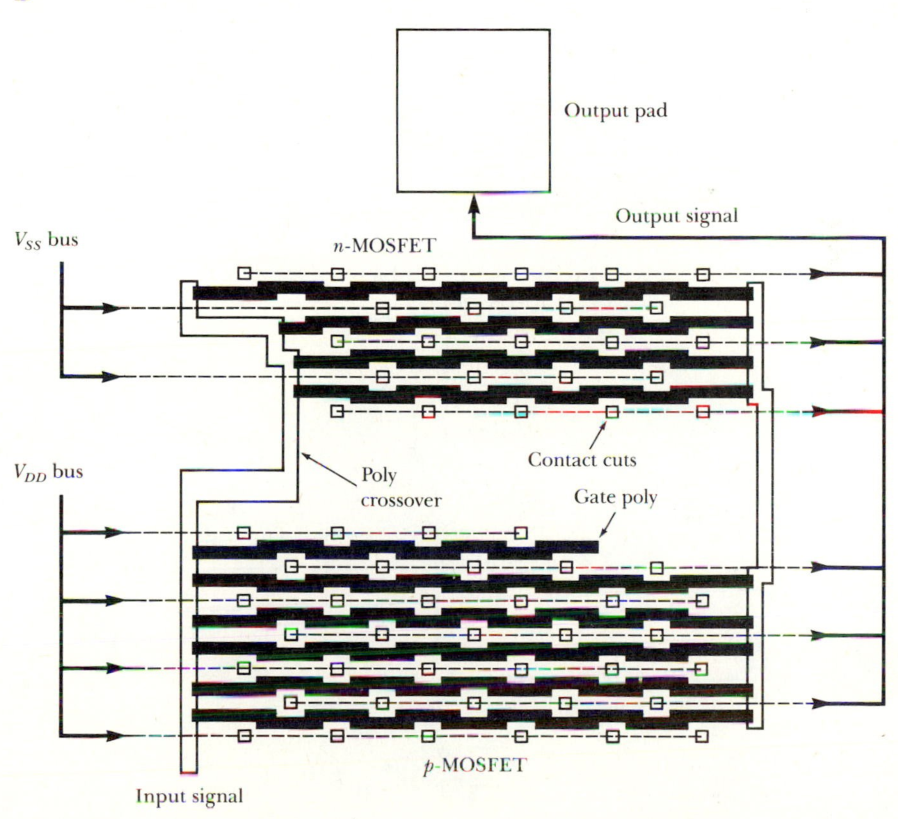

Figure 6.33 Tri-state pad. A. Block diagram schematic. B. NMOS realization.

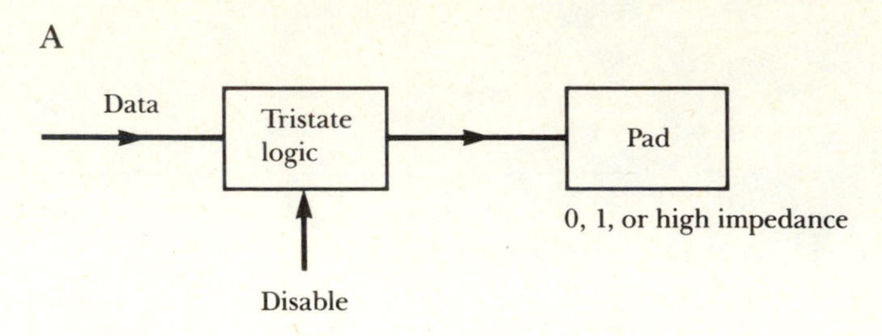

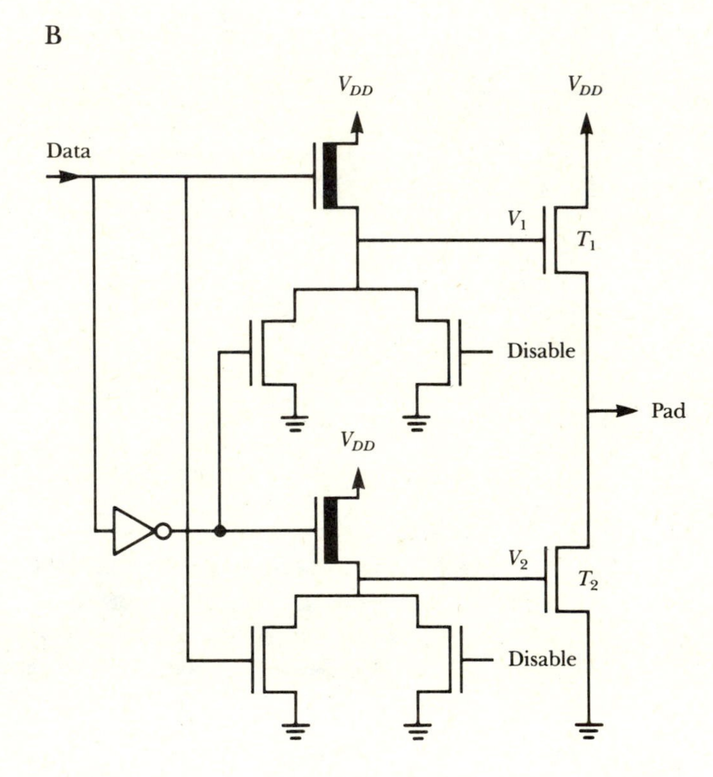

Data	Disable	V_1	V_2	Pad
0	0	0	1	0
1	0	1	0	1
X	1	0	0	High impedance

T_1 and T_2, are both enhancement-mode devices and are connected in a push-pull arrangement such that when one is being turned on, the other is being turned off. Since the T_1, T_2 pair form a ratioless configuration, the size of T_1 and T_2 can be individually adjusted to achieve symmetric rise and fall times.

6.5.2 MOS Input/Output-Protection Circuitry

The breakdown electric field in the gate oxide of a MOSFET is at best approximately 10^7 V/cm. If the gate oxide is 250 Å thick, then the corresponding gate breakdown voltage is 25 V. Including a safety margin to allow for variations in oxide quality, the gate voltage could be rated at about 10 V. This rating may seem fairly safe, considering that the power supply voltage is probably 5 V or less. However, a major concern for chip failure is static electricity. Consider, for example, a MOSFET with a 0.01 pF gate capacitance. The charge required to reach a 25 V breakdown voltage is given by the CV product to be only 0.25 pC! Of course an input/output buffer gate such as the one shown in Figure 6.32 will have a larger area and input capacitance and therefore will be more robust. Nevertheless, modest amounts of static charge can cause breakdown even in buffer gates. Therefore, on-chip protection circuitry is required to prevent excess voltages on input/output pins.

The basic philosophy behind input protection is to include devices that provide a path that will conduct to ground or to the supply before a charge reaches excessive voltage levels. Examples are avalanche breakdown diodes and punch-through MOSFETs. The latter are designed to allow source and drain space-charge-layer overlap at a predetermined breakdown voltage. Figure 6.34 presents a protection scheme showing a combination of protection methods in parallel. The breakdown diode provides a shunt to ground if the input voltage exceeds the avalanche voltage of the junction, as does the punch-through MOSFET. The diffused resistor, R, is a few hundred ohms to several thousand ohms and limits the current through the punch-through device. The combination of this series resistance and the node capacitance forms a low pass filter and limits the operating speed of this portion of the circuit. On the other hand, the diode does not have a series resistor and therefore can respond quickly to, for example, spikes on a power rail or ground line.

In addition to the punch-through device and the breakdown diode, device T_1 represents an enhancement-mode-transistor protection device whose threshold voltage is such that it begins to conduct if the pad voltage exceeds V_{DD} by a critical amount. Since V_{GS} for this device is zero, it is off so long as the pad voltage is less than V_{DD}. However, if the pad voltage exceeds V_{DD}, the role of source and drain are reversed and the device turns on when the pad voltage exceeds the power supply voltage by an amount equal to the device's threshold voltage. The W/L ratios of T_1 and T_2 should be large enough

Figure 6.34 Combination of input/output protection devices.

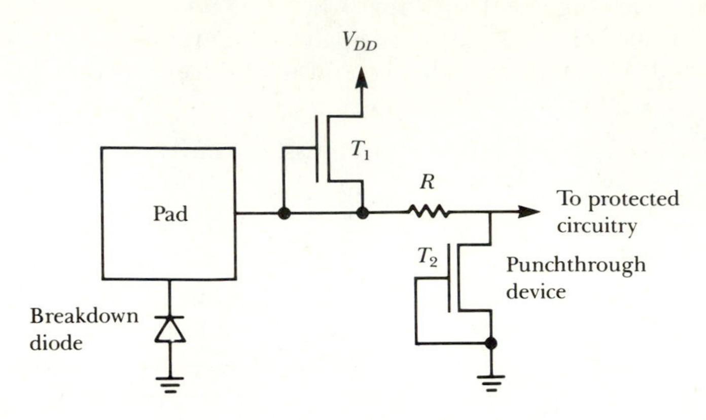

to provide an adequate path for the shunted current. Typical ratios run from about 10 to about 30.

6.6 Amplifier Example

Integrated circuit amplifiers may be realized in CMOS, NMOS, junction field-effect transistor (JFET), or bipolar configurations; the choice of technology depends on the application. For the sake of balance, this section considers a bipolar amplifier, since previous sections in this chapter focus on MOS circuits. Amplifiers may be combined with other analog signal-processing circuitry or, in some cases, with digital circuitry on the same chip. However, this section discusses a single-chip 741 operational amplifier. Although the 741 is an old design, it provides a useful setting for discussing many of the design opportunities and problems associated with monolithic amplifiers. The 741 amplifier was originally introduced by Fairchild but has since been produced by several manufacturers and is still widely used.

Bipolar operational amplifiers generally consist of three stages. A differential-amplifier input stage and a subsequent common-emitter amplifying stage provide the high voltage gain characteristic of this class of circuit, and a push-pull amplifier serves as the output stage. In addition, a fourth subcircuit provides the biasing voltages and currents. The 741 operational amplifier circuit is shown in Figure 6.35 [21].

An analysis of this circuit has appeared in several sources and is not repeated here [1, 21, 22]. However, note the following features, referring to the simplified circuit schematic in Figure 6.36. Transistor pairs T_1-T_2, T_3-T_4, and T_5-T_6, form the differential input stage. Since any offset voltages in the input stage will be highly amplified in the following stages, it is critical to

Figure 6.35 Circuit schematic for the 741 operational amplifier.

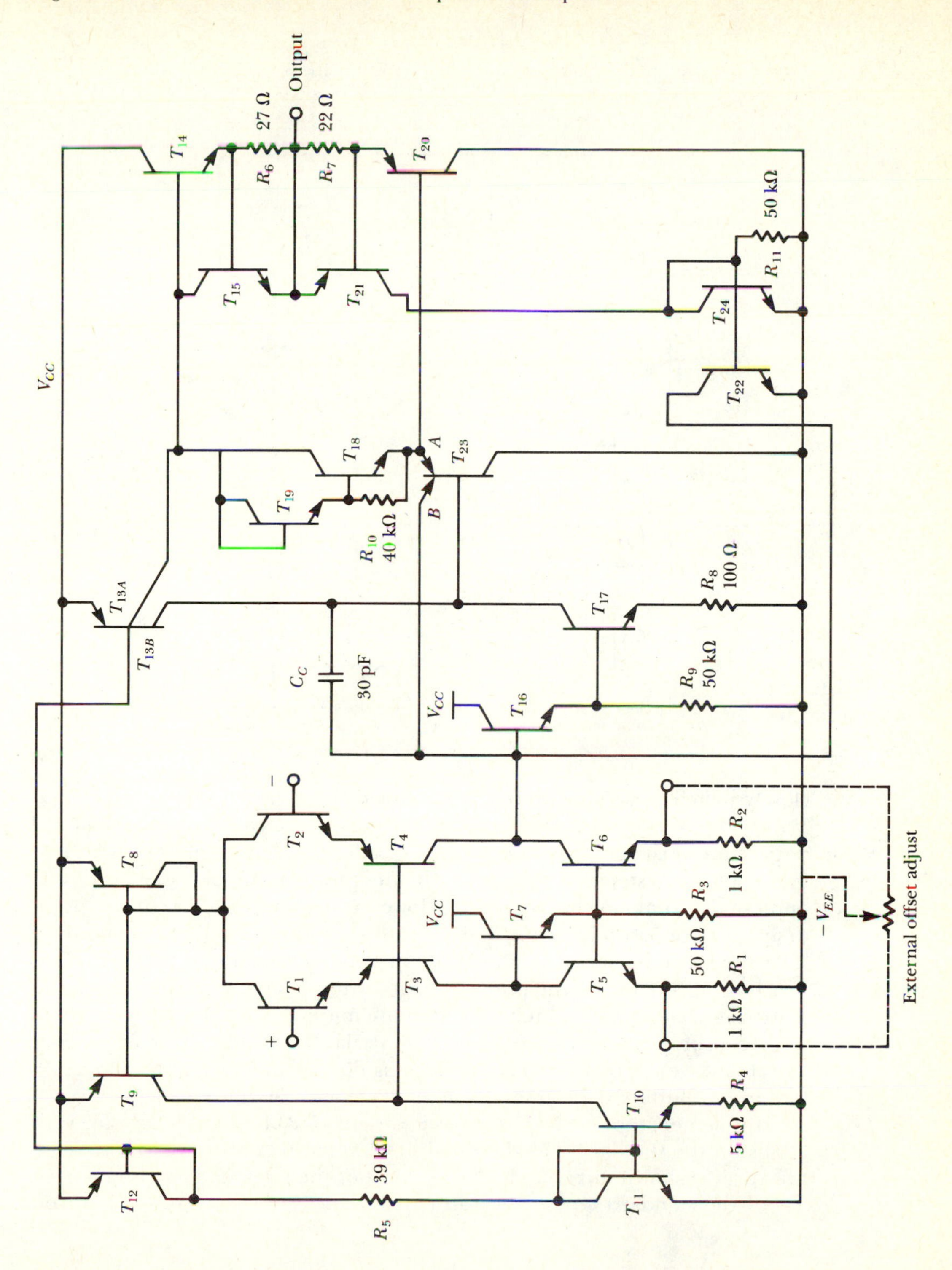

Figure 6.36 Simplified conceptual circuit model for the 741 operational amplifier.

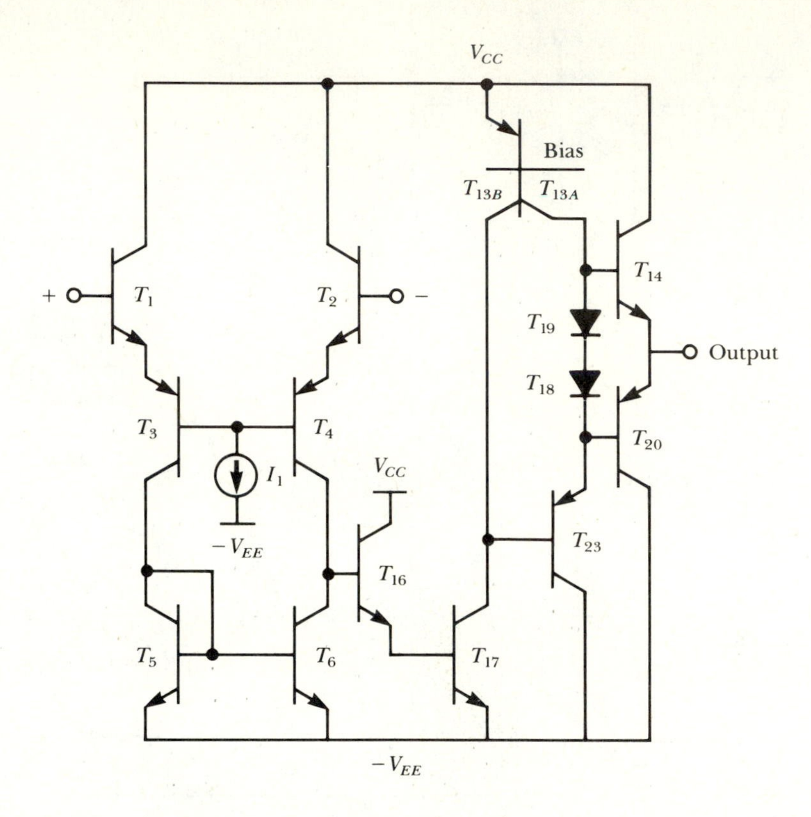

closely match these pairs so that 0 V applied at the input indeed results in 0 V out. Of course, one of the strengths of monolithic circuits is that matched devices are relatively easy to obtain, since adjacent devices go through identical processing steps. In that sense, the design draws on an inherent design opportunity in monolithic amplifiers. However, inappropriate layout can partly destroy this opportunity owing to temperature gradients, as discussed shortly.

The second stage, or high-gain common-emitter stage, is transistor T_{17}, which is buffered from the previous stage by the emitter follower, T_{16}. Transistor T_{13} is an active load for the common-emitter stage. Transistors T_{14} and T_{20} are a complementary pair that form the class *AB* output stage. Transistor T_{13} also serves to bias the output stage, as do T_{18} and T_{19}. The emitter follower, T_{23}, buffers the common-emitter stage and the output stage.

Implementation of the 741 uses a variety of BJT types, which are variations on the standard BJT discussed in Chapter 4. A set used by one manufacturer is shown in Figure 6.37 [22]. Most of the *n-p-n* and *p-n-p* transistors are of the standard types shown in parts A and C of this figure. However, the output stage requires larger transistors because of the current and power requirements for these devices. The large *n-p-n* output transistor shown in Figure 6.37B uses a double base stripe and a long collector stripe. The large complementary *p-n-p* output transistor shown in part D is a substrate *p-n-p* in which minority carriers are injected down from the *p*-diffused emitters through the *n*-type epitaxial layer to the *p*-type substrate collector. Since the current flow is vertical, the cross-sectional area of the emitter is much larger than is the case for lateral *p-n-p*'s, and the current rating is therefore larger, as discussed in Section 4.6.6. The double-collector transistor T_{13} is laid out as a dual-collector lateral *p-n-p,* as Figure 6.37E shows, and the double-emitter transistor T_{23} is realized as a dual-emitter substrate *p-n-p* shown in part F of Figure 6.37.

The amplifier is able to deliver approximately 100 mW to a load and, in the process of delivering this power, the output stage will internally dissipate comparable amounts of power. This dissipation causes the chip temperature to rise and produces a vertical heat flow, as discussed in Chapter 5. However, since heat is not generated uniformly on the chip, lateral heat flow also results and causes temperature gradients across the chip. Calculated thermal gradients across the 741 chip for a sample layout are shown in Figure 6.38 [23]. In this case, the output-stage transistors T_{14} and T_{20} and the biasing transistor T_{13} are located in such a way that temperature differences exist between the critical transistor pairs in the differential input stage.

The die photograph shown in Figure 6.39A shows a 741 layout corresponding to the isotherm example shown in Figure 6.38 [23]. An alternative layout by a different manufacturer is shown in Figure 6.39B [23]. In the second layout, the output transistors T_{14} and T_{20} are positioned symmetrically about the center line of the die, as are the critical input pairs. This arrangement helps to decouple temperature effects between the output and input stages and would appear to be a more optimal layout than the one in Figure 6.39A. However, T_{13} is still positioned along the edge of the chip, and power dissipation in this biasing transistor can also produce temperature differences between the input pairs. Also, the second layout contains an additional 100-Ω buffering resistor, R_{13}, in series with the output lead, which also contributes to input-pair temperature differences. The net result is that thermal effects in the two layouts distort the dc transfer characteristics of the amplifier by comparable magnitudes, although the shapes of the distorted transfer characteristics are quite different. For the first layout, the input voltage for $+10$ V out and a 1-kΩ load was measured to be approximately $+180$ μV. For the second layout, the corresponding input voltage was approximately -60 μV. These values can be compared to a computer-predicted value of $+50$ μV for the case of no thermal effects [23]. In both layouts it is interesting to note the large size of the 30-pF compensating capacitor relative to other circuit features.

Figure 6.37 Plan views of the 741 transistors. All dimensions are in mils. A. Small *n-p-n*. B. Large *n-p-n*. C. Lateral *p-n-p*. D. Large substrate *p-n-p*. E. Dual-collector lateral *p-n-p*. F. Dual-emitter substrate *p-n-p*.
SOURCE: From B. A. Wooley, S. J. Wong, D. O. Pederson, "A Computer-Aided Evaluation of the 741 Amplifier," *IEEE Journal of Solid-State Circuits,* SC-6 (1971), 357–366. © 1971 IEEE.

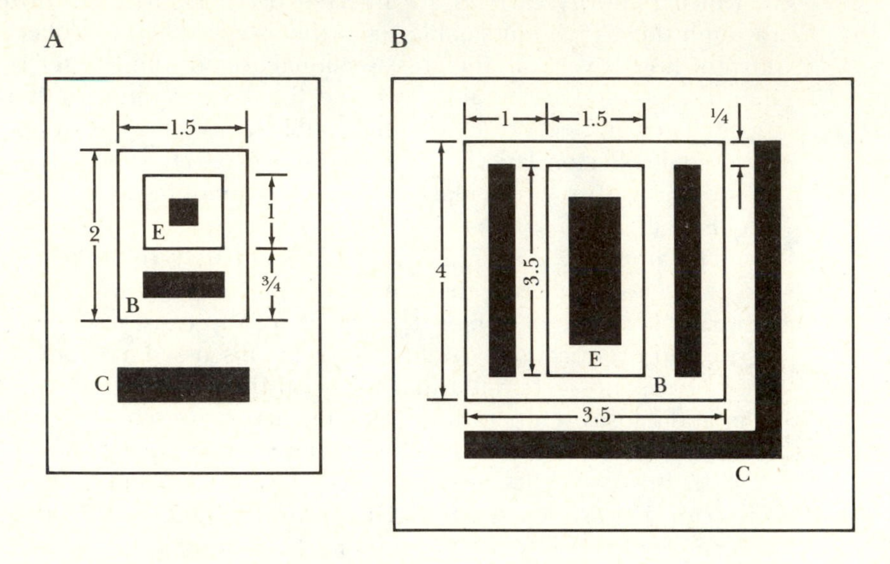

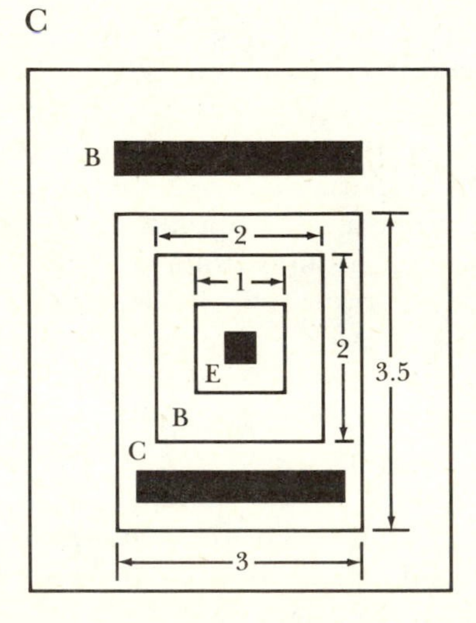

Figure 6.37 Plan views of the 741 transistors (*cont.*).

D

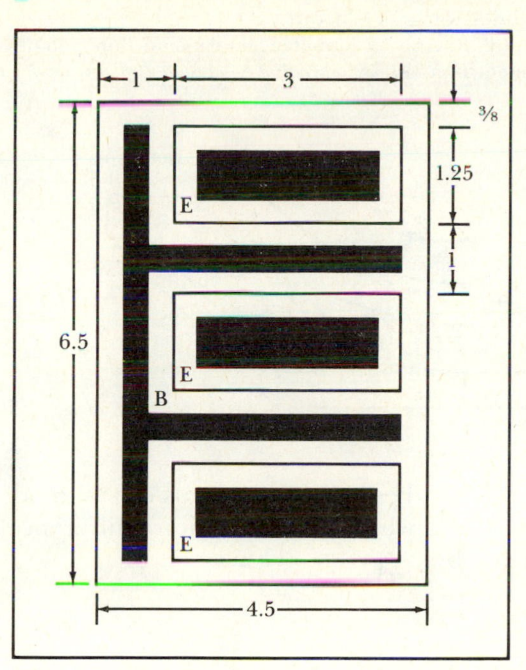

E

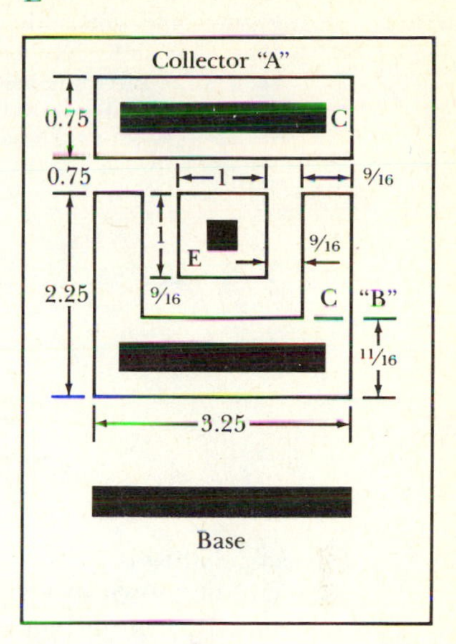

F

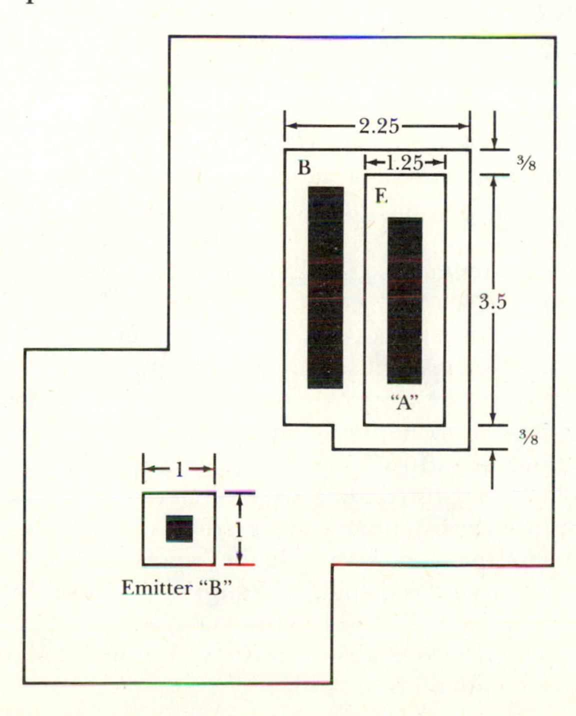

Figure 6.38 Die isotherm lines for a 741 layout.
SOURCE: From K. Fukahori and P. R. Gray, "Computer Simulation of Integrated Circuits in the Presence of Electrothermal Interaction," *IEEE Journal of Solid-State Circuits*, SC-11 (1976), 834–846. © 1976 IEEE.

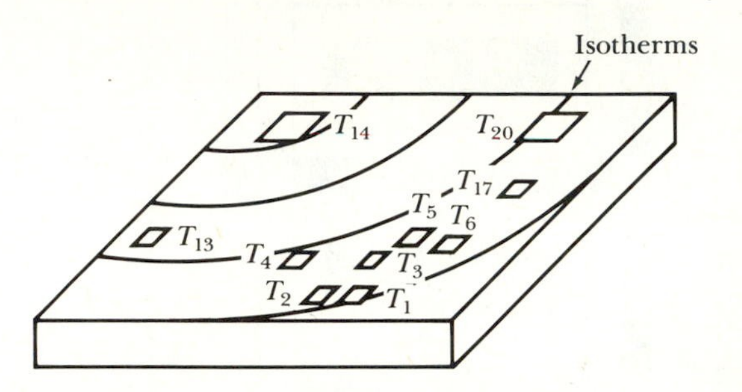

Conceptually, the effect of the thermal gradients can be modeled by the circuit shown in Figure 6.40, in which the temperature difference across the matched input pair is given by [24]

$$T_2 - T_1 \simeq \pm K_T P_D {}^\circ\text{C} \tag{6.16}$$

where P_D is the power dissipated in the output circuit and K_T is a constant with dimensions of °C/W. The ± sign indicates that the actual direction of the gradient depends on the layout in question and may, furthermore, actually change during the output swing as the dominant heat dissipation shifts from one transistor to another.

For a moderate-power bipolar integrated circuit amplifier with careful attention given to thermal considerations in the layout, a representative value of K_T is about 0.3°C/W. Since the separation between input pairs is roughly 100 μm, the corresponding lateral temperature gradient on the chip is about 3×10^{-3}°C/μm. On the other hand, for a high-power amplifier such as the 40-W single-chip audio amplifier discussed in Section 5.2, the maximum lateral temperature gradient is about 3×10^{-1}°C/μm.

In general, differential-input pair-mismatches fall into two classes. Systematic mismatches, such as those caused by thermal gradients and errors in the original layout, are the same from chip to chip. Random mismatches, on the other hand, may be due to nonuniform conditions in processing steps such as ion implantations or diffusions. Systematic mismatches can be dealt with by improved layouts or by including heating elements on the chip under feedback control to reduce temperature gradients. One way to reduce the effects of random mismatches is to split the input pair into a quad of input transistors with a common centroid, as Figure 6.41 shows. The original transistor T_1 is split into T_1 and T_4, and the original transistor T_2 is split into T_2 and T_3. In this way, a process gradient in the x or y direction affects the paired input transistors comparably [21].

Figure 6.39 Layout examples for the 741 amplifier. A. Die photograph 1. B. Die photograph 2.
SOURCE: From K. Fukahori and P. R. Gray, "Computer Simulation of Integrated Circuits in the Presence of Electrothermal Interaction," *IEEE Journal of Solid-State Circuits,* SC-11 (1976), 834–846. © 1976 IEEE.

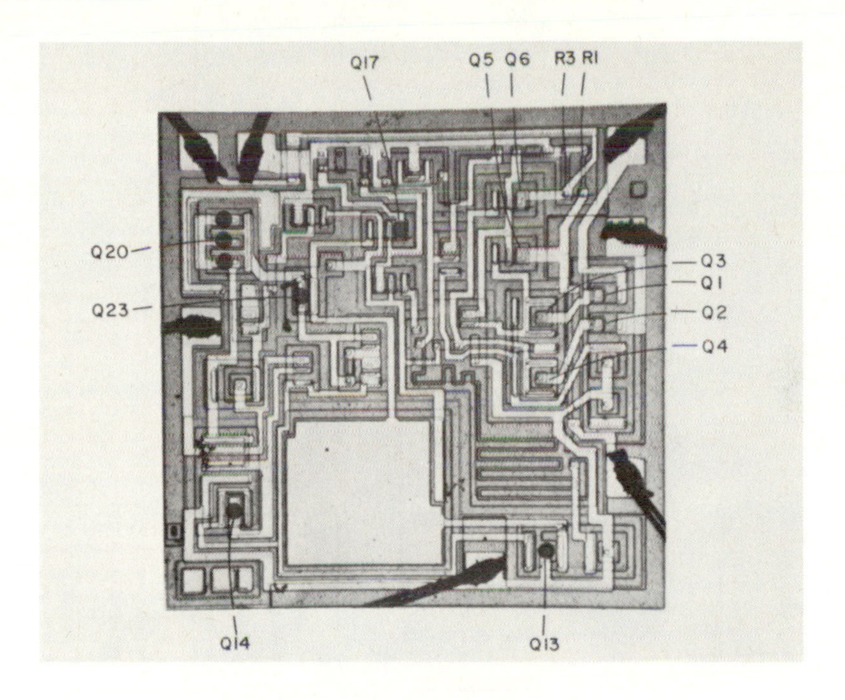

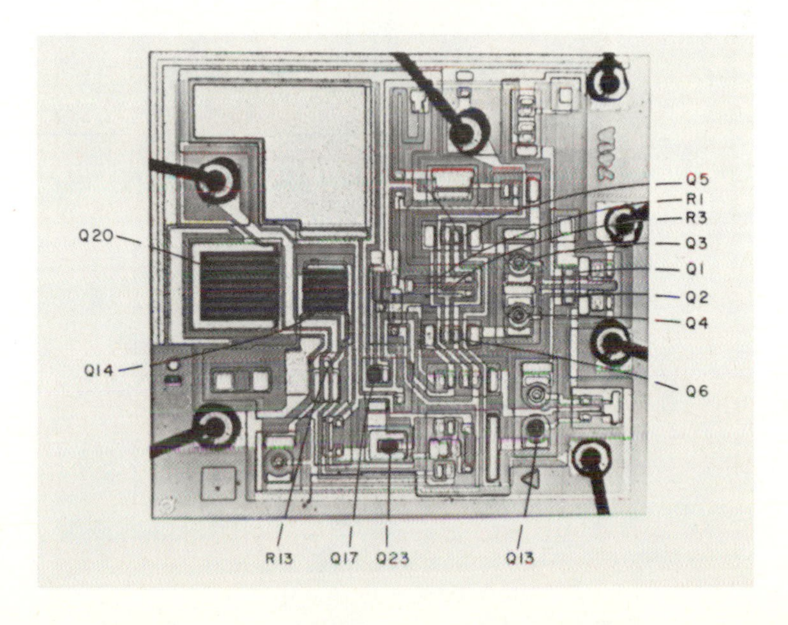

Figure 6.40 Model for thermal feedback in an integrated circuit operational amplifier with self-heating in the output stage.
SOURCE: Adapted from J. E. Solomon, "The Monolithic Op Amp: A Tutorial Study," *IEEE Journal of Solid-State Circuits*, SC-9 (1974), 314–332. © 1974 IEEE.

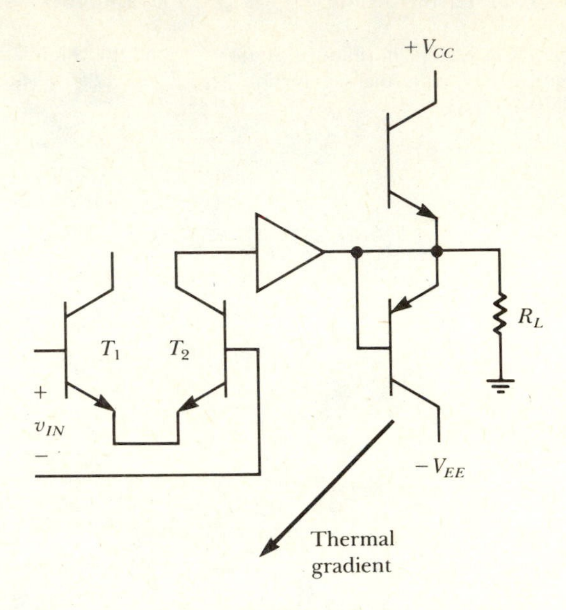

Figure 6.41 Quad of input transistors with a common centroid.

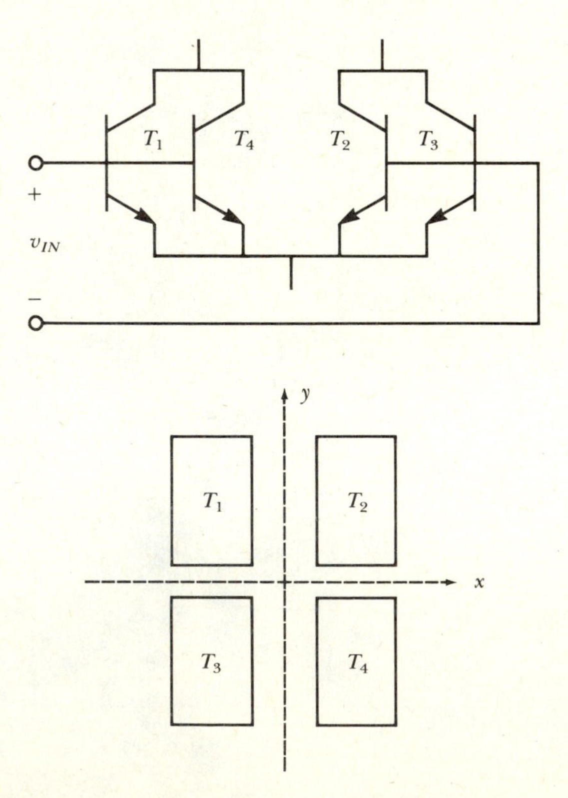

6.7 Analog-to-Digital Converter Example

Analog-to-digital converters (ADCs) produce a digital output signal that is proportional to an analog input signal. They provide an obviously necessary link in designing circuitry to bring the advantages of digital signal processing techniques to bear on situations where the original signal is analog. Two primary performance criteria of ADCs are the data sampling rate, or speed, and signal resolution. Resolution of the ADC is determined by the number of bits in the digital output; an N-bit ADC provides a resolution of 1 part in 2^N. Speed is determined by circuit design as well as by device and fabrication technology.

Many types of ADCs with quite different designs have been developed [25]. This section considers the fastest of the ADC options, namely the parallel-comparator ADC. High resolution, parallel-comparator ADCs require VLSI fabrication technology, since 2^N comparators are required for N-bit conversion.

Figure 6.42 shows the basic schematic of the parallel ADC. Each comparator has one input connected to the analog input bus and the other input

Figure 6.42 Conceptual schematic of an N-bit parallel analog-to-digital converter.

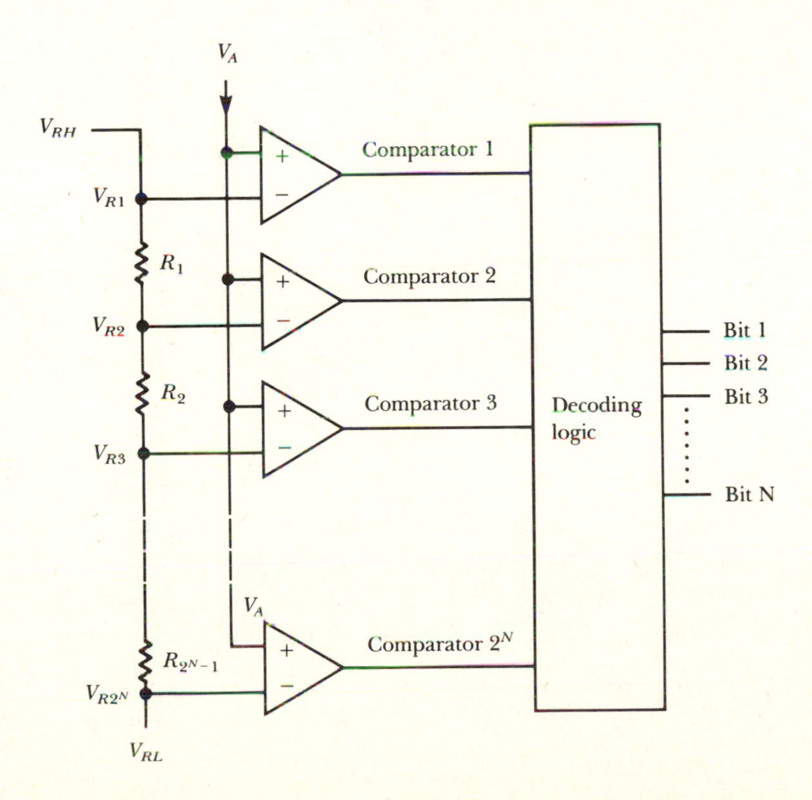

Figure 6.43 Monolithic 20 MHz analog-to-digital converter. A. Block diagram. B. Die photograph. The bipolar chip contains 40,000 devices and has dimensions of 9.2 mm by 9.8 mm.
SOURCE: From T. Takemoto, M. Inoue, H. Sadamatsu, A. Matsuzawa, K. Tsuji, "A Fully Parallel 10-bit A/D Converter with Video Speed," *IEEE Journal of Solid-State Circuits,* SC-17 (1982), 1133–1138. © 1982 IEEE.

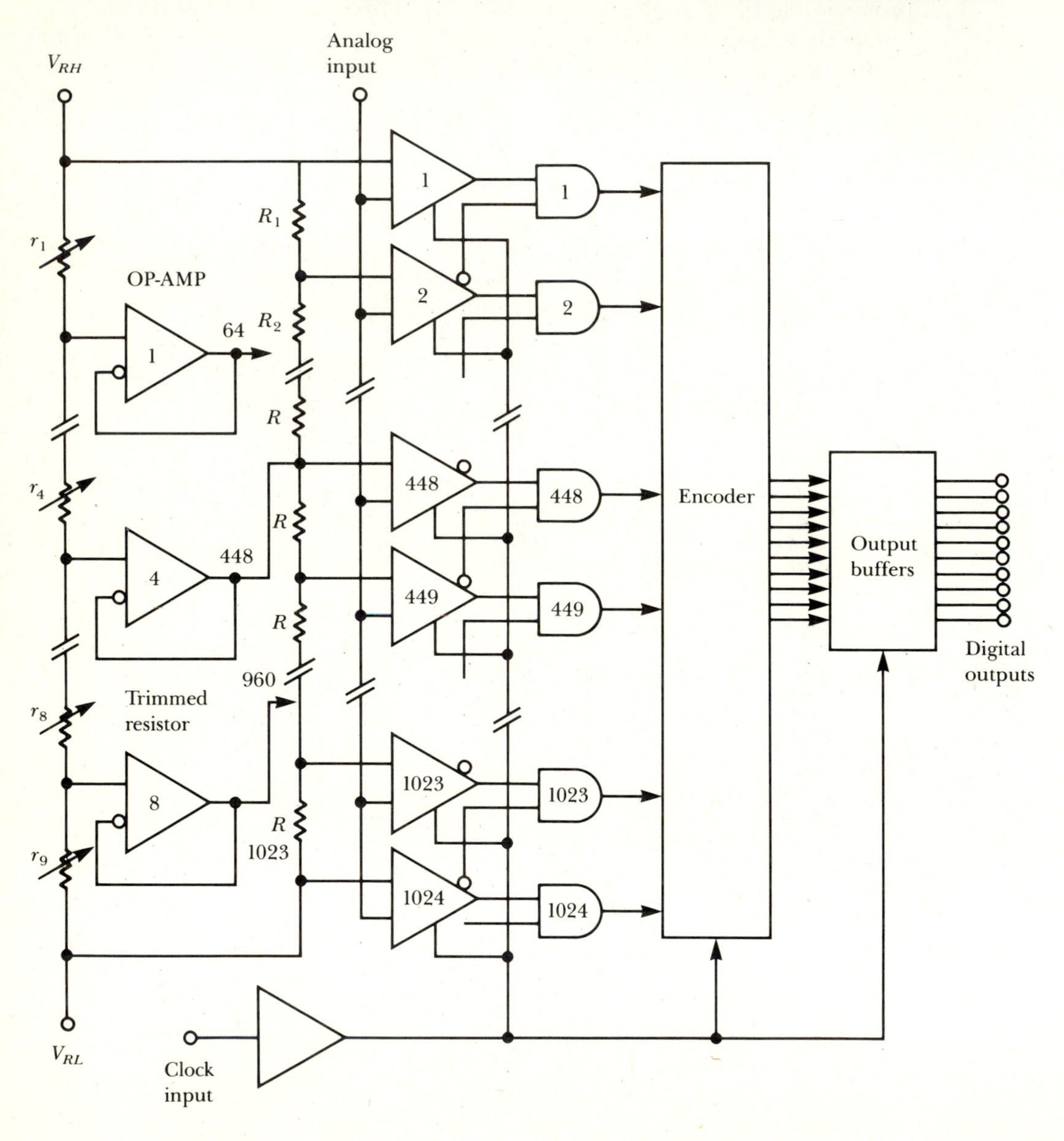

Figure 6.43 Monolithic 20 MHz analog-to-digital converter (*cont.*)

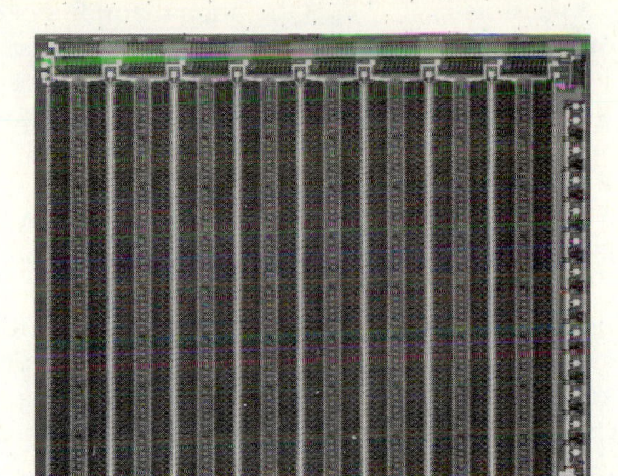

connected to a reference voltage derived by voltage division from a resistor ladder. The comparators are high-gain amplifiers, therefore all comparators with reference voltage V_{Rj} less than the analog signal V_A will have high outputs and will consequently be in the 1 state. For the circuit schematic shown in Figure 6.42, all comparators have a 0 output when V_A is less than the low reference voltage, V_{RL}, and all comparators have a 1 output when V_A is greater than V_{RH}. For values of V_A between these two limits, the jth comparator (and all higher numbered comparators) will be high if $V_A > V_{Rj}$, where

$$V_{Rj} = V_{RH} - \frac{(V_{RH} - V_{RL})}{(2^N - 1)}(j - 1) \tag{6.17}$$

The comparator outputs provide digital information about the analog input signal, but decoding is required to convert the 2^N comparator states to an N-bit digital output signal. The decoding logic can be in the form of custom logic circuitry or, alternatively, a PLA.

Figure 6.43 shows a block diagram of a 10-bit parallel ADC with a 20-MHz sampling frequency, implemented on a monolithic silicon integrated circuit [26]. The resistor chain, R_j, is fabricated with a thin aluminum film. The resistor ladder requires precision resistors to achieve the accuracy required for high conversion resolution. However, since the film thickness of sputtered or evaporated films may vary by 2 or 3% over the wafer, the aluminum-film resistors are not sufficiently uniform. Therefore component trimming is required.

To solve this problem, another series resistor chain has been added with variable resistors $r_1 \ldots r_9$. Operational amplifiers connected to the variable resistor ladder are attached, to provide compensation currents, at 8 intermediate points in the fixed resistor ladder consisting of resistors R_j. The variable resistors are adjusted by postprocessing laser trimming so that each V_{Rj} approaches the ideal value.

In this example, the variable resistors, r_i, are fabricated by depositing a resistive polysilicon film that is easily vaporized by a yttrium aluminum garnet (YAG) laser. Each operational amplifier output is connected to a monitoring pad, and while the resistor is being trimmed, the voltage between V_{RH} and the operational-amplifier-output pad is monitored so that the trimming operation is terminated when the desired value is reached. The operation is repeated sequentially beginning with resistor r_1 and ending with resistor r_9, and it provides voltage adjustments of approximately a few mv in the resistor chain, R_j.

The variable resistors are relatively large to accommodate the laser customizing process. The width of each trimming resistor is 95 μm, and the laser cuts a channel perpendicular to the current flow to a maximum distance of 20 μm.

The circuit in Figure 6.43 also differs from the conceptual schematic in Figure 6.42 in that a string of 2^N AND gates follow the comparators. By comparing the output of each comparator to that of the one immediately below it, these gates serve to determine the resistor segment in which V_A lies.

6.8 Microprocessor Floor-Plan Examples

When an integrated circuit contains several interconnected subcircuits, the arrangement and relative placement of the subcircuits on the chip are important in terms of both efficient area utilization and circuit performance. A case in point is the microprocessor, which generally contains an ALU, timing circuitry, and storage registers. In addition, the microprocessor may include on-chip RAM, ROM, and PLAs, as well as specialized circuitry for a given application. Careful choice of a floor plan reduces signal delays and simplifies the interconnect problem. This section considers the topological arrangement of two quite different microprocessors.

Figure 6.44 superimposes a die photograph on the floor plan of the major subsections of an NMOS 32-bit microprocessor developed by Hewlett-Packard [27]. This is a high-performance chip operating with a clock frequency of 18 MHz, dissipating approximately 4 W of power in the worst case, and containing about 450,000 transistors. The largest single subsection, in terms of both area and number of transistors, is the ROM, which stores the microcode supporting the instruction set for the microprocessor. This ROM is organized as 9216 words of 38 bits each and therefore contains over

Figure 6.44 Floor plan for a 32-bit NMOS microprocessor.
SOURCE: From J. W. Beyers, L. J. Dohse, J. P. Fucetola, R. L. Kochis, C. G. Lob, G. L. Taylor, and E. R. Zeller, "A 32-Bit VLSI CPU Chip," *IEEE Journal of Solid-State Circuits*, SC-16 (1981), 537–542. Photo courtesy of Hewlett-Packard Company.

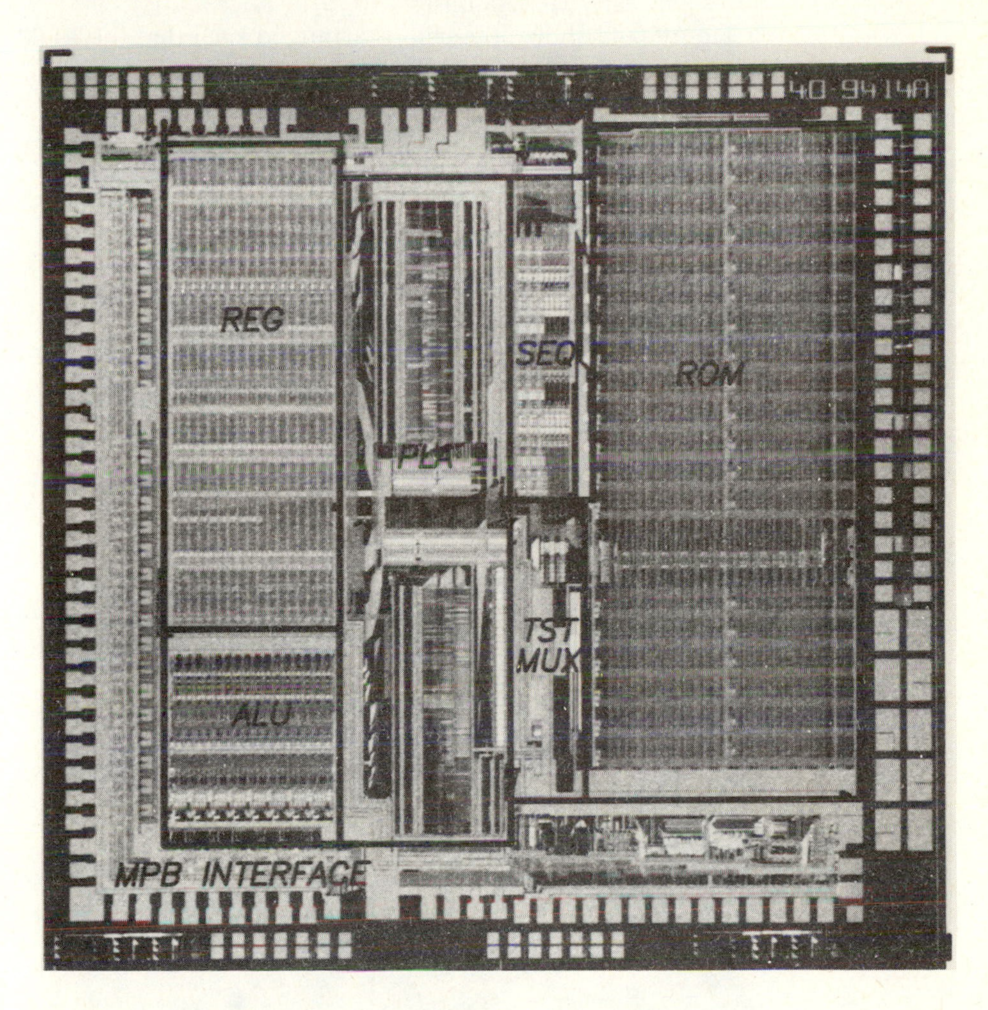

350,000 bit cells. Instructions accessed from the ROM are sent to the PLA, where they are decoded and then used to control the operation of the ALU and the register stack labeled REG on the die photograph. The register stack includes a set of 28 identical 32-bit-wide registers for storage of data, as well as top-of-stack and instruction registers.

The unit denoted SEQ is a sequencing machine that controls the flow of instructions to the PLA. It generates the starting address for the microcode routine that implements a given machine instruction. The unit labeled TST

MUX is a test-condition multiplexer that facilitates conditional jumps and skips in the microcode. Finally, the MPB subsection is a memory processor bus interface that provides the communication channel between internal chip data buses and external data buses.

As Figure 6.44 shows, the chip is arranged such that the ROM is a contiguous unit on one side. The ROM communicates with the PLA, SEQ, and TST MUX subsections which are therefore adjacent to the ROM. They are well placed in the middle of the chip, since the PLA must drive control lines for the ALU and register stack, which are on the opposite side of the chip from the ROM. The bus interface circuitry, logically, is near the periphery of the chip. As an interesting side comment, the entire chip was fabricated with 100 unique cells whose performance was simulated on desk-top computer systems.

Figure 6.45 shows the floor plan of a low-power, specialized 4-bit CMOS microprocessor [28]. Both of the microprocessors considered in this section are on 6 mm by 6 mm chips; however, the 32-bit chip contains 1-μm features and the 4-bit chip is based on 6-μm design rules. In contrast to the 4-W power dissipation of the 32-bit chip, the microprocessor in Figure 6.45 has an operating power of 100 μW at a 1.5-V supply voltage. In this case a user

Figure 6.45 Floor plan for a low-power, 4-bit CMOS microprocessor.
SOURCE: From G. Meusburger, R. C. Jones, G. Schönleber, and R. Mortimer, "A 1.5 V CMOS 4-Bit Microcomputer Needs only 100 μW," *IEEE Journal of Solid-State Circuits*, SC-18 (1983), 245–249. © 1983 IEEE.

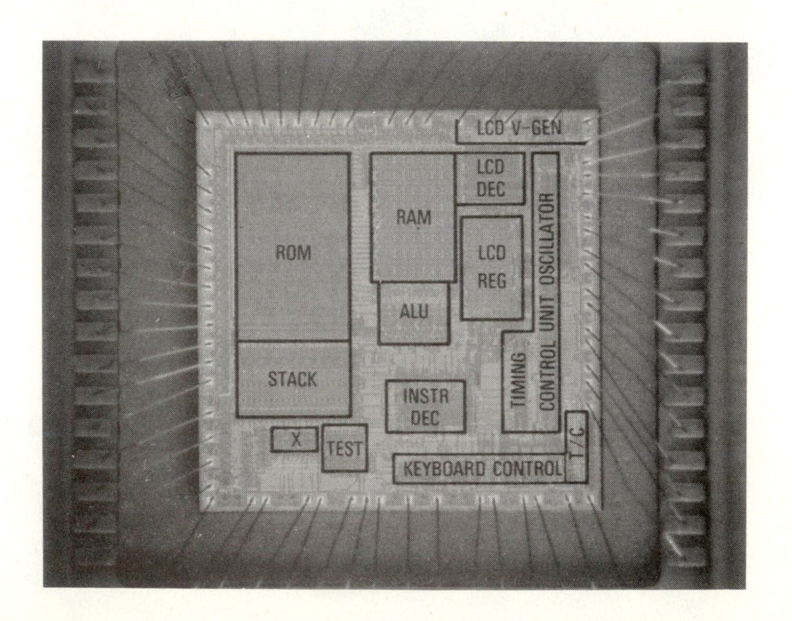

program is stored in a 10,240-bit ROM that is addressed by a 10-bit address from the program counter which, in turn, communicates with the stack, located adjacent to the ROM. The RAM subsection contains working registers and the index register and, as in the previous example, the registers are located next to the ALU. In addition an "X" register is used to facilitate addressing. Other subsections of interest include the instruction decoder, the timing/counter area denoted by TC, and the control unit oscillator. The internal oscillator can be disabled by a shutdown instruction, which halts microcomputer operations and reduces the power dissipation to virtually zero, making the device useful for battery-operated systems. Since the RAM uses a 6-transistor static cell, halting the microprocessor does not result in lost data. Because of the application orientation, the output circuitry is designed to drive a liquid crystal display (LCD) and includes a decoder to translate binary display-register data to appropriate display-segment patterns. This design is evidence that increased integration density makes possible not only larger and faster microprocessors but also the inclusion of special-purpose circuitry on processor chips. In some cases, analog-to-digital converters and digital-to-analog converters have been included with microprocessors on single-chip designs, such that digital signal processing is possible on single chips with analog interfaces.

6.9 Computer Aids

6.9.1 An Overview

At LSI levels of circuit complexity, a rule of thumb used during the mid-1970s to estimate the design and development cost for a fully-custom integrated circuit was about $100 per gate. At that rate, wide-scale and diverse development of the ultra-large-scale integrated circuits envisioned for the future, with 10^7 to 10^8 gates per chip, would be economically absurd. Clearly, increasing VLSI complexity has required not only improved fabrication facilities but also more efficient design procedures. In large part, the increase in design efficiency has been due to computer-aided design and to the adoption of highly structured design methodologies well suited to algorithm-based designs.

Computers are, of course, used in a variety of ways in integrated circuit design. In some cases, a completely automated integrated circuit layout is possible; an example is the design of random logic with PLAs. For a given technology, the basic architecture of the PLA is fixed. What varies from design to design is the number of inputs, outputs, and product terms. Given the Boolean functions to be realized, a variety of programs exist to generate the placement of cross-linking transistors, or fuse links, in the AND/OR planes. For fully-custom designs, interactive work stations with extensive computer-aided graphics capabilities are used to generate the basic patterns of subcells and subsystems and to tessellate them to form the mask patterns for larger circuits and systems. In addition, computer simulation programs form an important part of the design process in that they develop confidence in the design and help to identify problems before the design is committed to silicon.

In addition to making possible the design of complex VLSI chips, computer aids have also played a critical role in application-specific integrated circuits (ASICs). In contrast to more generic circuits, such as memories and microprocessors, the ASIC is designed to function in a specific application. For example, an ASIC may be designed to replace a printed circuit board in a particular piece of equipment, thereby taking advantage of improved fabrication technology to upgrade a system design. At a higher level of integration, an ASIC may represent a complete system on a chip. In any case, a particular ASIC is usually produced in smaller volume and on a tighter time scale than a general purpose chip. Consequently, computer aids that shorten design time and provide reliable simulations are crucial to ASICs. Standard cells, gate arrays, and PLAs are the design approaches most often used. As of 1987, ASICs account for about 15% of the total integrated circuit market, but this fraction is expected to increase.

6.9.2 The Silicon Compiler Concept

By drawing an analogy between computer programming and integrated circuit design, an interesting concept referred to as the *silicon compiler* emerges. Computer programmers routinely use compilers that translate a high-level language, or source language, into a machine language, or target language. Programming in FORTRAN, for example, allows a programmer to implement a series of commands and specifications without a detailed consideration of how the specifications are actually carried out in the computer. It is appealing to consider a similar situation in which a designer specifies a circuit design using high-level descriptions of the functions to be formed by a new integrated circuit. A silicon compiler then translates those specifications to the silicon "language," or the actual layout on the chip. Finally, the output of the compiler goes to a silicon foundry, which carries out the fabrication of the wafers containing the chips. In other words, the silicon compiler provides a link between a source language, which is the high-level description of the design, and a target language, which corresponds to the silicon layout [29]. Progress in this area is concerned not only with the compiler, but also with the development and refinement of the source languages that silicon compilers support.

A silicon compiler implies completely automated layout. In the most general sense, this characterization means the automated layout of a general-purpose chip that might include a variety of circuit types, such as random logic, storage registers, counters, clocking circuits, and on-chip memory. However, as one of the simplest of examples, consider some of the steps involved in the compilation of random logic into a PLA structure. The source language in this case consists of the Boolean equations that describe the functions to be generated. Once the equations have been generated by the designer and converted to a truth table, the number of inputs to the AND plane and outputs from the OR plane is apparent. Likewise, the number of product terms determines the number of links between the two planes. The compiler's job is to accept the input file of Boolean equations and to assemble the data into an area-efficient layout map showing the placement of transistors in the two planes.

An example of software for PLA synthesis is the set of University of Berkeley PLA design tools that run on UNIX and VMS operating systems [30, 31, 32]. This software package consists of several programs, each with a specific function. Considering the original input to the PLA package to be a set of logic equations, the program EQNTOTT generates a truth table that defines the PLA outputs in terms of the PLA inputs. Next, to achieve a compact PLA, it is desirable to perform a logic minimization. The program ESPRESSO accomplishes this procedure by producing a minimal equivalent Boolean representation. Reducing the number of product terms used to represent the original logic functions reduces the number of rows in the PLA.

Before proceeding to the layout stage of the compilation, the designer can verify the functional correctness of the PLA by use of the program SIMPLE, which is an interactive PLA simulator. Moreover, if topological minimization of area is desired, the program PLEASURE uses array-folding techniques, as described in Section 6.2.2, to optimize the silicon area occupied by the array. Finally the array is "assembled" into a silicon layout by PANDA, a program that implements a CMOS silicon layout based on the symbolic representation of the array provided by PLEASURE.

6.9.3 Interactive Graphics Work Station

An example of an interactive graphics terminal for computer-aided design of custom layouts and for generation of mask files is shown in Figure 6.46. Several such work stations may be time-shared with a single CPU and data storage unit. The intent of such a work station is to assist a designer in generating the set of patterns that correspond to each masking layer. At the most basic (and most flexible) level, a mask design is generated by creating a file consisting of geometric shapes such as rectangles, polygons, and circles. These shapes can be entered by keyboard entries, but a digitizing pad and pen, such as shown in Figure 6.46, or a mouse/menu combination greatly facilitates the process. Such procedures allow basic cells to be quickly generated and altered. Equally important is the ability to easily replicate structures, since many integrated circuit design methodologies involve basic cells that

Figure 6.46 A work station for interactive graphics and mask pattern file generation. Alternatively, the graphics tablet may be replaced by a mouse and screen menu.

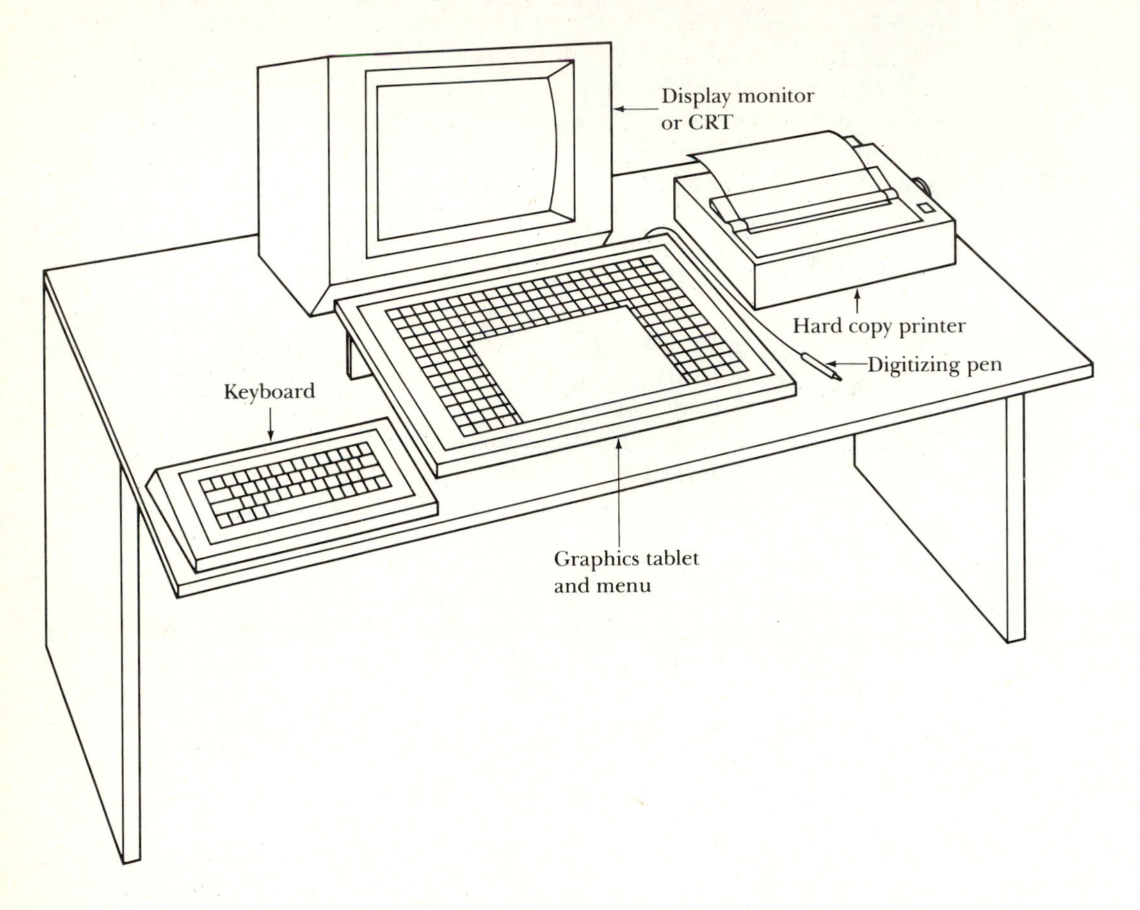

are combined to form subsystems and systems. The designer often draws on a library of predesigned circuits, or cells, which are the building blocks of the total circuit. This library can be supplemented by the addition of user-designed cells. Furthermore, many such work stations allow automatic checking of geometric design rules.

As an example of the interactive graphics ability of such a work station, consider the layout of an NMOS inverter and the subsequent tessellation of the inverter into a word shift register with pass transistors. To demonstrate the flexibility of pattern generation on an interactive graphics terminal, a library of cells will not be assumed to exist, so that all patterns must be defined by the user. The command examples are drawn from a specific integrated circuit software layout package [33]; however, they indicate the general

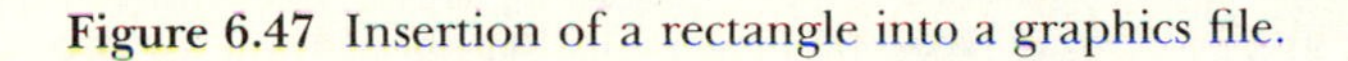

Figure 6.47 Insertion of a rectangle into a graphics file.

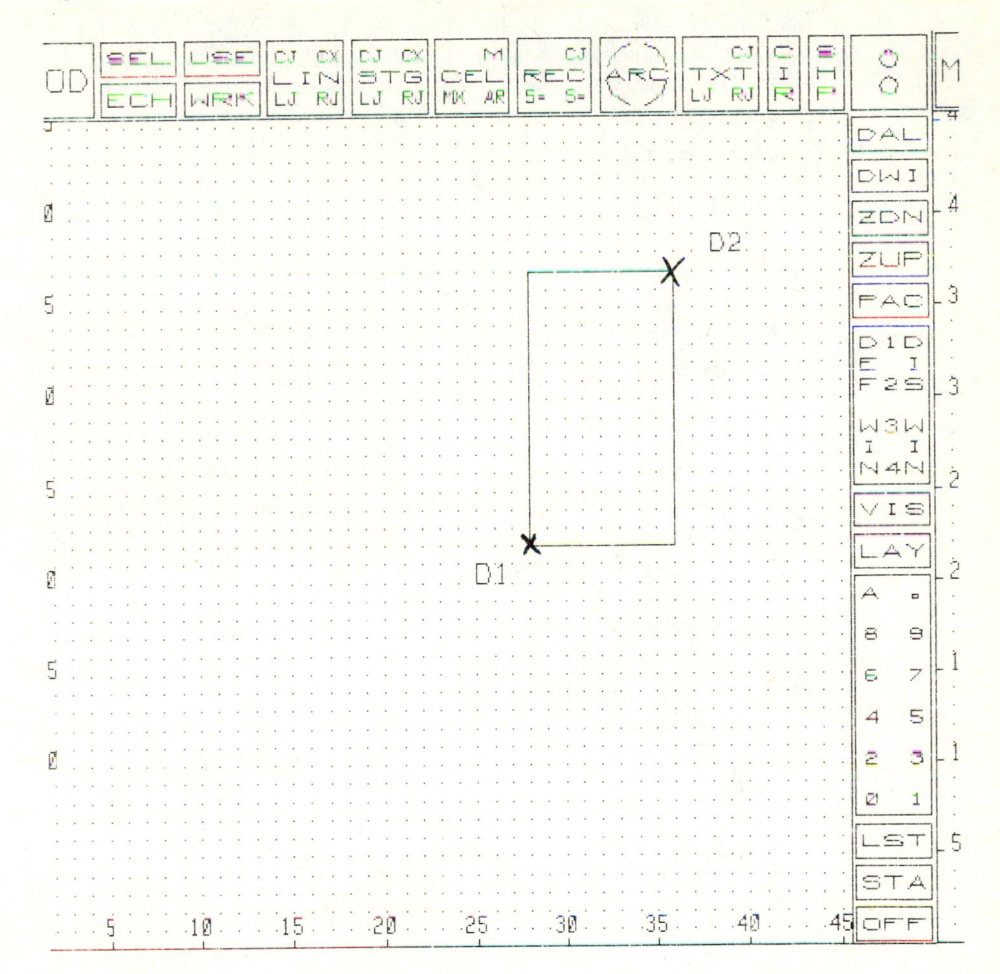

method of generating user-defined patterns on interactive graphics work stations. As a first step, several basic geometric shapes must be generated on the screen. These shapes are overlaid on a grid pattern, which corresponds to the fundamental unit of length used by the designer. This unit of length may be thought of as the λ of the design rules discussed in Chapter 5. A rectangle may be inserted into the grid, for example, by the command,

```
INS REC:d1d2
```

where d1 and d2 represent coordinates of the rectangle, as shown in Figure 6.47. A mouse can be used to implement the entire command; the selection of the command is from a menu on the screen and the coordinates d1 and d2 are entered by positioning cross-hairs or a cursor on the screen.

An interactive work station should allow a shape to be easily altered. Taking the grid dimension to be 1 μm, the rectangle shown in Figure 6.47 is 8 μm by 15 μm. The rectangle can be "stretched" by issuing the command

```
STR RIG ENT:d1d2
```

as shown in Figure 6.48. In the example shown, the new rectangle is actually shrunk, rather than stretched, to 6 μm by 15 μm. Again, the command can be entered entirely by using a mouse.

More powerful is the ability to convert the rectangle to another shape by using the "break" command. For example, Figure 6.49 shows the effect of the command

```
BRK CNV REC:d1d2d3d4
```

The combination of break and stretch commands allows flexible control over graphics entry. The NMOS inverter diffusion-mask pattern shown in Figure 6.50 was generated using a combination of these two commands. Similar command sequences generate patterns for the ion implant layer, the polysilicon layer, the contact cuts, and the metal layer.

Although the design is only active in one layer at a time, it is often desirable to display multiple layers at the same time to provide reference points for the layer being worked on. Viewing multiple layers of a layout simultaneously can be confusing, but there are two solutions to the problem. One way to distinguish layers is to use a different color for each layer. The second approach is to use a fonted fill, with a different font for each layer and fonts designed so that they can be seen through each other.

Figure 6.51 shows a combined layout for an NMOS inverter and pass transistor using an inverter layout design originated by Mead and Conway [3]. Note that this unit is the basic subcell of the shift register discussed in Section 6.2. The expansion of this subcell into a word shift register was shown by a modified stick diagram in Figure 6.5. Once the basic subcell has been defined as a graphics file, the layout for the word shift register is generated by replicating the subcell using mirroring and replicating commands. Figure 6.52 shows the diffusion pattern for a portion of the word shift register, and Figure 6.53 shows the five-layer combined layout. The busy appearance of Figure 6.53 shows the value of a zoom feature on the display monitor. This feature allows the screen display to vary from the entire structure to a single feature on a single device.

Once the entire pattern has been generated, checked, and stored as a file, the file is transferred by magnetic tape or a high-speed modem to a mask-making unit.

In the example in this section, the integrated circuit layout has been

Figure 6.48 Stretching of a rectangle. A. Screen display prior to command execution. B. New rectangle.

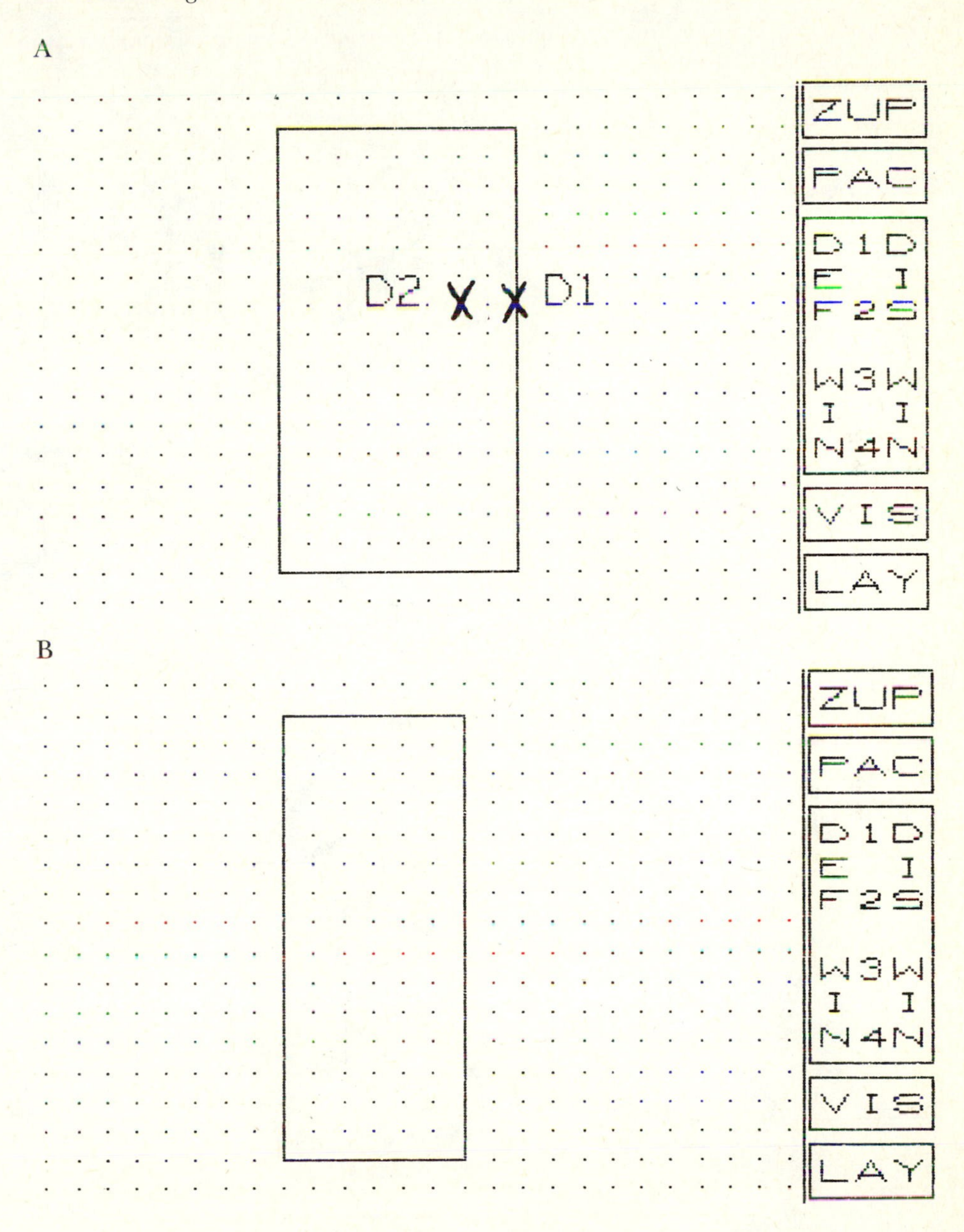

Figure 6.49 Conversion of the rectangle to another shape. A. Screen display prior to command execution. B. Result.

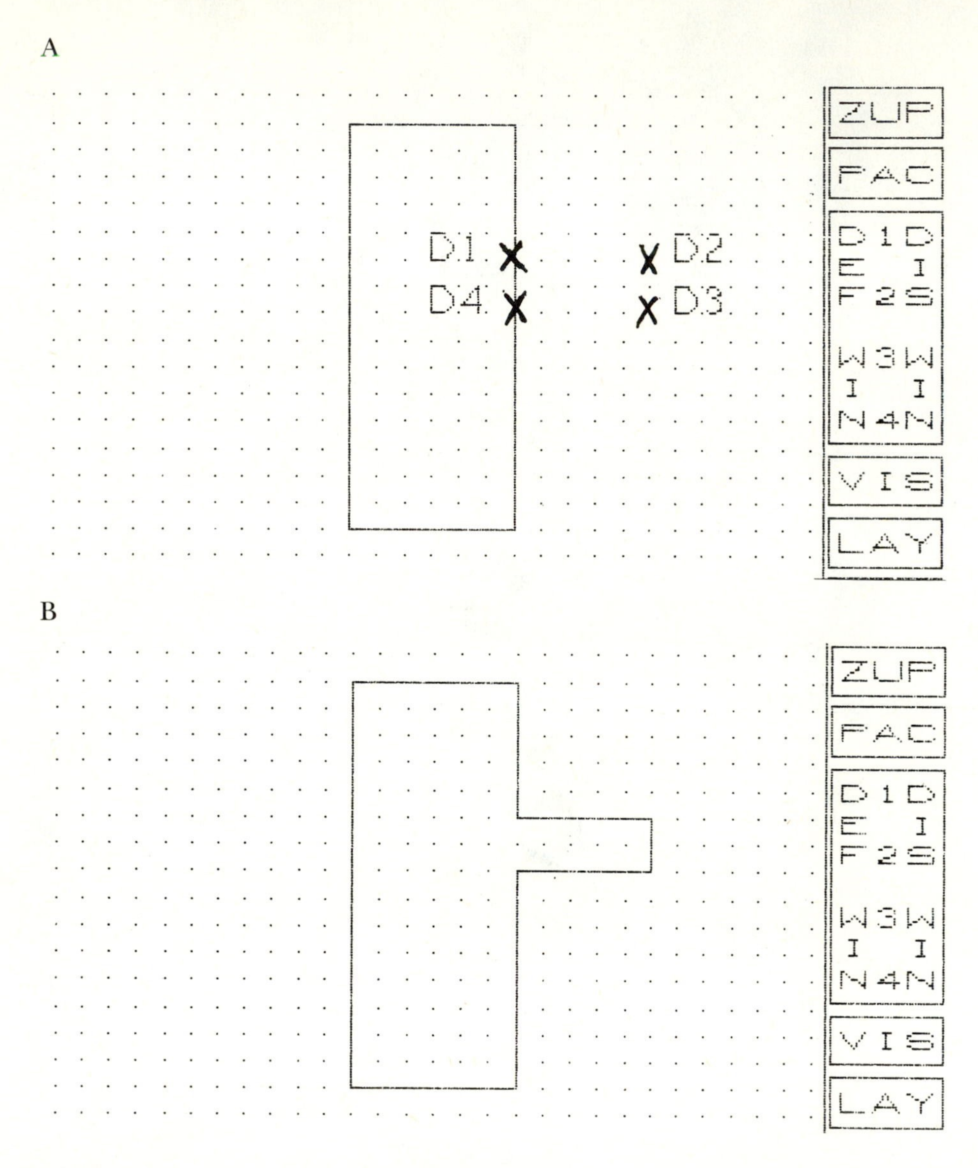

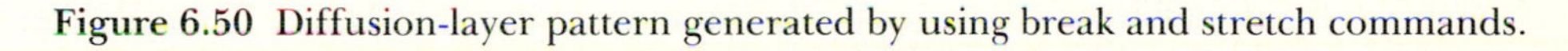

Figure 6.50 Diffusion-layer pattern generated by using break and stretch commands.

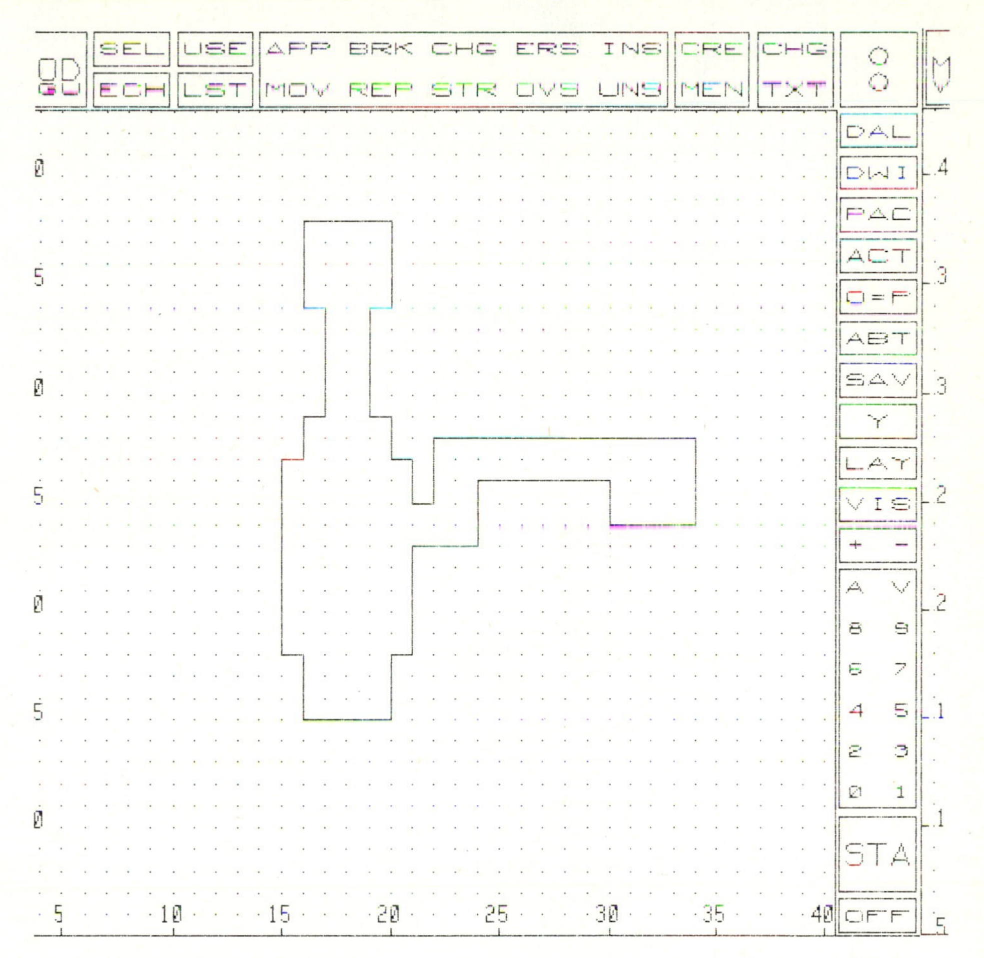

built from a combination of basic, user-defined, geometric shapes. A higher-level design approach, which offers greater ease but less overall flexibility to the designer, draws on a library of predesigned circuits or subcircuits. For example, Table 6.3 lists several entries from the standard-cell CMOS library included in the UW/NW VLSI Design System, which runs on a VAX with UNIX software [31, 34, 35]. Generation of mask artwork is facilitated by accessing cells from the library, placing the selected cells next to one another, and specifying connection points. In addition to the standard cells, custom cells may be designed by the user, provided that the new cells satisfy the design rules and interface standards associated with the standard cells. It has

Figure 6.51 Graphics screen display of the combined layout for a Mead and Conway inverter with a pass transistor.

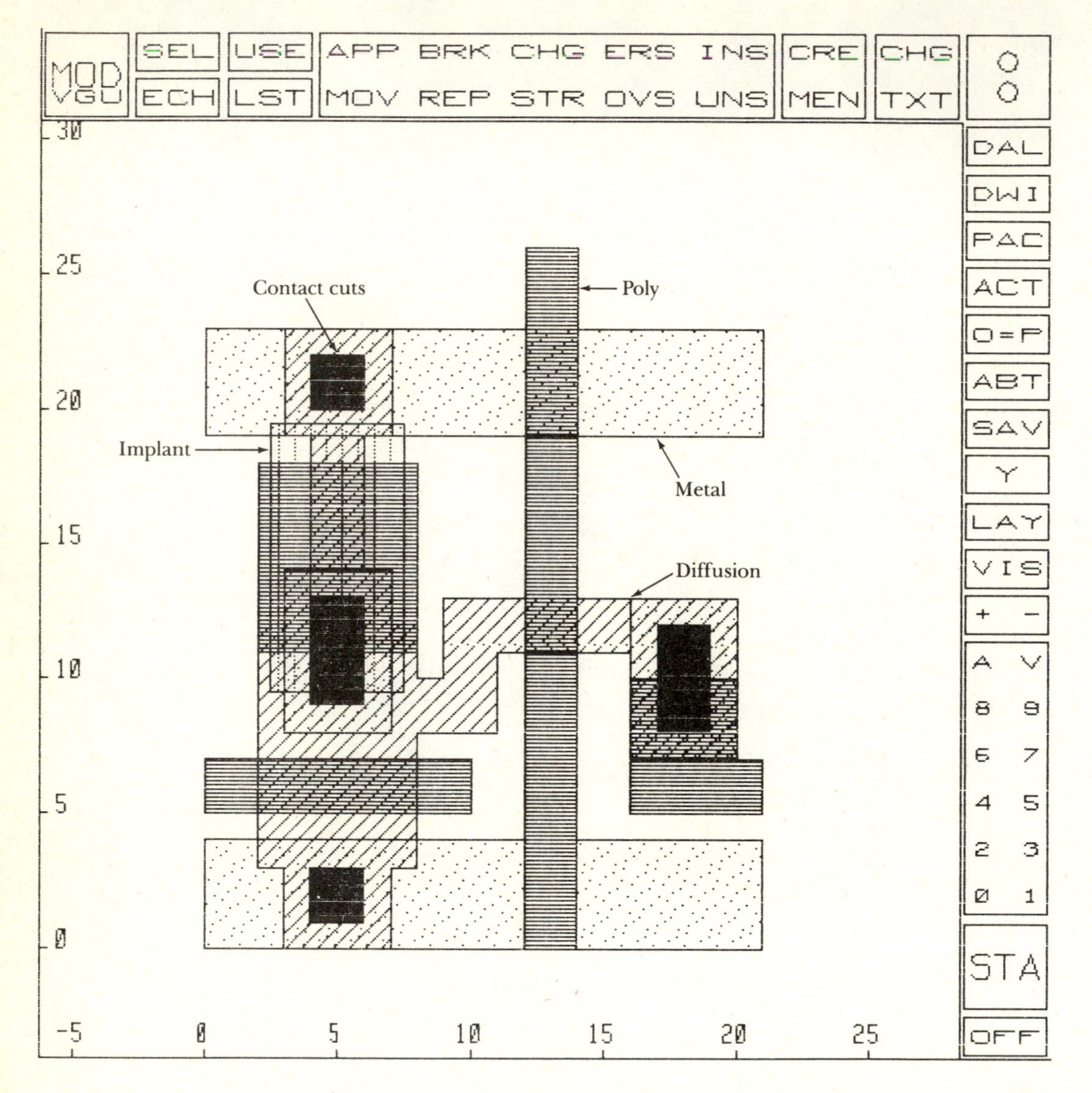

Figure 6.52 Graphics screen display of the tessellated diffusion layer for a 3 by 4 portion of a word shift register.

Table 6.3 Selected entries in the standard-cell CMOS library from the UW/NW VLSI design tools

Basic inverter	Input pad (three available sizes)
2-Input NAND gate	Buffered input pad (two available sizes)
3-Input NAND gate	Buffered TTL input pad (two available sizes)
4-Input NAND gate	Output pad (two available sizes)
2-Input NOR gate	TTL output pad
3-Input NOR gate	Tri-state pad
4-Input NOR gate	Ground pad (three available sizes)
2-Input exclusive-or gate	V_{DD} pad (three available sizes)
Clocked inverter	28-pin padframe (406 x 304 mm)
2-Input inverted selector	40-pin padframe (4.6 by 6.8 mm)
D-type latch	64-pin padframe (6.9 by 6.8 mm)
D-type latch with reset	84-pin padframe (7.9 by 9.2 mm)

Source: *VLSI Design Tool Reference Manual,* Release 3.0, UW/NW VLSI Consortium, University of Washington, Seattle, 1985.

Figure 6.53 Graphics screen display of the combined layout for a 3 by 4 portion of a word shift register.
SOURCE: From C. Mead and L. Conway, *Introduction to VLSI Systems.* © 1980, Addison-Wesley Publishing Company, Inc., Reading, Mass. Reprinted with permission.

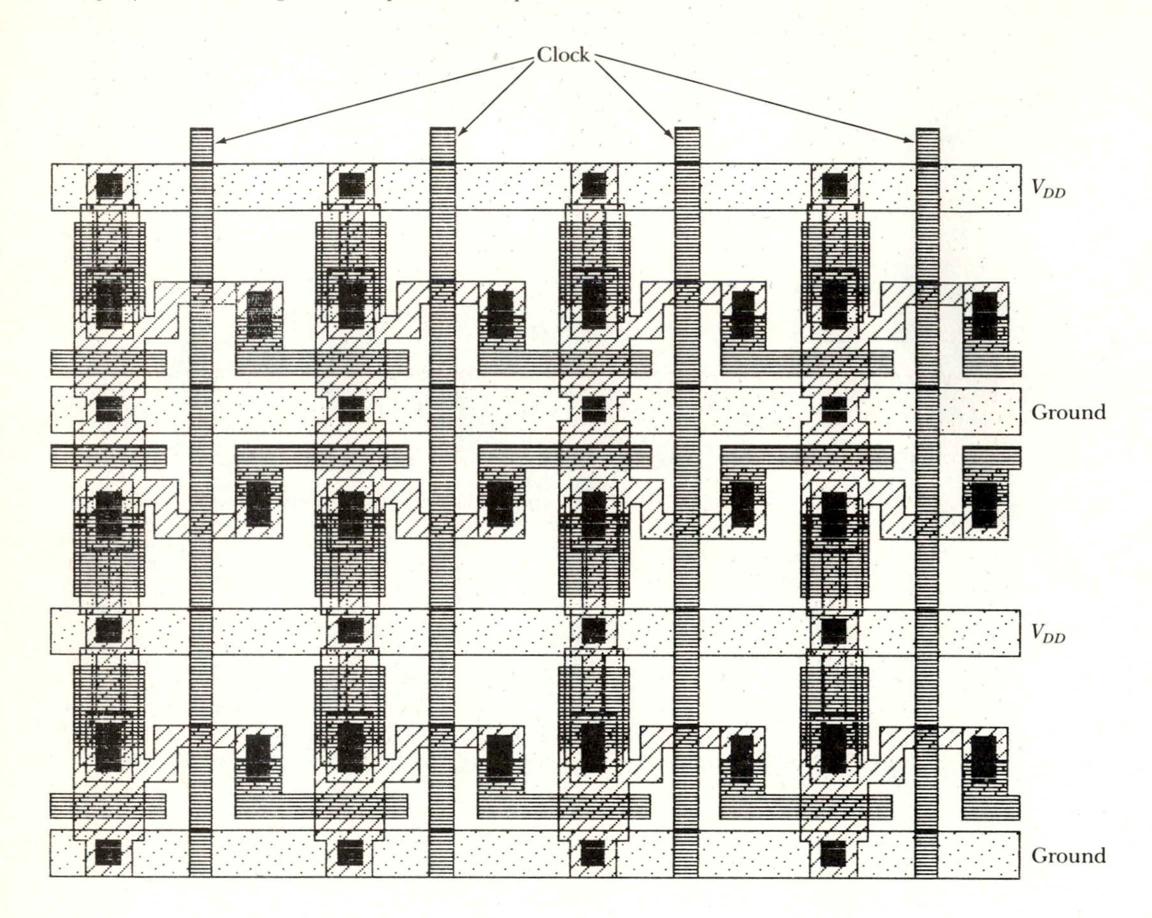

already been noted that cell libraries play a critical role in ASIC design. However, they are also important for VLSI designs that have a more general purpose; recall that the 32-bit microprocessor described in Section 6.8 was built-up from a library of 100 cells.

Another example of an interactive layout system that runs on UNIX is MAGIC, which is part of the Berkeley VLSI design tools [31, 36, 37]. MAGIC incorporates a design-rule checker that runs continuously in the background of the program so that when a layout is changed, the program records the area to be reverified. The design-rule checker rechecks these areas while the user proceeds with the design. As an example of the program's speed on a

Figure 6.54 The plowing operation in MAGIC. Layouts can be stretched or compacted while observing design rules and maintaining circuit structure. In this example, a vertical line is moved across the layout, pushing material out of its way.
SOURCE: From J. K. Ousterhout, G. T. Hamachi, R. N. Mayo, W. S. Scott, and G. S. Taylor, "The MAGIC VLSI Layout System," *IEEE Design and Test* (Feb. 1985), 19–30. © 1985 IEEE.

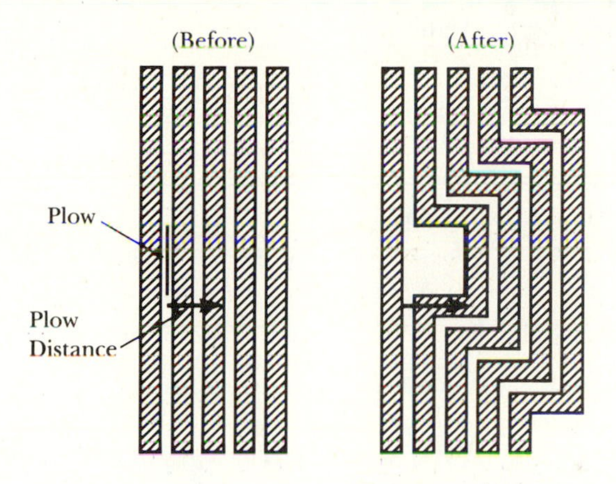

VAX-11/780, an entire recheck of design rules for a microprocessor layout with 40,000 transistors takes 18 minutes [36]. For small changes in layout, error information appears on the screen immediately.

Also available in MAGIC is an operation called *plowing,* which is illustrated in Figure 6.54. Plowing allows layouts to be stretched or compacted while the program maintains design rules. This operation is a useful editing device, since it makes it easy to open space for new circuitry. Furthermore, plowing allows a subcell portion of the circuit to be moved as a unit without changing its internal structure. MAGIC also provides routing tools for establishing interconnects. This feature facilitates the exploration of alternative layout designs and redesigns, since hand routing a large circuit is a formidable task.

In addition to the specific computer-aided layout systems discussed in this section, several others have been developed both commercially and by universities.

BIBLIOGRAPHY

For further reading on MOS digital integrated circuit design, see the following:

Glasser, L. A., and Dobberpuhl, D. W. *Design and Analysis of VLSI Circuits.* Reading, Mass.: Addison-Wesley, 1985.

Mavor, J., M. A. Jack, and P. B. Denyer. *Introduction to MOS LSI Design.* Reading, Mass.: Addison-Wesley, 1983.

Mead, C., and L. Conway. *Introduction to VLSI Systems.* Reading, Mass.: Addison-Wesley, 1980.

Mukherjee, A. *Introduction to NMOS and CMOS VLSI Systems Design.* Englewood Cliffs, N.J.: Prentice-Hall, 1986.

Newkirk, J. A., and R. Mathews. *The VLSI Designer's Library.* Reading, Mass.: Addison-Wesley, 1983.

Weste, N., and K. Eshraghian. *Principles of CMOS VLSI Design: A Systems Perspective.* Reading, Mass.: Addison-Wesley, 1985.

For further reading on analog integrated circuit design, see the following:

Gray, P. R., D. A. Hodges, and R. W. Broderson. *Analog MOS Integrated Circuits.* New York: IEEE Press, 1980.

Gray, P. R., and R. G. Meyer. *Analysis and Design of Analog Integrated Circuits.* 2nd ed. New York: Wiley, 1984.

Grebene, A. B. *Bipolar and MOS Analog Integrated Circuit Design.* New York: Wiley, 1984.

REFERENCES

1. A. S. Sedra and K. C. Smith, *Micro-Electronic Circuits* (New York: Holt, 1982).
2. H. Taub and D. Schilling, *Digital Integrated Circuits* (New York: McGraw-Hill, 1977).
3. C. Mead and L. Conway, *Introduction to VLSI Systems* (Reading, Mass.: Addison-Wesley, 1980).
4. J. Mavor, M. A. Jack, and P. B. Denyer, *Introduction to MOS LSI Design* (London: Addison-Wesley, 1983).
5. R. H. Dennard, "Field Effect Transistor Memory," U.S. Patent 3 387 286, June 4, 1968.
6. K. U. Stein, A. Sihling, and E. Doering, "Storage Array and Sense Refresh Circuit for Single-Transistor Memory Cells," *IEEE Journal of Solid State Circuits,* SC-7 (1972), 336–340.
7. J. J. Barnes and J. Y. Chan, "A High Performance Sense Amplifier for a 5 V Dynamic RAM," *IEEE Journal of Solid State Circuits,* SC-15 (1980), 831–839.
8. R. H. Dennard, "Evolution of the MOSFET RAM—A Personal View," *IEEE Transactions on Electron Devices,* ED-31 (1984), 1549–1555.
9. R. J. C. Chwang, M. Choi, D. Creek, S. Stern, P. H. Pelley, III, J. D. Schutz, P. A. Warkentin, M. T. Bohr, and K. Yu, "A 70 ns High Density 64K CMOS Dynamic RAM," *IEEE Journal of Solid State Circuits,* SC-18 (1983), 457–463.
10. R. T. Taylor and M. G. Johnson, "A 1-Mbit CMOS Dynamic RAM with a Divided Bit Line Matrix Architecture," *IEEE Journal of Solid State Circuits,* SC-20 (1985), 894–902.
11. L. F. Childs and R. T. Hirose, "An 18 ns 4K × 4 CMOS SRAM," *IEEE Journal of Solid State Circuits,* SC-19 (1984), 545–556.
12. J. Nokubo, T. Tamura, M. Nakamae, H. Shiraki, T. Ikushima, T. Akashi, H. Mayumi, T. Kubota, and T. Nakamura, "A 4.5 ns Access Time 1K × 4 Bit ECL RAM," *IEEE Journal of Solid State Circuits,* SC-18 (1983), 515–520.
13. L. R. Metzger, "A 16K CMOS PROM with Polysilicon Fusible Links," *IEEE Journal of Solid State Circuits,* SC-18 (1983), 562–567.
14. J.-I. Miyamoto, S. Saito, H. Momose, H. Shibata, K. Kanzaki, and T. Iizuka, "A

High-Speed 64K CMOS RAM with Bipolar Sense Amplifiers," *IEEE Journal of Solid State Circuits,* SC-19 (1984), 557–563.
15. S. Trimberger, "Automating Chip Layout," *IEEE Spectrum* (June 1982), 38–45.
16. X. Chen and S. L. Hurst, "A Comparison of Universal-Logic-Module Realizations and Their Application in the Synthesis of Combinatorial and Sequential Logic Networks," *IEEE Transactions on Computers,* SC-31 (1982), 140–147.
17. B. Alexander, "MOS and CMOS Arrays," in *Gate Arrays: Design Techniques and Application,* ed. J. W. Read, (New York: McGraw-Hill, 1985), Chap. 3.
18. W. Bräckelmann, H. Fritzsche, H. Ullrich, and A. Wieder, "A 150 ps 9000 gate ECL Masterslice," *IEEE Journal of Solid State Circuits,* SC-20 (1985), 1032–1035.
19. N. Toyoda, N. Uchitomi, Y. Kitaura, M. Mochizuki, K. Kanazawa, T. Terada, T. Ikawa, and A. Hojo, "A 2K-Gate GaAs Gate Array with a WN Gate Self-Alignment FET Process," *IEEE Journal of Solid State Circuits,* SC-20 (1985), 1043–1049.
20. M. L. Perrine, *Michigan State University Co-operative Education Report,* Sept. 1, 1983.
21. P. R. Gray and R. G. Meyer, *Analysis and Design of Analog Integrated Circuits,* 2nd ed. (New York: Wiley, 1984).
22. B. A. Wooley, S. J. Wong, D. O. Pederson, "A Computer-Aided Evaluation of the 741 Amplifier," *IEEE Journal of Solid State Circuits,* SC-6 (1971), 357–366.
23. K. Fukahori and P. R. Gray, "Computer Simulation of Integrated Circuits in the Presence of Electrothermal Interaction," *IEEE Journal of Solid State Circuits,* SC-11 (1976), 834–846.
24. J. E. Solomon, "The Monolithic Op Amp: A Tutorial Study," *IEEE Journal of Solid State Circuits,* SC-9 (1974), 314–332.
25. A. B. Grebene, *Bipolar and MOS Analog Integrated Circuit Design* (New York: Wiley-Interscience, 1984), Chap. 15.
26. T. Takemoto, M. Inoue, H. Sadamatsu, A. Matsuzawa, K. Tsuyi, "A Fully Parallel 10-bit A/D Converter with Video Speed," *IEEE Journal of Solid State Circuits,* SC-17 (1982), 1133–1138.
27. J. W. Beyers, L. J. Dohse, J. P. Fucetola, R. L. Kochis, C. G. Lob, G. L. Taylor, and E. R. Zeller, "A 32-Bit VLSI CPU Chip," *IEEE Journal of Solid State Circuits,* SC-16 (1981), 537–542.
28. G. Meusburger, R. C. Jones, G. Schönleber, and R. Mortimer, "A 1.5 V CMOS 4-Bit Microcomputer Needs only 100 μW," *IEEE Journal of Solid State Circuits,* SC-18 (1983), 245–249.
29. R. F. Ayres, *VLSI Silicon Compilation and the Art of Automatic Microchip Design* (Englewood Cliffs, N.J.: Prentice-Hall, 1983).
30. G. H. Mah, "PANDA; a PLA Generator for Multiply-Folded PLA's," Memorandum No. UCB/ERL M84/95, Electronics Research Laboratory, University of California, Berkeley, 1984.
31. UNIX is a trademark of Bell Laboratories.
32. VMS is a trademark of Digital Equipment Corporation.
33. Computervision Corporation, 201 Burlington Road, Bedford, MA 01730.
34. *VLSI Design Tools Reference Manual,* Release 3.0, UW/NW VLSI Consortium, University of Washington, Seattle, 1985.
35. VAX is a trademark of Digital Equipment Corporation.
36. J. K. Ousterhout, G. T. Hamachi, R. N. Mayo, W. S. Scott, and G. S. Taylor, "The Magic VLSI Layout System," *IEEE Design and Test* (Feb. 1985), 19–30.

37. W. S. Scott, G. Hamachi, J. Ousterhout, and R. N. Mayo, "1985 VLSI Tools: More Works by the Original Artists," Report No. UCB/CSD 85/225, Computer Science Division, University of California at Berkeley, Feb. 1985.

PROBLEMS

1. Consider the PLA circuit shown in Figure 6P.1. Assume that it is realized by a self-aligned polysilicon-gate NMOS process.

   a. List the masks required in the order that they are used.
   b. If $V_{DD} = 5$ V, and inputs A and B vary between 0 and 5 V, what would the maximum voltage be at Z_1 and Z_2?

Figure 6P.1 PLA circuit for Prob. 6.1.

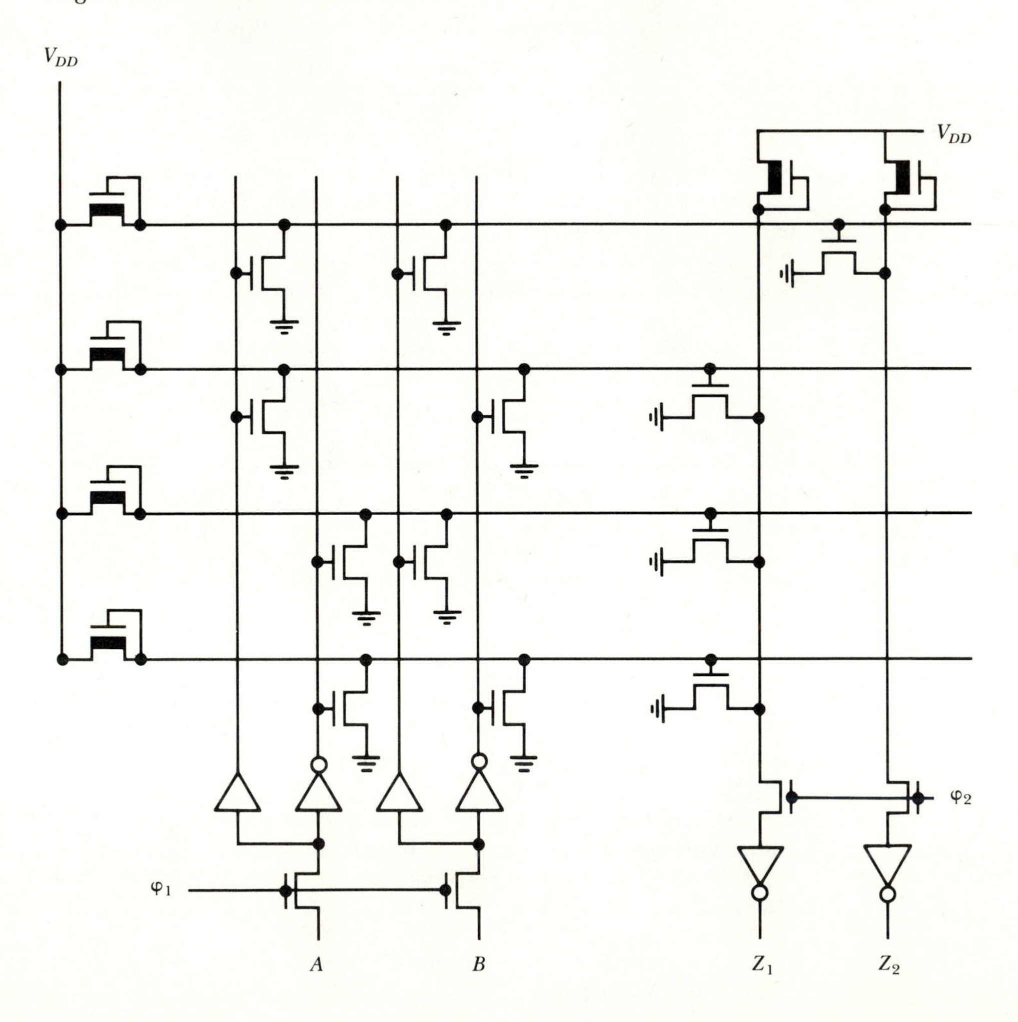

Table 6P.1 State-transition table for Prob. 6.2

A	B	$Y_{present}$	Y_{next}	Z_1	Z_2
1	0	0	0	1	0
1	0	1	1	0	1
1	1	0	1	1	0
1	1	1	0	1	1
0	x	0	0	1	1
0	x	1	1	1	0

c. Generate a logic truth table relating the outputs, Z_1 and Z_2, to the inputs, A and B.

2. A finite-state machine has 2 inputs, A and B, and two outputs, Z_1 and Z_2. Furthermore, there are two possible states as determined by a single feedback line, Y. Table 6P.1 shows the state-transition table.

   a. Generate a stick-diagram layout for this circuit, assuming a self-aligned NMOS polysilicon-gate process.
   b. Based on your stick diagram and the design rules in the second column of Table 5.1, estimate the area on the chip required by this circuit, assuming that $\lambda = 1\ \mu m$.

3. There are 16 logic "functions" associated with two Boolean variables A and B. (Actually, 2 of these functions are independent of both A and B, and 4 of the functions depend on only one variable.) Likewise, there are 16 control words, $S_3S_2S_1S_0$, for the selectable logic unit discussed in Section 6.2.3. Find the logic function associated with each control word.

4. The selectable-logic-unit stick diagram shown in Figure 6.13 uses depletion-mode transistors in every data path. When the gate inputs are low and the data input is high, the current-carrying ability of these transistors is limited, and speed may suffer as a result. Design a selectable-logic-unit layout, using stick diagrams, that does not use depletion-mode transistors. Is there a cost in the form of increased chip area?

5. The circuit depicted in Figure 6P.2 shows an NMOS gate structure that is an example of unstructured random logic.

   a. What is the logic relationship between the output and the six inputs?
   b. A PLA could realize the same function and offer the advantages of

Figure 6P.2 NMOS logic gate for Prob. 6.5.

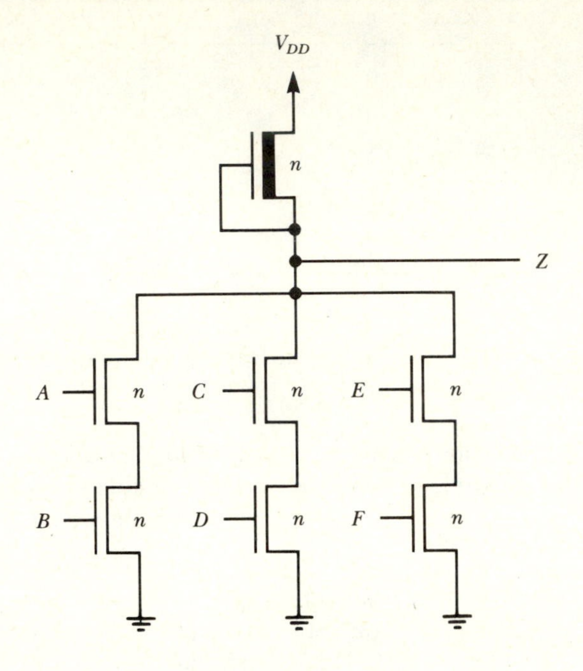

a highly structured layout. But how many transistors would be used by the PLA in comparison to the seven used in the NMOS gate?

6. Consider the circuit shown in Figure 6P.3. Find the relationship between the output and the two inputs. Is the quiescent power for this circuit negligible for all states?

7. The *D*-type flip-flop shown in Figure 6.2 uses 19 transistors if realized in *E*/*D* NMOS. Show that the two-phase-clock circuit of Figure 6P.4 functions as a *D* flip-flop with only 12 transistors.

8. Show that the circuit diagrammed in Figure 6P.5 functions as a storage register with automatic refresh. Specifically, show that the register can be loaded on command with a 1 or a 0 and will hold the stored data through any number of clock cycles until a new load command is issued. The load command, LD, and its complement are synchronized with the clock signal, φ_1. If a load command is to be issued, LD will be high during φ_1.

Figure 6P.3 MOS logic gate for Prob. 6.6.

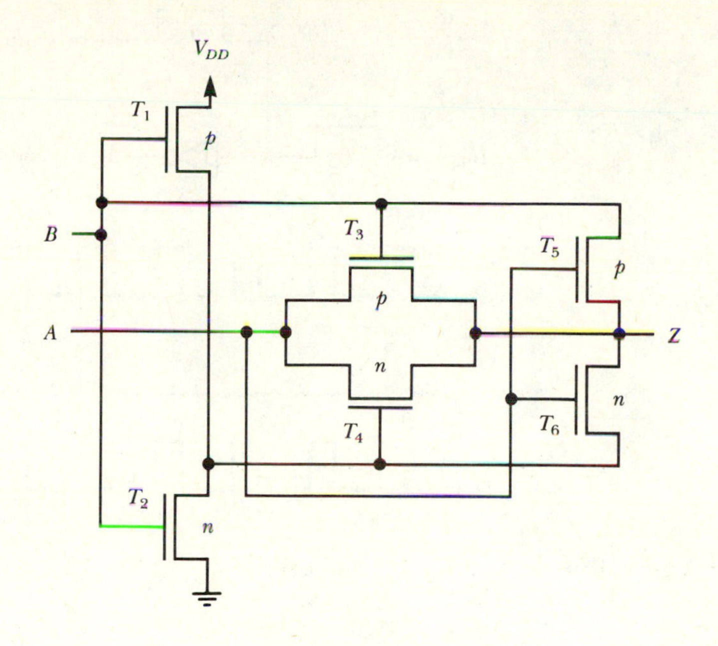

Figure 6P.4 Clocked type-D flip-flop for Prob. 6.7.

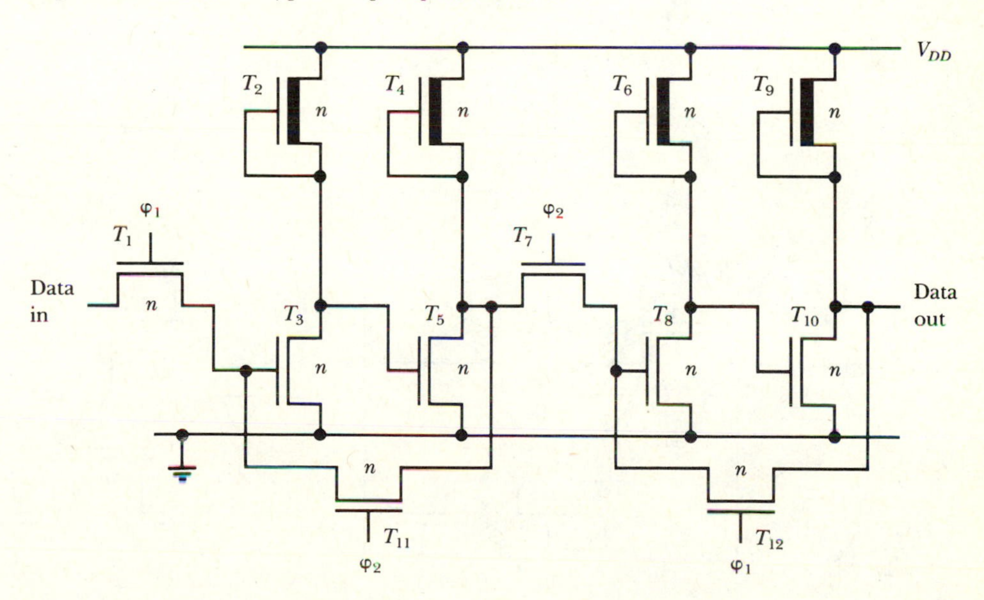

Figure 6P.5 Storage register with automatic refresh for Prob. 6.8.

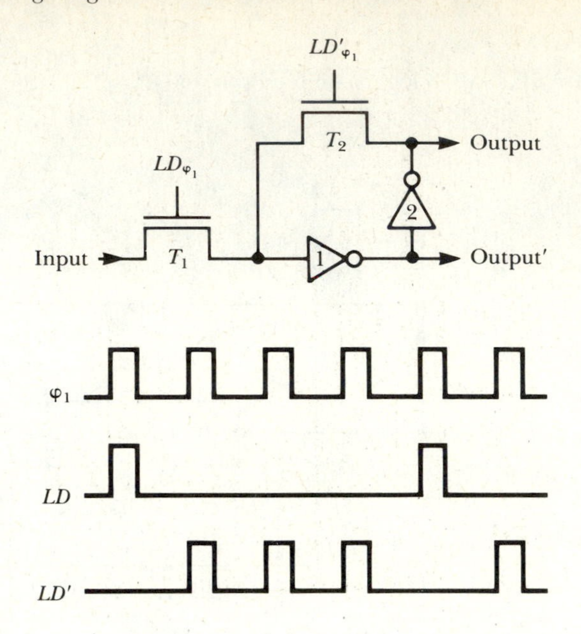

Figure 6P.6 Stack register for Prob. 6.9.

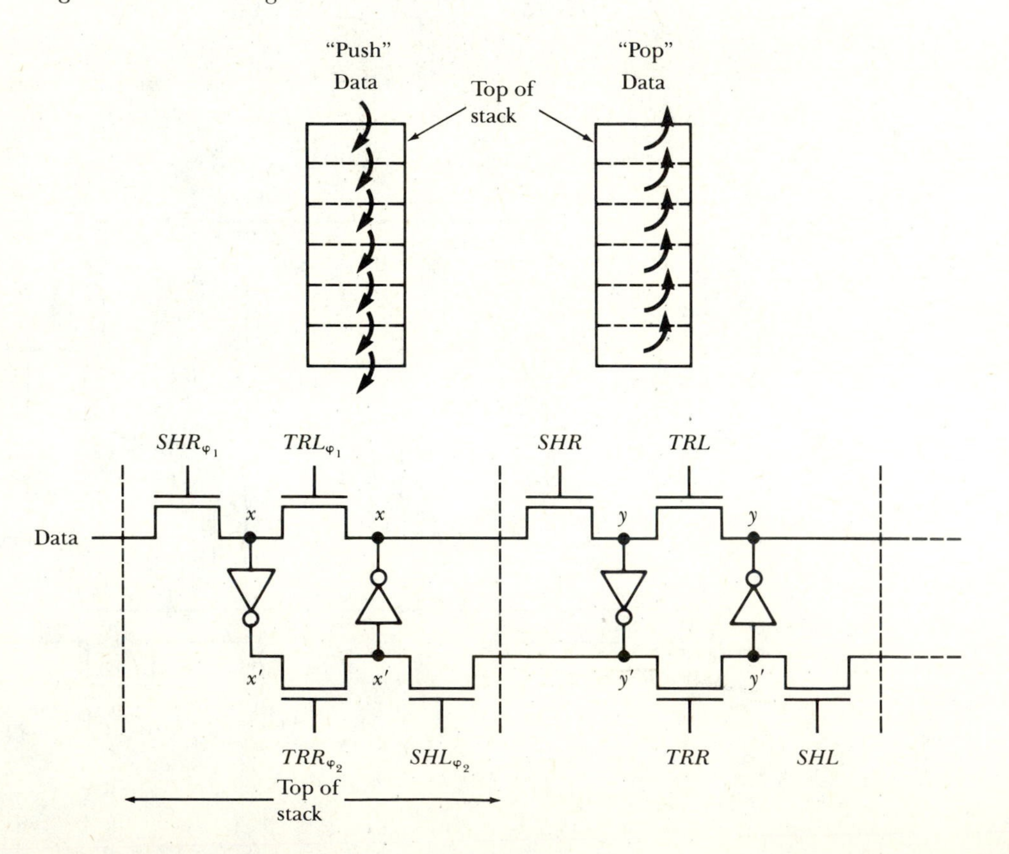

9. The circuit shown in Figure 6P.6 is a stack that functions as a first-in/last-out register. Data may be pushed into the stack, or popped out of the stack, or simply maintained by appropriate combinations of pass-transistor commands, SHR, SHL, TRR, and TRL. These commands are synchronized with the clock signals φ_1 and φ_2, as shown in the figure. Each stack operation requires two consecutive pass-transistor commands. Show that the sequence TRL high on φ_1, TRR high on φ_2 maintains the existing data. What sequences produce a push and a pop?

7 Packaging, Testing, and Yield Considerations

7.1 Introduction

The fabrication sequences discussed in Chapters 3 and 4 start with blank silicon wafers and end with processed wafers that contain arrays of dice, or chips, each of which contains a circuit the manufacturer would like to sell or use in a product. These fabrication sequences incorporate the processing technologies of Chapter 2 which, along with the design and layout methods of Chapters 5 and 6, play an important role in the economics of integrated circuit manufacturing. However, the processed wafer is obviously not the final product; to actually deliver a quality-assured product to the end user, the manufacturer must test and package the integrated circuits. These two issues are of considerable economic and technical importance because for some circuits, the cost of testing and packaging may exceed the cost of the die itself. Also, the cost per salable product is directly dependent on the yield, or the fraction of dice on the wafers that are good.

Packaging is important because the high performance and high density of a state-of-the-art chip are of little value if it cannot be interfaced, both electrically and thermally, to the next level of the system of which it is a part. Testing assures that a viable product is shipped and also helps to identify problems on the fabrication line. Indeed, testing actually begins during the process sequence and continues up to the point of the final packaged device. Yield is critical to the successful operation of a fabrication facility and is a function of many variables, including not only processing variables but also those associated with initial design and layout.

Packaging, testing, yield, and processing-related costs all combine to determine the production cost of an integrated circuit, and this chapter includes an exercise in estimating the production cost of a chip. Knowing which variables determine chip cost and how they do so is important to the designer who must decide how to partition a large system into chips. By allotting less circuitry per chip, the designer can make the chip area smaller and the yield higher. However, the more individual chips there are, the higher the testing and packaging costs are likely to be. Furthermore, more or larger circuit boards will be required. Therefore there is an economic balance between the drive to place more circuitry on a single chip and the decreasing chip yield with increasing chip area.

7.2 Process Testing and Evaluation

7.2.1 Junction Depth

An important part of integrated circuit processing is the establishment of *p-n* junctions at well-defined depths. Chapter 2 describes techniques for calculating predicted junction depths resulting from a diffusion or an ion implantation. It is often desirable to actually determine the junction depth by measurements on a test wafer. A common technique for direct junction-depth measurement is the lap and stain method illustrated in Figure 7.1, which uses an abrasive slurry, such as powdered Al_2O_3 and water, to remove the top few micrometers of silicon from a portion of the wafer. This polishing, or lapping operation is carried out at a slightly beveled angle. Next the wafer is placed in a staining bath, which may, for example, be HF and a small amount of HNO_3 [1]. As a result, the *p*-region darkens relative to the *n*-region.

Microscopic examination shows both the stained region and the onset of the bevel lap. Consequently, the lateral distance between these two lines can be measured using a microscope with a calibrated reticle in the eyepiece. For example, if the bevel angle is known to be 3 degrees and the lateral distance is found to be 95 μm, then the junction depth is 5 μm.

Another way to determine junction depth uses chemical-composition analytical methods, such as secondary-ion mass spectrometry, to measure the impurity profile as a function of depth. Section 7.2.4 describes this approach.

Figure 7.1 Lap and stain junction measurement. A. Beveled lapping operation. B. Junction measurement.

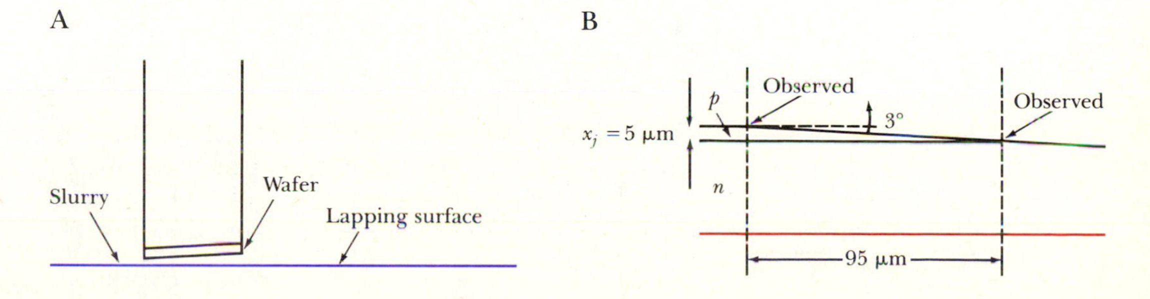

7.2.2 Irvin's Curves

Although the lap and stain provides information about the junction depth, it does not yield any further information about the actual profile of dopants between the wafer surface and the junction. A somewhat indirect indicator of the dopant profile $N(x)$ may be obtained by measuring the sheet resistance. The relationship between sheet resistance and the dopant profile is obtained from Eqs. (4.67) and (4.68), which are repeated here for convenience. The sheet resistance of a diffused or ion implanted region of semiconductor is given by

$$R_{\square} = 1/(\sigma_{\text{ave}})(x_j) \tag{7.1}$$

where x_J is the junction depth and the average conductivity is given by

$$\sigma_{\text{ave}} = (1/x_j) \int_0^{x_j} q\mu(x)[N(x) - N_B]dx \tag{7.2}$$

where μ is the majority carrier mobility and N_B is the background concentration.

If a dopant profile has been calculated using the theoretical methods of Chapter 2, then the expected sheet resistance can be calculated, using Eqs. (7.1) and (7.2). This method was adopted by Irvin to generate a set of curves for erfc and Gaussian $N(x)$ distributions; these curves relate the surface dopant concentration, $N(0)$, to the $R_{\square}x_j$ product for a given background doping level. Figures 7.2, 7.3, 7.4, and 7.5 show examples of Irvin's curves [2]. Given the conditions under which a predeposition or drive-in diffusion is carried out, these curves facilitate the calculation of the expected sheet resistance. For example, suppose the background doping of a uniformly doped wafer has been determined. Then, knowledge of the temperatures and times associated with the diffusion process allows the surface concentration and junction depth to be estimated, as described in Chapter 2. Irvin's curves then provide a value for the expected sheet resistance, and a subsequent test measurement of sheet resistance provides a comparison between actual results and expected results. Such an exercise is described in Section A.6.4 of Appendix A. Also, if the sheet resistance and the junction depth are measured, then Irvin's curves can be used to find the surface concentration.

EXAMPLE 7.1

As an example of the use of Irvin's curves, suppose that a p-type drive-in diffusion is carried out on an n-type wafer with a uniform background concentration of $N_B = 10^{14}$ cm^{-3}. A lap and stain measurement indicates that the junction depth is 3.0 μm and the sheet resistance is found to be 100 $\Omega/\square$. What is the surface concentration, $N(0)$?

Figure 7.2 Irvin's curves for an n-type erfc diffusion in uniformly doped p-type silicon.

SOURCE: Adapted by R. A. Colclaser, *Microelectronics Processing and Device Design* (New York: John Wiley & Sons, 1980), Chapter 9, copyright © by John Wiley and Sons, New York; from J. C. Irvin, "Resistivity of Bulk Silicon and of Diffused Layers in Silicon," *Bell System Technical Journal,* Vol. 41, No. 2, Part 1, March 1962. Reprinted with permission from the *AT&T Technical Journal,* copyright © 1962, AT&T.

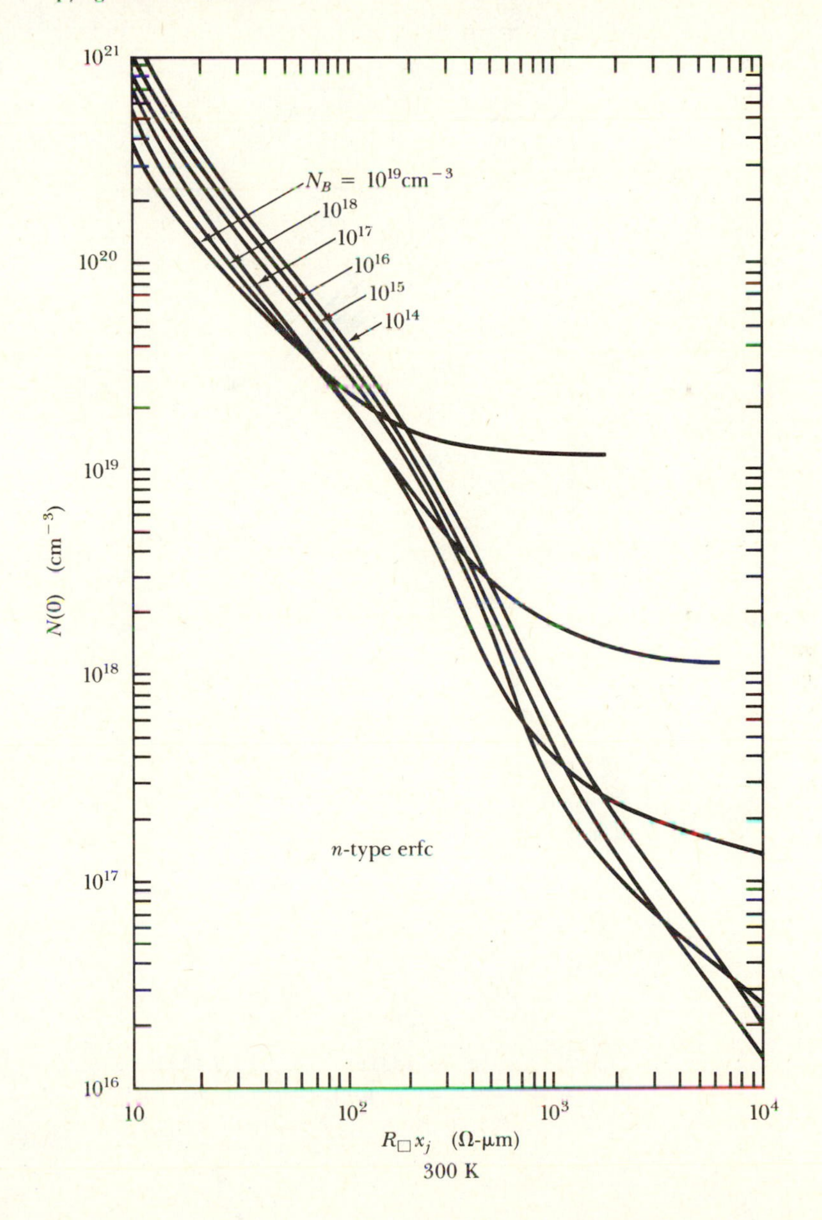

Figure 7.3 Irvin's curves for a *p*-type erfc diffusion in uniformly doped *n*-type silicon. SOURCE: Adapted by R. A. Colclaser, *Microelectronics Processing and Device Design* (New York: John Wiley & Sons, 1980), Chapter 9, copyright © by John Wiley and Sons, New York; from J. C. Irvin, "Resistivity of Bulk Silicon and of Diffused Layers in Silicon," *Bell System Technical Journal,* Vol. 41, No. 2, Part 1, March 1962. Reprinted with permission from the *AT&T Technical Journal,* copyright © 1962, AT&T.

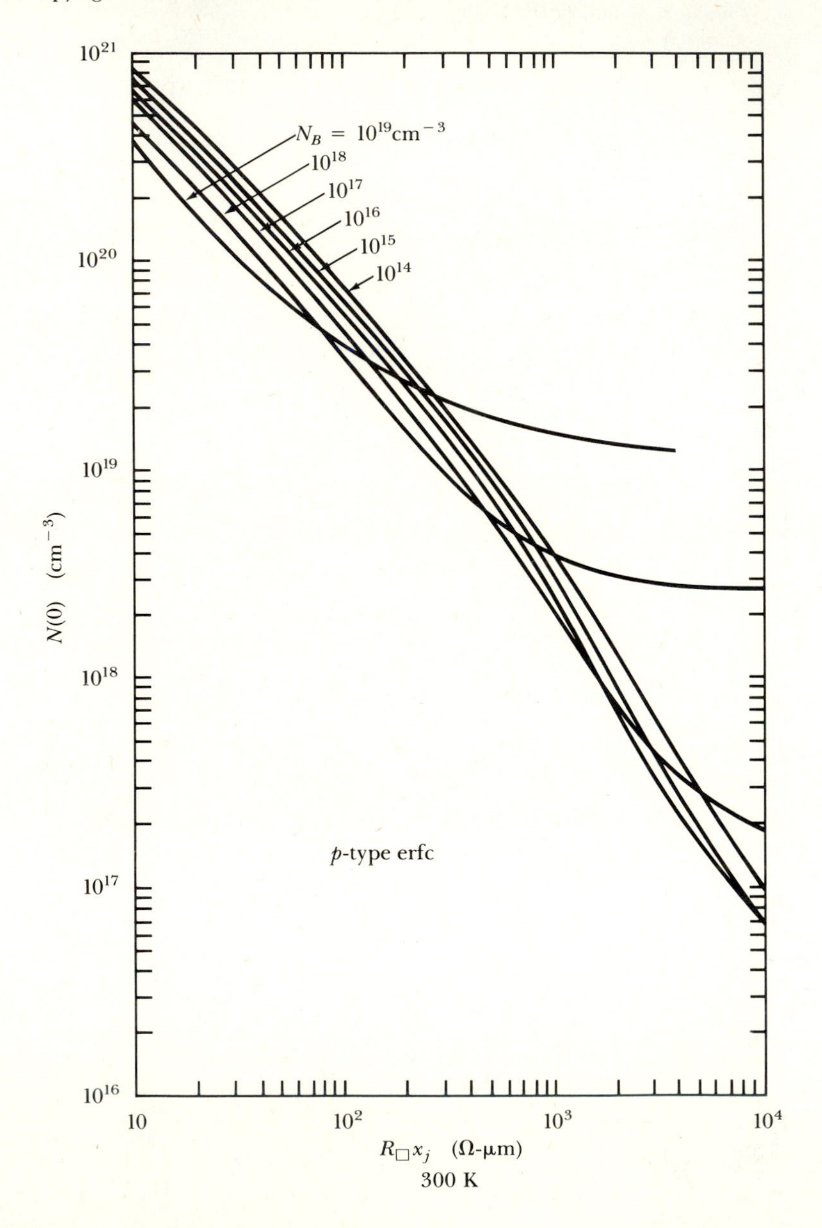

Figure 7.4 Irvin's curves for an n-type Gaussian diffusion in uniformly doped p-type silicon.

SOURCE: Adapted by R. A. Colclaser, *Microelectronics Processing and Device Design* (New York: John Wiley & Sons, 1980), Chapter 9, copyright © by John Wiley and Sons, New York; from J. C. Irvin, "Resistivity of Bulk Silicon and of Diffused Layers in Silicon," *Bell System Technical Journal,* Vol. 41, No. 2, Part 1, March 1962. Reprinted with permission from the *AT&T Technical Journal,* copyright © 1962, AT&T.

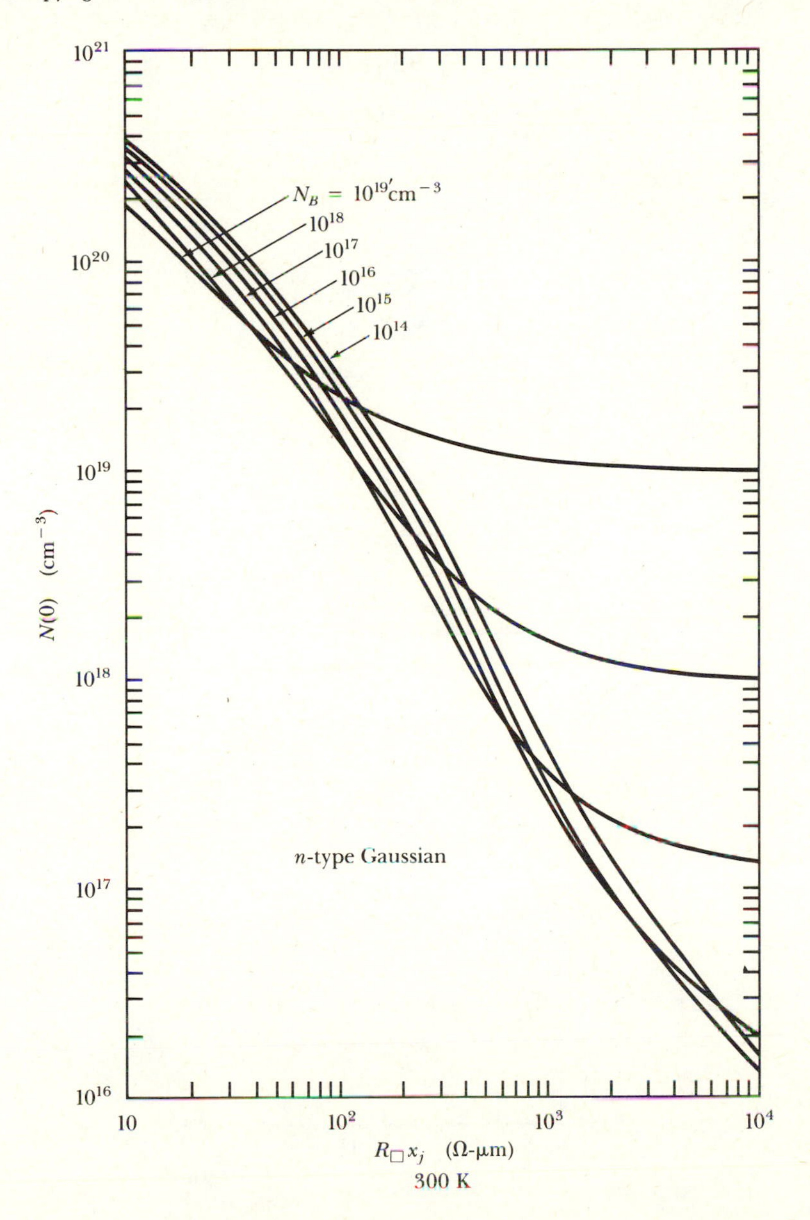

Figure 7.5 Irvin's curves for a *p*-type Gaussian diffusion in uniformly doped *n*-type silicon.

SOURCE: Adapted by R. A. Colclaser, *Microelectronics Processing and Device Design* (New York: John Wiley & Sons, 1980), Chapter 9, copyright © by John Wiley and Sons, New York; from J. C. Irvin, "Resistivity of Bulk Silicon and of Diffused Layers in Silicon," *Bell System Technical Journal,* Vol. 41, No. 2, Part 1, March 1962. Reprinted with permission from the *AT&T Technical Journal,* copyright © 1962, AT&T.

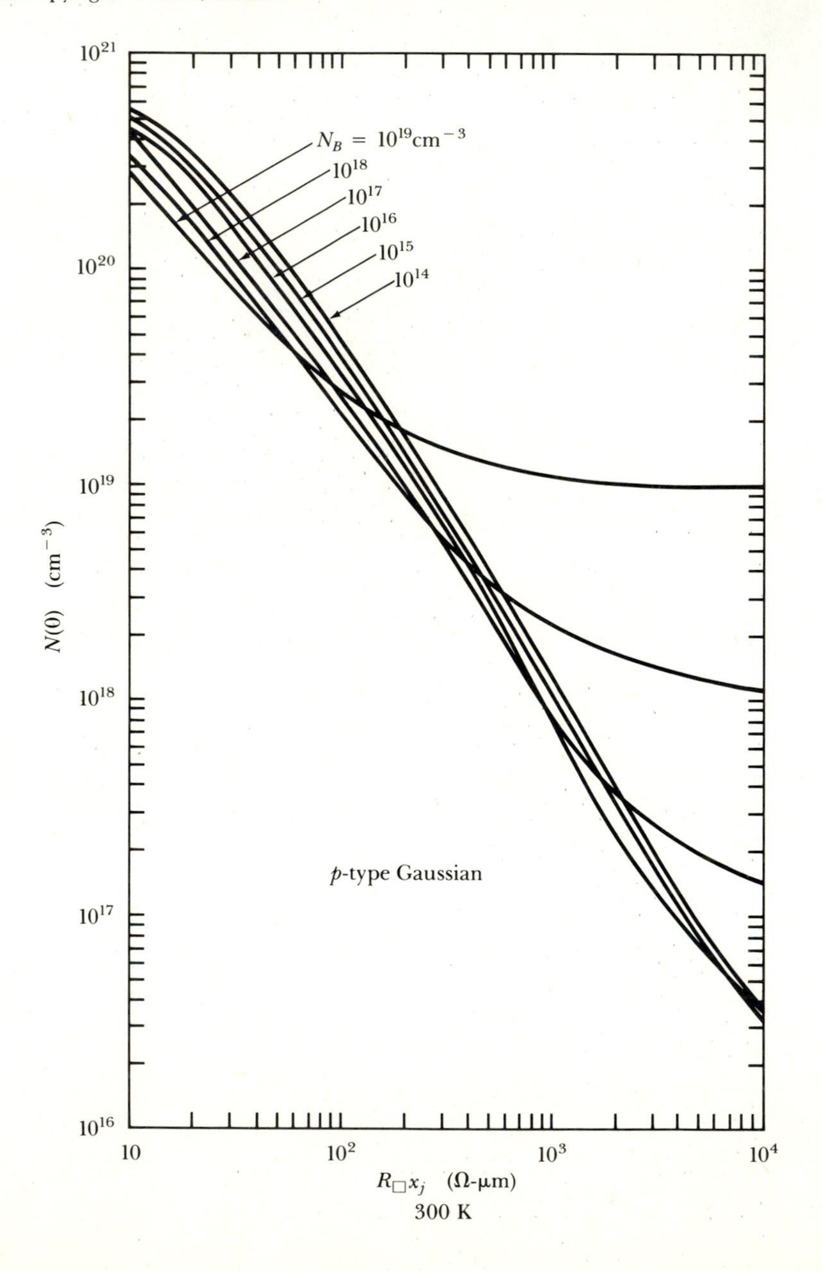

Solution Since the diffusion is a drive-in, the profile is Gaussian. From Figure 7.5, the value of $N(0)$ is found to be approximately 10^{19} cm^{-3}.

Irvin's curves, although useful, are limited in their application. For example, they cannot be used to evaluate the properties of the diffused emitter layer in a bipolar junction transistor (BJT) because the background concentration is not constant. More generally, note that the curves do not take into account the redistribution of impurities at the silicon-oxide interface. Therefore, measured values of the sheet resistance of drive-in boron-diffused layers tend to be less than predicted from Irvin's curves because of the depletion of boron at the interface, as described in Section 2.7. Also, the accuracy of Irvin's curves is limited in that real diffusion profiles deviate from the ideal erfc or Gaussian solutions since, as discussed in Section 2.5.3, the diffusion coefficient is to some extent a function of concentration.

A common method for measuring sheet resistance is the four-point-probe method shown in Figure 7.6. Four equally spaced point probes are brought into contact with the wafer surface and a known dc current is passed through the two outer probes. The inner two probes are connected to a high impedance voltmeter. This four-point arrangement largely eliminates contact resistance effects, since the voltage-measurement probes draw negligible current. If the diffused region dimensions a and d are much larger than the probe spacing s, then the relationship between V, I, and $R_\square$ is

$$R_\square = (\pi/\ln 2)(V/I) = 4.53(V/I) \tag{7.3}$$

Figure 7.6 The four-point-probe method for determining sheet resistance.

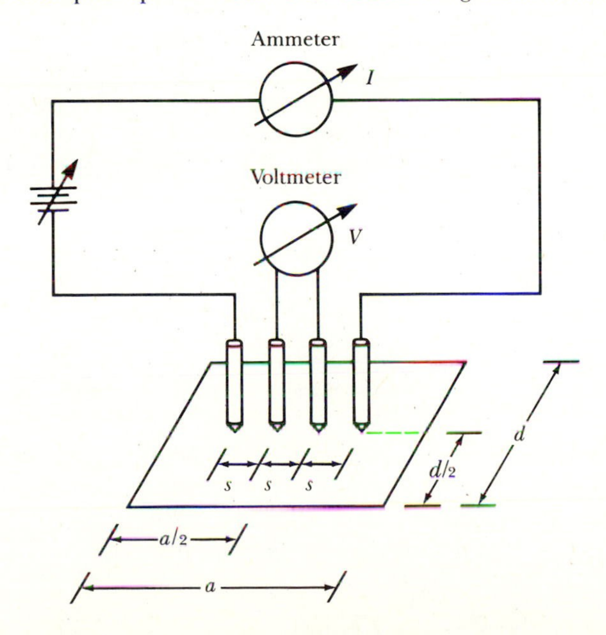

Table 7.1 Constant of proportionality between $R_\square$ and V/I for the four-point-probe structure of Figure 7.6. $R_\square = \zeta(V/I)$, where ζ is tabulated below.

d/s	$a/d = 1$	$a/d = 2$	$a/d = 3$	$a/d = 4$
3.0	2.4575	2.7000	2.7005	2.7005
4.0	3.1137	3.2246	3.2248	3.2248
5.0	3.5098	3.5749	3.5750	3.5750
7.5	4.0095	4.0361	4.0362	4.0362
10.0	4.2209	4.2357	4.2357	4.2357
15.0	4.3882	4.3947	4.3947	4.3947
20.0	4.4516	4.4553	4.4553	4.4553
40.0	4.5120	4.5129	4.5129	4.5129
∞	4.5324	4.5324	4.5325	4.5324

Source: From F. M. Smits, "Measurement of Sheet Resistivities with the Four-Point Probe," *Bell System Technical Journal,* 37, No. 3, Part 1, May 1958, pp. 711–718. Reprinted with permission from the *AT&T Technical Journal.*

For smaller-dimension samples, a different constant of proportionality must be used, as listed in Table 7.1 [3]. Alternatively, the sheet resistance may be determined by incorporating a four-probe test pattern into the wafer layout as a test site.

The four-point-probe is also useful for determining the background concentration of wafers prior to initiation of the fabrication process. Since the wafer doping is uniform, Eq. (7.1) can be used to find the resistivity by replacing x_J with the wafer thickness and thereby obtaining σ, which is the reciprocal of the resistivity. Since resistivity is tabulated as a function of doping for several semiconductors, the wafer doping is determined from the resistivity. Section A.5 of Appendix A documents this process for silicon wafers, and contains a table of ζ values for circular samples as well as a plot of resistivity versus doping for n- and p-type silicon.

7.2.3 Differential Capacitance Measurements of $N(x)$ [4]

Another experimental measure of $N(x)$ uses differential capacitance measurements in which the doping profile is related to the rate of change of a test-structure capacitance with voltage. In contrast to the sheet resistance technique described in the previous section, this technique does not assume a particular impurity profile. The capacitor test device may be a metal-oxide-semiconductor (MOS) structure, a p-n diode, or a Schottky barrier diode. This discussion assumes an MOS test device.

Figure 7.7 MOS capacitor probing of $N(x)$.

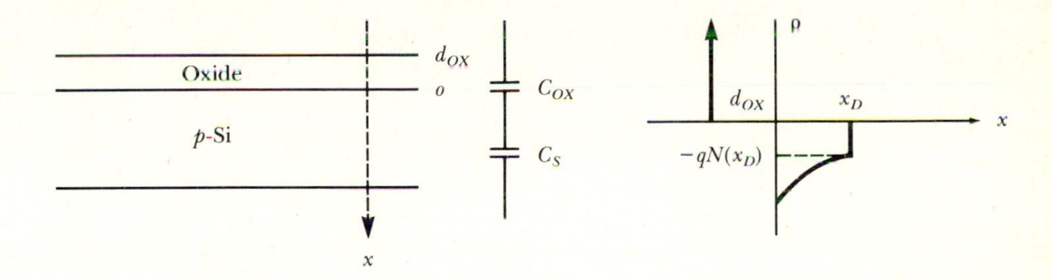

Provided the MOS capacitor is not in inversion or accumulation, a change in gate voltage causes a change in voltage across both the oxide and the semiconductor depletion layer, as discussed in Section 3.4 and indicated in Figure 7.7. Therefore the incremental gate voltage may be written as

$$dV_G = dV_{OX} + dV_S = \frac{dQ_S}{C_{OX}} + \frac{dQ_S}{C_S} \tag{7.4}$$

where Q_S is the space-charge density per unit area. The space-charge density is related to the width, x_D, of the semiconductor depletion region, according to the expression

$$dQ_S = -qN(x_D)dx_D \tag{7.5}$$

where $N(x_D)$ is the net dopant concentration at $x = x_D$, and the negative sign indicates that the dopants are negatively charged ionized acceptors for this analysis. Also,

$$C_{OX} = \epsilon_{OX}/d_{OX} \tag{7.6}$$

and

$$C_S = \epsilon_S/x_D \tag{7.7}$$

Eqs. (7.4) and (7.5) may be combined to yield

$$dQ_S = -qN(x_D)dx_D = CdV_G \tag{7.8}$$

where

$$C = [(1/C_{OX}) + (1/C_S)]^{-1} \tag{7.9}$$

The differential relationship between x_D and C_S may be expressed as

$$dx_D = \epsilon_S d(1/C_S) \tag{7.10}$$

and, since C_{OX} is a constant, Eq. (7.10) becomes

$$dx_D = \epsilon_S d(1/C) \tag{7.11}$$

Combining Eqs. (7.8) and (7.11) yields the following expression for $N(x_D)$:

$$N(x_D) = \left[\frac{q\epsilon_S}{2} \cdot \frac{d(1/C^2)}{dV_G}\right]^{-1} \tag{7.12}$$

Therefore, by plotting $1/C^2$ as a function of voltage, $N(x_D)$ can be obtained from the slope. By varying V_G, and therefore x_D, $N(x)$ may be mapped out up to $x = x_{D\max}$, which is the onset of inversion.

The differential capacitance method is useful for cases where the dopant profile is not constant, Gaussian, or erfc, and therefore for cases not amenable to simple sheet resistance analysis. The previous discussion assumes the existence of an MOS test structure on the wafer. However, a noninvasive method for carrying out differential capacitance analysis involves mercury Schottky diode probing, in which case the Schottky contact is due to a mercury drop that can later be removed by standard demetal cleaning procedures [5].

7.2.4 Surface-Analysis Techniques

A large variety of methods exist for both topological and chemical analysis of the surface or the near-surface region of an integrated circuit. These methods generally involve sophisticated observations of the interaction between the sample and an interrogating beam of either electrons, ions, neutrons, or photons (ranging from infrared to x-rays). A comprehensive review of these methods is beyond the scope of this discussion. Rather, this section briefly describes three techniques in fairly broad use, namely scanning electron microscopy (SEM), scanning Auger microanalysis (SAM), and secondary ion mass spectrometry (SIMS). This chapter's bibliography and references include sources for a more complete review of surface- and near-surface-analysis techniques.

The scanning electron microscope is a standard instrument in VLSI research laboratories because of the excellent detail provided in SEM examination of the topological details of an integrated circuit's surface. As illustrated in Figure 7.8, which shows a schematic drawing of a SEM, a small-diameter electron beam is raster-scanned across a sample mounted in a vacuum and an image is formed from secondary electrons liberated from the bombarded

Figure 7.8 Schematic drawing of a scanning electron microscope. See R. B. Marcus, "Diagnostic Techniques," in *VLSI Technology*, ed. S. M. Sze (New York: McGraw-Hill, 1983), Chap. 12, for further information.

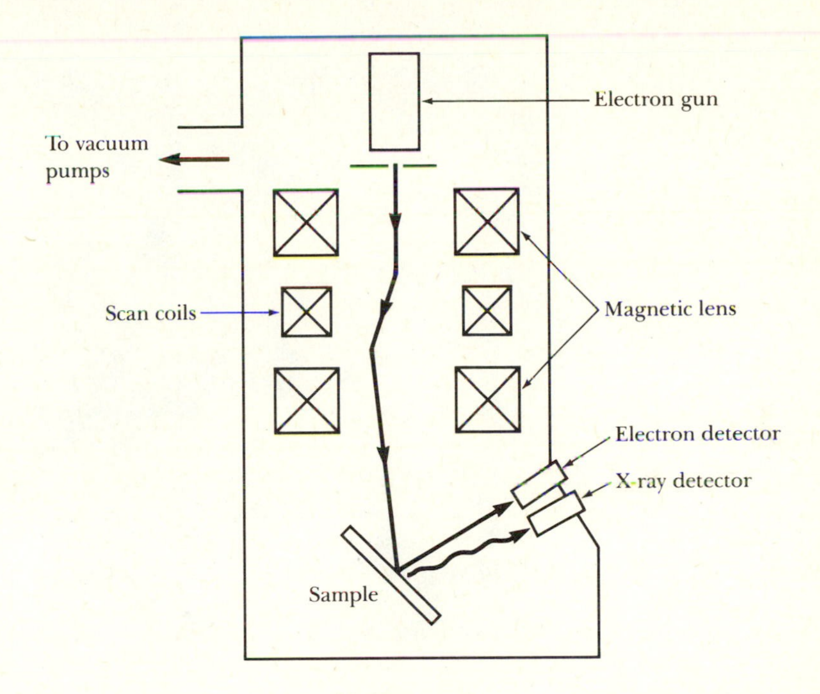

sample or, alternatively, from back-scattered electrons from the primary beam [6]. The former consist of fairly low-energy (several electron-volts) electrons, whereas back-scattered electrons have energies similar to that of the primary beam, which is several kilo-electron-volts.

The two electron populations form images with different and complementary information. Since the probability of an electron being back-scattered out of the specimen depends on the atomic number of the sample, the number of back-scattered electrons from a given portion of the sample depends on the average atomic number in that region. Consequently, the back-scattered electron image shows contrast between sample regions with different atomic numbers. The secondary electron yield, however, depends heavily on the work function of the sample region being interrogated and allows a strong distinction between metallic, oxide, and semiconductor regions based on work-function differences. Lateral resolution in SEM can be as small as 30 Å [7].

SEM is useful in inspecting for quality control, investigating new processing techniques, and identifying failure sites and mechanisms. For example, Figure 7.9 shows a SEM photograph of an integrated circuit aluminum

Figure 7.9 SEM photograph of an aluminum stripe integrated circuit conductor that has failed because of electromigration.
SOURCE: From J. R. Black, "Electromigration Failure Modes in Aluminum Metallization for Semiconductor Devices," *Proceedings of the IEEE,* 57 (1969), 1587. © 1969 IEEE.

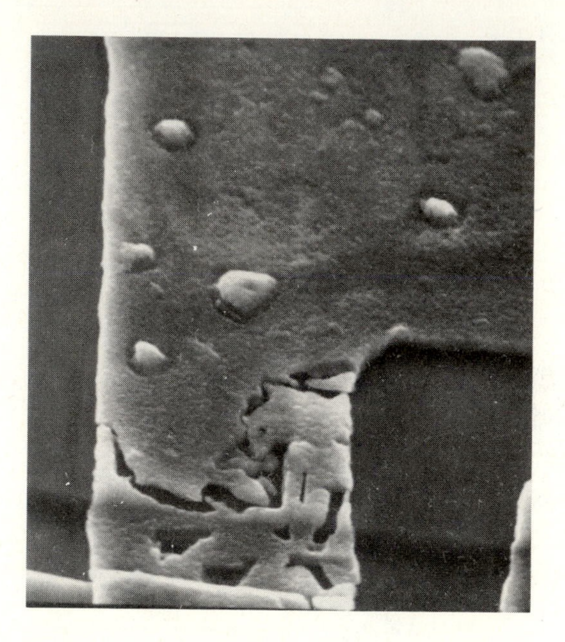

stripe that has failed because of electromigration [8]. Several voids appear in the aluminum film, and one extends across the entire width such that an open circuit resulted.

The SEM diagram in Figure 7.8 also shows an x-ray detector. Atoms in the near surface, approximately in the top micrometer of material, are stimulated by the SEM electron beam to emit x-rays. The resulting x-ray spectrum contains distinct peaks centered around specific energies that characterize the stimulated element, as well as a background continuum of bremstrahlung radiation resulting from the deacceleration of electrons. Since the x-ray peaks characterize the stimulated element, an analysis of the x-ray spectrum provides detailed information about the chemical make-up of the sample. These data complement the topographical information provided by the electron image. Because of lateral spreading of the emitted x-rays, the lateral resolution associated with the x-ray spectral information is only about 1 μm. Furthermore, since x-rays are received from a depth of about 1 μm as well, this method does not give a high degree of spatial resolution. Composition resolution is greatly enhanced by the Auger spectral analysis technique described next.

Figure 7.10 Block diagram of a SIMS apparatus. See R. B. Marcus, "Diagnostic Techniques," in *VLSI Technology*, ed. S. M. Sze (New York: McGraw-Hill, 1983), Chap. 12 for further information.

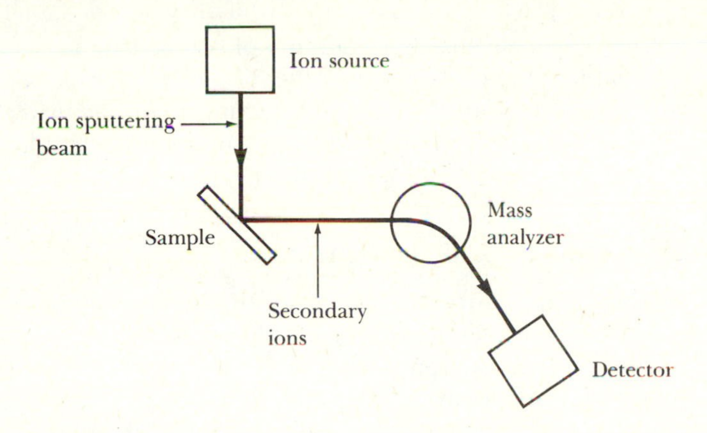

In addition to characteristic x-ray radiation, excited atoms also emit Auger electrons with energies characteristic of the excited atom. In energy, Auger electrons are intermediate between secondary electrons and backscattered electrons, and they correspond to electron emission from particular atomic shells. Unlike x-rays, Auger electrons have short ranges of tens of Angstroms, so that the Auger spectrum of emitted electrons is more representative of the surface than is the x-ray spectrum. Also, lateral resolution is approximately 0.1 μm, so the Auger technique provides compositional information from analyzed volumes as small as 10^{-17} cm^{-3}. Auger spectral analysis of a surface may also give compositional information as a function of depth if the method is combined with ion beam milling. Sequentially sputtering away the sample surface and measuring the Auger spectrum generates a depth profile of chemical composition.

The SIMS method also uses an ion beam to sputter away the surface, but in contrast to Auger spectral analysis, which interrogates the newly created surface, SIMS interrogates the material that is sputtered off. Some fraction of the sputtered material is ionized and can therefore be analyzed by a mass spectrometer, as Figure 7.10 shows. Since the SIMS method measures atomic mass, it is even possible to determine different isotopes of the same material. SIMS, therefore, peels the surface away layer by layer, and the ejected species provide direct information about the specimen's chemical composition as a function of depth. The method has proved a useful technique for direct measurement of ion implant or diffusion profiles. Figure 2.21 shows an example of a SIMS-measured dopant profile. The depth resolution of SIMS is about 80 Å, and lateral resolutions of 0.5 μm are achievable.

A method that far surpasses any of these techniques in terms of spatial resolution is transmission electron microscopy (TEM), in which an electron beam passes through a thin specimen and produces an image on a fluorescent screen. Resolutions on the order of 2 Å are achievable with this technique, which means that columns of atoms can be seen in images of thin crystalline samples. Because of the difficulty in preparing thin samples, TEM is not routinely used in semiconducting manufacturing for quality control. However, it is an important tool for fundamental studies, particularly of device and integrated circuit interfaces.

7.2.5 Film-Thickness Measurement

Another important process feature that needs to be documented and evaluated is the thickness of the various films used in the fabrication process. The thickness of transparent films can be determined by methods based on light-reflection phenomena at the top and bottom film surfaces. For example, oxide color, as observed under white-light illumination, is a useful and simple indicator of oxide-film thickness, and detailed tables of oxide color as a function of oxide thickness have been published. Color appears because destructive interference effects in the film cause an absence of certain frequencies in the reflected light. Oxide colors versus oxide thickness are tabulated in Table A.1 of Section A.5.4 in Appendix A. A more exact measure of oxide thickness is obtained by ellipsometry, which is based on measurement of the elliptical polarization that results when monochromatic, linearly polarized light is reflected from the oxide surface and the underlying silicon surface [4].

Opaque-film thickness measurements require a step in the film and can be accomplished by extremely sensitive stylus measurements in which a stylus tip coupled to an electromechanical transducer is scanned across the film step. This technique can measure film steps as small as 25 Å. Alternatively, optical interferometry can measure film thickness by measuring the shift in interference fringes produced by a step in a reflecting surface, as illustrated in Figure 7.11. Monochromatic light, when incident on the combination of an inclined, partially reflecting top reference surface and a bottom planar reflecting surface, produces interference fringes called Fizeau fringes, since the two reflecting surfaces are not parallel. If, furthermore, the bottom reflector has a film on it with a step of height t, then there will be a shift in the observed fringes by an amount x. It can be shown that the film step t is given by the expression

$$t = (x/L)(\lambda/2) \tag{7.13}$$

where λ is the wavelength of the reflected light and L is the fringe separation. For a sodium-vapor light source, the wavelength is 5892 Å. Practical restraints limit the best obtainable resolution to about 100 Å.

Figure 7.11 Interferometer measurement of film thickness.

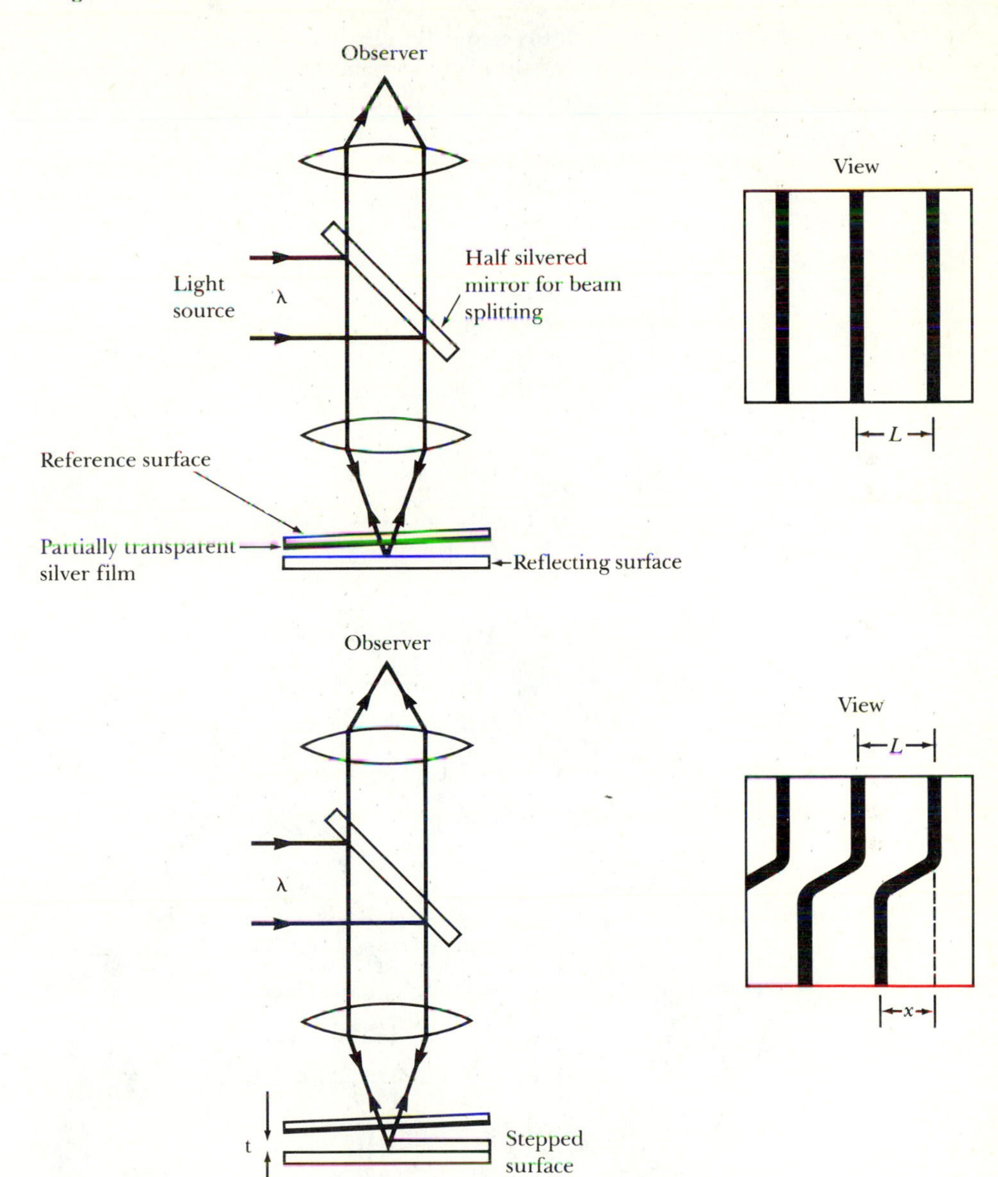

7.2.6 Test Patterns

Typically, a wafer contains some regions that serve as test areas for process evaluation so that process and device characteristics can be monitored from run to run. The test areas generally contain structures to evaluate specific devices such as capacitors, resistors, diodes, and transistors, as well as structures specifically designed to evaluate critical processing steps such as gate oxide integrity and interconnect continuity [9]. Test devices can also be included in the scribe channels, or kerf regions between the chips. As an example of how wafer area is employed for test purposes, Figure 7.12 shows the location of test areas on a 150-mm-diameter wafer with 256K random-access memory (RAM) chips.

Examples of process-related parameters that can be evaluated with on-wafer test structures include sheet resistance, contact resistance, oxide thick-

Figure 7.12 Example of wafer area allocation for test structures. In addition to the test areas noted, test devices are also included in the scribe channels. (The wafer is a 150-mm-diameter wafer with 256K RAM dice fabricated by Texas Instruments.)

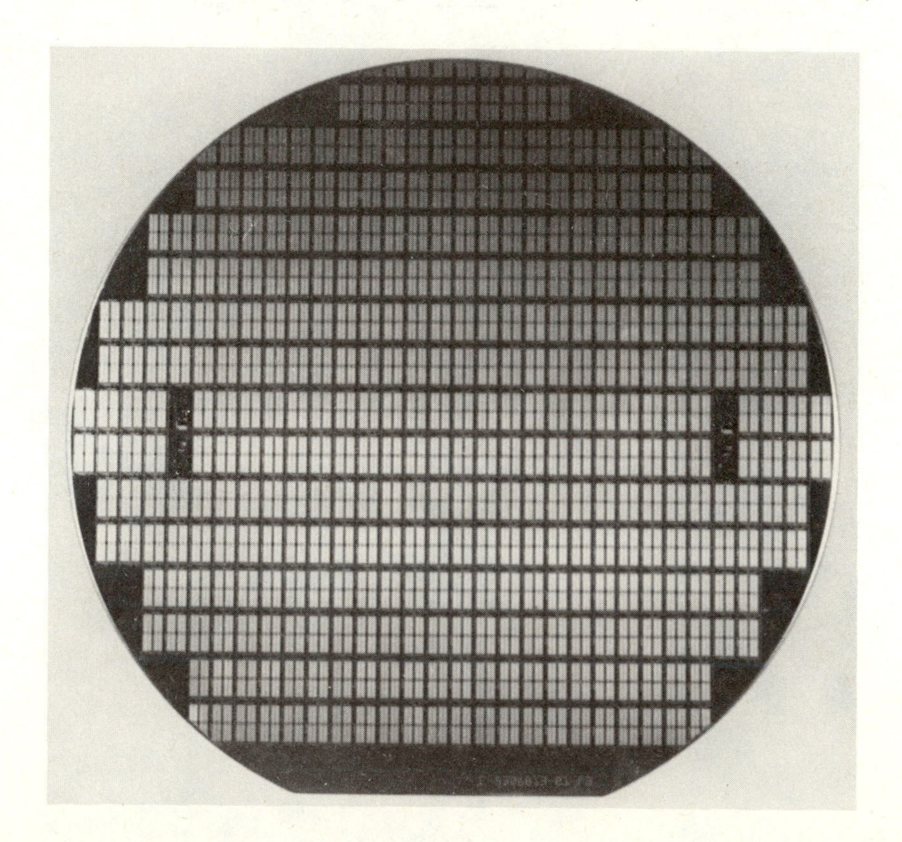

ness, flat band voltages, and junction breakdown voltages. Sheet resistance can be measured by forming resistors of well-defined length and width with four contacts to allow the accuracy of four-point measurements. To achieve high accuracy, these resistors should be straight to avoid the question of how to include edge squares in the number-of-squares calculation. Contact resistance can be monitored by including a contact chain with several hundred contacts in series. MOS capacitors with sufficiently large area to overcome parasitic effects provide capacitance voltage data from which flat band voltages and values of oxide interface charge can be determined. In addition, the oxide thickness can be determined because in accumulation, oxide thickness determines the capacitance. Interconnect layers can be checked for lead continuity and bridging by running two minimum-width interconnect paths in parallel, with minimum spacing, in a serpentine fashion. Continuity is checked by measuring end-to-end resistance for a given path, and bridging is checked by testing for an open circuit between the two paths.

Finally, test areas may include circuits or portions of circuits. For example, simple ring oscillator structures provide performance data on power and speed. In this case, buffer circuits need to be included on the test die to allow the actual testing of the oscillator. In addition, simple circuits may be included for verification of circuit simulation models.

7.3 Circuit Testing

Circuit testing of the finished die, although vital in integrated circuit manufacturing, is an area in which open questions remain. At one extreme, suppose the manufacturer did no testing. Every chip would be packaged and sold, and the customer would be given the cost, in dollars and time, of finding defective components. This arrangement would not likely lead to a good supplier-customer relationship. At the other extreme, suppose the manufacturer tried to exercise and check every conceivable circuit function for each chip. For even moderately complex logic circuits, this approach is not cost effective. Therefore, testing usually occupies a middle ground in which, for example, dc input/output characteristics, power consumption, and the response to a specific set of input vectors are determined to verify minimum performance specifications and check for certain classes of faults.

A variety of automatic test equipment (ATE) testing procedures have been developed. As one of many possible examples, consider the "walking-ones" test for RAMs. In this test, all cells are initially set to 0, then a 1 is written into the first cell. All cells are read and verified, and then the sequence is repeated for the next cell. This process continues until the memory is completely full of ones. Then the process is reversed, writing a 0 into a cell, reading and verifying all cells, and moving the 0 to the next cell until all cells again contain zeros. This test checks the decoding circuitry, the sense/refresh amplifiers, and the ability of each cell to receive and retain both logic states. To some extent it also checks the independence of cells by rechecking the entire contents of the memory after a write operation. It is possible to have a fault in which writing a 1 into one cell, for example, changes the contents of another cell.

The walking-ones test is time consuming, since it requires $2N^2$ tests where N is the size of the memory. If the ATE cycle time per test is 1 μs, for example, then the test time is approximately 34 s for a 4K RAM. However, as N approaches VLSI memory sizes, the test becomes unwieldy. For a 256K memory, the same 1-μs ATE cycle time would give a 38-hour test time. Obviously, such cases require more time-efficient testing, and in addition to higher-speed ATE, a variety of sophisticated RAM tests have been developed, including test patterns where the number of required tests goes as $N^{3/2}$ [10].

For complex logic circuits, fault checking is more difficult. Automatic test generation of input vectors may be used to check for stuck faults, in which a gate output is stuck at a 1 or a 0, by using a carefully chosen set of inputs. Of course, not all faults show up as stuck faults. On-chip diagnostic circuits help to alleviate the inherent problem associated with tens of thousands of logic circuits on a chip with only a relatively few access points (input/output pins). For analog circuits, the testing procedure depends on the nature and function of the circuit.

The testing process is facilitated if the testability of the circuit is considered in the original design. Nevertheless, the testing of VLSI-density integrated circuits is not a fully resolved issue. Rigorous fault testing may not always be cost effective, and in some cases manufacturers are accepting an increased risk of shipping defective products [11].

7.4 Yield Considerations

7.4.1 Controlling Contamination

Yields are directly related to the control of contamination. Sources of contamination in an integrated circuit fabrication facility include people, processing liquids, equipment, processing gases, and air, in approximately that order of importance [12]. Air is the least significant problem because it can be filtered as it is pumped in from outside the work area. A state-of-the-art high-efficiency particulate air (HEPA) filter can produce classs 1 air, which means that there is less than one particle, 0.2 μm or larger, per cubic foot of air. Other common air classifications for integrated circuit fabrication areas are class 100 and class 10, which correspond to less than 100 and less than 10 particles, respectively, of 0.5 μm or greater diameter. Because class 1 corresponds to a smaller particle size, class 1 is about 1000 times cleaner than class 100. The air supply to all critical fabrication areas is filtered to at least class 100, and access to these areas is carefully limited. By comparison, the particle count per cubic foot of air in a hospital operating room runs to 10,000 to 100,000, and ordinary room air has over a million particles, 0.5 μm or larger, per cubic foot. High vertical air velocity helps keep the clean room area swept clean, and at about 85 ft/min, the air flow becomes "laminar," so that a dust particle travels in a straight line from the top of the room to the bottom. A layout of the clean areas associated with the RCA Solid State Technology Center is representative of a process development facility and is shown in Figure 7.13 [13].

What happens to the air after it leaves the HEPA filter is up to the user. Sources of particles and contaminants always exist within the clean area, and the biggest source is definitely people. At rest, a person sheds roughly 100,000 particles per minute, including skin particles, lint, make-up, dried shaving cream, hair spray, and Flugge's droplets [14]. The latter moniker is essentially high-tech terminology for tiny saliva droplets, named for Carl Flugge, who did pioneering work on the subject in 1897 [15]. (Contamination of integrated circuits was not, obviously, a concern in the late nineteenth century, but the airborne spread of infectious diseases, particularly tuberculosis, by talking, coughing, and sneezing was.) The human particle-shedding rate increases with motion, and a sneeze can be a clean room disaster since over a million Flugge's droplets result.

If a contaminating particle lands on a wafer and sticks to a critical area prior to or during a processing step, failure of the entire chip can result for one of two reasons. First, the presence of a particle on the wafer may block the action of a processing step. For example, a dust particle may prevent etching of a critical area or cause a break in the deposition of an interconnect path. Second, and perhaps more importantly, the presence of a particle during a high-temperature processing step can cause electrochemical effects that alter the electrical characteristics of devices in the contaminated portion of the wafer. For example, contamination of the silicon-oxide interface may shift the threshold voltage of MOS field-effect transistors (MOSFETs) out of the range of useful circuit operation.

To reduce the human contribution to clean room contamination, stringent dress precautions require overclothing of tightly woven, monofilament nylon fabrics, caps of the same material to cover hair, and so on. However, an operator fully suited with a clean room outfit still contributes about 6000 particles per minute to an adjacent cubic foot of space [16]. Assuming that people are basically too dirty to clean up to class-10 or class-1 standards, the solution is to remove them from the working area entirely. This can be done by processing the wafers in laminar-flow benches or enclosed clean-air cabinets, with the people outside and the wafers and necessary equipment inside. Wafers can be transported from work station to work station in minimum volume, dustproof boxes or by other methods such as on an air stream within a clean-air tunnel. Glove box ports can allow necessary manipulation of

Figure 7.13. Layout example of an integrated circuit facility clean area.
SOURCE: From W. A. Bosenberg, "The SSTC Integrated Circuit Fabrication Facility," *Solid State Technology*, 25 (March 1982), 122–125. Reprinted with permission of *Solid State Technology*, published by Technical Publishing, a company of Dun & Bradstreet.

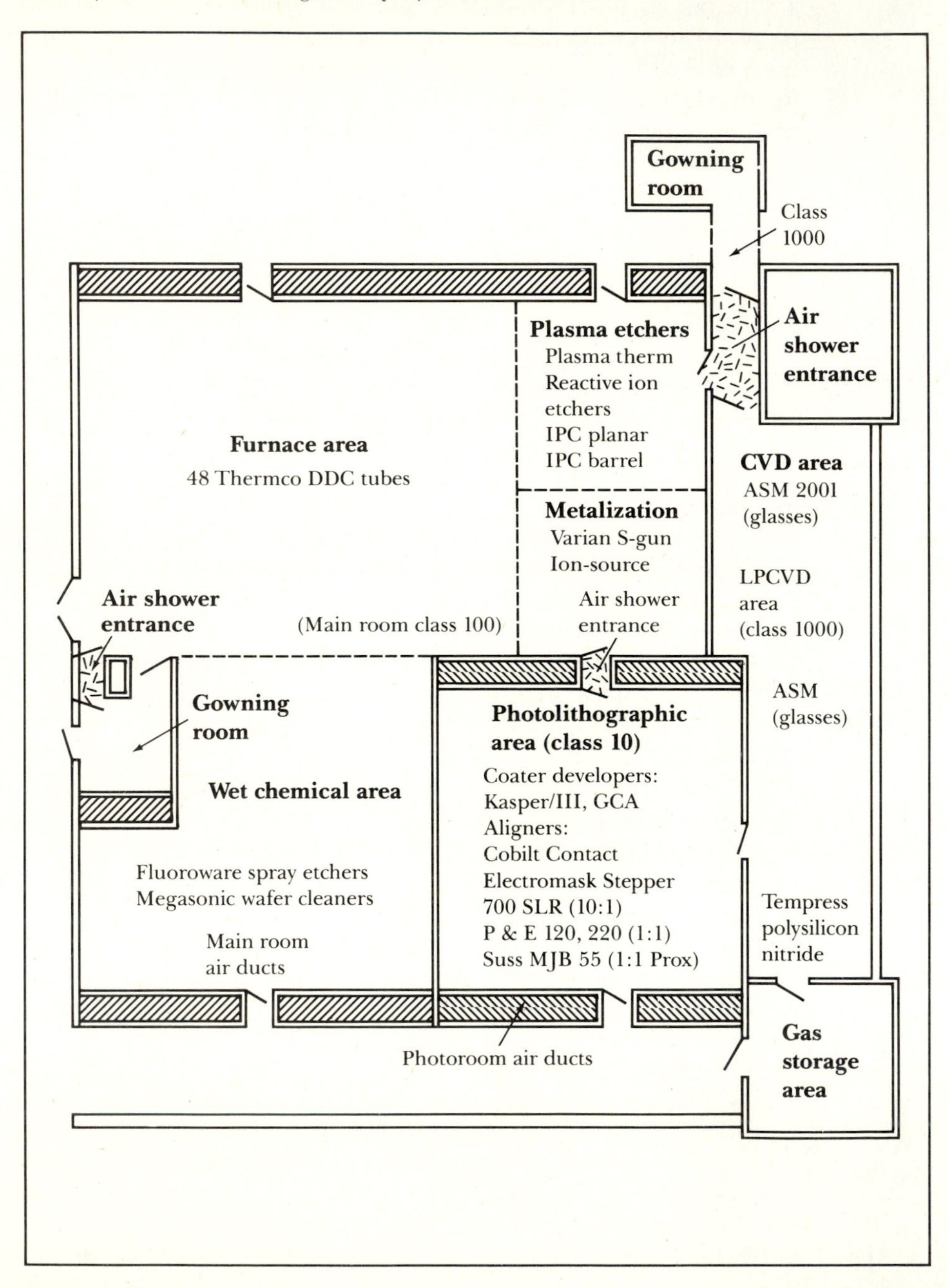

wafers and equipment; however, automated loading and handling equipment can minimize such intervention. This approach works well for dedicated production lines but is less suitable for research and development facilities where state-of-the-art processes are being developed and high flexibility is required. The clean-atmosphere requirements of VLSI provide a strong impetus for the development of process automation and for robotic engineering applications to wafer processing.

In addition to airborne contaminants, the liquids and gases that come into intimate contact with wafers are also a potential source of contamination. Semiconductor-grade chemicals typically have purity specifications expressed in parts per million (ppm), as do semiconductor-grade gases [17].

7.4.2 Yield Statistics

Several factors contribute to yield problems, but the sources fall into three broad classes [18]. The first class consists of that fraction of wafers that never reach the final test site because of handling breakage or out-of-order or missing processing steps. The discussion in this section mainly concerns the second class of yield problems, namely, the loss resulting from localized process defects as measured at the final wafer probe testing procedure prior to packaging. The third class of yield problems has to do with packaging: a chip that passed the wafer probe test, prior to die separation, may not work after packaging because of problems such as those associated with wire-bonding accuracy.

As Stapper, Armstrong, and Saji point out, the prepackaging wafer probe yield can be further divided into two subdivisions [18]. The gross, or parametric yield is associated with problems that affect entire wafers or entire portions of wafers, as opposed to random chips on the wafer. As a specific example of the parametric yield, MOSFET threshold voltages depend on process temperatures, timings, and levels of chemical contamination. Excessive variations in any of these factors may cause the value of V_T to stray beyond the allowed window for an acceptable product. Consequently, entire wafers may fail the test procedure, or in marginal cases, portions of the wafer may have parameters outside the acceptable limits. Figure 7.14 shows an example of parametric variation, specifically in n-channel MOS (NMOS) V_T values as measured on kerf test devices during a two-month manufacturing period.

The second class of yield problems is related to random defects caused by particulate contamination, as described in the previous section. For typical manufacturing environments, the parametric yield could account for 10% to 20% of the prepackaging test yield problems and random defects could cause the remaining 80% to 90%, although these figures depend a great deal on the chip area and the exact fabrication process.

There is considerable interest in developing predictive yield formulas.

Figure 7.14 Distribution of MOSFET threshold voltages during a two month manufacturing period. Devices outside the process window correspond to failing chips. SOURCE: From C. H. Stapper, F. M. Armstrong, and K. Saji, "Integrated Circuit Yield Statistics," *Proceedings of the IEEE,* 71 (1983), 453–470. © 1983 IEEE.

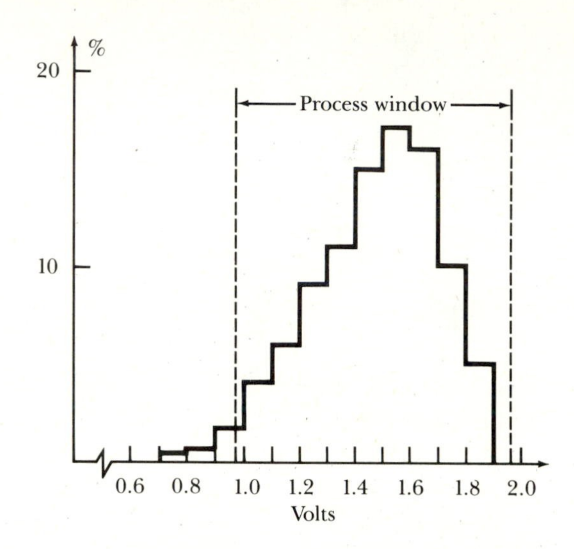

The statistical problem is not trivial, but it is instructive to consider some simple yield models to appreciate the relationship of yield to chip area. Consider a process characterized by a fatal defect density, $\mathcal{D}$. In fact, there are several types of defects, including insulator pin holes, scratch marks, dirt particles, photoresist defects, and so on, each with their own density, $\mathcal{D}_i$. However, for this discussion all are lumped into one combined defect density. If we assume that the random defects are independent—that is, that the probability of a fault defect occurring on a chip is independent of the number of defects already on that chip—then Poisson statistics apply and the yield, Y, is given by

$$Y = \exp(-A\mathcal{D}) \tag{7.14}$$

where $\mathcal{D}$ is the defect density per unit area and A is the active area of the chip.

In fact, however, some defects are correlated; the evidence is that defective chips on wafers tend to cluster. One reason is that a portion of the yield problem is parametric, and entire portions of wafers, or entire wafers are affected. Another reason is that contaminating particles can approach wafers from only one side in some cases, as for example when the wafers are stored in containers between processing steps. Therefore, once a chip has some dust particles on it, it is likely to get more, since it is probably preferentially exposed to contamination from, say, the open side of the container. A chip without any particles, is more probably protected from contamination and is less likely to pick up particles. Because of the correlation of defects, the Poisson yield expression in Eq. (7.14) is usually too pessimistic when used to predict the area dependence of yield. The failure of the simple Poisson model has been clearly demonstrated by first measuring the yield on a chip-by-chip basis in which the area, A, is the area of one chip, and then considering the yield from the point of superchips—in clusters of 2, 4, 6, and so on, adjacent chips—with a correspondingly larger value of A. If the value of $\mathcal{D}$ derived for the single-chip yield is used for the clustered chips, then the Poisson distribution underestimates the yield of clustered chips.

Since yield analysis is important for quality control of existing lines as well as for predicting the yields of products in development, a number of approaches have been developed to predict yields more accurately than the simple Poisson model can. Some of the resulting equations used by companies have a statistical basis, and others, being mainly curve-fitting equations, do not. One simple modification to the Poisson equation is to combine the concepts of parametric yield and random yield in the following equation:

$$Y = Y_0 \exp(-A\mathcal{D}) \tag{7.15}$$

where Y_0 is the parametric, or gross yield [19]. This approach, which has intuitive appeal, assumes that a parametric problem causes a fraction $(1 - Y_0)$ of the chips to be nonfunctional and a fraction Y_0 to be functional. Furthermore, it assumes that Y_0 is independent of the chip area. Random defects are assumed to be independent and therefore described by a Poisson distribution. Combining both sources of yield loss gives Eq. (7.15). Using data from actual wafer runs, Eq. (7.15) has been successfully used to extrapolate the yield of single chips so as to accurately predict the yield of bunched superchips [19].

EXAMPLE 7.2

Suppose a 64K RAM process is characterized by a parametric yield of $Y_0 = 0.85$ and a defect density of $\mathcal{D} = 2\ \text{cm}^{-2}$. What is the expected yield owing to the combined effect of parametric and random faults if the active chip area is $0.25\ \text{cm}^2$?

Next, suppose the chip is expanded to a 256K memory by brute force, using the same design rules and minimum dimensions, such that the area grows to $1\ \text{cm}^2$. Now what would be the yield? Repeat for the case of a 1-Mbit memory with an area of $4\ \text{cm}^2$.

Solution Applying Eq. (7.15) for the 64K memory yields

$$Y = 0.85\ \exp[-2(0.25)] = 0.516$$

for a yield percentage of 51.6%. For the 256K memory, the same equation yields

$$Y = 0.85 \exp(-2) = 0.115$$

for a yield percentage of 11.5%. Finally, for the 1-Mbit memory,

$$Y = 0.85 \exp[-2(4)] = 0.00029$$

which indicates a yield percentage of only 0.029%.

This example shows rather dramatically that improved lithography, and therefore smaller minimum dimensions, are key to the achievement of practical high-density VLSI circuits. Furthermore, die sizes can be reduced by new processes and layout schemes. For example, vertical capacitors and vertical transistors have been incorporated in memory cells to reduce the planar area per cell by fabricating both the capacitor and transistor of a RAM cell on the sidewalls of a silicon trench.

A more sophisticated yield analysis would take into account the different types of defect densities, $\mathcal{D}_i$, as well as that there is usually at least some correlation in the random defects. Also, every product shows a learning-curve enhancement in yield, as demonstrated by Figure 7.15, which shows the steady increase of yield for a 64K dynamic RAM over a four-year period. Such an increase can be expected because of experiential learning and improvements in equipment, processes, and clean rooms.

7.5 Assembly and Packaging

7.5.1 Die Separation

The scribe channels (also called scribe alleys, scribe streets, and kerf regions) that separate chips on the wafer run along natural cleavage planes in the silicon crystal. This configuration is accomplished by orienting the photolithographic patterns for the scribe lines perpendicular and parallel to a "flat" ground onto a wafer and aligned with a specific crystal plane of the wafer, as Figure 7.12 shows. The individual dice are separated by through-cutting or by a scribe-and-break technique. Scribing may be accomplished by a diamond-tipped stylus, a laser beam, or a diamond-coated saw blade. After the scribing operation, the wafer is placed, scribed side down, on a slightly flexible support, and a roller with applied pressure passes over it. This causes the wafer to break along the scribe lines parallel to the roller. Then the wafer is rotated 90 degrees, and the operation is repeated. However, the best yields

Figure 7.15 Actual and planned yields of a 64K dynamic RAM chip.
SOURCE: From C. H. Stapper, F. M. Armstrong, and K. Saji, "Integrated Circuit Yield Statistics," *Proceedings of the IEEE*, 71 (1983), 453–470. © 1983 IEEE.

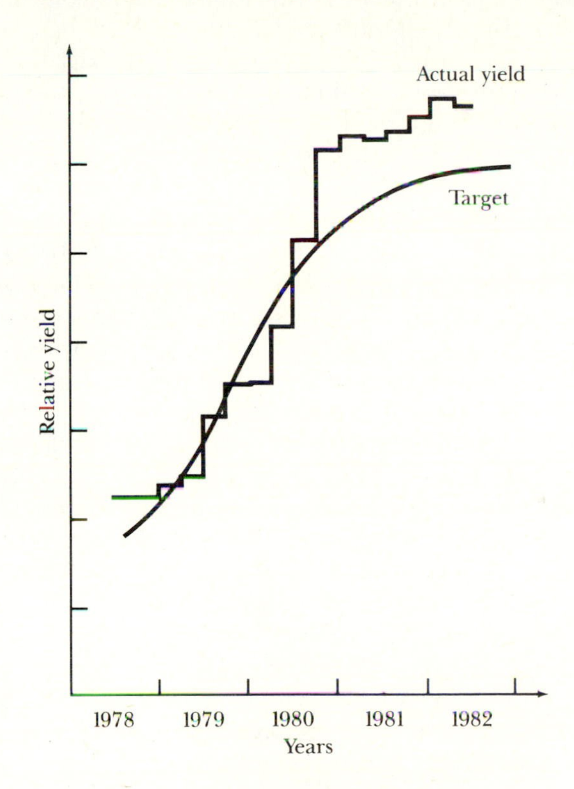

for die separation are obtained by using a diamond-coated saw blade to cut through the wafer, thereby eliminating the breaking step.

At this point the dice are separated from one another. Often the wafer is bound to an adhesive plastic sheet prior to the separation operation, so that the dice of a given wafer are still attached to a common plastic carrier. A vacuum chuck may be used to pick off the dice that have passed the pre-packaging test. Those that have not passed the test are marked with an ink dot. In an automated line, the die pickup is performed automatically with a tool that is in communication with the testing station and knows the location of the good chips.

7.5.2 Bonding and Packaging

It has been noted that "packaging begins where the chip ends," that is, at the metallic bonding pads on the surface of the chip [20]. From a systems designer's point of view, there are three levels of packaging. The first level is the packaging around a single chip, the second is the mounting of several chips on a card (circuit board) or, in some cases, on a laminate structure with multiple layers of wiring, and the third level is the interconnection of the cards into a total system. Most of this section concerns the first level, namely, the packaging of the individual chip.

One of the requirements of the chip package is that it must allow an electrical path between the connecting pins on the integrated circuit package and the input/output pads on the chip. Doing so involves a bonding procedure between the chip bonding pads and the surrounding package. There are essentially variations on two main approaches: the wire bonding technique and the flip-chip technique.

The wire bonding method uses thin gold or aluminum wire to connect the chip bonding pads to the package leads. First, using either an epoxy or a gold-silicon eutectic alloy, the chip is attached to a substrate that is part of the integrated circuit package. Next, a wire approximately 20 to 40 μm in diameter is welded to a chip bonding pad using either thermal-compression bonding or ultrasonic bonding.

Thermal-compression bonding welds a gold wire to the chip bonding pad and the package bonding post using a combination of heat and pressure, following the sequence shown in Figure 7.16. Prior to the weld to the chip bonding pad, a ball is formed on the end of the gold wire via a brief exposure to a hydrogen flame kindled by a short release of hydrogen gas and a simultaneous electric-spark ignition. The chip-substrate combination is heated to around 220° to 250°C, and the capillary tip is often heated as well. As the capillary is lowered onto the chip bonding pad, the combination of heat and pressure deforms the ball into a "nail-head" bond and welds the gold wire to the pad. Next, the capillary tip is raised to form a loop in the gold wire and the tip is brought down again on a gold-plated lead post on the bonding package. This bond is referred to as a stretch bond. Finally, the tip is raised again, and the wire is severed by being clamped in the feed mechanism so that the upward pull breaks the wire at a weak point after the stretch bond.

Ultrasonic bonding forms the weld with a burst of ultrasonic energy. In this method it is not necessary to heat the chip and aluminum wire is usually used instead of gold. Ultrasonic bonding is a viable alternative to thermal-compression bonding, although the bonds are somewhat weaker. In either case, wire bonding has developed into a highly automated process in which automatic feed mechanisms provide the die to the bonder, and pattern-recognition mechanisms automatically align the die for the bonding operations. Automatic bonders achieve rates of over 10,000 bonds per hour. A modern wire bonding operation on standard chips may involve several machines busily bonding away while one operator monitors the operation via video-displayed microscope views of the bonding procedure. Figure 7.17 shows a 3000-gate Honeywell chip with 144 wire bonds [21].

The flip-chip approach eliminates the wire bond between the chip bonding sites and the package leads; instead, the chip is placed face down such

Figure 7.16 The thermal-compression wire bonding sequence. A. Wire and capillary tip. B. Ball formation. C. Thermal-compression nail-head bond. D. Loop formation. E. Stretched bond on package substrate. F. Wire sever.

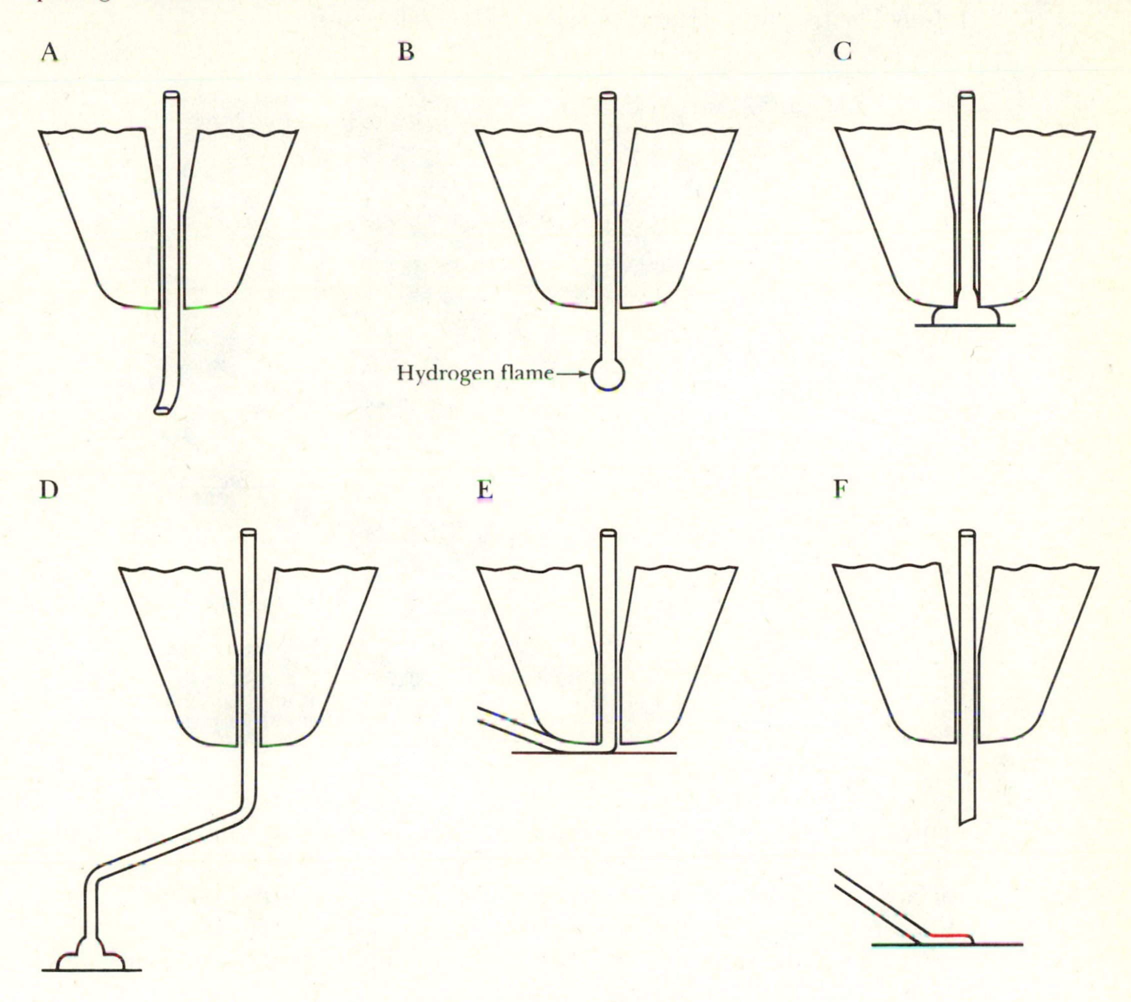

that the chip bonding sites line up with corresponding package lead sites. IBM has used this approach extensively. In the IBM version of this technology, an array of solder bumps cover the surface of the chip and make input/output connections to the underlying circuitry, as shown in Figure 7.18 [22]. This configuration contrasts the wire bonding format, in which input/output bonding pads are arranged on the periphery of the chip. Clearly the array approach lends itself to a larger number of input/output connections per chip, and flip chips with 354 solder-bump connections have been reported [23]. In the bonding operation, the chip is flipped and mated with

Figure 7.17 A 3000-Gate Honeywell chip with 144 wire bonds.
SOURCE: From A. H. Mones and R. K. Spielberger, "Interconnecting and Packaging VLSI Chips," *Solid State Technology*, 26 (Jan. 1984), 119–122. Reprinted by permission of *Solid State Technology*, published by Technical Publishing, a company of Dun & Bradstreet.

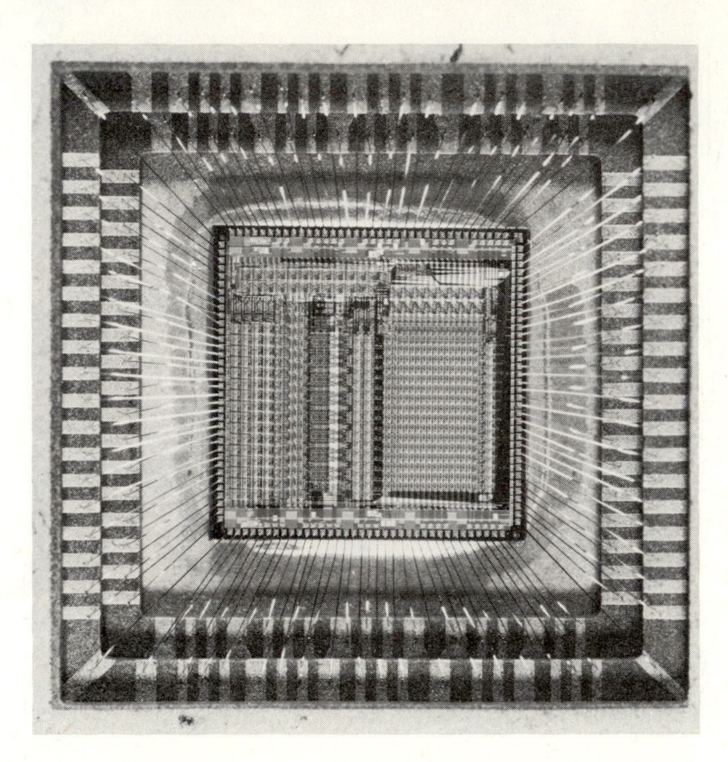

matching solder bumps on a laminated ceramic substrate having multiple metal paths to the package leads. During the bonding operation, the mating surfaces are heated so that the solder bumps on both surfaces are molten. When the chip pads and their corresponding substrate "footprints" touch, metal surface tensions provide a certain amount of self-alignment such that misaligned chips are pulled into place.

In addition to electrical connections, a package must also provide encapsulation, protecting the chip from the environment and allowing handling of the integrated circuit. Individual chip packages are usually either plastic or ceramic. A ceramic package offers the ability to hermetically seal the chip in a dry, inert atmosphere, which is an advantage in hostile environments such as those encountered in automotive, space, and military applications. General purpose applications often use the injected-plastic dual-in-line package (DIP), a cut-away view of which is shown in Figure 7.19.

The DIP package is effective when there are relatively few leads. Since leads are only located on two sides of the package, however, the size of the

Figure 7.18 Solder-ball connections for flip-chip bonding. A. Solder-ball connection to chip via triple metal layer. B. VLSI chip with 184 solder-ball terminals.
SOURCES: Part A from L. J. Fried, J. Havas, J. S. Lechaton, J. S. Logan, G. Paal, P. A. Totta, "A VLSI Bipolar Metalization Design with Three-Level Wiring and Area Array Solder Connections," *IBM Journal of Research and Development,* 26 (1982), 362–371. Copyright © 1982 by International Business Machines Corporation; reprinted with permission; part B from A. H. Mones and R. K. Spielberger, "Interconnecting and Packaging VLSI Chips," *Solid State Technology,* 26 (Jan. 1984), 119–122. Reprinted with permission of *Solid State Technology,* published by Technical Publishing, a company of Dun & Bradstreet.

A

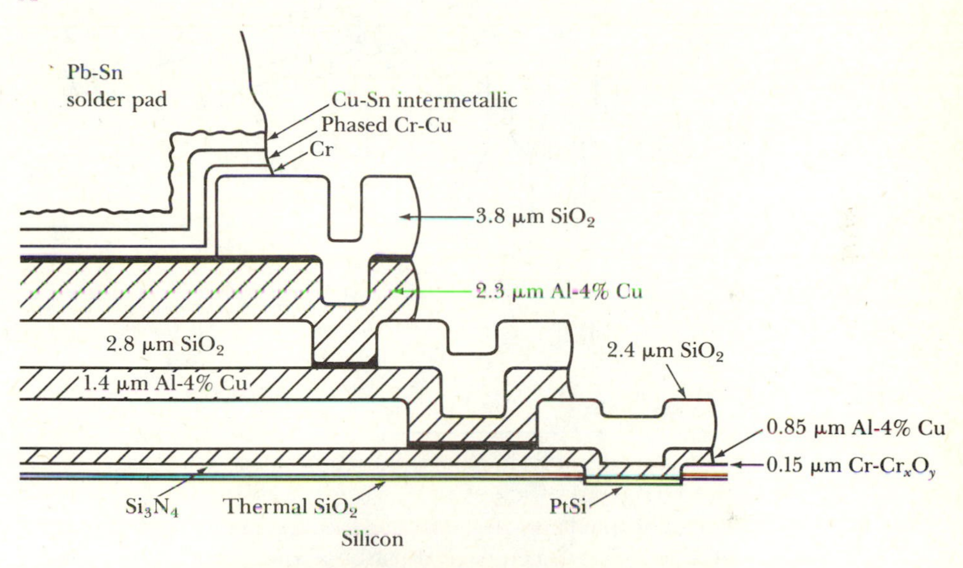

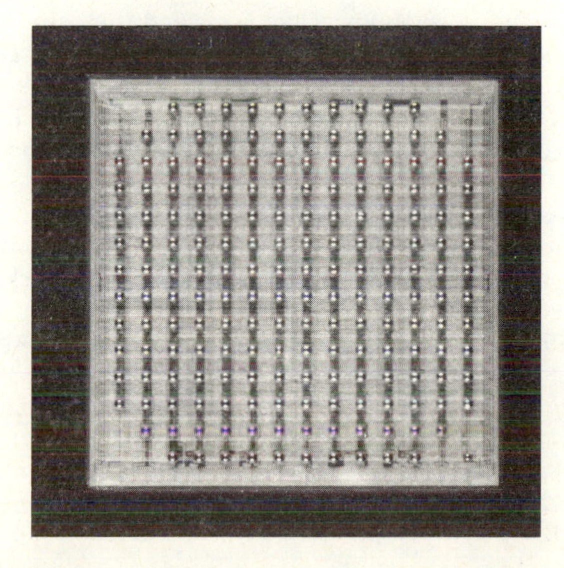

Figure 7.19 Cut-away view of the DIP package.
SOURCE: C. Mitchell and H. M. Berg, "Thermal Studies of a Plastic Dual-in-Line Package," *IEEE Transactions on Components, Hybrids, and Manufacturing Technology,* CHM-2 (1979), 501. © 1979 IEEE.

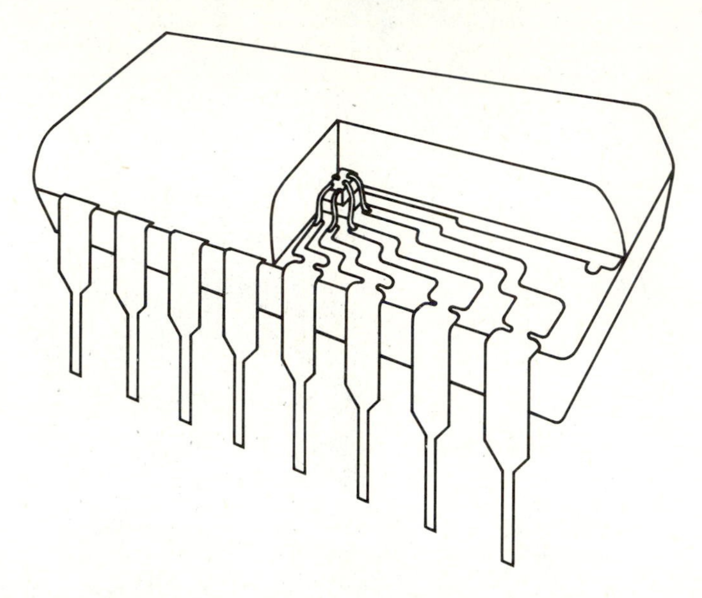

DIP grows considerably as the number of leads increases. As Figure 7.20 shows, the DIP is an efficient package for a 16-pin circuit, but not for a 64-pin circuit. The package not only gets longer as the pin count goes up but must also get wider since space must be allowed inside the package for connections between the package leads and the chip. Figure 7.20 compares the footprints of a variety of chip carriers, including DIPs, square chip carriers with contacts on all four sides, and chip carriers with an array of contact pins. Obviously, the array is most space efficient. In fact, 64 pins in an 8 by 8 array with 0.05-in. spacings takes up less area on a board than a standard 16-pin DIP.

The second level of packaging concerns the interconnection of single chips on a circuit card or board, most typically, on a printed circuit board. As mentioned in Chapter 5, an air-cooled DIP package can dissipate about 2 W maximum when board mounted. In some high-performance designs, this limitation is unacceptable. For example, the LSI chips on the Control Data Cyber 205 supercomputer dissipate about 5 W each, on average [24]. Furthermore, each printed circuit board holds 150 arrays, so that the power dissipated per board is 750 W. To maintain junction temperatures at 55°C, it was necessary to provide liquid cooling to the chips. The solution in this case was to run a liquid-freon coolant tube under each ceramic chip carrier.

IBM has developed a "thermal conduction module" (TCM) for its high-

Figure 7.20 A comparison of chip carrier footprints with 0.1 inch and 0.05 inch pin spacing.
SOURCE: From A. J. Blodgett, Jr., "Microelectronic Packaging," *Scientific American*, 249 (July 1983), 86–96. © 1983 Scientific American, Inc.

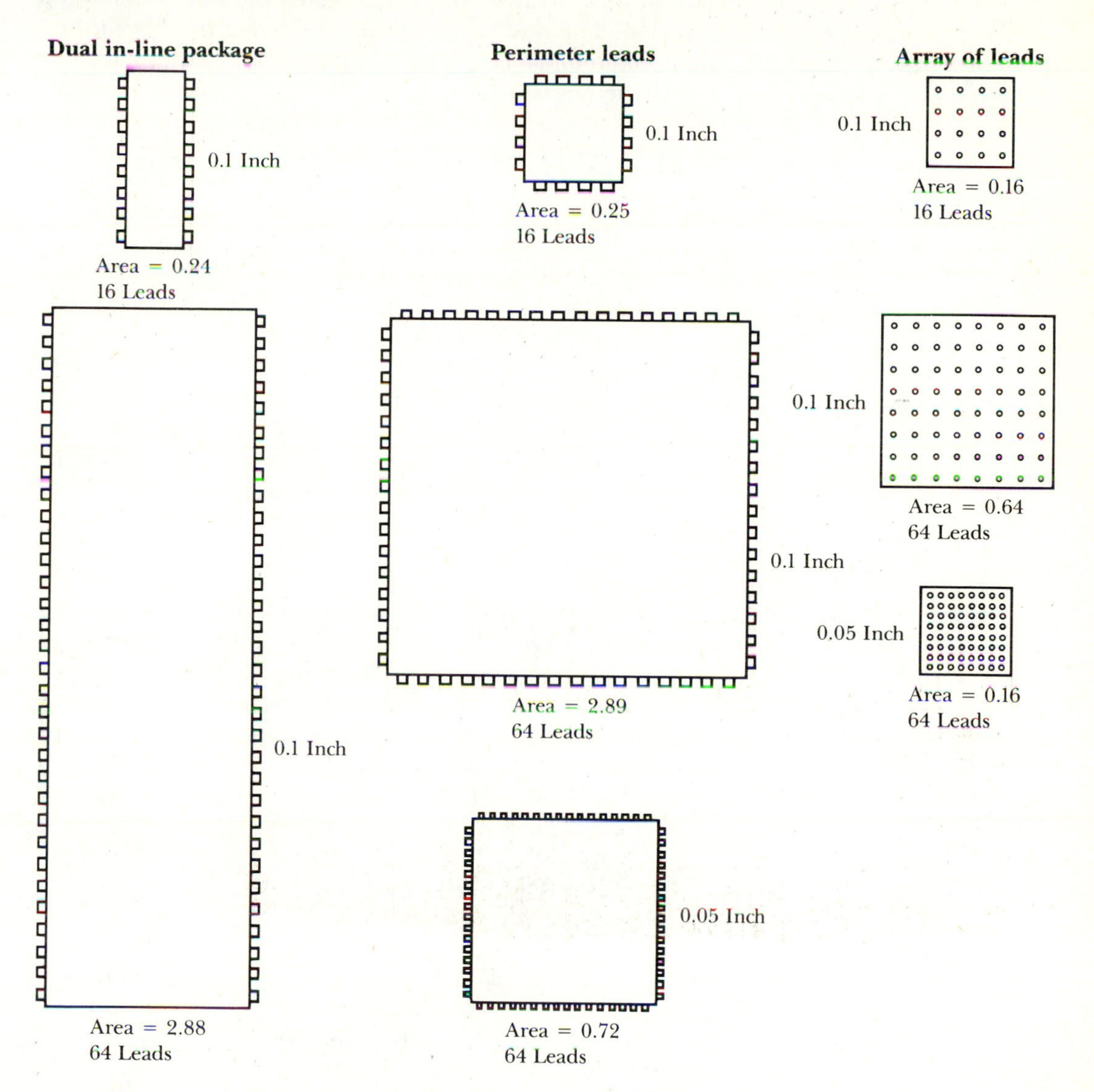

speed computing systems, to some extent removing the boundary, referred to earlier in this section, between the first and second level of packaging. In this technique, a close-packed array of flip chips is mounted on a common ceramic laminate substrate. The multilayer ceramic substrate provides up to 33 layers of conductors, interconnected by more than 350,000 vias that allow interconnections among the chips. A typical substrate has a 10 by 10 array of chips and 130 m of connecting wires in the multilaminate structure. To conserve space, both the contact pads, or solder bumps on the chips and the pins on the TCM are organized as two-dimensional arrays. The total number of chip contact pads is over 10,000, and 1800 pins on the bottom of the substrate carry both power to the chip array and input/output signals.

Since a typical chip in the array dissipates about 3 W, the TCM's total power dissipation is approximately 300 W. Again, special cooling is required, but this situation is different from that of standard packages because the silicon substrate of a flip chip is not in contact with the package substrate; the solder bumps are in contact with the substrate, but the back of the chip is pointing away. The TCM turns this potential difficulty into an advantage: the back of the chip is directly contacted by a spring-loaded piston that transfers heat from the chip to a water-cooled plate. Figure 7.21 shows details of this arrangement. Spring-loaded aluminum pistons mounted in the hat assembly act as thermal conduction paths to the water-cooled cold plate, so that at 4 W of chip power the device temperature rises to 44°C above the water temperature. Figure 7.22 shows a view of the entire TCM.

Packaging at all levels is a crucial part of VLSI technology. The high-performance and high-density advantages of a state-of-the-art chip are of little use to a systems designer if heat cannot be dissipated effectively or if signal-time delays from one chip to another are prohibitive. A variety of development efforts at several companies have given the systems designer a wide choice of packaging approaches.

7.6 Economic Considerations

The production cost of a chip depends on many variables, including yield, test and packaging costs, wages, material costs, and depreciation of processing equipment. This section illustrates a method for estimating the approximate production cost of a chip at three points in the fabrication, testing, and packaging sequence.

Consider, first, the cost per die after the wafer probe testing step, prior to wafer dicing and separation of the individual dice. Letting $P =$ the cost per die at this point,

$$P = \frac{\text{wafer cost} + \text{wafer probe cost}}{\#\ \text{dice/wafer} \times \text{yield}} \tag{7.16}$$

Figure 7.21 Heat sinking of a flip chip in the IBM thermal conduction module.
SOURCE: From A. J. Blodgett, Jr., "Microelectronic Packaging," *Scientific American*, 249 (July 1983), 86–96. © 1983 Scientific American, Inc.

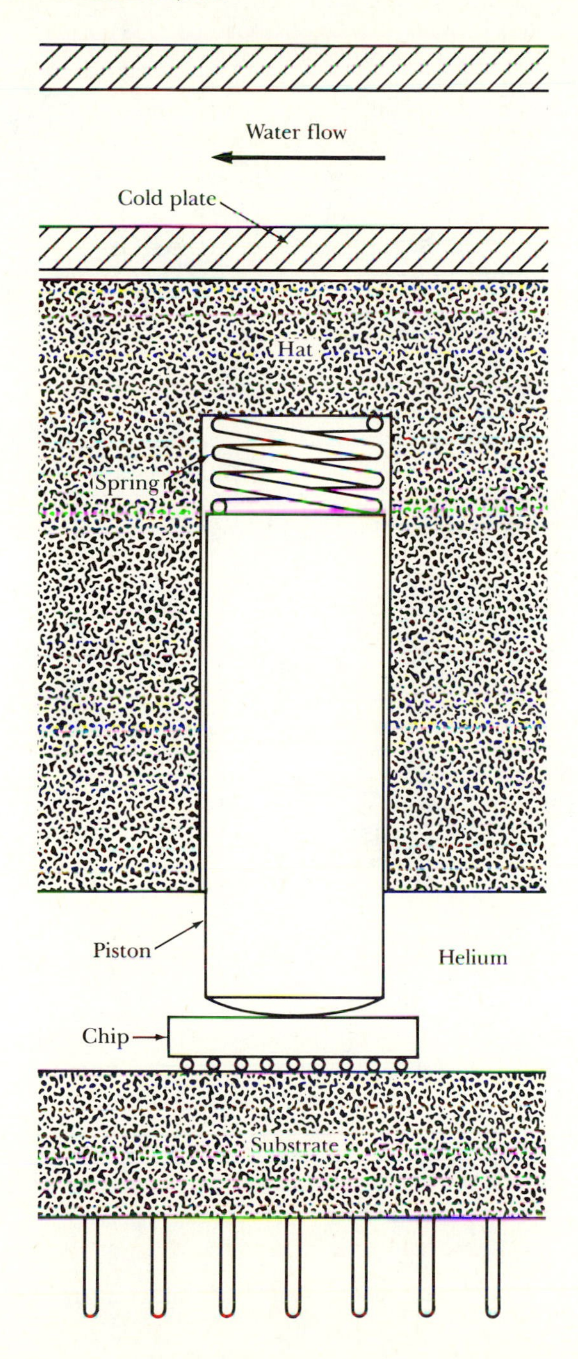

Figure 7.22 Exploded view of the IBM thermal conduction module.
SOURCE: From A. J. Blodgett, Jr., "Microelectronic Packaging," *Scientific American,* 249 (July 1983), 86–96. © 1983 Scientific American, Inc.

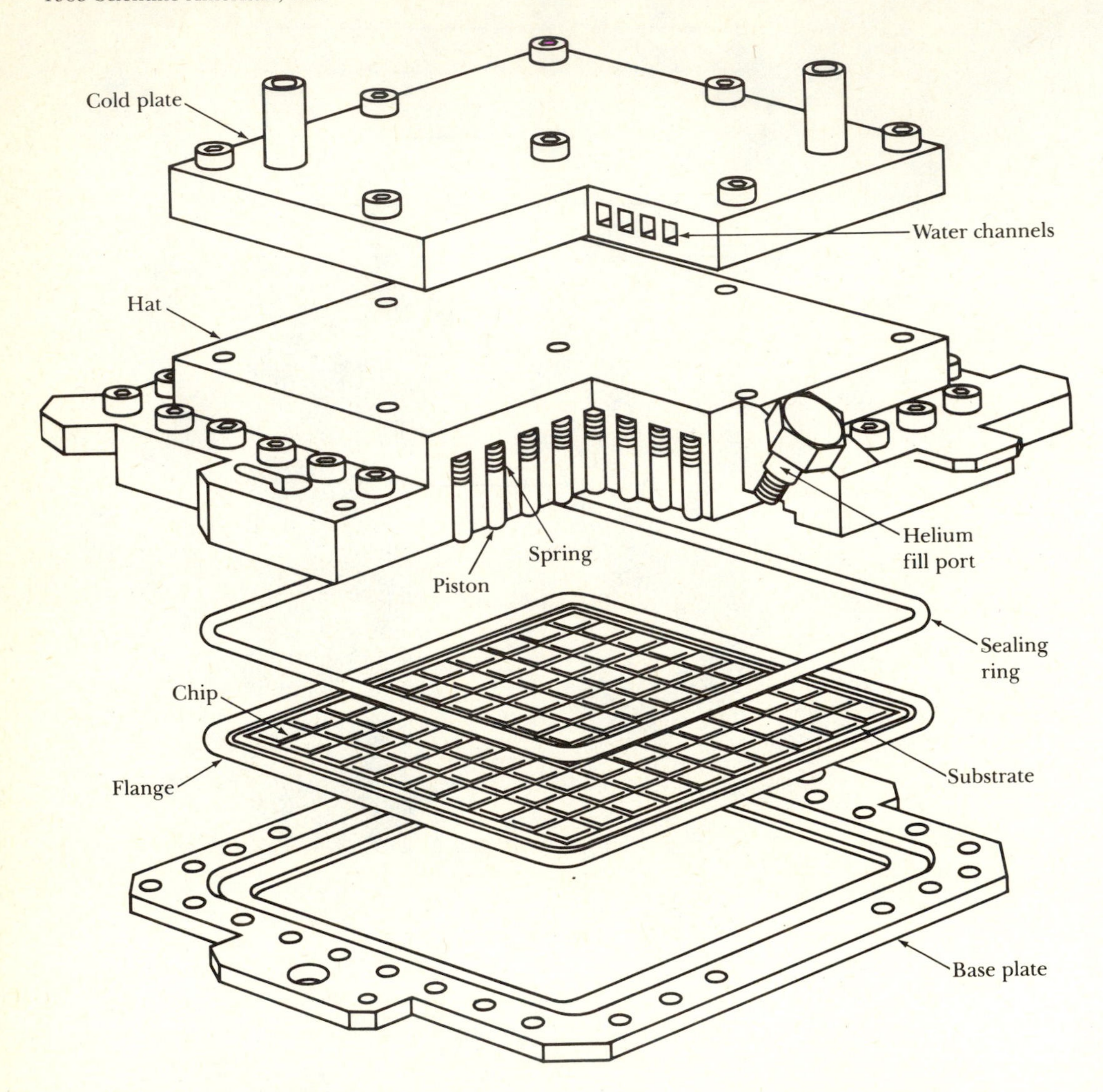

where the wafer cost represents all the costs associated with wafer processing steps up to the point of wafer probing, including the loss of wafers from breakage and other line yield problems. After the wafer probe, the next step is to scribe the wafer and separate the good dice from the bad. Letting $S =$ the cost per die after the scribe and separation step,

$$S = \frac{P + \text{scribe and separation cost per die}}{\text{scribe and separation yield}} \tag{7.17}$$

Finally, the chip cost after packaging and final testing may be expressed as follows:

$$F = \frac{S + \text{assembly cost} + \text{packaging cost} + \text{final test cost}}{\text{final test yield}} \tag{7.18}$$

where the assembly cost includes such steps as connecting the good dice to the package substrate and wire bonding.

It is interesting to put some numbers into these cost equations, but such an exercise must be qualified by noting that the value of the various costs and yields will vary considerably from product to product and, in some cases, from facility to facility. First, consider the cost per die after wafer probing. The initial cost of a blank silicon wafer depends on the wafer size and on the presence or absence of an epitaxial layer. However, as mentioned, the cost associated with each wafer at the point of probe testing must also include several factors such as labor, materials, depreciation, and other overhead, including utilities. A 1984 study of cost per wafer based on a factory model with five thousand 125 mm wafer starts per week produced the following breakdown [25].

Labor costs associated with individuals involved with the direct processing of wafers totaled $605,000 per month. This figure is based on 280 direct operators with a pay scale of $12/hour at 180 hours/month. In addition, there are salaries associated with indirect operators such as engineers and managers. Assuming 70 indirect operators at $3000 per month, the monthly cost is $210,000. Combining this figure with the cost of direct operators, the total labor and benefits cost is $815,000 per month.

Assuming a $36 million equipment investment, the monthly depreciation costs associated with equipment is $750,000 per month. A $15 million plant investment contributes a depreciation cost of $125,000 per month, for a total monthly depreciation cost of $875,000. Utilities and other operating overhead are figured at $290,000 per month, and the monthly materials cost for this model was taken to be $1,540,000.

Therefore the total monthly cost, as tabulated in Table 7.2, is $3,520,000. Consequently, the cost per wafer is about $160. Assuming that the line yield

Table 7.2 Wafer cost breakdown for 5000 wafer starts per week

	Cost ($1000/month)	%	Yielded cost per wafer ($)
Labor, benefits	815	23	46
Materials	1,540	44	88
Depreciation	875	25	50
Utilities and other overhead	290	8	16
Total	3,520	100	200

Source: From J. G. Harper and L. B. Bailey, "Flexible Material Handling Automation in Wafer Fabrication," *Solid State Technology*, 27 (July 1984), 89–98. Reprinted with permission of *Solid State Technology*, published by Technical Publishing, a company of Dun & Bradstreet.

rate is about 80%—owing to handling breakage, incorrect processing, and so on—the cost per wafer at the probe test station is about \$200.

Now suppose that the product in question is the 64K RAM from Ex. 7.2 with a defect density of $\mathcal{D} = 2\ \text{cm}^{-2}$ and a parametric yield of 85%. The expected wafer probe test yield is therefore 51.6%. Since the chip area is $0.25\ \text{cm}^2$, the number of dice per 125-mm diameter wafer is about 430 and the expected number of good dice per wafer is approximately 222. The wafer probe cost depends on the time it takes to test the wafer and on the depreciation cost of test equipment. This example take a value of \$20. Using this information, the cost per good die after the wafer test probe is, from Eq. (7.16),

$$P = \frac{\$200 + \$20}{0.516(430)} \tag{7.19}$$

or $P = \$0.99$ per die.

The cost associated with die separation and inspection is strongly dependent on the degree of automation associated with this step. Assuming that the scribe and separation cost per die is \$0.10 and that the scribe and separation yield is 95%, the cost per die after the scribe and separation steps, from Eq. (7.17), is

$$S = \frac{\$0.99 + \$0.10}{0.95} \tag{7.20}$$

which is \$1.15 per die.

Assembly and packaging costs vary over a wide range, depending on the type of package. A simple plastic DIP package may contribute only about \$0.15 to the cost, whereas a more sophisticated package can contribute several dollars. This example assumes a 16-pin DIP package, as well as highly automated assembly and final test procedures, such that the combined assembly, packaging, and final test cost is estimated to be \$0.40 per die. Then, if the final postpackaging test yield is 95%, the final cost is calculated from Eq. (7.18) to be

$$F = \frac{\$1.15 + \$0.40}{0.95} \tag{7.21}$$

or \$1.63 for the final packaged and tested product.

In this exercise, the cost per good die, P, was an appreciable portion of the final cost. However, if a chip requires complex and lengthy testing or nonstandard packaging, then the cost of packaging and testing can exceed the cost of the die. This can also be the case if the chip is small and the yield high, such that P is very small.

EXAMPLE 7.3

Reconsider the preceding cost analysis for the 64K RAM, assuming that two improvements have been made. Specifically, assume that an improved layout and tighter design rules have reduced the chip area by 20%, from 0.25 cm^2 to 0.20 cm^2. Also assume that improved processing conditions have reduced the defect density, $\mathcal{D}$, from 2 cm^{-2} to 1 cm^{-2} and increased the parametric yield from 85% to 95%. What would be the new cost per good die after the wafer probe test, assuming all other factors remain the same?

Solution Because of the changes in A, Y_0, and $\mathcal{D}$, a new yield must be calculated. The new yield at the wafer probe test is calculated from Eq. (7.15) as

$$Y = 0.95 \exp[-1(0.20)] = 0.778$$

or 77.8%. The new cost per die after the wafer probe test is

$$P = \frac{200 + 20}{0.778(430)} = \$0.66.$$

Therefore, the cost per good die has been reduced by one-third.

Example 7.3 shows that yield is critical in maintaining a successful wafer fabrication facility. However, yield is a function of many variables. It hinges on the initial circuit design and layout, particularly in that it depends greatly on chip area. Also, a good parametric yield requires tight control of the various wafer processing steps. And obviously, particulate contamination is a strong variable that relates directly to how the fabrication facility is designed and operated.

BIBLIOGRAPHY

For further reading on integrated circuit yield analysis, see the following:

C. H. Stapper, F. M. Armstrong, and K. Saji. "Integrated Circuit Yield Statistics." *Proceedings of the IEEE,* 71 (1983), 453–470.

For further information on integrated circuit testing and diagnostics, see the following sources:

Graf, M. C. "Testing—A Major Concern for VLSI." *Solid State Technology,* 27 (Jan. 1984), 101–107.

Healy, J. T. *Automatic Testing and Evaluation of Digital Integrated Circuits.* Reston, Va.: Reston Publishing, 1981.

Marcus, R. B. "Diagnostic Techniques." In *VLSI Technology.* Ed. S. M. Sze. New York: McGraw-Hill, 1983, Chap. 12.

For further reading on integrated circuit assembly and packaging techniques, see the following sources:

Blodgett, A. J. "Microelectronic Packaging." *Scientific American,* 249 (July 1983), 86–96.

Steidel, C. A. "Assembly Techniques and Packaging." In *VLSI Technology.* Ed. S. M. Sze. New York: McGraw-Hill, 1983, Chap. 13.

REFERENCES

1. R. A. Colclaser, *Microwave Electronics Processing and Device Design* (New York: Wiley, 1980).
2. J. C. Irvin, "Resistivity of Bulk Silicon and of Diffused Layers in Silicon," *Bell System Technical Journal,* 41 (1962), 387–410.
3. F. M. Smits, "Measurement of Sheet Resistivities with the Four-Point Probe," *Bell System Technical Journal,* 37 (1958), 711–718.
4. A. B. Glaser and G. E. Subak-Sharpe, *Integrated Circuit Engineering* (Reading, Mass.: Addison-Wesley, 1979).
5. P. S. Schaffer and T. R. Lally, "Silicon Epitaxial Wafer Profiling Using the Mercury-Silicon Schottky Diode Differential Capacitance Method," *Solid State Technology,* 26 (April 1983), 229–233.
6. R. B. Marcus, "Diagnostic Techniques," in *VLSI Technology,* ed. S. M. Sze (New York: McGraw-Hill, 1983), Chap. 12.
7. J. A. Buono, A. W. Wisniewski, and W. S. Andrus, "Surface Science Analysis Techniques," *Solid State Technology,* 25 (Feb. 1982), 95–101.
8. J. R. Black, "Electromigration Failure Modes in Aluminum Metallization for Semiconductor Devices," *Proceedings of the IEEE,* 57 (1969), 1587.
9. J. Toker, "Test Bar Design Using the Cellular Approach," *Texas Instruments Engineering Journal,* 1 (1984), 15–26.
10. J. T. Healy, *Automatic Testing and Evaluation of Digital Integrated Circuits* (Reston, Va.: Reston Publishing, 1981).

11. M. C. Graf, "Testing—A Major Concern for VLSI," *Solid State Technology*, 27 (Jan. 1984), 101–107.
12. J. Berger, "Sources of Contamination in VLSI Processing: A User's Point of View," *Microcontamination*, 3 (1985), 16–18.
13. W. A. Bosenberg, "The SSTC Integrated Circuit Fabrication Facility," *Solid State Technology*, 25 (March 1982), 122–125.
14. E. Larson, "Why You Shouldn't Sing 'La Marseillaise' in Computer Factory," *Wall Street Journal*, Dec. 28, 1984, p. 1.
15. C. Flugge, "Über Luftinfection," *Zeitschrift für Hygiene ünd Infektionskrankheiten*, 25 (1897), 179–224.
16. M. Parikh and U. Kaempf, "SMIF: A Technology for Wafer Cassette Transfer in VLSI Manufacturing," *Solid State Technology*, (July 1984), 111–115.
17. E. M. Juleff, W. J. McLeod, E. A. Hulse, and S. Fawcett, "Advances in Contamination Control of Processing Chemicals in VLSI," *Solid State Technology*, 25 (Sept. 1982), 82–86.
18. C. H. Stapper, F. M. Armstrong, and K. Saji, "Integrated Circuit Yield Statistics," *Proceedings of the IEEE*, 71 (1983), 453–470.
19. C. L. Mallory, D. S. Perloff, T. F. Hasan, R. M. Stanley, "Spatial Yield Analysis in Integrated Circuit Manufacturing," *Solid State Technology*, 26 (Nov. 1983), 121–127.
20. A. J. Blodgett, Jr., "Microelectronic Packaging," *Scientific American*, 249 (July 1983), 86–96.
21. A. H. Mones and R. K. Spielberger, "Interconnecting and Packaging VLSI Chips," *Solid State Technology*, 26 (Jan. 1984), 119–122.
22. L. J. Fried, J. Havas, J. S. Lechaton, J. S. Logan, G. Paal, P. A. Totta, "A VLSI Bipolar Metalization Design with Three-Level Wiring and Area Array Solder Connections," *IBM Journal of Research and Development*, 26 (1982), 362–371.
23. A. H. Dansky, "Bipolar Circuit Design for a 5000 Circuit VLSI Gate Array," *IBM Journal of Research and Development*, 25 (1981), 116.
24. N. R. Lincoln, "Technology and Design Tradeoffs in the Creation of a Modern Supercomputer," *IEEE Transactions on Computers*, SC-13 (1984), 349–362.
25. J. G. Harper and L. B. Bailey, "Flexible Material Handling Automation in Wafer Fabrication," *Solid State Technology*, 27 (July 1984), 89–98.

PROBLEMS

1. A boron drive-in diffusion is carried out in an n-type epitaxial layer and produces a Gaussian distribution of acceptor atoms. Before the diffusion, the epitaxial layer was uniformly doped with 10^{16} cm^{-3} donors. The intent of the diffusion is to establish a p-type region with a sheet resistance of 200 $\Omega/\square$ and a junction depth of 2.5 μm. If the furnace temperature is such that the effective boron diffusion coefficient is 10^{-12} cm^2/s, find the time required for the drive-in diffusion. This problem requires use of Irvin's curves.

2. A uniformly doped n-epitaxial layer has 10^{14} donors per cubic centimeter. A p-type base drive-in is carried out for 1 hour with a diffusion

coefficient $D = 10^{-13}$ cm^2/s. A lap and stain measurement on a test wafer shows the junction depth to be 1 μm. The wafers are returned to the drive-in furnace for another 2 hours. What is the final sheet resistance of the base? Repeat the problem with a doping in the epitaxial layer of 10^{15} cm^{-3}. Again, Irvin's curves are required.

3. To document the background doping level, a four-point-probe measurement is made on a p-type silicon wafer that is uniformly doped. The probe spacing is 62.5 mils, the wafer diameter is 100 mm, and the wafer thickness is 0.5 mm. The probe measurement results in 44 mV and 2 mA. Refer to the data in Figure A.2 of Appendix A to find the resistivity of the silicon, and refer to Figure A.3 to find the dopant concentration. (It is useful to have some experience with both metric and English units, and they are intentionally mixed in this exercise. In the United States the units of mils and inches are still used in portions of the semiconducting industry; however, the general trend is toward increased use of metric units.)

4. An MOS C-V measurement on a silicon wafer is used to probe $N(x)$. The following data result:

   $V_{G1} = 2.0$ V; $C_1 = 10.5 \times 10^{-9}$ F/cm^2
   $V_{G2} = 2.5$ V; $C_2 = 8.2 \times 10^{-9}$ F/cm^2

   The oxide thickness of the MOS capacitor is known to be 500 Å. Approximately what value of x is being probed? In other words, what value of x_D corresponds to the average of the above voltages? What is $N(x_D)$?

5. A proposed circuit has a chip area of 40,000 mils^2. Assume that it is to be fabricated on a 125-mm diameter wafer with costs and yields as for the example in Section 7.6, namely, a wafer cost of \$200, a wafer probe cost of \$20, a scribe and separation cost of \$0.10 per die, a scribe and separation yield of 95%, and a packaging and final testing yield of 95%. The cost of assembly, packaging, and final testing is taken to be \$1.00. Assuming a parametric yield of 80% and 4 defects per square centimeter, what is the final cost per chip if there are 430 dice per wafer.

6. It is suggested that the chip in Prob. 5 be divided into two chips to improve yield. Assume that it is possible to simply divide the integrated circuit into two 20,000-mils^2 chips. Also assume the same cost and yield parameters as in Prob. 2, and that the cost of assembly, packaging, and final test per chip is still \$1.00. What is the final cost of the two-chip set?

7. Consider again the wafer in Prob. 5. Now advanced processing improvements are suggested that can substantially reduce the die size. However, the requirement of new processing equipment will add $50 to the wafer cost. Furthermore, since the new process is more sensitive to contaminants, the parametric yield is reduced to 75%, and the effective defect density is increased to 6 defects per square centimeter. Assume all other cost and yield parameters remain the same. How much reduction in die size is required to make the new process attractive in terms of reducing the final cost per chip?

Appendix A
Laboratory Experiments

A.1 Introduction

This appendix describes laboratory experiments that allow the student to learn several of the concepts in the text chapters in a learning-by-doing environment. Topics addressed include doping principles, diffusion, oxidation, metalization, photolithographic pattern definition, etching, the metal-oxide-semiconductor (MOS) interface, threshold-voltage considerations, MOS field-effect transistor current-voltage (MOSFET) *(I-V)* characteristics, MOS capacitor diagnostics, diffused resistors, and diffused junctions. In addition, the student is provided with a unique opportunity to compare theoretical predictions with actual device and circuit performance, since he or she has carried out the fabrication and knows exactly what went into the making of the chip.

Three experiments are outlined and discussed. The first simply involves basic wafer characterization and the oxidation process. The next two involve a series of mask steps in which a variety of devices and simple circuits are fabricated. In the Diffused Devices Experiment, the MOS fabrication process is reduced to the simplest possible form. Only three masks are used in its most basic version: one to establish *p*-type diffusion patterns, including resistors, junctions, and FET sources and drains; one to make contact cuts; and one to define the metalization pattern. No gate oxidation patterning is done. Consequently there is no distinction between a field oxide and a gate oxide. The resulting chip contains several test devices, including *p*-type enhancement-mode MOSFETs, MOS capacitors, diffused resistors, and a *p-n* junction diode. By adding an optional fourth mask for an n^+ diffusion, a bipolar junction transistor (BJT) is also included. During the testing and analysis portion of the experiment, properties of the MOS system, diffused layers, *p-n* junctions, and transistors of different geometries are studied. Therefore, in spite of the simplicity of the fabrication steps, the student obtains hands-on experience with several key issues. PMOS is used rather than NMOS because it is a more forgiving technology; enhancement-mode transistors are easier to obtain. Although this experiment refers to a specific mask set, the processing procedures presented in this appendix could also be applied to other mask sets developed by users of this text.

The third experiment, on basic PMOS logic, uses five masks and includes gate oxide definition and an *n*-type diffusion. For the specific die layout that is used to illustrate this experiment, two choices are available for the metalization pattern. In the first, transistors are interconnected to form several simple logic gates. In the second, the interconnection forms a selectable logic unit. Again, however, the processing procedures for the five-mask sequence could be applied to user-designed masks.

The first two experiments have formed the standard portion of a combination lecture and laboratory course offered at Michigan State University for the past several years, primarily to seniors in electrical engineering. The third experiment has been used for special individual or group projects.

A.2 Equipment and Time Requirements

One academic quarter, with the laboratory section meeting once a week for three hours, provides sufficient time to finish the Oxidation Experiment and the Diffused Devices Experiment (without the bipolar option). Experience shows that the following is a suitable minimum-progress schedule:

Week 1 Wafer characterization and oxidation
Week 2 Diffused devices mask 1: photoresist exposure and development
Week 3 Oxide etch and boron predeposition
Week 4 Boron drive-in
Week 5 Mask 2: photoresist exposure and development
Week 6 Oxide etch and metalization
Week 7 Mask 3: photoresist exposure and development; metal etch and anneal
Week 8 Testing and analysis
Week 9 Testing and analysis

As outlined in Sections A.6, A.7, and A.8, there are natural break points in the fabrication sequence where the process can safely be stopped and the wafers put into storage until the next laboratory session.

At Michigan State University, the student laboratory is equipped with four furnaces for diffusions and oxidations, a photoresist spinner, three photoresist ovens, two mask aligners, an aluminum evaporator, a de-ionized water supply, a four-point resistivity probe, and a wafer probe station. An incident-light microscope with an instant film camera provides magnification levels from 65X to 400X. Also, a capacitance bridge and a curve tracer are used with the wafer probe station. As outlined in the following sections, each of these items of equipment plays an important role in the experiments.

All of the equipment described, except for the diffusion furnaces and the evaporator, is housed in a room under positive pressure with class 100 filtered air. However, only a few basic precautions are taken to maintain clean room conditions. Students wear lab coats while working in the area, and if shoes are dirty, disposable shoe covers are available. No gowns, hair covers, or face masks are used, so class 100 conditions do not exist in the working area. However, the simple PMOS fabrication process used in these experiments and the fairly large dimensions (the smallest critical dimension is 10 μm) combine to provide a fairly robust process. The diffusion furnaces open into a laminar-flow hood with class 100 air. This hood also has a sink, and most of the wet chemistry is performed in this hood.

For safety's sake, both n- and p-type diffusions are carried out with commercial solid planar sources in which the active dopant component is dispersed in an inert, refractory matrix. Toxic gases are not required in these experiments.

Students usually work in groups of two or three, and there are three groups in each section. Each group has its own wafer and is responsible for carrying the wafer through all the fabrication steps. However, at some points in the fabrication sequence a batch process involves all the wafers in a section. For example, each diffusion furnace is dedicated to a given purpose: one is used only for oxidations; one is used only for boron predepositions; one is used only for boron drive-ins; and the fourth is used for both aluminum annealing and the n^+ diffusion. Therefore, although each group is individually responsible for the handling, cleaning, photolithography, mask alignment, etching, and testing of its wafer, it is not practical for each group to separately do a boron drive-in, for example. Therefore, the section's activities are planned such that all the wafers go into the drive-in at the same time. Usually all furnace operations, as well as the metalization, are done in a batch process. Near the end of the term, chip testing is done on an open-lab basis.

Michigan State University offers the laboratory course in three terms a year. Consequently, even though the individual laboratory sections are small, about 50 students a year take the laboratory portion of the course.

A.3 Safety Considerations

A.3.1 General Considerations

These laboratory experiments involve hazards not normally encountered in an electrical engineering laboratory. Principally, the hazards involve the use of chemicals including strong acids, solvents, and bases. With proper precautions, these chemicals can be used quite safely. But without due respect for the chemicals, serious problems could result. It is important that the student be aware of the characteristics of each chemical used in the laboratory. Also, to be prepared in the unlikely event of an accident, he or she should be aware of a variety of emergency information such as that listed here.

Safety Precautions

1. **Know the institution's emergency phone number and the location of a nearby phone that is always available when the laboratory is in session.**
2. **An emergency shower, an eye-wash fountain, a fire extinguisher, a respirator mask, and a first aid kit should be available near the work area. Know their locations.**
3. **In any accident, get assistance. If acid or base splashes into the eye, flush the eye profusely with cold water for 15 to 20 minutes. The first few seconds are critical. An instructor or lab partner should make sure that the injured person's eyes are open during the rinse and should hold them open if necessary.**
4. **If acid or base spills or splashes onto the skin, rinse immediately with cold water for 15 to 20 minutes.**
5. **If an acid or base spills on clothing, remove the clothing.**
6. **No smoking or eating is allowed in the laboratory.**

A.3.2 General Handling of Laboratory Chemicals

Several basic guidelines follow for the general handling of chemicals like the ones used in these exercises.

Safety Precautions

1. **Always wear laboratory coats. When handling solvents, acids, or bases, always wear rubber gloves. When working with acids and etches, always wear a face shield or goggles. A face shield gives the best protection.**
2. **Acids, bases, and strong solvents should always be used in an exhausted fume hood. Photoresist-related chemicals should be used in well-ventilated areas.**
3. **All chemicals should be disposed of according to the guidelines of the institution. Chemicals such as solvents, acids, and bases usually require special pickup. In any event, *never* pour a solvent down the drain immediately after an acid, or vice-versa, because quite aside from ground water contamination, an explosion may result.**
4. **Generally, it should not be necessary for a student to mix any of the chemicals. The mixing will be done by staff before the laboratory starts. Likewise, the disposal of chemicals is normally the responsibility of the staff. In special cases where the student is given and accepts the responsibility of preparing and mixing chemical solutions, see the detailed instructions in Section A.3.7, Solution-Mixing Guidelines.**

5. **When in doubt about the proper procedure, never guess. Always ask your instructor.**
6. **As a general rule, use a minimum amount of chemicals. About 2 cm of fluid in a beaker is plenty to cover a wafer in a basket holder.**

A.3.3 Handling Solvents

Solvents used in these exercises include trichlorethylene (TCE), methanol (methyl alcohol), acetone, 2-propanol (isopropyl alcohol), and xylene. Some particular comments pertain to solvents.

Safety Precautions

1. **Rubber gloves are important when using solvents, not necessarily because they are hazardous to the skin, but because they can be absorbed through the skin. There are reports, for example, that chronic exposure to TCE may cause liver damage.**
2. **Solvents are volatile, and the fumes they give off are highly flammable. Except for TCE, solvents are not heated above room temperature in these experiments. Clearly, flames should not be allowed in the laboratory. If fumes from solvent in a beaker should happen to ignite as a result of mishandling such as overheating, cover the top of the beaker to extinguish the flames.**
3. **Use solvents in a well-ventilated area, preferably under an exhaust hood. Diligently avoid breathing the vapors.**
4. ***Never* mix an acid and a solvent. *Never* heat an acid and a solvent on the same hot plate because the fumes can react violently in certain cases.**

Some characteristics of the solvents used in these experiments follow. For additional information, see *Dangerous Properties of Industrial Materials,* by Sax [1].

Safety Information

***Acetone* (CH_3COCH_3) is a mild skin irritant, and prolonged inhalation can cause headaches, but its primary hazard is that it is very flammable when exposed to heat or flames. Also it can react violently with certain acids.**

***Isopropyl alcohol* ($CH_3CHOHCH_3$), or 2-propanol, is in most medicine cabinets as the main ingredient of rubbing alcohol. In spite of its familiarity, be aware that it can cause eye irritation and eye damage. It also is flammable and can react violently with acids.**

Methyl alcohol **(CH_3OH), or methanol, is flammable and reacts violently with several acids. On very severe exposure, methanol is toxic to the nervous system, especially the optic nerves.**

Trichlorethylene **($CHClCCl_2$) can cause headache and drowsiness if inhaled in moderate quantities. Chronic exposure may damage internal organs, including the liver.**

Xylene **($C_6H_4(CH_3)_2$) is highly flammable and can react with oxidizing materials.**

A.3.4 Handling Acids and Bases

Strong acids and bases may represent the most serious hazard in the laboratory. As a general rule, always wear rubber gloves (without holes), lab coats or aprons, and goggles or a face shield when working with any acid or base. Again, always work under the exhaust hood. Some particular dangers of the acids and bases used in these experiments follow. Again, for more information, see *Dangerous Properties of Industrial Materials,* by Sax [1].

Safety Information

Sulfuric acid **(H_2SO_4) is extremely corrosive and toxic to body tissues. It will rapidly attack the skin and can cause severe burns and blistering.**

Nitric acid **(HNO_3) can also destroy tissue and cause burns, but it is not quite as destructive as sulfuric acid. Nitric acid leaves a characteristic yellow stain on the skin.**

Hydrofluoric acid **(HF) can produce severe skin and deep tissue burns that are slow in healing. It is all the more dangerous in that burns may not be noticed until some time after exposure. Therefore, HF should be handled with great caution. It is a good idea to rinse your hands carefully after using HF even if you did wear rubber gloves. Avoid the fumes.**

Hydrochloric acid **(HCl) is strongly corrosive and causes burns on contact with the skin.**

Phosphoric acid **(H_3PO_4) should not be heated to the boiling point since the PO_x fumes are toxic. It will cause burns if it contacts the skin.**

Ammonium hydroxide **(NH_4OH) is a powerful base that can cause severe burns. It should always be used with adequate ventilation.**

Generally these acids and bases will be encountered in solutions containing more than one chemical. Some solutions contain hydrogen peroxide:

Hydrogen peroxide (H_2O_2) **is also found in many household medicine cabinets in dilute solution and is used as a topical antiseptic. However, it is used in concentrated form in these experiments. Solutions of H_2O_2 of 35% by weight and over can easily blister the skin. In solution with acids, hydrogen peroxide can be highly volatile.**

A.3.5 Furnace Safety

Furnace temperatures may be as high as 1100°C in the flat-zone region inside the tube. Observe the following procedures when working with a furnace:

Safety Precautions

1. **Although the sides of the furnace are generally cool to the touch, the quartz ends may be quite hot. When removing an end cap, use a thick cloth or glove for protection. When a push rod is removed from the furnace, the end will be too hot to touch for about 5 minutes.**
2. **Glass components on the furnace setups are actually quartz, which means they are expensive. Handle them carefully. Do not apply a torque; the push rods are especially fragile. And when using lint-free gloves, the quartz may be slippery, so be certain of your grip.**
3. **Gases flow continuously through the furnaces. Gas cylinders may be under pressures of over 2000 psi and are potential torpedoes if they are not properly secured and handled. Generally the student should not adjust flowmeter or regulator controls unless supervised by the instructor.**
4. **Scavenge the output of the furnace tubes to an external exhaust.**

A.3.6 Electrical Safety

The electrical hazards in this laboratory are comparable to those of other electrical engineering laboratories:

Safety Precautions

1. **Most hazardous is the familiar curve tracer, which can produce several hundred volts and fatal shocks! Keep the curve tracer voltage at zero until the leads are safely connected to the device or circuit being tested.**
2. **Equipment such as furnaces and mask aligners may have high-voltage circuitry behind protective panels. This circuitry is not to be worked with during the laboratory.**

A.3.7 Solution-Mixing Guidelines

A student who is authorized to mix solutions should observe the following guidelines:

Safety Precautions

1. **Use a face shield rather than goggles. Always wear rubber gloves and a lab coat. Mix chemicals under an exhaust hood. *Never* peer into the top of a beaker—always view it from the side.**
2. **Pour chemicals slowly. Make sure you have the right chemicals. Always pour acid into water, *never* water into acid. Use Teflon or an appropriate plastic beaker for HF solutions.**
3. ***Never* mix acids and solvents or heat them on the same hot plate.**
4. **Always label beakers. Do not let used chemicals accumulate in beakers. When you are finished, dispose of the solution unless someone else will be using it during that lab period. (An exception to this rule is photoresist strip solution.)**
5. **Rinse empty beakers thoroughly by placing them in a sink under a running faucet for several minutes.**
6. **Always start with clean beakers that have been stored top down in a clean area. If beakers must be cleaned, an adequate procedure for purposes of these experiments is as follows:**

   a. Scrub in hot water and Alconox [2] with a brush.
   b. Rinse in hot water.
   c. Rinse in distilled water.
   d. Rinse in de-ionized water.
   e. Bake out glass beakers in clean area until dry. Dry plastic and Teflon [3] beakers at room temperature.

   For higher-purity procedures, chromic acid is used to clean glassware.

   Some commonly used solutions in amounts appropriate for single-wafer, beaker processing follow. In all cases where water is used, de-ionized water is assumed.

Degrease etch ($5H_2O:1H_2O_2:1NH_4OH$)

75 ml H_2O
15 ml H_2O_2 (30% solution)
15 ml NH_4OH

Demetal etch ($8H_2O:2H_2O_2:1HCl$)

80 ml H_2O
20 ml H_2O_2 (30% solution)
10 ml HCl

Aluminum etch ($2H_2O{:}16H_3PO_4{:}1HNO_3$:1 acetic acid)

80 ml H_3PO_4
5 ml HNO_3
5 ml acetic acid
10 ml H_2O

Buffered oxide etch (mixture of HF and ammonium fluoride)

7 parts 40% NH_4F to 1 part 50% HF (this ratio may be varied for faster or slower etch rates)

Borosilicate etch ($1H_2SO_4{:}1HNO_3$)

50 ml H_2SO_4
50 ml HNO_3

A.4 Laboratory Reports and Records

It is recommended that each student be individually responsible for a laboratory notebook and for lab reports. The laboratory notebook should contain a daily record of laboratory work and observations. All data, records of experiments, and observations during the lab should go directly into the notebook.

In addition to and separate from the notebook, three laboratory reports are suggested, two for the Diffused Devices Experiment and one for the Oxidation Experiment. The reports should show the pertinent data, answers to questions, calculations, observations, and comments, as discussed later in the Analysis of Results sections of this appendix.

A.5 Silicon Oxidation Experiment

A.5.1 Introduction

The purpose of this experiment is to gain experience with wafer handling, cleaning, and characterization techniques, and to grow a silicon dioxide layer using wet and dry thermal oxidations. In later experiments, this oxide layer will be used as a diffusion mask.

A.5.2 Wafer Handling

In the steps that follow, you will be handling the wafer with tweezers and carrying it from one work area to another. Clean the tweezers at the beginning of the lab by rinsing in hot TCE, then in acetone, methanol, and deionized water, in that order. Dry the tweezers with N_2 gas. When carrying the wafer, always hold it over a piece of filter paper, or a petri dish, in case the wafer slips from the tweezers. If you don't catch it and it lands on the floor or some other nonclean surface, it will be necessary to reclean the wafer using either the Initial Wafer Cleaning Procedure or the Prephotoresist Cleaning Procedure (see Section A.9), depending on where the wafer is in the fabrication sequence.

During these experiments, never touch with bare hands the wafer or anything that will come in contact with it, such as the quartz furnace components (push rods, boats, and so on). When working with dry wafers or quartz furnace components, use lint-free gloves. When working with wet processes, use rubber gloves. Small amounts of impurities, particularly sodium ions from the skin or from impure water, will wreck an MOS process. Always make sure beakers are clean. One improperly cleaned beaker or inadvertent contact with tap water can send threshold voltages into the stratosphere.

A.5.3 Wafer Documentation

The starting point for this experiment is a uniformly doped, polished, *n*-type silicon wafer whose background doping is reasonable for basic MOS and junction devices; an appropriate range is 10^{15} to 5×10^{15} cm^{-3}. It will be necessary to know the actual carrier concentration in later processing steps. Also, the carrier type should be verified and documented, since it would be unfortunate to find several weeks into the term that somehow a *p*-type wafer was used instead of an *n*-type wafer. In this experiment, type and concentration of background doping will be determined by two measurements, the hot probe and the four-point probe.

Given a semiconductor, a quick and easy method of determining whether the majority carriers are holes or electrons is the hot probe technique, which proceeds as follows. Using a voltmeter, heat one of the two voltmeter probes and then place (do not clip) both probes on the wafer surface. If the meter reads a positive voltage, roughly several mV, for the hot probe relative to the cold one, then the wafer is *n*-type. Contrarily, if the hot probe is negative, the sample is *p*-type. It is a good idea to reverse the heating of the probes and make sure that the polarity of the measured voltage indeed changes. For this experiment, the wafer should be *n*-type, and the purpose of this measurement is simply to check and document that indeed you are starting with the correct wafer type.

A rigorous analysis of this thermoelectric phenomenon is complex, but it is useful to consider the hot probe effect from a nonmathematical point of view. Since the thermal velocity of a carrier increases with temperature, there will be a velocity gradient corresponding to the thermal gradient produced by the hot probe. Considering *p*-type material, holes near the hot probe will move faster than holes far from the hot probe. Referring to Figure A.1,

Figure A.1 The hot probe measurement. The meter should be capable of measurements in the mV range.

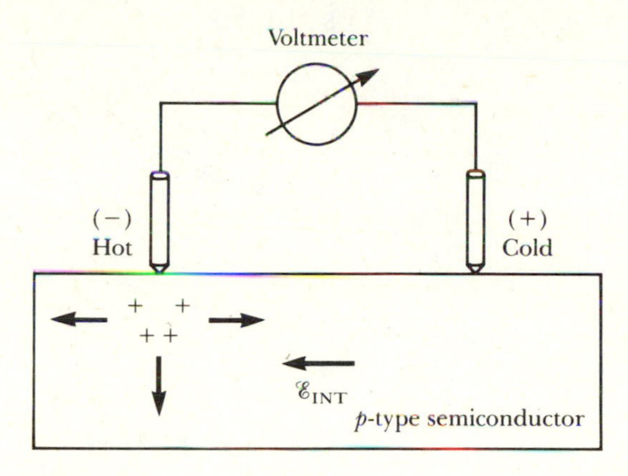

holes will tend to move away from the hot probe, which would correspond to a current flow. However, because the voltmeter draws essentially no current, an internal electric field must be set up inside the semiconductor to oppose this flow. In p-type material, this situation means the hot probe will be negative with respect to the cold probe. For n-type material, the electron flow must be counteracted by an internal field causing opposite polarity for the voltmeter reading.

Prior to the four-point probe experiment, review Section 7.2.2. Recall that the sheet resistance is proportional to the voltage-to-current ratio according to

$$R_{\square} = \zeta(V/I) \tag{A.1}$$

where V is the voltage obtained from the inner voltage-sensing probes and I is the current obtained from the outer current-driving probes shown in Figure A.2, and ζ is a tabulated proportionality factor. When the wafer size is much larger than the probe spacing, the constant of proportionality, ζ, is nearly equal to 4.53, as discussed in Section 7.2.2. For example, the Signatone model S-301 probe spacing is specified by the manufacturer to be 62.5 mils [4]. Using this probe and a 3-inch-diameter wafer, the d/s ratio is 48 and the correction factor from Figure A.2 is about 4.51 [5].

When carrying out the four-point probe measurement, note that the probe tips are rather fragile. *Once the probes are in contact, the wafer should not be moved.* After the sheet resistance has been experimentally determined, the wafer resistivity is found from

$$\rho = R_{\square}t \tag{A.2}$$

Figure A.2 The four-point probe measurement. Metering is by digital multimeters. (See Figure 7.6 and Table 7.1 for rectangular samples.)
SOURCE: From F. M. Smits, "Measurement of Sheet Resistivities with the Four Point Probe," *Bell System Technical Journal* (May 1958), 711–718.

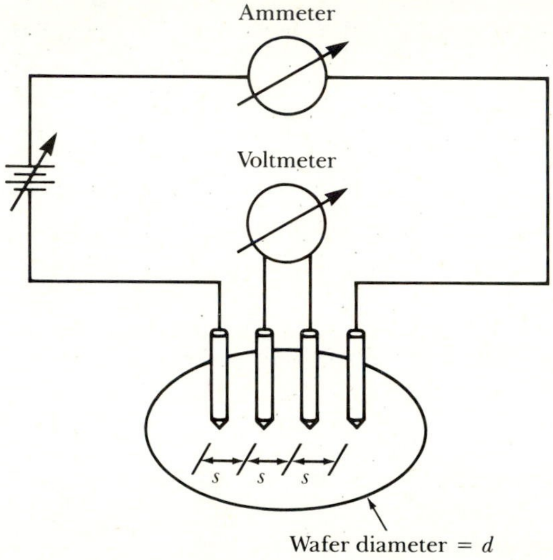

$$R_{\square} = \zeta(V/I)$$

d/s	ζ
3.0	2.2662
4.0	2.9289
5.0	3.3625
7.5	3.9273
10.0	4.1716
15.0	4.3646
20.0	4.4363
40.0	4.5076
∞	4.5324

where t is the wafer thickness. To ensure a representative reading, it is a good idea to measure the sheet resistance more than once, at slightly different wafer locations, and to take readings at more than one current setting.

A.5.4 Oxidation

Next, an oxide layer is to be grown on the wafer according to the procedure that follows. In this experiment, the oxide is grown partly by a dry technique in which the oxygen flows directly into the furnace tube, and partly by a wet technique in which the oxygen is bubbled through a heated quartz flask containing de-ionized water [6].

1. Clean the wafer, using the Initial Wafer Cleaning Procedure from Section A.9.
2. Place the wafer in the quartz boat of the oxidation furnace. Do not force the wafer into a slot or apply force perpendicular to the wafer surface. (This is one step where wafers are sometimes broken.) Use a heavy cloth glove to remove the quartz end cap from the furnace tube and use the push rod dedicated to that furnace to pull the boat out of the tube onto the boat holder. Place the wafer in the boat and record its position. Use the push rod to slide the boat back into the front part of the furnace tube.
3. Check that the furnace temperature is at 1100°C and that the oxygen flow rate is approximately 500 sccm (standard cubic centimeters per minute). The oxygen flow should have been on for about an hour, depending on the size of the furnace tube, prior to this point to establish an oxygen atmosphere in the furnace tube.
4. Slide the boat into the furnace slowly. The rate should be about 5 cm per 10 seconds. Place the boat near the center of the furnace. Pull the push rod out again.

Safety Precaution

The end of the push rod will be hot and should not be touched for 5 minutes.

5. Do a dry oxidation for 10 minutes.
6. Switch the O_2 flow through the de-ionized water bubbler. The water temperature should be between 95 and 98°C, and the oxygen flow rate should be such as to cause bubbling in the water, but not violent splashing. A typical flow-rate range is 200 to 300 sccm. Do a wet oxidation for 15 minutes.
7. Switch back to a dry oxidation for 10 minutes.
8. Pull the boat out at the same rate it was pushed in.
9. Let the boat cool at the end of the tube for 5 minutes.
10. Pull the boat out onto the boat holder and let it cool another 5 minutes.
11. Remove the wafers, inspect them, and estimate the oxide thickness based on color by referring to Table A.1 [7]. Calculate the expected thickness using information from Section 2.7. Compare the calculation of thickness and the experimental observation of color.
12. Store the wafer in a petri dish, or other suitable container, in a desiccator kept in a clean area.
13. The furnace gas flow should be returned to the standby condition of 30

Table A.1 Color chart for thermally grown oxide films observed perpendicularly under daylight fluorescent lighting

Film thickness (μm)	Order (5450 Å)	Color and comments
0.050		Tan
0.075		Brown
0.100		Dark violet to red-violet
0.125		Royal blue
0.150		Light blue to metallic blue
0.175	I	Metallic to very light yellow-green
0.200		Light gold to yellow—slightly metallic
0.225		Gold with slight yellow-orange
0.250		Orange to melon
0.275		Red-violet
0.300		Blue to violet-blue
0.310		Blue
0.325		Blue to blue-green
0.345		Light green
0.350		Green to yellow-green
0.365	II	Yellow-green
0.375		Green-yellow
0.390		Yellow
0.412		Light orange
0.426		Carnation pink
0.443		Violet-red
0.465		Red-violet
0.476		Violet
0.480		Blue-violet
0.493		Blue
0.502		Blue-green
0.520		Green (broad)
0.540		Yellow-green
0.560	III	Green-yellow
0.574		Yellow to "yellowish" (not yellow but is in the position where yellow is to be expected; at times it appears to be light creamy gray or metallic)
0.585		Light-orange or yellow to pink borderline
0.600		Carnation pink
0.630		Violet-red
0.680		"Bluish" (not blue but borderline between violet and blue-green; it appears more like a mixture between violet-red

Table A.1 Color chart for thermally grown oxide films observed perpendicularly under daylight fluorescent lighting (*cont.*)

Film thickness (μm)	Order (5450 Å)	Color and comments
		and blue-green, and overall looks grayish)
0.72	IV	Blue-green to green (quite broad)
0.77		"Yellowish"
0.80		Orange (rather broad for orange)
0.82		Salmon
0.85		Dull, light red-violet
0.86		Violet
0.87		Blue-violet
0.89		Blue
0.92	V	Blue-green
0.95		Dull yellow-green
0.97		Yellow to "yellowish"
0.99		Orange
1.00		Carnation pink
1.02		Violet-red
1.05		Red-violet
1.06		Violet
1.07		Blue-violet
1.10		Green
1.11		Yellow-green
1.12	VI	Green
1.18		Violet
1.19		Red-violet
1.21		Violet-red
1.24		Carnation pink to salmon
1.25		Orange
1.28		"Yellowish"
1.32	VII	Sky blue to green-blue
1.40		Orange
1.45		Violet
1.46		Blue-violet
1.50	VIII	Blue
1.54		Dull yellow-green

From W. A. Pliskin and E. E. Conrad, "Nondestructive Determination of Thickness and Refractive Index of Transparent Films," *IBM Journal of Research and Development*, 8 (1964), 43–51.

sccm nitrogen. This nitrogen purge is used in each of the furnaces to prevent contamination from room air.

A.5.5 Analysis of Results

1. What were the actual times, flow rates, and temperatures of interest for the wafer oxidation?
2. If the process is indeed dry-wet-dry, what is the predicted oxide thickness?
3. What would be the thickness and mass of silicon consumed, based on the answer to question 2?
4. What would be the expected color? What was the actual color? Comment.
5. Actually, the wet oxidation was not a true steam oxidation since a portion of the 15 minutes was just used to establish a wet oxygen ambient in the furnace tube. As a lower limit, assume the entire oxidation was dry. Now what is calculated for the oxide thickness, consumed silicon, and expected color?

Figure A.3 Silicon resistivity as a function of impurity concentration.
SOURCE: From S. M. Sze and J. C. Irvin, "Resistivity, Mobility, and Impurity Levels in GaAs, Ge, and Si at 300°K," reprinted with permission from *Solid State Electronics,* vol. 11. Copyright © 1968, Pergamon Press, Ltd., pages 599–602.

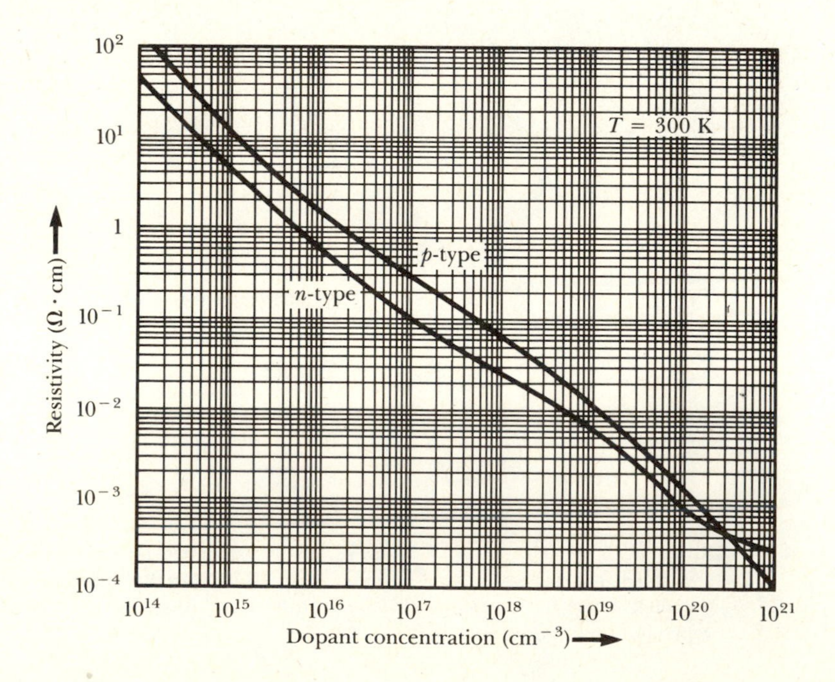

Figure A.4 Excess carrier lifetime as a function of silicon resistivity.

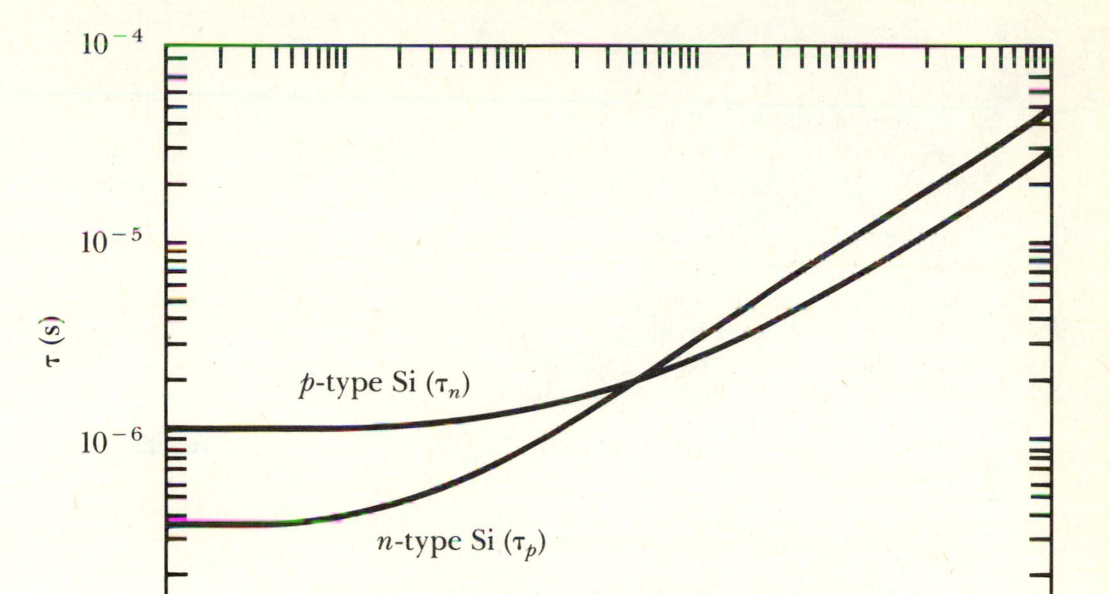

6. What is your best guess for the actual oxide thickness at this point? What are your reasons?
7. What were the voltage values measured with the hot probe? What do you conclude from the hot probe measurement?
8. Based on the four-point probe measurement, what is the sheet resistance for the wafer? What is the resistivity? Estimate the uncertainty associated with these values. What are the origins of the uncertainties, and how much does each origin contribute to the total uncertainty?
9. Referring to Figures A.3 and A.4, what are the background doping concentration, the majority carrier mobility, and the excess carrier lifetime for the wafer [8,9]?

A.6 Diffused Devices Experiment

A.6.1 Introduction

This experiment continues processing on the wafer from the oxidation experiment with the fabrication of diffused resistors, MOS capacitors, *p-n* junction diodes, and MOSFET transistors. A BJT is optional. There are five major parts to the experiment.

1. Patterning and etching of windows in the oxide for the source and drain diffused regions, diode anode diffused regions, and resistors
2. Boron predeposition diffusion and drive-in diffusion
3. Patterning and etching of contact cuts in the oxide
4. Patterning and etching of aluminum contact pads
5. Testing and evaluation of devices

The next section gives a detailed breakdown of the laboratory fabrication procedure.

A.6.2 Diffused Devices Fabrication Sequence

The steps in the fabrication sequence in this section refer to the procedures described in Section A.9. To help schedule the time in the laboratory, dotted lines indicate points in the sequence where the process can be stopped and resumed again later. At such break points, the wafers should be stored in a clean, dry environment. A standard laboratory desiccator is a suitable container.

In the steps that follow DI refers to de-ionized water,

1. Prephotoresist Cleaning Procedure, unless going directly from the oxidation furnace to the spinner, in which case this step can be skipped.
2. Photoresist Coating Procedure.
3. Photoresist Exposure Procedure with mask 1, "BASIC PMOS P-DIFFUSION," for diffusion oxide patterning.
4. Photoresist Develop Procedure.
5. Microscope inspection of the wafer.

- -

6. Photoresist Postbake Procedure.
7. Oxide Etch Procedure.
8. Microscope inspection of the wafer.

- -

9. Photoresist Strip Procedure
10. Microscope inspection of wafer.
11. Boron Predeposition Procedure.

- -

12. Boron-Skin Removal Procedure.

- -

13. Test-area evaluation (hot probe measurement and four-point probe measurement).

- -

14. Rinse with running DI for 2 minutes, dry with N_2
15. Boron Drive-In Procedure.

- -

16. Prephotoresist Cleaning Procedure, unless going directly from the diffusion furnace to the spinner, in which case this step can be skipped.
17. Photoresist Coating Procedure.
18. Photoresist Exposure Procedure using mask 2, "BASIC PMOS CONT," for contact cut patterning.
19. Photoresist Develop Procedure.
20. Microscope inspection of the wafer.

- -

21. Photoresist Postbake Procedure.
22. Oxide Window Etch Procedure.
23. Microscope inspection of the wafer.

- -

24. Photoresist Strip Procedure.
25. Microscope inspection of the wafer.

- -

26. Test-area evaluation (hot probe measurement and four-point probe measurement).

- -

27. Preevaporation Cleaning Procedure.
28. Aluminum Evaporation Procedure on front of wafer.

- -

29. Photoresist Coating Procedure.
30. Photoresist Exposure Procedure using mask 3, "BASIC PMOS METAL," for metalization patterning.
31. Photoresist Develop Procedure.
32. Microscope inspection of the wafer.

- -

33. Photoresist Postbake Procedure.
34. Aluminum Etch Procedure.
35. Microscope inspection of the wafer.

- -

36. Photoresist Strip Procedure.
37. Microscope inspection of the wafer.

- -

38. Aluminum Evaporation Procedure on back of the wafer (optional).
39. Aluminum Anneal Procedure.

- -

40. Device testing and evaluation.

A.6.3 Diffused Devices Mask Set [10]

The mask set used at Michigan State University for the Diffused Devices Experiment provides patterning for two MOSFETs with different geometries, two MOS capacitors of different sizes, two different diffused resistors, and a *p-n* junction diode. In addition, four test areas are provided on the wafer that are large enough to allow four-point probing of sheet resistance and hot probe testing to determine carrier type. An optional fourth mask for an *n*-type diffusion establishes an emitter region for an eighth device, an *n-p-n* BJT.

Figure A.5 shows the relative position of the test areas on the wafer. These test regions are opened during the *p*-type diffusion and are present for evaluating the results of the boron predeposition and drive-in. Figures A.6, A.7, and A.8 show photographs of the three basic masks, and Figure A.9 shows details of the chip layout for each mask.

A.6.4 Analysis of Results: Part 1

Note: In answering the questions that follow, use times and temperatures specific to the conditions actually experienced by the wafer in the laboratory, and show all calculations.

1. Treat the boron predeposition as a constant-source, solid-solubility-limited diffusion. What would you expect for $N(0)$, x_j, and the sheet resistance, $R_{\square}$, based on the theoretical principles developed in Chapter 2? Since D is a function of N, and therefore of x, a simulation such as SUPREM is in order. Alternatively, and more approximately, an "effective" diffusion coefficient, D_{CEF}, for the complementary-error-function diffusion is given by the expression [11]

$$D_{CEF} \simeq \left(\frac{1.225}{z}\right)^2 \left[\frac{N(0)}{n_i}\right] D_i \qquad \text{(A.3)}$$

where D_i is the diffusion coefficient calculated for intrinsic silicon and where, as in Chapter 2,

Figure A.5 Wafer test-area positions for the diffused devices experiment.

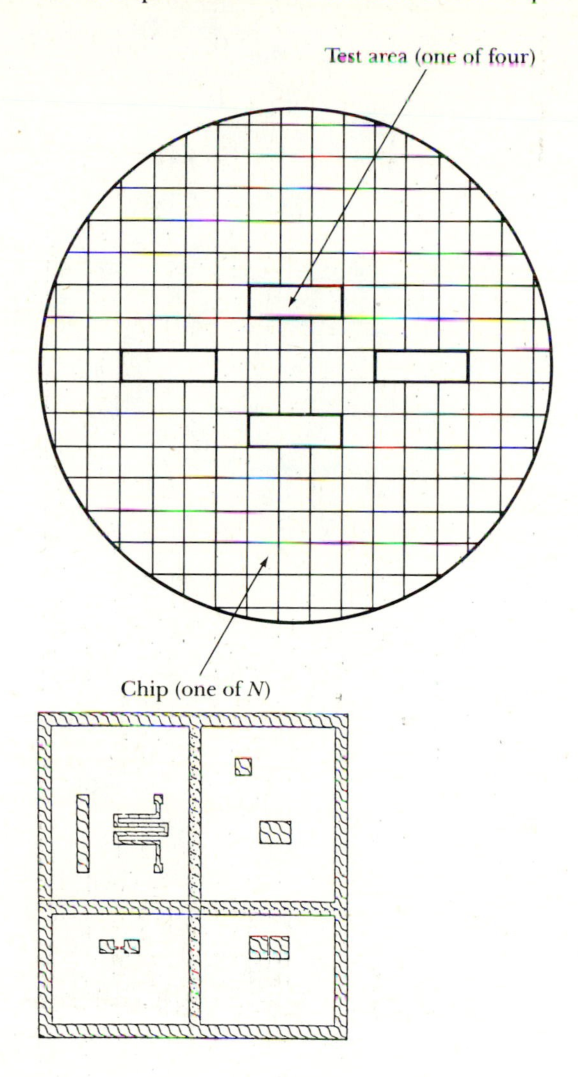

$$z = \frac{x_j}{2(Dt)^{1/2}} \tag{A.4}$$

and the rest of the terms also have the same meaning as in Chapter 2. For a given ratio, $N(0)/N_B$, z is determined from the erfc curve in Figure 2.12; then D_{CEF} is calculated. Finally, x_j is calculated by using D_{CEF} in Eq. (2.38) for the erfc profile. The Irvin's curves of Chapter 7 can then be used to find the approximate sheet resistance.

Figure A.6 Photograph of mask 1, "BASIC PMOS DIFF," for the diffused devices experiment.

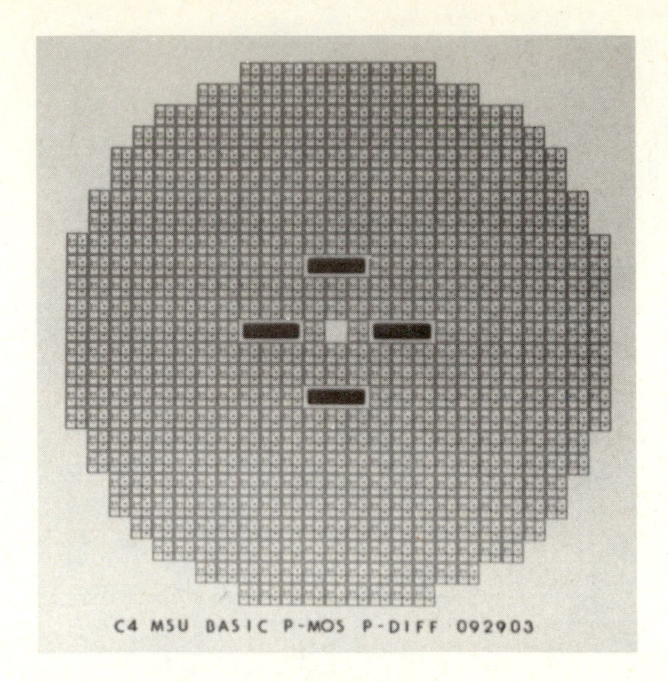

Figure A.7 Photograph of mask 2, "BASIC PMOS CONT," for the diffused devices experiment.

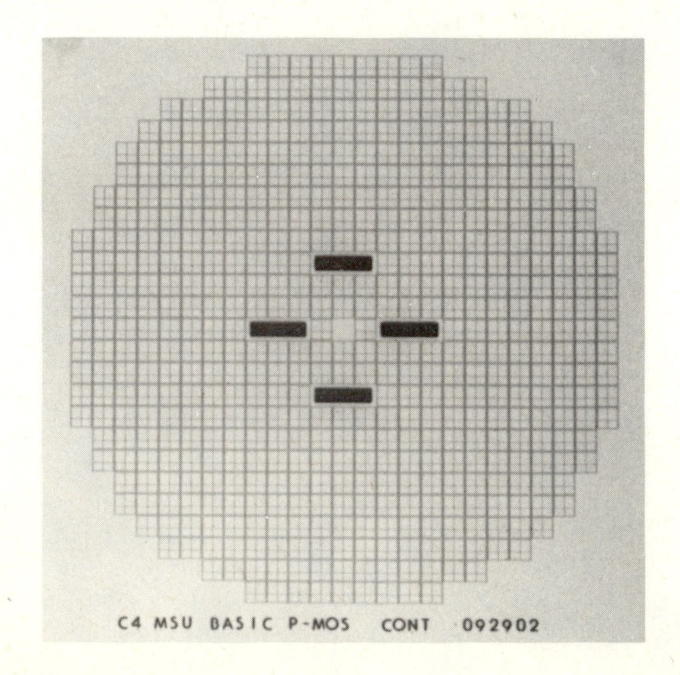

Figure A.8 Photograph of mask 3, "BASIC PMOS METAL," for the diffused devices experiment.

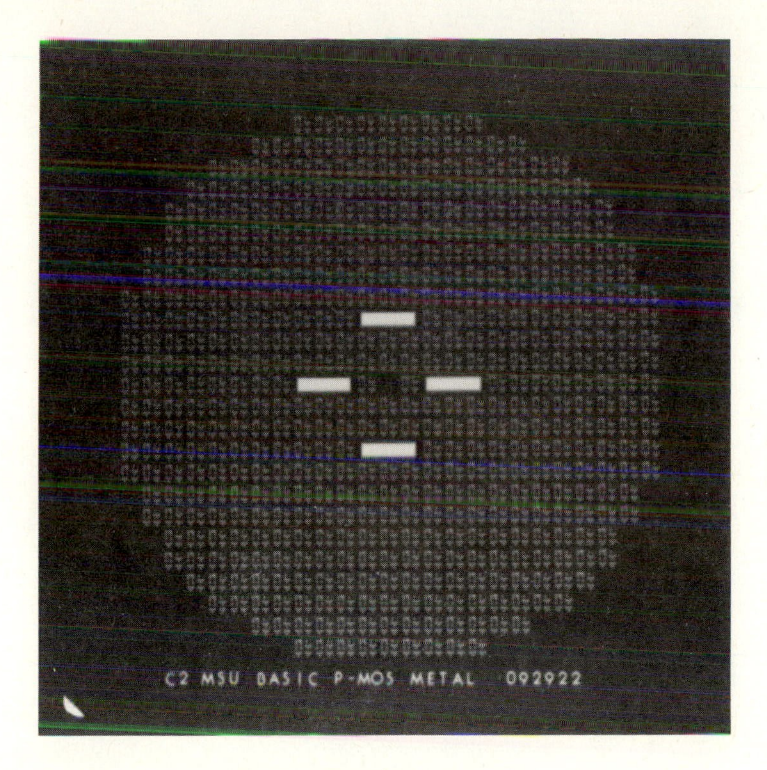

2. Based on the manufacturer's data from Figure A.17 in the Boron Predeposition Procedure in Section A.9 (or a comparable figure if different sources are used), what is expected for the sheet resistance?
3. What is the measured value of $R_\square$ in each of the four test areas. Estimate the uncertainty in each case. How do the measurements of sheet resistance compare with the predictions of questions 2 and 3? Comment on the differences. What can you say about the doping uniformity?
4. What do you calculate for the theoretical values of $N(0)$, x_j, and $R_\square$ after the drive-in? Again, an approximate effective diffusion coefficient that accounts for the fact that D changes with x may be used in the Gaussian expression, namely [11],

$$D_{\text{GAUSSIAN}} \simeq 2D_i \qquad \text{(A.5)}$$

5. What are the measured values of $R_\square$ in the test areas after the drive-in? Again, how uniform does the doping appear to be?

Figure A.9 (A–E) Chip layout for the diffused devices experiment. A. Diffusion pattern for mask 1, "BASIC PMOS P-DIFFUSION."

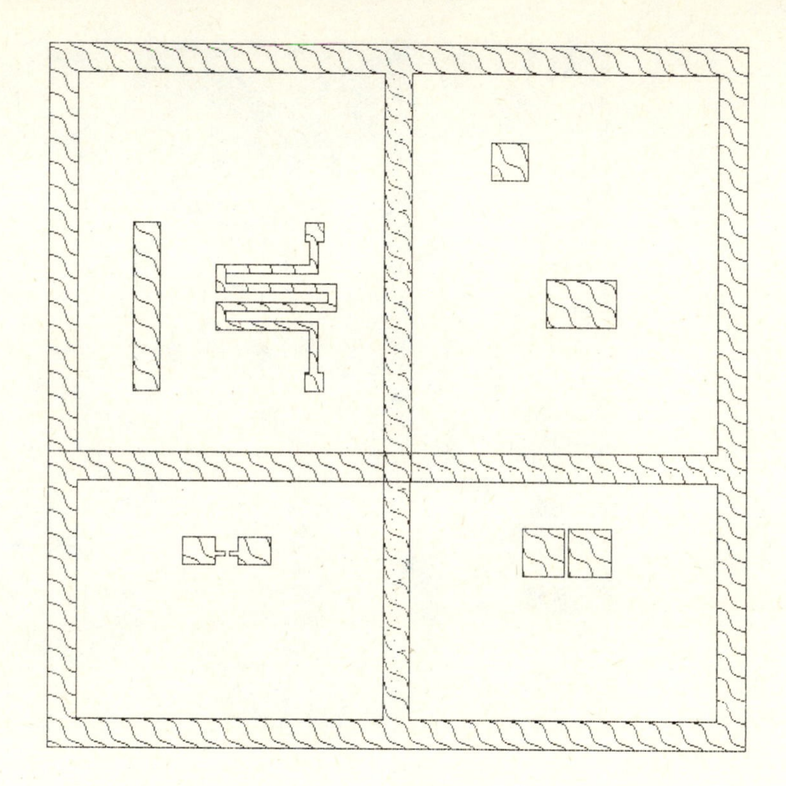

6. Compare the measured values of $R_\square$ after the drive-in to the predicted values. The theory that leads to a Gaussian distribution says that $R_\square$ should decrease. Why? But experiment probably shows that $R_\square$ actually increases. Why?
7. How would you go about actually measuring x_j if that were required? Give a step-by-step procedure.

A.6.5 Analysis of Results: Part 2

Note: Make device measurements by contacting the metal pads associated with each device with a wafer prober. To avoid photocurrents, measurements should be without incident light. The primary test instruments are a curve tracer and a high-frequency (100 kHz or higher) capacitance bridge.

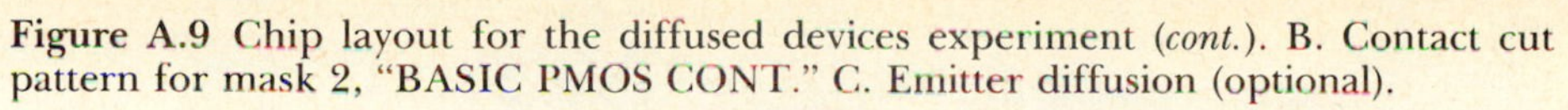

Figure A.9 Chip layout for the diffused devices experiment (*cont.*). B. Contact cut pattern for mask 2, "BASIC PMOS CONT." C. Emitter diffusion (optional).

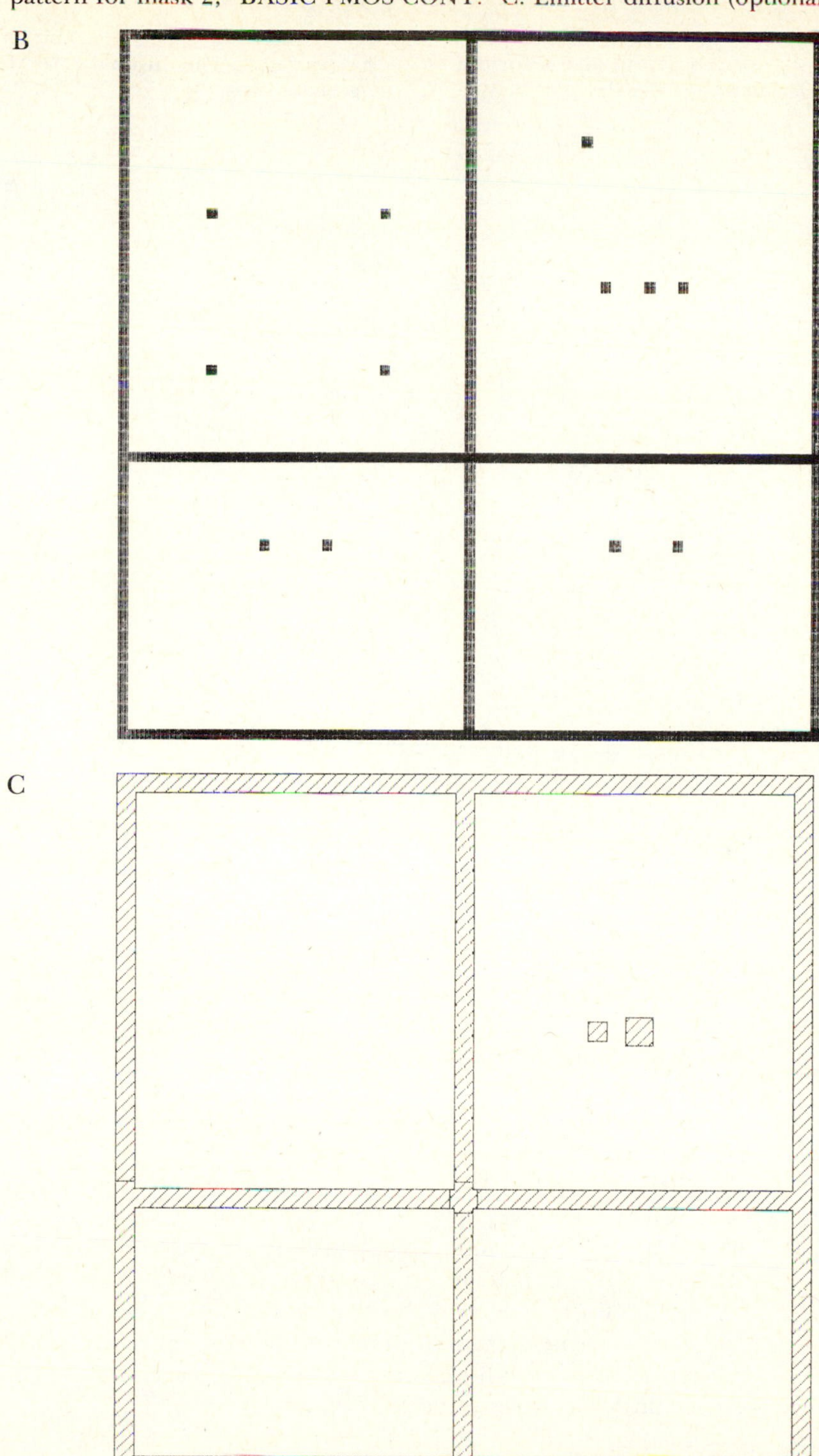

Figure A.9 Chip layout for the diffused devices experiment (*cont.*). D. Metal pattern for mask 3, "BASIC PMOS METAL" (negative of mask).

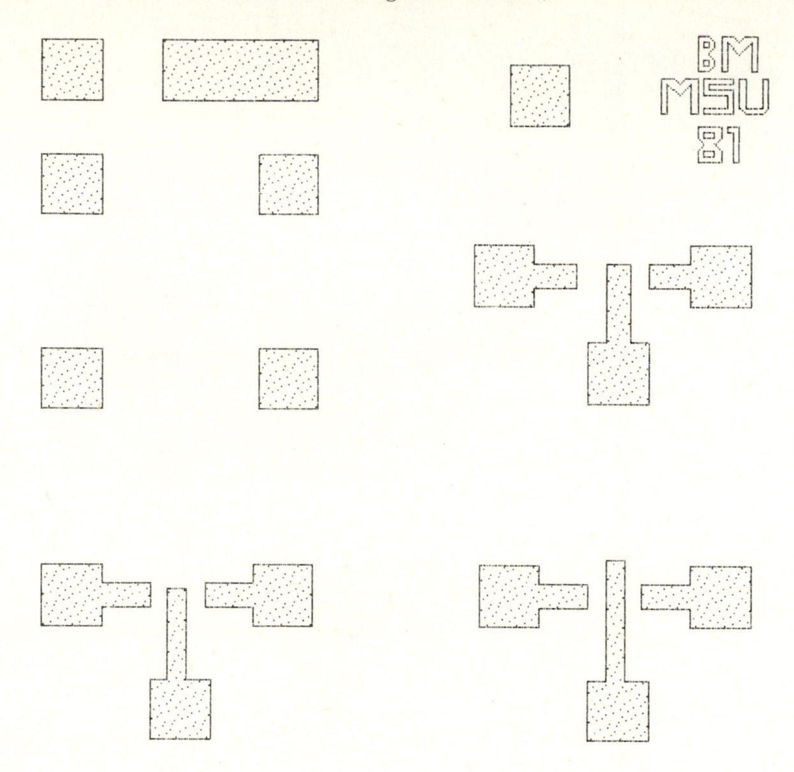

1. Take a microscope photograph of the two resistors and another photograph of the MOSFET with $W/L = 1$. Your lab partner should include one in his or her report and you should include the other in your report.
2. Count the number of squares in each resistor, using a microscope with a calibrated reticle in the eyepiece. Refer to Section 4.6.2 for the weighting of corner squares and terminations.
3. Check for resistors that appear to be "good" on visual inspection with the microscope. Measure at least five short resistors and five long resistors. Record their resistance values and check for linearity of current with voltage. Choose devices near the wafer edge as well as near the wafer center. What can you deduce about doping uniformity from these measurements? Calculate the average resistor value and standard deviation for both the long and short resistors.
4. Given your earlier test-area measurements of sheet resistance, compare the resistor values with expected values using the number of squares and the test-area sheet resistance.

Figure A.9 Chip layout for the diffused device experiment (*cont.*). E. Combined chip layout.

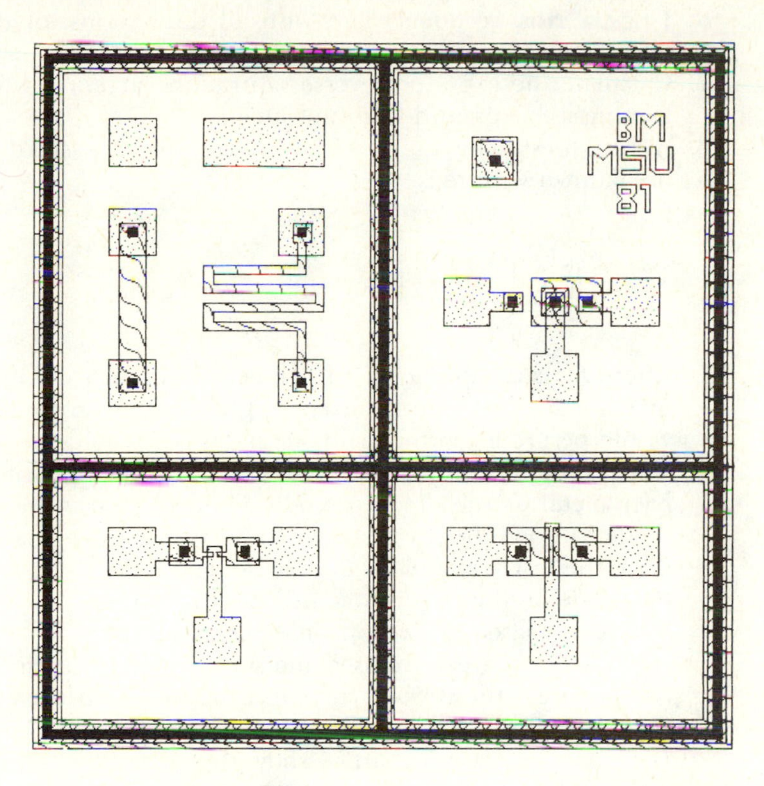

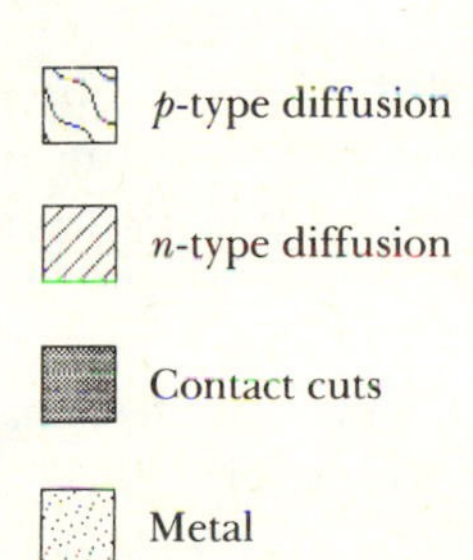

5. Check a few resistors that appear to be "bad" on microscope inspection, owing to scratches, blemishes, overetching, and the like.
6. Draw a cross-sectional view, with all dimensions, of the diode on the chip.
7. Calculate the expected reverse saturation current, I_S, for the diode, assuming it is a one-sided abrupt junction.
8. A rough-estimate universal expression for one-sided abrupt junction breakdown voltage is

$$V_B \simeq 60\left[\frac{E_G}{1.1}\right]^{3/2}\left[\frac{10^{16}}{N_B}\right]^{3/4} \tag{A.6}$$

where E_G is the energy gap in units of electron-volts and N_B is the background doping in units of cm^{-3} [12]. Based on this expression, what would be predicted for the diode breakdown voltage?
9. For the diode testing, if the back of the wafer was metalized, use the back-metal as the diode cathode. Otherwise, use the test areas as the diode cathode. In any case, you are actually testing back-to-back diodes, since the test areas and wafer back-surface are also *p-n* junctions. When the diode on the chip is forward biased, the diode at the cathode contact is reverse biased. However, since the cathode area is about 1 cm^2 if the combined test areas are used and several square centimeters if the wafer back is used, the reverse current associated with the cathode contact is fractions of a μA or larger. At higher currents, the chip diode's forward *I-V* characteristic will start to show the effect of the nonohmic cathode contact, but good diode characteristics can be measured at lower currents.

    Test several diodes and observe the variations in I_S, V_B, and turn-on voltage. For purposes of this question, take I_S to be the current at 1 V reverse bias and define V_B to be the reverse voltage giving 10 μA of leakage current. Take the turn-on voltage to be the forward-bias voltage at 1 μA forward current.
10. For one of the better diodes, make a careful *I-V* measurement using several current scales on a curve tracer. Plot the forward- and reverse-bias results on both a semi-log plot and a linear plot, letting the forward current range from 0.1 μA to about 10 μA. Compare results with theory, and comment.

    The forward diode characteristics are often modeled by the expression

$$I = I_o \exp\left[\frac{q(V - IR_S)}{\eta kT}\right] \tag{A.7}$$

where η is the diode quality factor and R_S is a series resistance. Since there are three unknowns, three measurements should in principle provide values for I_o, η, and R_S using the three equations that follow.

$$R_S \simeq \frac{(V_3 - V_2)\ln(I_2/I_1) - (V_2 - V_1)\ln(I_3/I_2)}{(I_3 - I_2)\ln(I_2/I_1) - (I_2 - I_1)\ln(I_3/I_2)} \tag{A.8}$$

$$\eta \simeq \frac{q(V_2 - V_1) - qR_S(I_2 - I_1)}{kT\,\ln(I_2/I_1)} \tag{A.9}$$

$$I_o \simeq I_x \exp\left[\frac{(-qV_x - R_S I_x)}{\eta kT}\right] \tag{A.10}$$

where $V_3 > V_2 > V_1$ and $x = 1$, 2, or 3 [13]. What are I_o, R_S, and η for the diode? How do these values change with your choice of three data points? For example, how do they change as current increases? For best results, use current values of less than 1 μA.

11. Measure capacitance as a function of voltage for a large and a small MOS capacitor. Obtain a plot of the results and identify the three regions of interest, namely, accumulation, depletion, and inversion. What value of oxide thickness is indicated by the capacitance results in the accumulation region? How does this value compare to what you would expect from the oxide color?
12. Referring to Section 3.2 and using the value of oxide thickness, d_{OX}, obtained from question 11, calculate the theoretical value of the flat band voltage, V_{FB}, for an aluminum gate and for the background doping of your wafer.
13. Referring to Section 3.4, calculate the extrinsic Debye length, L_D, for your wafer.
14. Now calculate the flat band capacitance, C_{FB}, using the oxide thickness, d_{OX}, and the Debye length, L_D.
15. Based on both the value of C_{FB} obtained in question 14 and the measured capacitance as a function of voltage, find the value of V_{FB} indicated by measurement. Comparing this with the calculated V_{FB} of question 12, what may be concluded about the combined effect of oxide charges Q_f, Q_m, and Q_{ot}?
16. Calculate the expected value of MOSFET threshold voltage based on the theoretical value of V_{FB}.
17. Calculate the expected value of MOSFET threshold voltage based on the measured value of flat band voltage.
18. Note that one of the FETs has a large W/L ratio characteristic of a driving transistor, and that the other FET has a small W/L ratio characteristic of a load transistor. Assume that $V_{GS} - V_T = 2$ V and calculate the

expected transconductance, g_m, in saturation for the load FET and the driver FET.

19. Measure the *I-V* characteristic of a good driving MOSFET and a good load MOSFET. How do the values of V_T deduced from these transistor measurements compare with values of V_T calculated in questions 16 and 17? How do the measured values of transconductance compare with the calculated value in question 18?
20. Take photographs of the *I-V* family of curves for both MOSFETs. One should go in your lab report and the other in your lab partner's report.
21. Use SPICE to computer-generate a set of *I-V* characteristics for the load transistor and the driving transistor. For SPICE parameters such as oxide thickness, substrate doping, gate dimensions, and so forth, use the values you have determined for your particular wafer. Compare the SPICE output with your measurements. Include the SPICE output with your lab report.

A.7 MOS Logic Experiment: Option 1

A.7.1 Introduction

This experiment adds two more masks, one for an additional diffusion and another for gate oxide definition. The use of a field oxide allows self-isolation of the transistors so that the devices can be interconnected in circuits. For the mask set discussed here each chip contains four PMOS inverting gates and a two-input PMOS NAND gate. Two of the inverters are of the *p*-enhancement-load-saturated (PELS) design, with an enhancement-mode load operating in the saturated region, and two of the inverters are of the *p*-enhancement-load-triode (PELT) design, with an enhancement load operating in the triode region.

In addition to the logic circuits, each wafer has four test areas containing a variety of test devices. When aligning each mask, after mask 1, *make sure the test areas are in the right place.* A detailed breakdown of the laboratory procedure follows, with the dotted lines again indicating appropriate break points. As earlier, procedures are from Section A.9.

A.7.2 MOS Logic Option 1 Fabrication Sequence

1. Four-point probe measurement, as in the oxidation experiment.
2. Hot probe measurement, as in the oxidation experiment.

- -

3. Initial Wafer Cleaning Procedure.
4. Wafer oxidation at 1100°C, as in the oxidation experiment, but with 10 minutes dry, 1 hour wet, and 10 minutes dry.

- -

5. Prephotoresist Cleaning Procedure, unless going directly from the oxidation furnace to the spinner, in which case this step can be skipped.
6. Photoresist Coating Procedure.
7. Photoresist Exposure Procedure with mask 1, "PMOS LOGIC P P DIFF," for boron diffusion oxide cuts.
8. Photoresist Develop Procedure.
9. Microscope inspection of the wafer.

---

10. Photoresist Postbake Procedure.
11. Oxide Window Etch Procedure.
12. Microscope inspection of the wafer.

---

13. Photoresist Strip Procedure.
14. Microscope inspection of the wafer.
15. Boron Predeposition Procedure.

---

16. Boron-Skin Removal Procedure.
17. Boron Drive-In Procedure.

---

18. Prephotoresist Cleaning Procedure, unless going directly from the diffusion furnace to the spinner, in which case this step can be skipped.
19. Photoresist Coating Procedure.
20. Photoresist Exposure Procedure using mask 2, "PMOS LOGIC N P DIFF," for phosphorus diffusion oxide cuts; this establishes the substrate contact.
21. Photoresist Develop Procedure.
22. Microscope inspection of the wafer.

---

23. Photoresist Postbake Procedure.
24. Oxide Window Etch Procedure.
25. Microscope inspection of the wafer.

---

26. Photoresist Strip Procedure.
27. Microscope inspection of the wafer.
28. Phosphorus Predeposition Procedure.

---

29. Phosphosilicate Glass Removal Procedure.
30. Bake out at 200°C for 30 minutes.
31. Photoresist Coating Procedure.
32. Photoresist Exposure Procedure using mask 3, "PMOS LOGIC GATE OX," for gate oxide definition.

33. Photoresist Develop Procedure.
34. Microscope inspection of the wafer.

---

35. Photoresist Postbake Procedure.
36. Oxide Window Etch Procedure.
37. Microscope inspection of the wafer.

---

38. Photoresist Strip Procedure.
39. Microscope inspection of the wafer.
40. Gate Oxidation Procedure.

---

41. Prephotoresist Cleaning Procedure, unless going directly from the furnace to the spinner.
42. Photoresist Coating Procedure.
43. Photoresist Exposure Procedure using mask 4, "PMOS LOGIC CONT," for contact cut definition.
44. Photoresist Develop Procedure.
45. Microscope inspection of the wafer.

---

46. Photoresist Postbake Procedure.
47. Oxide Window Etch Procedure.
48. Microscope inspection of the wafer.

---

49. Photoresist Strip Procedure.
50. Microscope inspection of the wafer.

---

51. Preevaporation Cleaning Procedure.
52. Aluminum Evaporation Procedure on front of wafer.

---

53. Photoresist Coating Procedure.
54. Photoresist Exposure Procedure using mask 5, "PMOS LOGIC MET-1," for metalization patterning.
55. Photoresist Develop Procedure.
56. Microscope inspection of the wafer.

---

57. Photoresist Postbake Procedure.
58. Aluminum Etch Procedure.
59. Microscope inspection of the wafer.

---

60. Photoresist Strip Procedure.
61. Microscope inspection of the wafer.

---

62. Aluminum Anneal Procedure.

---

63. Circuit testing and evaluation.

A.7.3 MOS Logic Option 1 Mask Set and Pin Assignment [14]

The mask set for the logic chips in this experiment and for the test areas are shown in Figures A.10 and A.11, respectively. The pin assignments for the logic chip and for the test area are shown in Figures A.12 and A.13, respectively. Details of the pin assignments are shown in Tables A.2 and A.3.

Figure A.10 (A–H) Mask set for PMOS logic option 1 and option 2. A. Boron diffusion mask, "PMOS LOGIC P P DIFF."

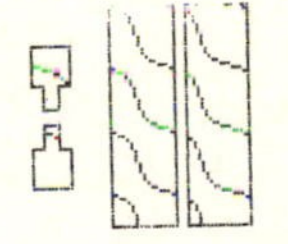

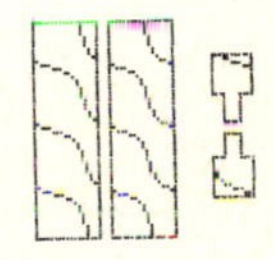

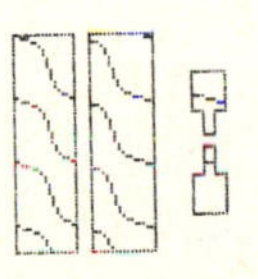

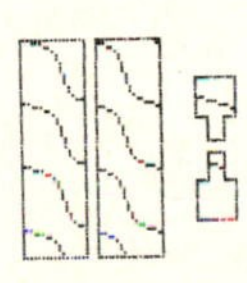

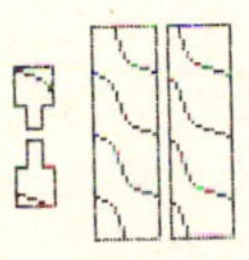

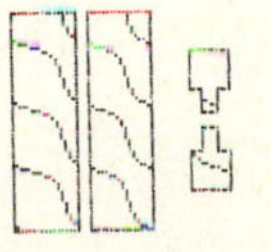

Figure A.10 Mask set for PMOS logic option 1 and option 2 (*cont.*) B. Phosphorus/boron diffusion mask, "PMOS LOGIC N P DIFF." C. Gate oxide mask, "PMOS LOGIC GATE OX."

B

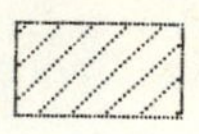

C

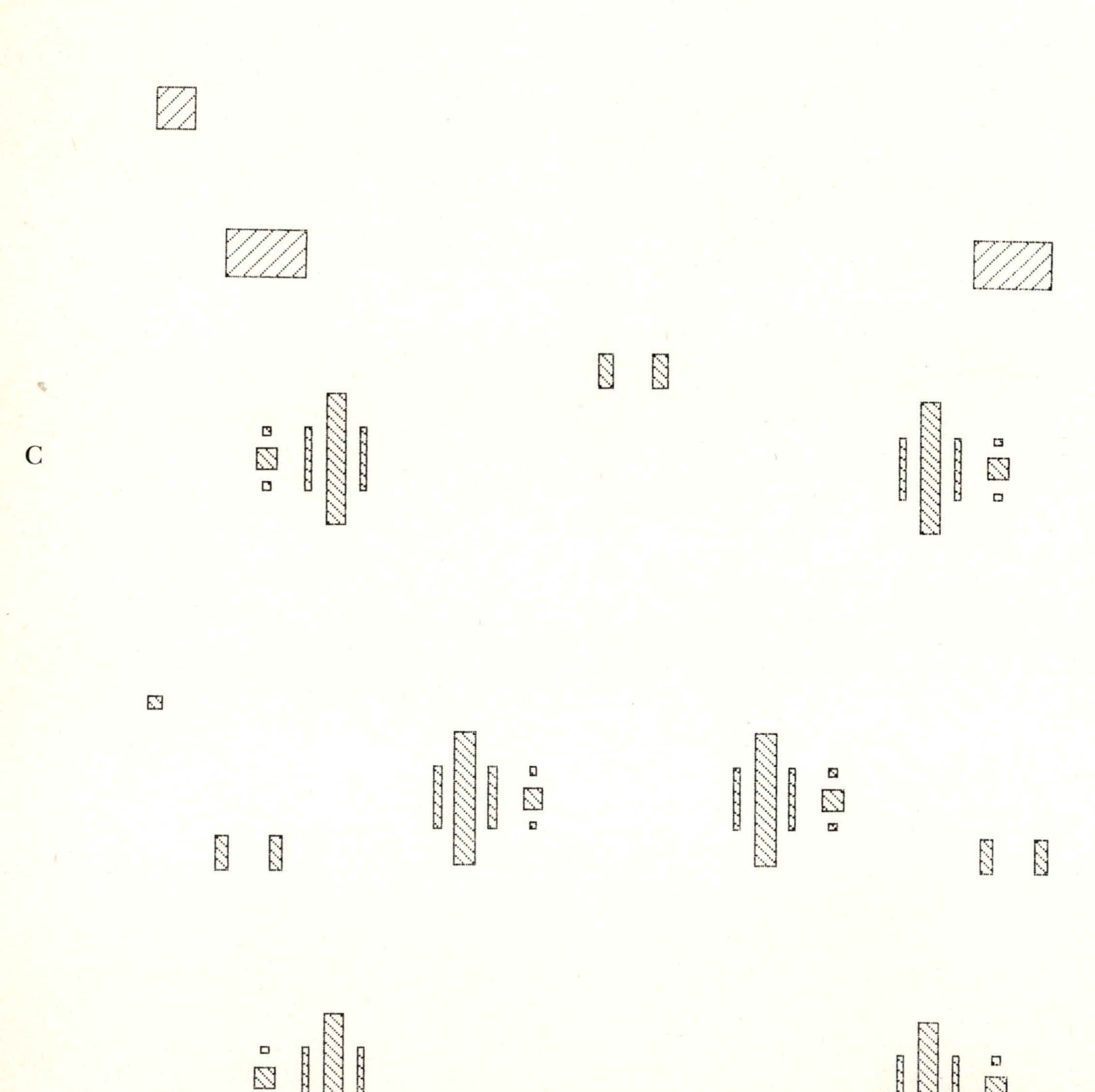

Figure A.10 Mask set for PMOS logic option 1 and option 2 (*cont.*) D. Contact cut mask, "PMOS LOGIC CONT."

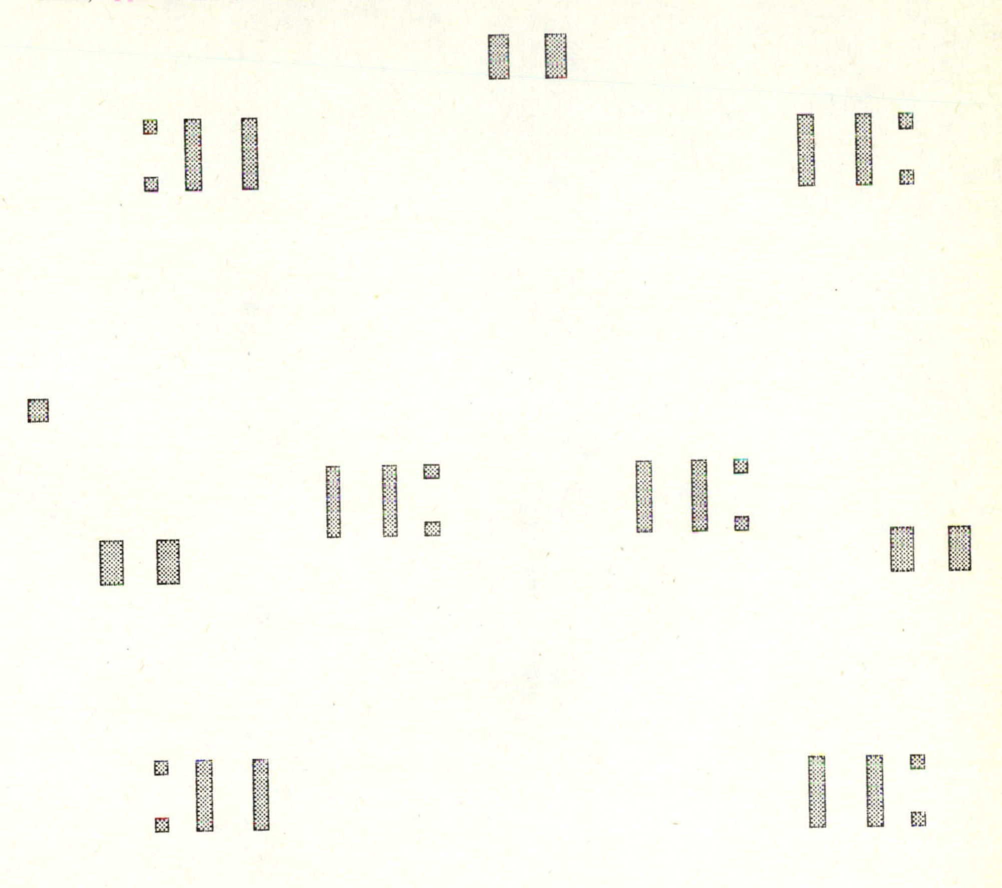

Figure A.10 Mask set for PMOS logic option 1 and option 2 (*cont.*) E. Option 1 metal mask, "PMOS LOGIC MET-1." F. Option 2 metal mask, "PMOS LOGIC MET-2."

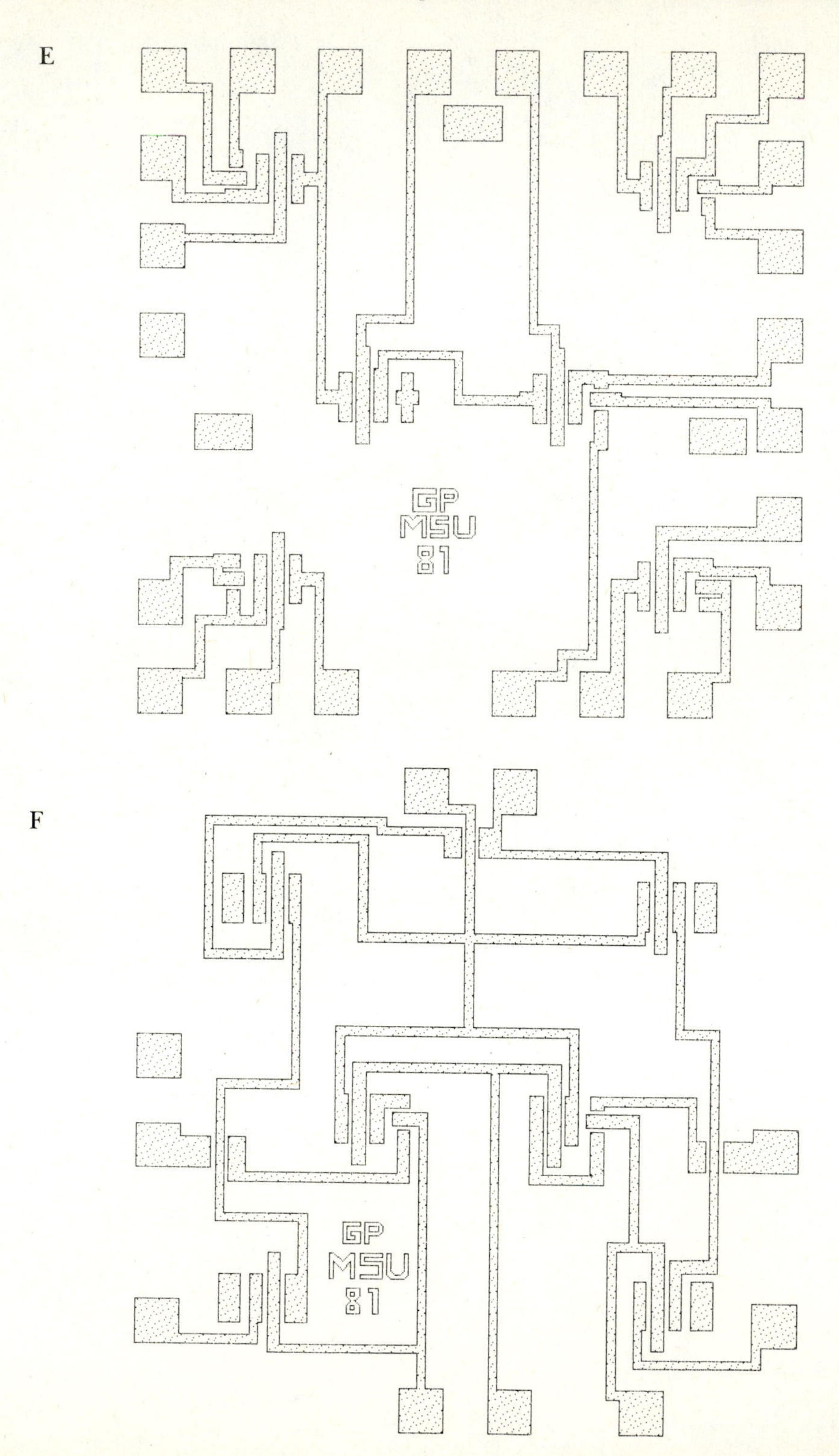

Figure A.10 Mask set for PMOS logic option 1 and option 2 (*cont.*) G. Combined layout for option 1.

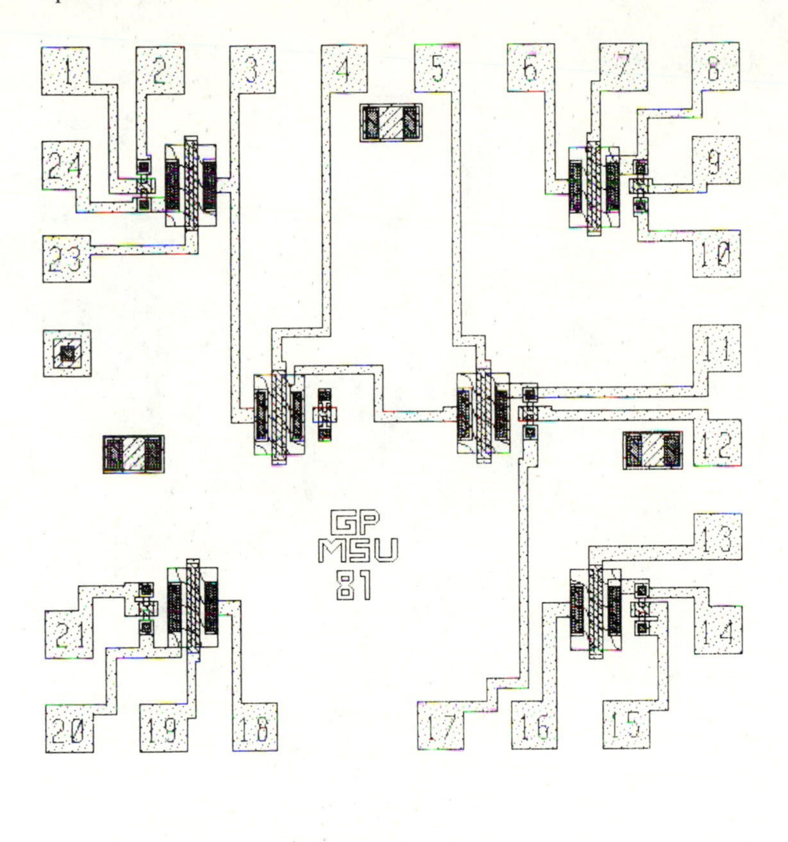

p-type diffusion

n-type diffusion

Gate oxide

Contact cuts

Metal

Figure A.10 Mask set for PMOS logic option 1 and option 2 (*cont.*) H. Combined layout for option 2.

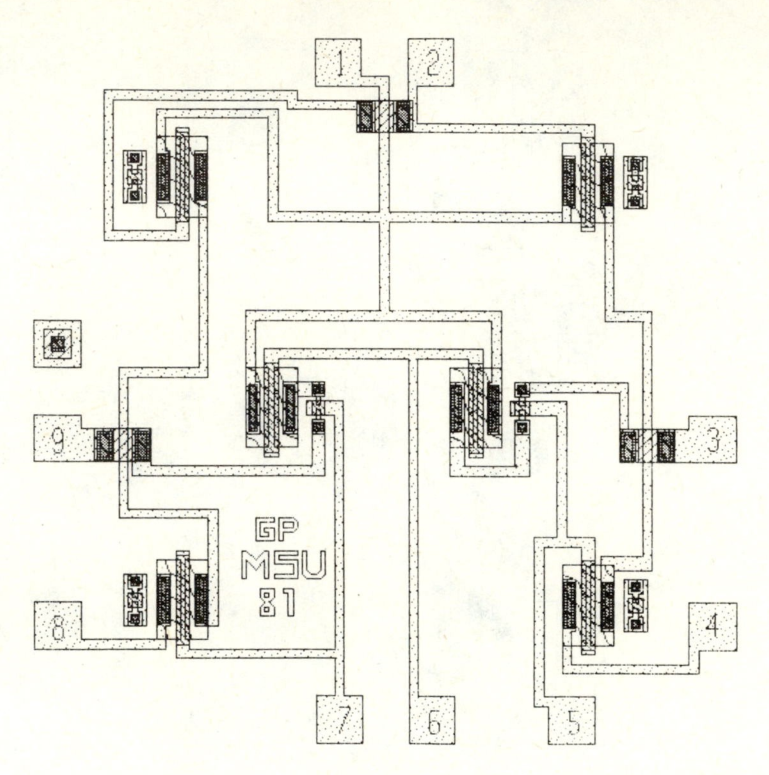

p-type diffusion

Second *p*-type diffusion

Gate oxide

Contact cuts

Metal

Figure A.11 Test-area layout for PMOS logic option 1 and option 2.

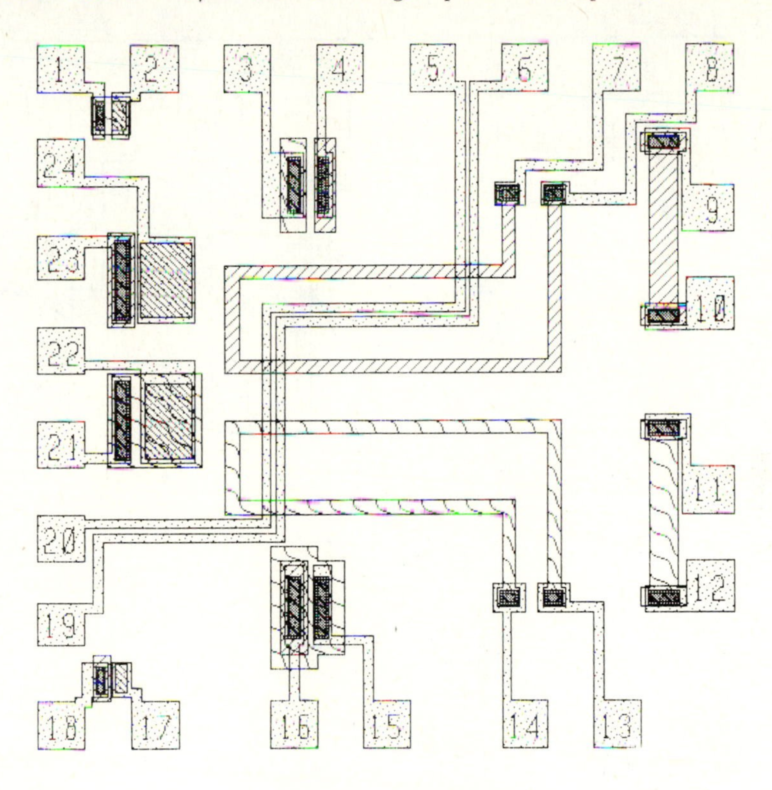

p-type diffusion

n-type, or second *p*-type, diffusion

Gate oxide

Contact cuts

Metal

Figure A.12 Pin assignment for PMOS logic experiment option 1.

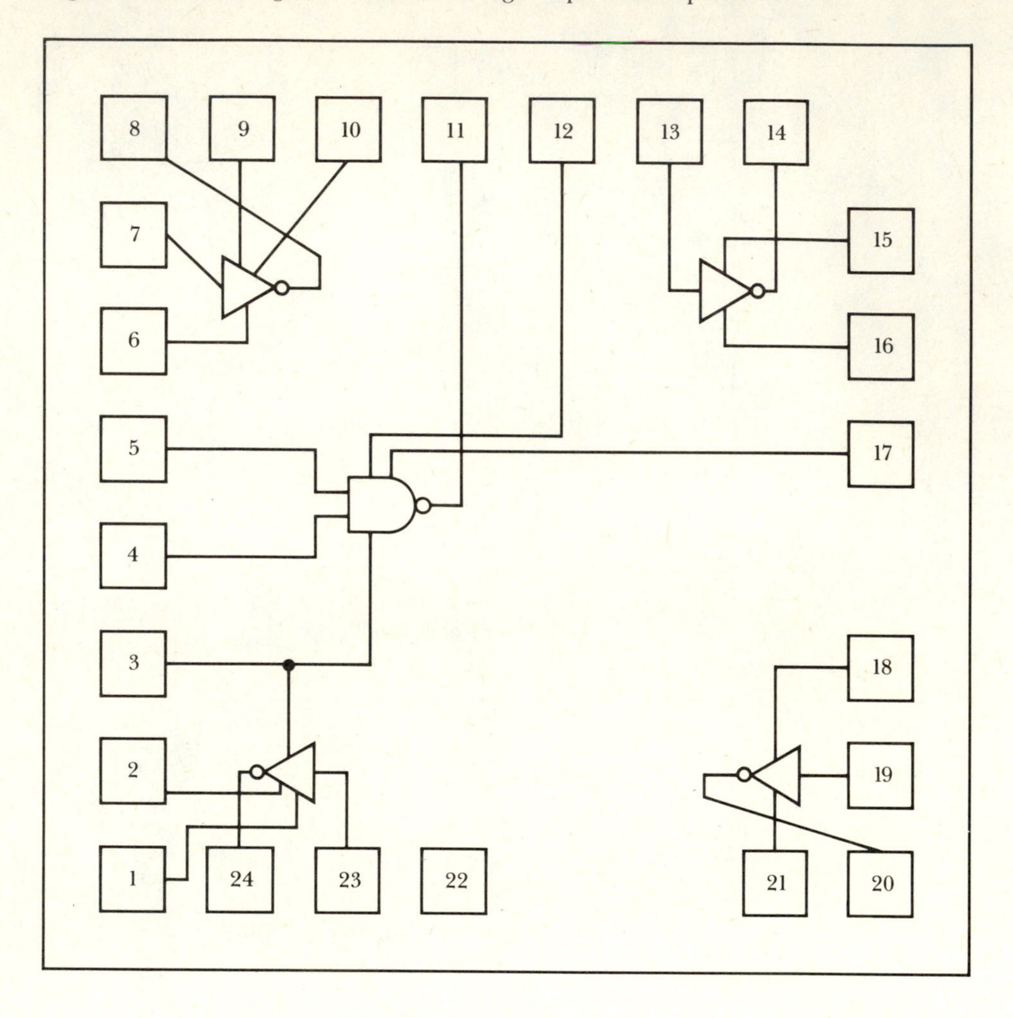

Figure A.13 Pin assignment for PMOS logic test area for option 1 and option 2.

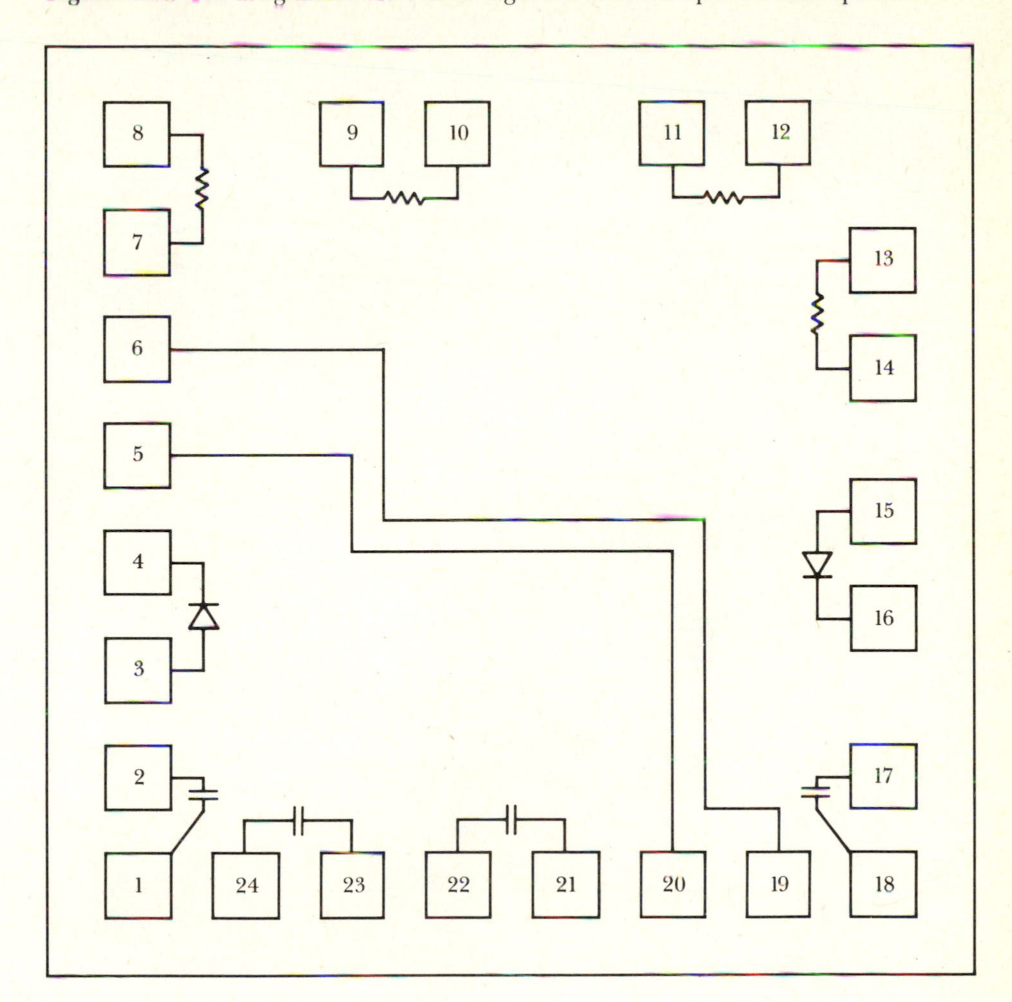

Table A.2 MOS logic option 1 pin assignment for the logic chip

Pin	Function
First PELT Inverter	
2	$-V_{DD}$
1	$-V_{GG}$
24	Output
23	Input
3	Ground
22	Substrate contact (for all gates)
NAND Gate	
4	Input 1
5	Input 2
3	Ground
17	$-V_{DD}$
12	$-V_{GG}$
11	Output
Second PELT Inverter	
6	Ground
10	$-V_{DD}$
9	$-V_{GG}$
8	Output
7	Input
First PELS Inverter	
15	$-V_{DD}$
14	Output
13	Input
16	Ground
Second PELS Inverter	
18	Ground
21	$-V_{DD}$
20	Output
19	Input

Table A.3 MOS logic test-area pin assignment (*Note:* The same test area is used for both option 1 and option 2. However, some test devices are only functional in one of the two options.)

Pins	Test device
1,2	Small MOS capacitor. Gate oxide over first diffusion. Pin 2 is the top electrode.
3,4	Diode resulting from the junction between the first diffusion and the substrate. Pin 3 is connected to the first diffusion.
5,6,19,20	Metal test. There should be a short circuit between 5 and 20 and between 6 and 19. There should be an open circuit between 5 and 6 and between 19 and 20.
7,8	Resistor with 68 squares formed from the second diffusion. Note that for option 1, this is not a useful test device.
9,10	Resistor with 5.8 squares formed from the second diffusion. Note that for option 1, this is not a useful test device.
11,12	Resistor with 5.8 squares formed from the first diffusion.
13,14	Resistor with 68 squares formed from the first diffusion.
15,16	Diode formed by the junction between the first and second diffusions. Pin 15 is the first diffusion. Note that this is not functional for option 2.
17,18	Small MOS capacitor resulting from the gate oxide over the silicon substrate. Pin 17 is the gate electrode.
21,22	Large MOS capacitor resulting from the gate oxide over the first diffusion. Pin 22 is the gate electrode.
23,24	Large MOS capacitor resulting from the gate oxide over the silicon substrate. Pin 24 is the gate electrode.

A.8 MOS Logic Experiment: Option 2

A.8.1 Introduction

In this experiment, each chip contains a general logic function block, the circuit of which is shown in Figure A.14. The operation of this circuit is as discussed in Section 6.2.3; however, the implementation in this experiment does not use ion-implanted depletion-mode transistor technology. All of the pass transistors are of the driver type except for two that are of the load type. These two are indicated with the letter *L* on the circuit diagram. For this option, the differing transistor geometries serve no useful function. Two load-type transistors are used because there are not enough drivers on the

Figure A.14 Circuit for PMOS logic option 2.

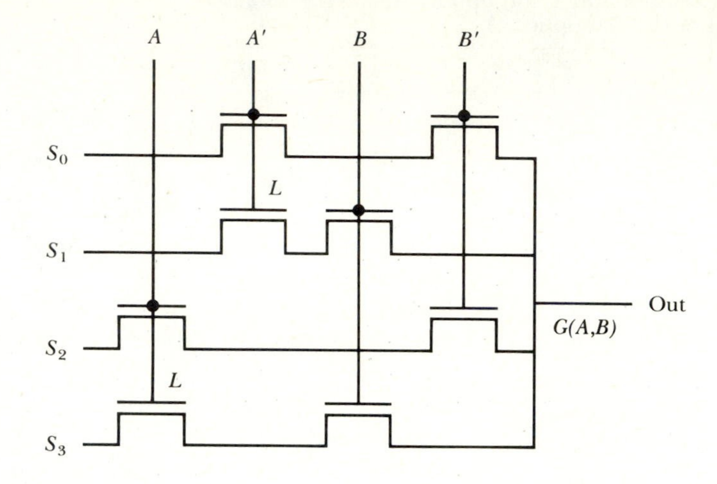

Table A.4 PMOS logic experiment option 2 functions

Control Word				$G(A,B)$	
S_3	S_2	S_1	S_0	Logic	Arithmetic/Logic
0	0	0	0	0	0
0	0	0	1	$(A + B)'$	A plus $(A + B')$
0	0	1	0	$A'B$	A plus $(A + B)$
0	0	1	1	A'	A plus 1
0	1	0	0	AB'	A plus (AB)
0	1	0	1	B'	$(A + B')$ plus (AB)
0	1	1	0	$A \oplus B$	A plus B
0	1	1	1	$(AB)'$	(AB) plus 1
1	0	0	0	AB	A plus (AB')
1	0	0	1	$(A \oplus B)'$	A plus B plus 1
1	0	1	0	B	$(A + B)$ plus (AB')
1	0	1	1	$A' + B$	(AB') plus 1
1	1	0	0	A	A' plus 1
1	1	0	1	$A + B'$	$(A'B)$ plus 1
1	1	1	0	$A + B$	$(A + B)'$ plus 1
1	1	1	1	1	1

chip for this circuit. As a review, it should be verified that the relationship between the 4-bit control word, $S_3S_2S_1S_0$, and the logic/arithmetic function of this block is as shown in Table A.4.

The fabrication sequence is quite similar to option 1's, and four of the five masks are common to both options. The chief differences in option 2 are that the second diffusion, which is used to establish crossovers, is *p*-type rather than *n*-type, and that a different metalization mask is used. As with option 1, there are four test areas on the wafer, and care should be taken that these are in the right location for each mask.

A.8.2 MOS Logic Option 2 Fabrication Sequence

Again the dotted lines indicate appropriate stopping points, and procedures are from Section A.9.

1. Four-point probe measurement, as in the oxidation experiment.
2. Hot probe measurement, as in the oxidation experiment.

- -

3. Initial Wafer Cleaning Procedure.
4. Wafer oxidation at 1100°C, as in the oxidation experiment, but with 10 minutes dry, 1 hour wet, and 10 minutes dry.
5. Prephotoresist Cleaning Procedure, unless going directly from the oxidation furnace to the spinner, in which case this step can be skipped.
6. Photoresist Coating Procedure.
7. Photoresist Exposure Procedure with mask 1, "PMOS LOGIC P P DIFF," for boron diffusion oxide cuts.
8. Photoresist Develop Procedure.
9. Microscope inspection of the wafer.

- -

10. Photoresist Postbake Procedure.
11. Oxide Window Etch Procedure.
12. Microscope inspection of the wafer.

- -

13. Photoresist Strip Procedure.
14. Microscope inspection of the wafer.
15. Boron Predeposition Procedure.

- -

16. Boron-Skin Removal Procedure.
17. Boron Drive-In Procedure.

- -

18. Prephotoresist Cleaning Procedure, unless going directly from the diffusion furnace to the spinner, in which case this step can be skipped.
19. Photoresist Coating Procedure.
20. Photoresist Exposure Procedure using mask 2, "PMOS LOGIC N P DIFF," for second boron diffusion oxide cuts.
21. Photoresist Develop Procedure.
22. Microscope inspection of the wafer.

- -

23. Photoresist Postbake Procedure.
24. Oxide Window Etch Procedure.
25. Microscope inspection of the wafer.

- -

26. Photoresist Strip Procedure.
27. Microscope inspection of the wafer.
28. Boron Predeposition Procedure.

- -

29. Boron-Skin Removal Procedure.
30. Boron Drive-In Procedure.

- -

31. Prephotoresist Cleaning Procedure, unless going directly from the diffusion furnace to the spinner, in which case this step can be skipped.
32. Photoresist Coating Procedure.
33. Photoresist Exposure Procedure using mask 3, "PMOS LOGIC GATE OX," for gate oxide definition.
34. Photoresist Develop Procedure.
35. Microscope inspection of the wafer.

- -

36. Photoresist Postbake Procedure.
37. Oxide Window Etch Procedure.
38. Microscope inspection of the wafer.

- -

39. Photoresist Strip Procedure.
40. Microscope inspection of the wafer.
41. Gate Oxidation Procedure.

- -

42. Prephotoresist Cleaning Procedure, unless going directly from the furnace to the spinner.
43. Photoresist Coating Procedure.

44. Photoresist Exposure Procedure using mask 4, "PMOS LOGIC CONT," for contact cut definition.
45. Photoresist Develop Procedure.
46. Microscope inspection of the wafer.

---

47. Photoresist Postbake Procedure.
48. Oxide Window Etch Procedure.
49. Microscope inspection of the wafer.

---

50. Photoresist Strip Procedure.
51. Microscope inspection of the wafer.

---

52. Preevaporation Cleaning Procedure.
53. Aluminum Evaporation Procedure on front of wafer.

---

54. Photoresist Coating Procedure.
55. Photoresist Exposure Procedure using mask 5, "PMOS LOGIC MET-2," for metalization patterning.
56. Photoresist Develop Procedure.
57. Microscope inspection of the wafer.

---

58. Photoresist Postbake Procedure.
59. Aluminum Etch Procedure.
60. Microscope inspection of the wafer.

---

61. Photoresist Strip Procedure.
62. Microscope inspection of the wafer.

---

63. Aluminum Evaporation Procedure on back of the wafer. In option 2, the substrate contact is the back of the wafer.
64. Aluminum Anneal Procedure.

---

65. Circuit testing and evaluation.

A.8.3 MOS Logic Option 2 Mask Set and Pin Assignment

The mask set for option 2 differs from option 1's only in the metalization mask, as shown in Figure A.10. Figure A.15 shows the pin assignment.

Figure A.15 Pin assignment for PMOS logic experiment option 2.

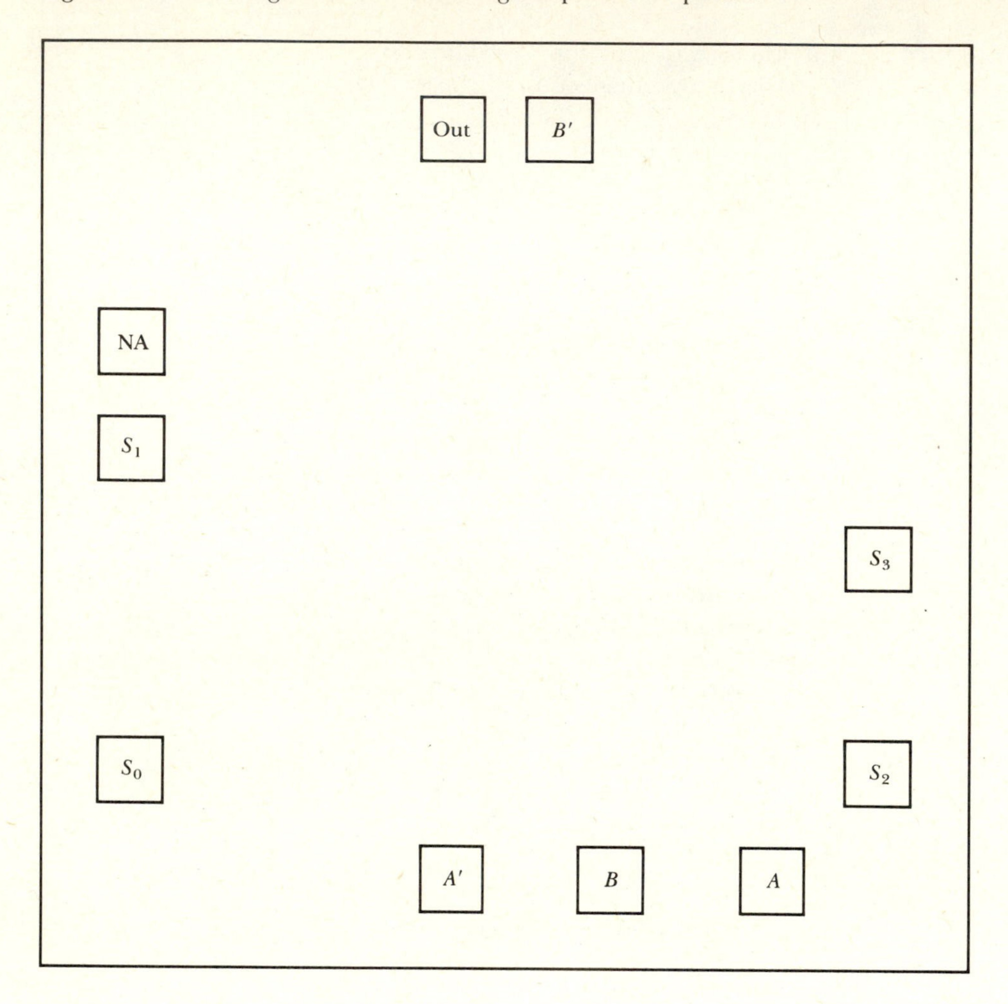

A.9 Laboratory Procedures

This section lists a number of laboratory procedures that are used, several times in some cases, during the fabrication sequences described in the preceding sections. They are not necessarily identical to procedures used in other laboratories or in industry; from one facility to another, considerable variation is often found in the details of how a given fabrication procedure is carried out. However, techniques used at other institutions have served as helpful guidelines here [15, 16]. Since the following procedures are intended for a student laboratory setting, they generally tend toward simplicity. In a setting where a very high degree of purity is required, a more stringent set of steps would be followed for some of the procedures.

De-ionized water is again referred to as DI in the following procedures. Satisfactory results have been obtained using reagent-grade solvents and acids, which are available from most standard chemical stock suppliers. However, more consistent results can be expected with chemicals that meet the specifications of the Semiconductor Equipment and Materials Institute (SEMI). SEMI specifications limit trace impurities of over 20 elements to values of a part per million or less. The price difference between reagent-grade and SEMI-grade chemicals is not great. For nitrogen gas, which is used at several points in the fabrication sequence, the 99.9%-purity grade, which is readily available at a few dollars per 220-ft^3 cylinder, is acceptable for these exercises. Suggested flow rates are appropriate for 3-inch diameter furnace tubes. For the oxidation and drive-in furnaces, the undergraduate lab at Michigan State University has used 99.5%-purity oxygen gas, which is also available at a few dollars per cylinder. This grade is rather substandard; nevertheless, MOS characteristics have been acceptable for purposes of the teaching lab. For FET's with fairly thick gate oxides, most students have measured PMOS threshold voltages in the range of -3 to -8 V, with the actual value depending on whether forming gas was used for the annealing process and on the cleanliness of the processing for the particular wafer. Differences between theory and experiment are usually large enough to appreciate the importance of nonideal phenomena, but small enough to appreciate the basic validity of the theoretical analysis.

The wet chemistry steps are carried out in glass beakers, unless the solution contains HF, in which case Teflon is used [3]. Wafer handling is greatly simplified by using wafer holders and wafer baskets [17]. It is advisable to use separate wafer holding baskets for each type of chemical solution. An exception is the photoresist development procedure, where timing is critical and chemicals are compatible. It is always important to use clean utensils to handle the wafers.

Most of the procedures in this section are not equipment specific and should transfer well to other laboratories. However, a few procedures involve specialized equipment, such as mask aligners. In those cases it is necessary for the user to supplement the text material with operating information specific to the particular equipment being used.

Finally, it is repeated for the sake of emphasis that safety issues associated with the various chemicals should be well in mind before carrying out a procedure.

Initial Wafer Cleaning Procedure

1. If the wafer has obvious dirt or fingerprints, clean it by scrubbing with TCE and a cotton-tip swab. (Lint-free swabs are also available.) TCE is an effective solvent and degreaser.

2. Immerse the wafer and basket in a boiling TCE solution for 3 minutes. Don't let the TCE dry on the wafer because it may leave stains that will persist during the rest of this procedure. If such stains do occur, they can be removed by a TCE scrub as in step 1.

Safety Precaution

Never heat TCE, or any solvent, near heated solutions containing hydrogen peroxide such as degrease etch and demetal etch. Solvent vapors are flammable and H_2O_2 produces large amounts of oxygen. The combination can create an explosion hazard.

3. Rinse immediately in running acetone, from a squeeze bottle, for several seconds. This rinse removes the TCE residue and acts as a further cleaning solvent.
4. Rinse in running methanol for several seconds. This rinse removes the acetone residue.
5. Rinse in running DI for 2 minutes.
6. Dry with N_2.
7. Immerse in freshly prepared degrease etch (see Section A.3.7) for 10 minutes at $70 \pm 5°C$ to remove residual organic contamination left over from the solvent cleaning. H_2O_2 plays a critical role in degrease etch, but the half-lifetime of H_2O_2 in this heated solution is only about 30 minutes.
8. Rinse in DI for 2 minutes.
9. Dry with N_2.
10. Immerse in freshly prepared demetal etch (see Section A.3.7) for 10 minutes at $70 \pm 5°C$ to remove ionic and metallic contaminants. Again, the half-lifetime of H_2O_2 in this heated solution is only about 30 minutes.
11. Rinse in DI for 3 minutes.
12. Dry with N_2.
13. The wafer should be spotless at this point, with no stains or blemishes. Proceed to the next fabrication step.

Prephotoresist Cleaning Procedure

When taking the wafer straight from the oxidation or diffusion furnace to the photoresist procedures, this step can be skipped.

1. Immerse the wafer in boiling TCE for 3 minutes.
2. Rinse in running acetone for several seconds.
3. Rinse in running methanol for several seconds.
4. Rinse in running DI for 3 minutes.
5. Dry with N_2.

6. Bake out the silicon wafer in a clean room oven for 30 minutes at $200 \pm 5°C$.

Proceed directly to *Photoresist Coating*.

Photoresist Coating Procedure

The steps here may be supplemented by information specific to the photoresist and photoresist spinner model being used.

Note: The wafer and photoresist must not be exposed to light other than long-wavelength yellow light.

1. Allow the wafer to cool down from the bake-out.
2. If the spinner chuck that holds the wafer has residual resist on it, clean the chuck with xylene. Center the wafer on the chuck. Spin the wafer and blow it clean with N_2 to remove dust particles.
3. Cover the entire wafer with negative resist. Wait for about 5 seconds.
4. Spin the wafer such that the resulting photoresist layer is about 1 μm thick. For Waycoat Negative HR Resist 100, 200, and 300, the thickness versus spin speed is shown in Figure A.16 [18]. As the wafer spins, the thickness of the photoresist changes as can be seen from the moving color fringes. When the color becomes uniform across the wafer, the photoresist has reached a final thickness. An appropriate spin/speed combination for Waycoat 200 resist is 3000 rpm and 30 seconds.
5. With the help of tweezers, slide the wafer off the chuck onto filter paper. Ideally, the wafer should show a uniformly colored layer of photoresist. However, some degree of nonuniform color will not wreck the experiment. Also, there may be a "comet" on the wafer owing to a dust particle that has caused local streaking of the resist. This flaw is also tolerable since any yield loss will be fairly localized.
6. Place the wafer in an oven, under a glass petri dish, and prebake at the time and temperature recommended for the particular photoresist being used. For Waycoat HR-200, the procedure is 20 minutes at $65 \pm 10°C$. The purpose of the prebake is to solidify the resist.

Proceed directly to *Photoresist Exposure*.

Photoresist Exposure Procedure

This information is to be supplemented by information specific to the mask aligner being used. The following are some generic instructions.

1. Study the mask aligner operating sequence carefully since it is possible to damage both the wafer and the mask if the correct procedures are not followed.

Figure A.16 Resist thickness as a function of spin speed.
SOURCE: Waycoat® resist is a registered trademark for several photoresist-related products from Olin Hunt Specialty Products, Inc., West Paterson, N.J. 07494.

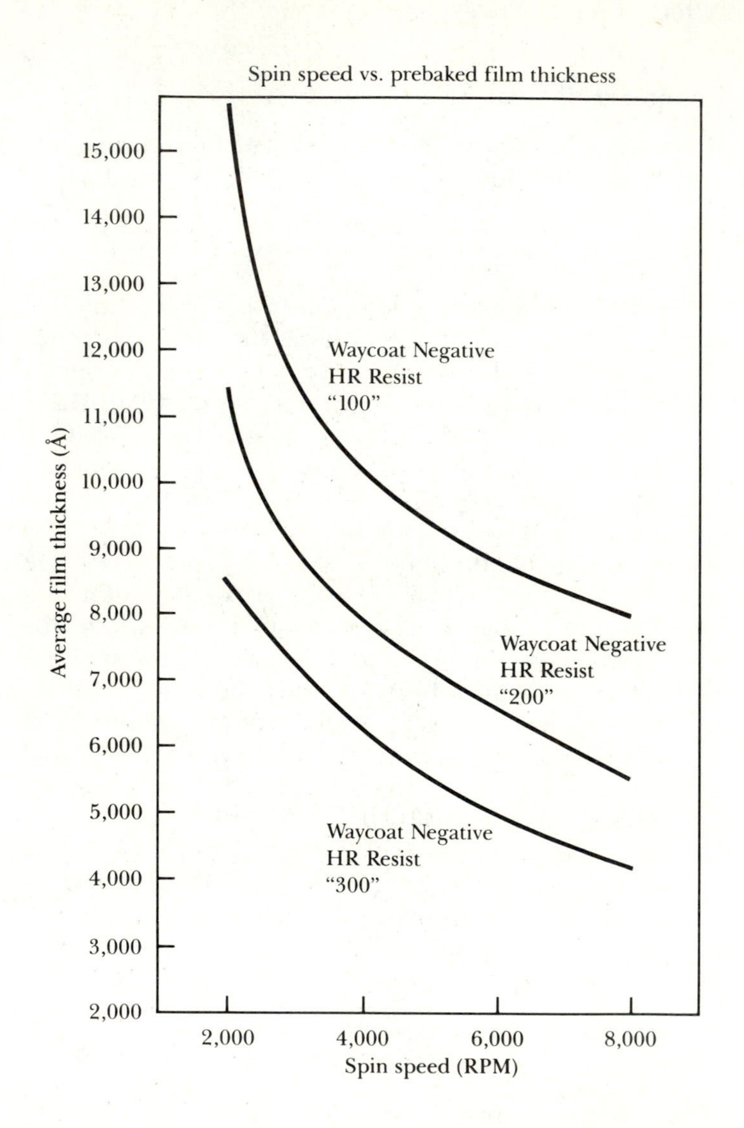

2. Make sure the ultraviolet lamp is on and that all vacuum and pressure lines are functioning. To allow warm-up time, wait 5 minutes after the lamp has been turned on before making an exposure. Usually, the lamp is on a power supply separate from that for the logic circuitry associated with the mask aligner controls. It is good to have the logic power supply turned off when igniting the lamp to avoid damage from high-voltage spikes.
3. Place the wafer on the holding chuck and the mask in the mask holder. Verify that both are held securely and that the correct mask is being used.
4. Align the mask relative to the previous patterning on the wafer. If it is the first mask, make sure that the wafer is approximately centered under the mask and that the scribe lines on the mask are parallel and perpendicular to the flat marked on the wafer. If it is a later mask, use the translate and rotate features of the wafer holder to achieve alignment. Generally the microscope viewer is fitted with a split-field feature such that two sides of the wafer are viewed simultaneously. Scan across the wafer to double-check alignment.
5. If a contact aligner is being used, bring the wafer into contact with the mask and expose the wafer to the ultraviolet light. Typically the exposure time is several seconds; however, the appropriate exposure time for a particular combination of lamp strength and photoresist should be established by experiment for a given setup.
6. Disengage the wafer from the mask and remove the wafer from the mask aligner. If the wafer sticks to the mask, one possible explanation is that the prebake temperature or time was too low.

Proceed directly to *Photoresist Develop.*

Photoresist Develop Procedure

A variety of available developing solutions act as a solvent for the unexposed, and therefore nonpolymerized negative photoresist. Companies that manufacture photoresist also make developing solutions. Stoddard's industrial solvent may also be used as a developer. The instructions that follow are for Waycoat negative photoresist and developer [18], but a similar sequence is used for other developers.

1. Using tweezers, place the wafer in a Teflon basket.
2. Immerse the wafer in the developer, agitating continuously so that a continual supply of fresh developer solution is available at the wafer surface. A typical development time is approximately 100 seconds. The developer acts as a solvent, removing the unexposed, nonpolymerized negative resist.
3. Remove the basket and wafer and shake off the excess developer. Before

the wafer has a chance to dry and develop hard-to-remove residues, immerse the basket and wafer in xylene for 15 to 20 seconds, again with agitation. The xylene removes the remaining developer and stops the development process. Remove the wafer from the bath and rinse with running xylene for a few seconds.

4. After rinsing the wafer with xylene and before the wafer can dry, immerse the basket and wafer in isopropyl alcohol for 15 to 20 seconds, with agitation, to remove the xylene. Remove the wafer from the bath and rinse it in running isopropyl alcohol for a few seconds.
5. Use tweezers to transfer the wafer to a wafer holder to facilitate drying and blow dry with N_2. Do so immediately, before the wafer dries in the air, since stains may result otherwise.

The wafer is now ready for the Postbake Procedure and is no longer sensitive to room light.

Photoresist Postbake Procedure

The postbaking step further hardens the photoresist and increases the robustness of the resist as a barrier to etching solutions. As a result of some degree of plastic flow, this procedure may also eliminate small pin holes in the resist. The following specific times and temperature are for Waycoat 200 negative photoresist.

1. Place the wafers face up in glass petri dishes in an oven.
2. Bake at 135 ± 5°C for 15 minutes.
3. Remove the wafers and let cool.

Proceed directly to the appropriate etch procedure.

Oxide Window Etch Procedure

This procedure uses three beakers. One contains the buffered oxide etch (see Section A.3.7), and the other two contain DI for sequential rinses and are referred to as DI-1 and DI-2.

1. Immerse the wafer and basket into the buffered oxide etch. The etch rate is typically about 1000 Å/min. Since extreme degrees of overetching should be avoided, it is a good idea to initially leave the wafer in the etch for somewhat less than the estimated required time based on your oxide thickness.
2. Remove the wafer from the etch and immerse it in DI-1 for a few seconds, to stop the etching procedure, and then in DI-2. Observe the back surface. If it is wet, then SiO_2 still remains since the oxide is hydrophilic. However, if the SiO_2 is completely etched away, there will be only a few isolated drops since silicon is hydrophobic.
3. If SiO_2 remains, immerse the wafer in the buffered etch again for an-

other 30 seconds. Check the back of the wafer as before. Continue until the oxide is removed. A microscope inspection of the etched windows is also helpful in determining whether the oxide is gone.

4. Rinse in running DI.
5. Dry with N_2.
6. As a final check, view the etched areas of the wafer under the microscope; you should not see oxide, which has colors as listed in Table A.1, but instead silicon, which appears rather white under a high-power microscope with vertically incident light.

Particularly for contact cuts, it is *extremely important* that all the oxide in the windows be removed. Otherwise, electrical contact between the metal and semiconductor will be prevented and device and circuit testing will be impossible. Also, the effect of the boron diffusion is reduced by a residual oxide layer. A thin oxide layer is invisible since the last few hundred Angstroms of oxide do not show any color. So it is better to err on the side of overetching. It is much better to etch the oxide 30 seconds longer than necessary than 30 seconds less than necessary.

Photoresist Strip Procedure

The intent of this procedure is to completely remove the resist from the wafer after it has served the intended purpose of selectively exposing portions of the wafer to an etching medium. Again, companies that supply photoresist often supply a photoresist stripping solution as well. The procedure below is based on Waycoat Microstrip, which is a resist-removal solution [18]. The procedure makes use of a two-bath system. It is not necessary to use fresh Microstrip each time.

1. Immerse the wafer in a basket in hot Microstrip bath number 1 for 5 minutes at 90 to 100°C. Repeat for 5 minutes in Microstrip bath number 2 for 5 minutes at 90 to 100°C.
2. Rinse with methanol.
3. Rinse with DI.
4. Dry with N_2.
5. If residue remains, repeat the procedure. If a problem persists, it may be time to replace the stripping solution. Let the old bath number 2 become the new bath number 1, and pour a fresh solution for the new bath number 2.

Boron Predeposition Procedure

As discussed in Chapter 2, a variety of techniques are available for carrying out boron diffusions, including using gas, liquid, and solid sources of boron. From the perspective of safety and simplicity, which is the underlying theme of these procedures, planar boron sources that contain a heavy concentration of B_2O_3 in a glass-ceramic matrix or on a boron-nitride wafer are very suitable. These dopant sources are in the form of wafers, and are designed to be interleaved with the silicon wafers in the diffusion furnace using a quartz boat, as shown in Figure 2.17. The times and temperatures that follow are for GS 139 Boron$^+$ Planar Dopant Sources from Electro Oxide, which provide B_2O_3 distributed in a glass carrier instead of on a boron-nitride carrier [19]. Other boron sources may require somewhat different times, temperatures, and procedures for appropriate results.

1. The silicon wafer should appear clean on microscope inspection. If it does not, do a DI rinse followed by an N_2 dry. If it is still dirty, go through the TCE, acetone, methanol, DI, N_2-dry sequence. For more rigorous cleaning, or if the wafer has been dropped on the floor or touched with bare hands, steps 1 through 8 of the Preevaporation Cleaning Procedure can be used. This procedure provides for a degrease and demetal etch followed by a mild oxide etch.
2. The wafers should be placed in the quartz diffusion boat, with the polished front side of the wafer toward the nearest boron source. Silicon wafers are interspersed between dopant wafers with one dopant wafer, then two silicon wafers, then one dopant wafer, and so on. Use an N_2 stream to remove any dust particles from the wafers after they are loaded in the boat. Record each wafer's position in the boat.
3. Check that the furnace temperature is 925°C and that the flow rate is approximately 500 cc/min of N_2.
4. Slide the boat into the furnace slowly, at about 5 cm per 10 seconds. Remove the push rod.

Safety Precaution

Remember that the end of the rod is hot and will take about 5 minutes to cool.

5. A typical predeposition time is 30 minutes. Note the relationship, shown in Figure A.17, between sheet resistance and predeposition time.
6. Pull the boat out at the same rate it was pushed in.
7. Let it cool at the end of the tube for about 5 minutes.
8. Pull the boat out onto the boat holder and let it cool another 5 minutes before removing the wafer.
9. Switch the furnace gas flow back to a 30-sccm-nitrogen standby condition.

Boron-Skin Removal Procedure

Use separate wafer baskets for each chemical solution.

1. Place the wafer in a basket and immerse the wafer in a 9:1 H_2O:HF solution for 40 seconds.

Figure A.17 Sheet resistance as a function of deposition time for B_2O_3-loaded glass-ceramic wafers.
SOURCE: Data are from Electro Oxide Division of Solitron Devices, Inc., Riviera Beach, FL.

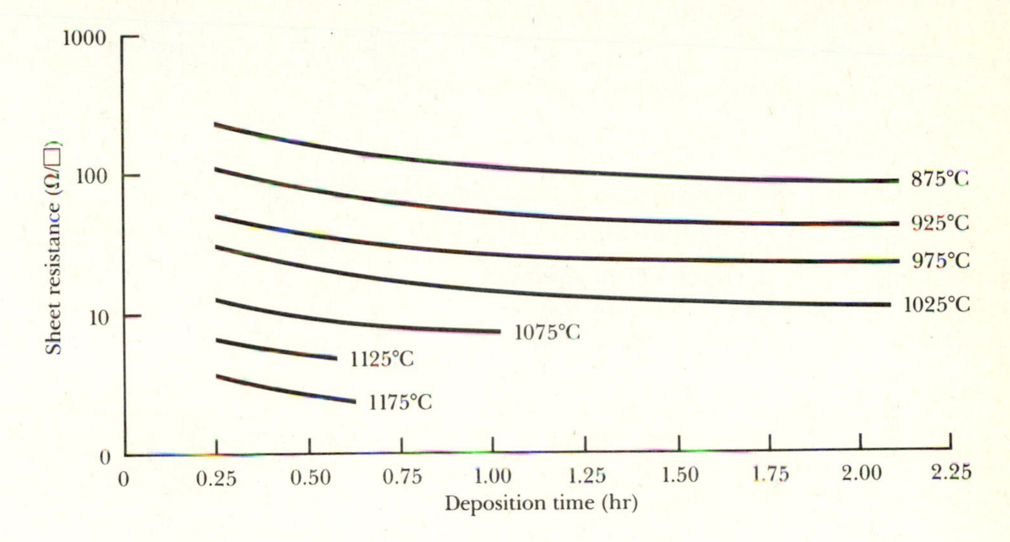

2. Rinse in running DI and dry with N_2.
3. Immerse the wafer in borosilicate etch (see Section A.3.7) for 7 minutes.
4. Rinse well in running DI and dry with N_2.
5. Immerse the wafer in a 50:1 H_2O:HF solution for 30 seconds.
6. Rinse in running DI and dry with N_2.

Boron Drive-In Procedure

The main purpose of this step for the MOS experiments in this appendix is to establish an oxide layer for later photolithographic steps. That the junction is driven deeper is somewhat a secondary consideration for these particular experiments; so the desired oxide thickness mainly determines the choice of drive-in time at this point. In a bipolar experiment, on the other hand, junction depths are more critical and drive-in times must be calculated to satisfy junction-depth and base-width requirements.

1. The wafer should appear clean on microscope inspection. If it does not, do a DI rinse followed by an N_2 dry. If it is still dirty, go through the TCE, acetone, methanol, DI, N_2-dry sequence. Steps 1 through 8 of the Preevaporation Cleaning Procedure are appropriate for more rigorous cleaning.
2. Place the wafers in the quartz boat of the *p*-type drive-in furnace. Use a nitrogen stream to dislodge dust particles from the wafers once all are loaded into the boat.

3. Check that the furnace temperature is 1100°C and that the flow rate is approximately 500 cc/min in a dry O_2 flow. The oxygen flow should have been on for about an hour depending on the size of the furnace tube, prior to this point.
4. Slide the wafer in slowly, at about 5 cm per 10 seconds. Pull out the push rod.

Safety Precaution

Again, remember that the end will be hot, and let the rod cool for about 5 minutes.

5. A typical drive-in/oxidation sequence for the diffused devices experiment is 10 minutes dry, 10 minutes wet, and 10 minutes dry. For the wet oxidation portion of the cycle, switch the oxygen flow through the de-ionized water bubbler. The water temperature should be between 95 and 98°C, and the oxygen flow rate should be such that it causes bubbling in the water, but not violent splashing. A typical flow rate is 200 to 300 sccm.
6. Pull the boat out at the same rate it was pushed in.
7. Let the boat cool at the end of the furnace tube for about 5 minutes.
8. Pull the boat out onto the boat holder and let it cool for another 5 minutes before removing the wafer.
9. Switch the furnace gas flow back to the 30-sccm-nitrogen standby condition.

Phosphorus Predeposition Procedure

Use separate wafer baskets for each chemical solution. As for the boron diffusion, a solid planar diffusion source is used as the phosphorus source. In this case, SiP_2O_7 is dispersed in an inert refractory matrix. The times and temperatures for the procedure that follows are for PH-1000 Planar Diffusion Sources from The Standard Oil (Ohio) Engineered Materials Company [20].

1. Prepare the PH-1000 dopant wafers by heating them at 925°C for 25 minutes in the *n*-type predeposition furnace. The N_2 flow should be about 500 sccm.
2. The silicon wafer should appear clean on microscope inspection. If it does not, do a DI rinse followed by an N_2 dry. If the wafer is still not clean, go through the TCE, acetone, methanol, DI, N_2-dry sequence. Again, steps 1 through 8 of the Preevaporation Cleaning Procedure are appropriate for more rigorous cleaning.
3. Place the silicon wafers in a quartz diffusion boat. The silicon wafers should be interspersed between dopant wafers with one dopant wafer,

two silicon wafers, one dopant wafer, and so on. Each silicon wafer should have the front side toward a dopant wafer. Use a nitrogen stream to clean dust particles from the loaded wafers. Record each wafer's position in the boat.

4. Check that the furnace temperature is 925°C and that the N_2 flow rate is 500 sccm.
5. Slide the boat in slowly, at about 5 cm per 10 seconds. Pull the push rod out slowly.

Safety Precaution

Again, remember that the end will be hot, and let it cool for about 5 minutes.

6. A typical predeposition time is 30 minutes. Note the relationship, shown in Figure A.18, between sheet resistance and time.
7. Pull the boat out at the same rate it was pushed in.
8. Let it cool at the end of the tube for approximately 5 minutes.
9. Pull the boat out onto the boat holder and let it cool another 5 minutes before removing the wafer.
10. Switch the furnace gas flow back to the 30-sccm-nitrogen standby condition.

Phosphosilicate Glass Etch Procedure

1. Dip the wafer and basket into a beaker of fresh DI.
2. Immerse the wafer for 5 seconds in either buffered oxide etch (see Section A.3.7) or in 9:1 H_2O:HF.
3. Rinse with running DI for 3 minutes.
4. Dry with N_2.

Gate Oxidation Procedure

1. It is advisable to first clean the wafer with the hydrogen peroxide–based degrease and demetal etches. Toward this end, steps 1 through 8 of the Preevaporation Cleaning Procedure are also appropriate for preparing the wafer for gate oxidation.
2. Use the same technique as in the oxidation experiment, except that only a dry oxidation is used.
3. Check that the temperature is 1100°C and that the O_2 flow rate is approximately 500 sccm.
4. Choose an oxidation time to give the desired gate oxide thickness. To avoid pin-hole problems, a conservative value of about 1200 Å is appropriate for the procedures described in this appendix.

Figure A.18 Sheet resistance as a function of deposition time for SiP_2O_7-loaded planar sources. The spacing between the source and silicon wafers is 0.08 inches and the gas flow is 500 to 1000 sccm of N_2.
SOURCE: Data are from The Standard Oil (Ohio) Engineered Materials Company, Semiconductor Products Division, P.O. Box 664, Niagara Falls, NY 14302.

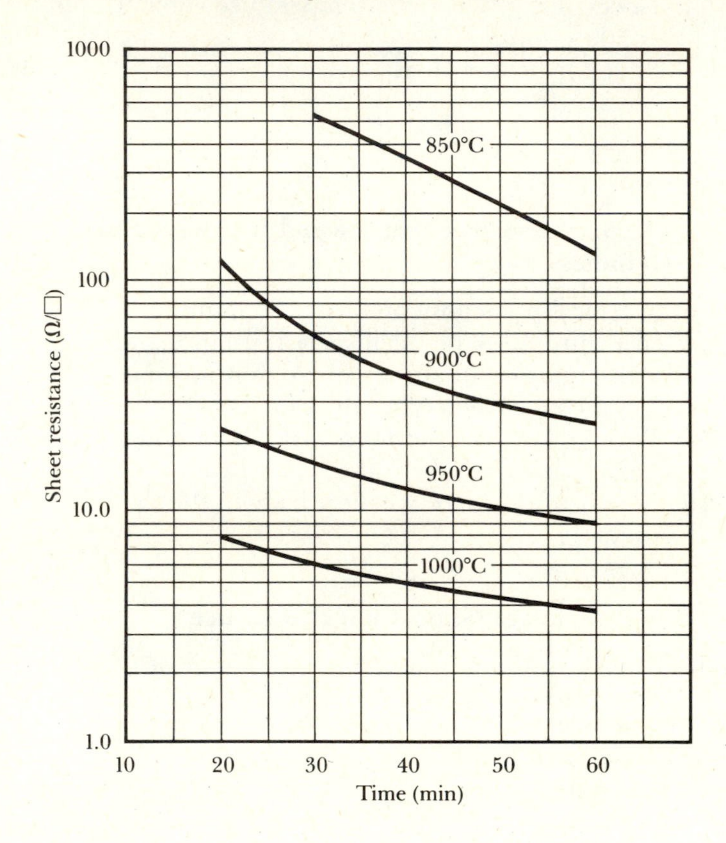

As-grown dry oxide layers have a fairly high interface-state density, which can be reduced by an annealing procedure. The Aluminum Anneal Procedure describes a low-temperature hydrogen anneal for this purpose.

Preevaporation Cleaning Procedure

1. Dip the wafer and basket in a beaker containing fresh DI.
2. Immerse the wafer for 10 minutes in degrease etch (see Section A.3.7) heated to $70 \pm 5°C$. Remember that the degrease and demetal etch solutions have a short life-time because the hydrogen peroxide leaves the heated solution fairly rapidly.

3. Rinse in running DI for 2 minutes.
4. Immerse the wafer for 10 minutes in demetal etch (see Section A.3.7) heated to $70 \pm 5°C$.
5. Rinse in running DI for 2 minutes.
6. Immerse the wafer in 50:1 H_2O:HF for 30 seconds.
7. Rinse in running DI for 3 minutes.
8. Dry with N_2.
9. Place the wafer in the anneal furnace at 525°C for 15 minutes with a 500-sccm-N_2 flow.
10. Proceed immediately to the evaporation procedure.

Safety Precaution

The degrease etch and the demetal etch should not be heated on the same hot plate as a solvent.

Aluminum Evaporation Procedure

This information is to be supplemented by information specific to the evaporator and pumping station being used. The notes below are generic for a thermal-filament evaporation unit.

Depending on time constraints, this procedure may be carried out by the staff between lab sessions.

1. Break the vacuum, if necessary, and raise the bell jar.
2. Using clean lint-free gloves and tweezers, secure the wafer(s) to the evaporator unit's substrate holder and note the distance between the filament and the wafer.
3. Load the filament with aluminum clips that have been rinsed in methanol and dried with N_2. The number of aluminum clips should be sufficient to produce a film about 0.5 μm thick. Refer to Section 2.10.2 to calculate the required number of clips. Both the filament and the aluminum should be low-sodium-content materials to avoid ionic contamination of the MOS system.
4. Place a clean glass slide over the observation port. Lower the bell jar, making sure the gasket mates well with a clean base-plate surface. Follow the prescribed pump-down procedure for the pumping station.
5. When the pressure is in the mid-10^{-6}-torr range, the evaporation can begin. Turn the filament on and gradually increase the current through the coil. When the aluminum clips start to melt, close the high-vacuum valve to the diffusion pump. Increase the current until evaporation is complete.
6. Let the system cool down, then break the vacuum, open the bell jar, and remove the wafers. Lower the bell jar again and evacuate to minimize exposure to room air.

Aluminum Etch Procedure

1. Rinse the wafer in running DI for about 30 seconds.
2. Immerse the wafer and basket in aluminum etch (see Section A.3.7) at 40 to 50°C. Agitate by gently swirling the beaker. It is advisable to use a separate beaker of metal etch for each wafer.
3. Continue until the pattern is completely developed to the eye, then proceed for another 15 seconds. The etch rate is highly temperature dependent.
4. Rinse in gentle running DI for 3 minutes.
5. Dry with N_2.

Note: When stripping the photoresist off the aluminum, make sure the strip temperature is not more than 100°C to avoid aluminum lift-off.

Aluminum Anneal Procedure

A heat treatment is necessary to form a good bond between the aluminum and silicon in the contact openings. One alternative is as follows:

1. Rinse wafers in running DI for 3 minutes.
2. Dry with N_2.
3. Anneal at 525°C for 15 minutes with a 500-sccm-N_2 flow.

Safety Information

It is usually possible to achieve lower values of flat band voltage, and more nearly ideal values of threshold voltage, if a forming gas (95% N_2 and 5% H_2) is used for the anneal. However, obvious caution is required if H_2 is introduced into the gas stream because hydrogen is highly flammable. Mixtures of hydrogen and oxygen present a severe explosion hazard when exposed to heat or flame, so the furnace must be properly vented to an external exhaust.

An annealing procedure that uses hydrogen to neutralize most of the interface trapped charge is as follows.

1. Rinse wafers in running DI for 3 minutes.
2. Dry with N_2.
3. Anneal at 450°C for 1 hour in a furnace atmosphere of 5% hydrogen and 95% nitrogen.

REFERENCES

1. N. I. Sax, *Dangerous Properties of Industrial Materials,* 5th ed. (New York: Van Nostrand Reinhold, 1979).
2. Alconox is a trademark of Alconox, Inc. Alconox detergent is distributed by VWR Scientific Company, P.O. Box 3200, San Francisco, CA 94119.

3. Teflon is a trademark of E. I. Du Pont de Nemours Co., Inc.
4. Signatone, 3687 Enochs Street, Santa Clara, CA 95051.
5. F. M. Smits, "Measurement of Sheet Resistivities with the Four Point Probe," *Bell System Technical Journal* (May 1958), 711–718.
6. Suppliers of diffusion and oxidation furnace quartzware include Berkeley Glasslab, 1279 Quarry Lane, Pleasanton, CA 94566, and United States Fused Quartz Company, Inc., 17 Madison Road, Fairfield, NJ 07006.
7. W. A. Pliskin and E. E. Conrad, "Nondestructive Determination of Thickness and Refractive Index of Transparent Films," *IBM Journal of Research and Development,* 8 (1964), 43–51.
8. S. M. Sze and J. C. Irvin, "Resistivity, Mobility, and Impurity Levels in GaAs, Ge, and Si at 300°K," *Solid State Electronics,* Pergamon Press Ltd. 11 (1968), 599–602.
9. H. F. Wolf, *Silicon Semiconductor Data* (New York: Pergamon, 1969), p. 501.
10. The Diffused Devices mask set was designed by Bob Matthews, MSEE, Michigan State University, 1981. The masks were fabricated by Delco Electronics, a division of General Motors Corporation.
11. R. A. Colclaser, *Microelectronics Processing and Device Design* (New York: Wiley, 1980), Chap. 7.
12. S. M. Sze, *Physics of Semiconductor Devices,* 2nd ed. (New York: Wiley, 1981), p. 104.
13. L. J. Giacoletto, "Solid State Devices," in *Electronics Designer's Handbook,* 2nd ed. (New York: McGraw-Hill, 1977), Chap. 10.
14. The PMOS logic mask set was designed by Gordon Priebe, BSEE, Michigan State University, 1981. The masks were fabricated by Delco Electronics, a division of General Motors Corporation.
15. K. P. Roenker and T. Kohli, *Standard Operating Procedures for Solid State Microelectronics Laboratory,* Department of Electrical and Computer Engineering, University of Cincinnati, 1980.
16. *EE 344 Laboratory Manual,* University of Illinois, Department of Electrical Engineering, 1977.
17. Labware for handling wafers is supplied, for example, by Fluoroware, Jonathan Industrial Center, Chaska, MN 55318.
18. Waycoat® is a trademark for several photoresist-related products from Olin Hunt Specialty Products, Palisades Park, NJ 07494.
19. Electro Oxide Corporation, Division of Solitron Devices Inc., Riviera Beach, FL.
20. The Standard Oil (Ohio) Engineered Materials Company, Semiconductor Products Division, P.O. Box 664, Niagara Falls, NY 14302.

Appendix B
Physical Constants

Quantity	Symbol and value
Electron charge	$q \simeq 1.602 \times 10^{-19}$ C
Electron rest mass	$m_0 \simeq 9.109 \times 10^{-31}$ kg
Electron-volt	1 eV $\simeq 1.602 \times 10^{-19}$ J
Boltzmann's constant	$k \simeq 1.381 \times 10^{-23}$ J/K
	$k \simeq 8.619 \times 10^{-5}$ eV/K
Avogadro's constant	$N_A \simeq 6.022 \times 10^{23}$ mole^{-1}
Angstrom	1 Å $= 10^{-8}$ cm $= 10^{-4}$ μm
Vacuum permittivity	$\epsilon_0 \simeq 8.854 \times 10^{-14}$ F/cm
Vacuum permeability	$\mu_0 = 4\pi \times 10^{-9}$ H/cm
Planck's constant	$h \simeq 6.626 \times 10^{-34}$ J·s
Reduced Planck's constant	$\hbar \simeq 1.055 \times 10^{-34}$ J·s
Speed of light in vacuum	$c \simeq 2.998 \times 10^{10}$ cm/s

Appendix C
Material Properties at 300 K

Table C.1 Silicon and gallium arsenide [1]

Property	Si	GaAs
Atoms/cm^3	5.0×10^{22}	4.42×10^{22}
Lattice constant (Å)	5.43	5.65
Density (g/cm^3)	2.33	5.32
Dielectric constant	11.9	13.1
Effective density of states in the conduction band, N_C (cm^{-3}) [2]	2.8×10^{19}	4.7×10^{17}
Effective density of states in the valence band, N_V (cm^{-3}) [2]	1.04×10^{19}	7.0×10^{18}
Energy gap, E_G (eV)	1.12	1.42
Intrinsic carrier concentration n_i (cm^{-3}) [2]	1.45×10^{10}	1.79×10^{6}
Thermal conductivity (W/cm·°C)	1.5	0.46

Table C.2 SiO_2 and Si_3Ni_4 [1]

Property	SiO_2	Si_3N_4
Density (g/cm^3)	2.2	3.1
Dielectric constant	3.9	7.5
Dielectric strength (V/cm)	10^7	10^7
Energy gap (eV)	9	5
dc resistivity (Ω·cm)	10^{14}–10^{16}	10^{14}

Table C.3 Aluminum

Density (g/cm^3)	2.70
dc resistivity (Ω·cm)	2.67×10^{-6}

REFERENCES

1. S. M. Sze, *Physics of Semiconductor Devices,* 2nd ed. (New York: Wiley, 1981), pp. 848–852.
2. The values of N_C, N_V, and n_i are from Sze. However, there is a considerable spread in the reported values of these quantities. Experimental discrepancies of up to 25% in the literature are noted for n_i, and reported measurements of the effective masses from which N_C and N_V are calculated vary up to about 50%. The situation is complicated in part because effective masses are temperature dependent and, at high doping, also depend on doping concentration. Furthermore, n_i measurements tend to be done on very pure material, with carrier concentrations of about 10^{13} cm^{-3}, while effective mass measurements at room temperature tend to be done on very heavily doped material (10^{17}–10^{20} cm^{-3}).

   For room temperature calculations of n and p, equations based on n_i will generally give more accurate results than equations based on N_C and N_V.

Index